D0939068

FORENSIC DNA TYPING

FORENSIC DNA TYPING

BIOLOGY, TECHNOLOGY, AND GENETICS OF STR MARKERS
Second edition

John M. Butler

ELSEVIER
ACADEMIC
PRESS

Amsterdam Boston Heidelberg London New York Oxford
Paris San Diego San Francisco Singapore Sydney Tokyo

Elsevier Academic Press
30 Corporate Drive, Suite 400, Burlington, MA 01803, USA
http://www.elsevier.com

Elsevier Academic Press
84 Theobald's Road, London WC1X 8RR, UK
http://www.elsevier.com

Library of Congress Catalog Number: 2004114214

British Library Cataloguing in Publication Data
A catalogue record for this book is available from the British Library

ISBN 0-12-147952-8

Acquisitions Editor: *Mark Listewnik*
Associate Acquisitions Editor: *Jennifer Soucy*
Developmental Editor: *Pamela Chester*
Marketing Manager: *Christian Nolin*
Cover Designer: *Eric DeCicco*

05   06   07   08   09   8   7   6   5   4   3   2   1

CONTENTS

> *High-profile cases and other interesting information are included as D.N.A. Boxes (Data, Notes, and Applications) scattered throughout the book in the chapter pertaining to a particular subject.*

FOREWORD

Forensic DNA Typing charts the progress and development of DNA applied to criminal forensics, providing vivid demonstrations of the amazing potential of the method, not only to convict the guilty but also to exonerate the innocent. John Butler has created a text that caters to all audiences, covering the basics of DNA structure and function and describing in detail how the techniques are used. In addition, the extensive use of D.N.A. (Data, Notes, and Application) Boxes in the text enables the reader to dip in and out as he or she pleases.

Probably the most important development of recent years is the universal use of polymerase chain reaction (PCR) to replicate DNA molecules *in vitro*. This has led to the rapid development of new platforms and biochemistry that have revolutionized the methods used to carry out DNA analysis. These new technologies are clearly explained in great detail in this book, with lavish illustrations. The culmination of recent advances has led to the instigation of massive National DNA offender databases using short tandem repeat (STR) loci. For example, since its inception in 1995, the England and Wales National DNA database (NDNADB) now has more than 2.75 million reference DNA profiles from suspects and offenders alike, against which all crime stains are routinely compared. Many more countries throughout the world have since followed suit. The social benefits of such databases are considerable – individuals who commit major crimes such as murder usually already have a criminal record. UK policy enables the collection of DNA profiles from all offenders regardless of the seriousness of the crime. Consequently, those who re-offend can be quickly identified and apprehended. In the US thirteen different STRs are combined together into one or two-tube reactions known as multiplexes to provide data for the Combined DNA Index System (CODIS). When a complete DNA profile is obtained the probability of a chance match with a randomly chosen individual is usually less than one in one trillion using these 13 CODIS loci.

Other areas are explored in detail including mitochondrial, Y chromosomal DNA and use of forensic science in wildlife crime such as poaching. Recently, as a result of terrorist attacks, new areas of forensic DNA profiling have arisen in response. Foremost amongst these is the field of microbial forensics, which is used to identify pathogens such as anthrax.

Although it is very difficult to anticipate all future developments, STRs are probably the system of choice for the foreseeable future although other systems, especially single nucleotide polymorphisms (SNPs), have been suggested. SNPs may find a special niche to analyze very highly degraded material and may play a valuable role in future mass disasters. However, there is no doubt that the utility of both STRs and SNPs will benefit from new biochemistry and new platforms such as microchips. Automation, miniaturization and expert systems will all play an increasingly important role over the coming years. The main aims of the new technology can be summarized: to enable faster processing; to reduce costs; to improve sensitivity; to produce portable instruments; to de-skill and to automate the interpretation process; to improve success rates; to improve quality of the result and to standardize processes. The next few years will probably see a new revolution as this new technology comes of age and becomes widely available.

John Butler reviews these new innovations in great detail – he is to be congratulated for preparing such a readable book that will appeal to everyone from the layperson, the lawyer and the scientist alike.

PETER GILL, Ph.D.
Birmingham, UK
December 2004

INTRODUCTION

An expert is one who knows more and more about less and less until they know absolutely everything about nothing...

(First part by Nicholas Butler, Bartlett's 585:10)

The work is its own reward.

(Sherlock Holmes, *The Adventure of the Norwood Builder*)

Several significant things have happened since the first edition of *Forensic DNA Typing* was published in January 2001. The Human Genome Project published a draft sequence of the human genome in February 2001 and completed the 'finished' reference sequence in April 2003. In addition, human mitochondrial DNA population genomics is underway and more than a thousand full mitochondrial genomes have been published. Technology for DNA sequencing and typing continues to advance as does our understanding of genetic variation in various population groups around the world. These milestones are a tribute to the progress of science and will benefit the field of forensic DNA typing.

The literature on the short tandem repeat (STR) markers used in forensic DNA testing has more than doubled in the four years since writing the first edition. More than 2000 publications now detail the technology and report the allele frequencies for forensically-informative STR loci. Hundreds of different population groups have been studied; new technologies for rapidly typing DNA samples have been developed, and standard protocols have been validated in laboratories worldwide. Yet DNA results are still sometimes challenged in court – not usually because of the technology, which is sound – but rather the ability of practitioners to perform the tests carefully and correctly. A major purpose of this book is to help in the training of professionals in the field of forensic DNA testing. The knowledge of forensic scientists, lawyers, and students coming into the field will be enhanced by careful review of the materials found herein.

The advent of modern DNA technology has resulted in the increased ability to perform human identity testing. Individual identification is desirable in a number of situations including the determination of perpetrators of violent crime such as murder and rape, resolving unestablished paternity, and identifying remains of missing persons or victims of mass disasters.

In the past few years, the general public has become more familiar with the power of DNA typing as the media has covered efforts in identifying remains from victims of the World Trade Center twin towers collapse following the terrorist attacks of 11 September 2001, the O.J. Simpson murder trial, the President Clinton–Monica Lewinsky scandal, and the identification of the remains in the Tomb of the Unknown Soldier. In addition, our perceptions of history have been changed with DNA evidence that revealed Thomas Jefferson may have fathered a child by one of his slaves.

These cases have certainly attracted widespread media attention in recent years, however, they are only a small fraction of the thousands of forensic DNA and paternity cases that are conducted each year by public and private laboratories around the world. The technology for performing DNA typing has evolved rapidly since the 1990s to the point where it is now possible to obtain results in a few hours on samples with only the smallest amount of biological material.

This book will examine the science of current forensic DNA typing methods by focusing on the biology, technology, and genetic interpretation of short tandem repeat (STR) markers, which encompass the most common forensic DNA analysis methods used today. The materials in this book are intended primarily for two audiences: forensic scientists who want to gain a better understanding of STRs and professionals in the law enforcement and legal communities who find it hard to comprehend the complexities of DNA profiling. This text should also directly benefit college students learning more about forensic DNA analysis in an academic environment. The references cited at the end of each chapter provide a fairly comprehensive view of this dynamic field.

This book is also intended to aid forensic DNA laboratories in meeting the training requirements stated in the DNA Advisory Board Quality Assurance Standards. These standards are striving to improve the quality of work performed in forensic laboratories by requiring technical managers and DNA examiners to have training in biochemistry, genetics and molecular biology in order to gain a basic understanding of the foundation of forensic DNA analysis. See Standard 5.2.1 and 5.3.1 in Appendix IV of this book.

NEW MATERIAL IN THIS SECOND EDITION

Since the first edition was written in the winter months of 2000, the published literature has grown dramatically on the topic of STR typing and its use in forensic DNA testing. With more than 2000 papers now available describing STR markers, technology for typing these STRs, and allele frequencies in various populations around the world, the scientific basis for forensic DNA typing is sound. The basic foundational material in the first edition is still relevant and thus has remained essentially unchanged. However, ten new chapters have been added to accommodate the explosion of new information since the turn of the century.

New topics such as single nucleotide polymorphisms (SNPs) and Y chromosome testing have gained greater acceptance within the forensic community since 2000 and therefore have become areas of expansion in this edition. A very comprehensive look at mitochondrial DNA and its application to forensic DNA analysis is included in Chapter 10. There is updated information on new DNA extraction procedures, real-time PCR for DNA quantification, multi-capillary electrophoresis instruments, and 5-dye chemistries that are now used in many forensic DNA laboratories. Citations have expanded to include more than 500 new literature references enabling readers to find original source material or to conduct extensive background research on the various topics covered herein and more than 50 new figures and 45 new tables containing helpful information have been added in this second edition of *Forensic DNA Typing*.

Statistical issues with data analysis and interpretation that were missing in the first edition are covered in Chapters 19–23 in this new edition. Extensive examples are provided for each equation discussed and corresponding population data can be found in Appendix II to enable readers to review the source of conclusions reached. Another appendix includes the description of a hypothetical case from start to finish in an attempt to bring together the information discussed throughout the book and to aid in training students and professionals in the field.

In this edition, we utilize Data, Notes, and Applications (D.N.A.) Boxes to cover specific topics of general interest. Many of the high-profile cases included in the last chapter of the first edition, such as the O.J. Simpson trial, are now scattered throughout the book near the sections dealing with the science or issues behind these cases. It is hoped that these D.N.A. Boxes will help readers see the practical value of forensic DNA typing.

AN OVERVIEW OF THE BOOK CHAPTERS

The book has been divided into three primary sections covering the biology, technology, and statistical analysis (genetics) of STR markers. Within each section, the chapters progress from basic introductory information to on-going 'cutting-edge' research. The first few chapters in particular are meant as introductory material for those readers who might be less familiar with DNA or as a review of useful materials for more advanced readers. The biology section is contained in Chapters 2 through 11, the technology section involves Chapters 12 through 18, and the genetics section may be found in Chapters 19 to 23. The final chapter examines the use of DNA testing in mass disaster victim identification efforts, which include the greatest national tragedy in U.S. history, the events of 11 September 2001.

BIOLOGY SECTION

The book begins with an overview and history of DNA and its use in human identification. An actual criminal investigation where DNA evidence proved crucial is used to illustrate the value of this technology to law enforcement. Chapter 2 provides some basic information on DNA structure and function while Chapter 3 covers the processes involved in preparing samples for DNA amplification via the polymerase chain reaction, which is discussed further in Chapter 4. Chapter 5 focuses on the 13 commonly used STR markers in the United States today with details about naming of alleles and unique characteristics of each marker. Chapter 6 goes into the biology of STR markers including stutter products, non-template addition, microvariants, and null alleles. These aspects can complicate data interpretation if they are not understood properly. Chapter 7 discusses issues that are unique to the forensic DNA community, namely mixtures, degraded DNA samples, PCR inhibition, and contamination, all of which impact forensic casework since many samples do not come from a pristine, controlled environment. What was previously Chapter 8 in the first edition that discussed additional markers used in conjunction with STRs to aid in human identification has now been expanded upon in four additional chapters. The new Chapter 8 covers single nucleotide polymorphisms (SNPs) and technologies for typing them. Chapter 9 reviews Y chromosome markers for specifically identifying the male contributor of a sample and Chapter 10 discusses maternally inherited mitochondrial DNA, the use of which often provides results in situations involving highly degraded DNA. Finally, we touch on the use of non-human DNA to aid forensic investigations in Chapter 11 through a discussion of animal, plant, and microbial DNA testing.

TECHNOLOGY SECTION

The technology portion of the book begins in Chapter 12 with a discussion of DNA separations using slab gel and capillary electrophoresis. Fluorescent detection methods are the primary topic of Chapter 13. This chapter has a number of colorful figures featuring the fluorescent dyes in use today. A description of the most widely used DNA analysis instruments in modern forensic laboratories is presented in Chapter 14. This chapter covers the ABI Prism 310 Genetic Analyzer, the 16-capillary array ABI Prism 3100, and reviews the Hitachi FMBIO II Fluorescence Imaging System used in conjunction with slab gel electrophoresis. Issues surrounding genotyping of STR results are the focus of Chapter 15 and Chapter 16 reviews laboratory validation and quality assurance of DNA analysis.

Alternative DNA analysis technologies such as mass spectrometry and microchips are reviewed in Chapter 17 along with robotics and expert systems for automated data analysis. The final chapter in the technology section, Chapter 18, discusses the use of computer DNA databases to solve crimes. Large national DNA databases will continue to benefit law enforcement for many years to come by connecting violent crimes and serial criminal activity with otherwise unknown perpetrators.

GENETICS AND STATISTICAL ANALYSIS SECTION

The genetics and statistical analysis section begins with a review of genetic principles and statistics in Chapter 19. Chapter 20 discusses Hardy–Weinberg and linkage equilibrium and the role of these calculations in checking performance of genetic markers in population databases. Calculations for DNA profile frequency estimates including random match probabilities and likelihood ratios are covered in Chapter 21. Chapter 22 discusses approaches to interpreting mixtures or partial profiles resulting from degraded DNA. Finally, Chapter 23 deals with kinship and paternity testing situations. Throughout all of these sections clear examples are used to demonstrate how equations are applied to the calculations required.

MASS DISASTER VICTIM IDENTIFICATION: UTILIZING BIOLOGY, TECHNOLOGY, AND GENETICS

Forensic DNA laboratories may be called upon to assist in victim identification following mass disasters such as airplane crashes or terrorist attacks. Situations where remains are highly fragmented prevent the use of fingerprints or dental records to rapidly determine the identity of each victim and DNA analysis often becomes the only method to bring closure to the chaos of such an event. To recognize the increasing role that DNA information is playing in mass disaster victim identification, we discuss here the application of STRs, mitochondrial DNA, and single nucleotide polymorphisms in the identification of individuals who died in the terrorist attacks of 11 September 2001 at the World Trade Center twin towers, the Pentagon, and Shanksville, Pennsylvania. The original information from the first edition on the Waco Branch Davidian fire (April 1993) and the airline crash of Swissair Flight 111 (September 1998) are retained to provide historical perspective.

APPENDICES

There are seven appendices at the back of the book that provide valuable supplemental material. Appendix I describes all reported alleles for the 13 CODIS

STR loci as of January 2004. Sequence information, where available, has been included along with the reference that first described the noted allele. As most laboratories now use either a Promega GenePrint® STR kit or an Applied Biosystems AmpF*l*STR® kit for PCR amplification, we have listed the expected size for each allele based on the sequence information. Appendix II lists some STR allele frequency information from U.S. populations of African-Americans, Caucasians, and Hispanics. This information is used for all statistical calculations performed in the book. Appendix III is a compilation of companies and organizations that are suppliers of DNA analysis equipment, products, and services. Approximately 100 companies are listed along with their addresses, phone numbers, internet web pages, and a brief description of their products and/or services. Appendix IV contains the DNA Advisory Board (DAB) Quality Assurance Standards that pertain to forensic DNA testing laboratories and convicted offender DNA databasing laboratories in the United States. These standards are important for laboratory validation and maintaining high quality results as DNA testing becomes more prevalent. Appendix V includes the DAB recommendations on statistics. Appendix VI reviews the National Research Council's *The Evaluation of Forensic DNA Evidence*, better known as NRC II, and the application of its recommendations to STR typing. Finally, Appendix VII provides two example forensic cases in an attempt to put the information contained in this book within a proper context.

ACKNOWLEDGMENTS

I express a special thanks to colleagues and fellow researchers who kindly provided important information and supplied some of the figures for this book. These individuals include Martin Bill, George Carmody, Mike Coble, David Duewer, Dan Ehrlich, Nicky Fildes, Lisa Forman, Ron Fourney, Lee Fraser, Chip Harding, Debbie Hobson, Bill Hudlow, Alice Isenberg, Margaret Kline, Carll Ladd, Demris Lee, Steve Lee, Bruce McCord, Steve Niezgoda, Richard Schoske, Jim Schumm, Bob Shaler, Melissa Smrz, Amanda Sozer, Kevin Sullivan, and Lois Tully. I am indebted to the dedicated project team members, past and present, who work with me at the U.S. National Institute of Standards and Technology: Peter Vallone, Margaret Kline, Janette Redman, Mike Coble, David Duewer, Jill Appleby, Amy Decker, Christian Ruitberg, and Richard Schoske. It is a pleasure to work with such supportive and hard-working scientists.

Several other people deserve specific recognition for their support of this endeavor. The information reported in this book was in large measure made possible by a comprehensive collection of references on the STR markers used in forensic DNA typing. For this collection now numbering more than 2000 references, I am indebted to the initial work of Christian Ruitberg for tirelessly collecting and cataloging these papers and the steady efforts of Janette Redman to monthly update this STR reference database. A complete listing of these references may be found at http://www.cstl.nist.gov/biotech/strbase. George Carmody reviewed the statistical materials in Chapters 19–24 and provided many valuable comments.

My wife Terilynne, who carefully reviewed the manuscript and made helpful suggestions, was always a constant support in the many hours that this project took away from my family. As the initial editor of all my written materials, Terilynne helped make the book more coherent and readable.

Since I was first exposed to forensic DNA typing in 1990 when a friend gave me a copy of Joseph Wambaugh's *The Blooding* to read, I have watched with wonder as the forensic DNA community has rapidly evolved. DNA testing that once took weeks can now be performed in a matter of hours. I enjoy being a part of the developments in this field and hope that this book will help many others come to better understand the principles behind the biology, technology, and genetics of STR markers.

John Marshall Butler grew up in the Midwest and enjoying science and law decided to pursue a career in forensic science at an early age. After completing an undergraduate education at Brigham Young University in chemistry, he moved east to pursue his graduate studies at the University of Virginia. While a graduate student, he enjoyed the unique opportunity of serving as an FBI Honors Intern and guest researcher for more than two years in the FBI Laboratory's Forensic Science Research Unit. His Ph.D. dissertation research, which was conducted at the FBI Academy in Quantico, Virginia, involved pioneering work in applying capillary electrophoresis to STR typing. After receiving his Ph.D. in 1995, Dr. Butler obtained a prestigious National Research Council postdoctoral fellowship to the National Institute of Standards and Technology (NIST). While a postdoc at NIST, he designed and built STRBase, the widely used Short Tandem Repeat Internet Database (http://www.cstl.nist. gov/biotech/strbase) that contains a wealth of standardized information on STRs used in human identity applications. Dr. Butler then went to California for several years to work as a staff scientist and project leader at a startup company named GeneTrace System to develop rapid DNA analysis technologies involving time-of-flight mass spectrometry. In the fall of 1999, he returned to NIST to lead their efforts in human identity testing with funding from the National Institute of Justice.

Dr. Butler received the Presidential Early Career Award for Scientists and Engineers from President George W. Bush in a White House ceremony held in July 2002. In September 2003, he was awarded the Scientific Prize of the International Society of Forensic Genetics, the first American to be given this honor by his scientific peers. Following the terrorist attacks of 11 September, 2001, Dr. Butler's expertise was sought to aid the DNA identification efforts, and he served as part of the distinguished World Trade Center Kinship and Data Analysis Panel (WTC KADAP). He is also a regular invited guest and participant in the semi-annual meetings of the FBI's Scientific Working Group on DNA Analysis Methods (SWGDAM). In addition, he serves on the Department of Defense Quality Assurance Oversight Committee for DNA Analysis and as a guest editor for the *Journal of Forensic Sciences*. His more than 65 publications in the field make him one of the most prolific active authors in the field with articles appearing regularly in every major forensic science journal. He has been an invited speaker to numerous national and international forensic DNA meetings and in the past few years has spoken in Germany, France, England, Portugal, Cyprus, and Australia. He is well-qualified to present the information found in this book, much of which has come from his own research efforts over the past decade. In addition to his busy scientific career, Dr. Butler and his wife serve in their community and church and are the parents of five children, all of which have been proven to be theirs through the power of STR typing.

To my wife Terilynne and our children
Amanda, Marshall, Katy, Emma, and Ethan

CHAPTER 1

OVERVIEW AND HISTORY OF DNA TYPING

DNA testing is to justice what the telescope is for the stars; not a lesson in biochemistry, not a display of the wonders of magnifying glass, but a way to see things as they really are.

(Barry Scheck and Peter Neufeld, *Actual Innocence*)

In the darkness of the early morning hours of 26 August 1999, a young University of Virginia student awoke to find a gun pointed at her head. The assailant forced her and a male friend spending the night to roll over on their stomachs. Terrorized, they obeyed their attacker. After robbing the man of some cash, the intruder put a pillow over the man's head and raped the female student. The female was blindfolded with her own shirt and led around the house while the intruder searched for other items to steal.

Throughout the entire ordeal, the intruder kept his gun to the back of the male student's head, daring him to look at him and telling him if he tried he would blow his head off. The assailant forced the young woman to take a shower in the hope that any evidence of the crime would be washed away. After helping himself to a can of beer, the attacker left before dawn taking with him the cash, the confidence, and the sense of safety of his victims. Even though the assailant had tried to be careful and clean up after the sexual assault, he had left behind enough of his personal body fluids to link him to this violent crime.

The police investigating the crime collected some saliva from the beer can. In addition, evidence technicians found some small traces of semen on the bed sheets that could not be seen with the naked eye. These samples were submitted to the Virginia Department of Forensic Sciences in Richmond along with control samples from other occupants of the residence where the crime occurred. The DNA profiles from the beer can and the bed sheets matched each other, but no suspect had been developed yet. Because of intense darkness and then the blindfold, the only description police had from the victims was that the suspect was black, medium height, and felt heavy set.

A suspect list was developed by the Charlottesville Police Department that contained over 40 individuals, some from the sex offender registry and some with extensive criminal histories who were stopped late at night in the area of the home invasion. Unfortunately, no further leads were available leaving the victims as well as other University of Virginia students and their parents suspicious and fearful. The police were at the end of their rope and considered asking many of the people on the suspect list to voluntarily donate blood samples for purposes of a DNA comparison. The top suspects were systematically eliminated by DNA evidence leaving the police frustrated.

Then on 5 October, six long weeks after the crime had been committed, the lead detective on the case, Lieutenant J.E. 'Chip' Harding of the Charlottesville Police Department, received a call that he describes as being 'one of the most exciting phone calls in my 22 years of law enforcement.' A match had been obtained from the crime scene samples to a convicted offender sample submitted to the Virginia DNA Database several years before. The DNA sample for Montaret D. Davis of Norfolk, Virginia was among 8000 samples added to the Virginia DNA Database at the beginning of October 1999. (Since 1989, a Virginia state law has required all felons and juveniles 14 and older convicted of serious crimes to provide blood samples for DNA testing.)

A quick check for the whereabouts of Mr. Davis found him in the Albemarle-Charlottesville Regional Jail. Ironically, because of a parole violation, he had been court ordered weeks before to report to jail on what turned out to be the same day as the rape. Amazingly enough he had turned himself in at 6 p.m. just 14 hours after committing the sexual assault! Unless he would have bragged about his crime, it is doubtful that Mr. Davis would ever have made it on the suspect list without the power of DNA testing and an expanding DNA database. At his jury trial in April 2000, Mr. Davis was found guilty of rape, forcible sodomy, and abduction among other charges and sentenced to a 90-year prison term.

DNA typing, since it was introduced in the mid-1980s, has revolutionized forensic science and the ability of law enforcement to match perpetrators with crime scenes. Thousands of cases have been closed and innocent suspects freed with guilty ones punished because of the power of a silent biological witness at the crime scene. This book will explore the science behind DNA typing and the biology, technology, and genetics that make DNA typing the most useful investigative tool to law enforcement since the development of fingerprinting over 100 years ago.

HISTORY OF FORENSIC DNA ANALYSIS

'DNA fingerprinting' or DNA typing (profiling) as it is now known, was first described in 1985 by an English geneticist named Alec Jeffreys. Dr. Jeffreys found that certain regions of DNA contained DNA sequences that were repeated over and over again next to each other. He also discovered that the number of repeated sections present in a sample could differ from individual to individual. By developing a technique to examine the length variation of these DNA repeat sequences, Dr. Jeffreys created the ability to perform human identity tests.

These DNA repeat regions became known as VNTRs, which stands for variable number of tandem repeats. The technique used by Dr. Jeffreys to examine the VNTRs was called restriction fragment length polymorphism (RFLP) because it involved the use of a restriction enzyme to cut the regions of DNA surrounding the VNTRs. This RFLP method was first used to help in an English immigration case and shortly thereafter to solve a double homicide case (see D.N.A. Box 1.1). Since that time, human identity testing using DNA typing methods has been widespread. The past 15 years have seen tremendous growth in the use of DNA evidence in crime scene investigations as well as paternity testing. Today over 150 public forensic laboratories and several dozen private paternity testing laboratories conduct hundreds of thousands of DNA tests annually in the United States. In addition, most countries in Europe and Asia

---

D.N.A. Box 1.1

First use of forensic DNA testing

The first use of DNA testing in a forensic setting came in 1986. Two young girls, Lynda Mann and Dawn Ashworth, were sexually assaulted and then left brutally murdered in 1983 and 1986. Both murders occurred near the village of Narborough in Leicestershire, England with similar features leading the police to suspect that the same man had committed the crimes. A local man confessed to killing one of the girls and his blood was compared to semen recovered from the crime scenes. The man did not match evidence from either crime! Thus, the first use of DNA was to demonstrate innocence of someone who might otherwise have been convicted.

A mass screen to collect blood for DNA testing from all adult men in three local villages was conducted in a thorough search for the killer. Over 4000 men were tested without a match. About a year later a woman at a bar overheard someone bragging about how he had given a blood sample for a friend named Colin Pitchfork. The police interviewed Mr. Pitchfork, collected a blood sample from him, and found that his DNA profile matched semen from both murder scenes. He was subsequently convicted and sentenced to life in prison.

The story behind the first application of forensic DNA typing or genetic fingerprinting, as it was then called, has been well told in Joseph Wambaugh's *The Blooding*. The DNA typing methods used were Alec Jeffrey's multi-locus RFLP probes, which he first described in 1985. Since it was first used almost 20 years ago, DNA testing has progressed to become a sensitive and effective tool to aid in bringing the guilty to justice and in exonerating the innocent.

Source:
Joseph Wambaugh (1989) *The Blooding*. New York: Bantam Books;
   http://www.forensic.gov.uk

have forensic DNA programs. The number of laboratories around the world conducting DNA testing will continue to grow as the technique gains in popularity within the law enforcement community.

COMPARISON OF DNA TYPING METHODS

Technologies used for performing forensic DNA analysis differ in their ability to differentiate two individuals and in the speed with which results can be obtained. The speed of analysis has dramatically improved for forensic DNA analysis. DNA testing that previously took 6 or 8 weeks can now be performed in a few hours.

The human identity testing community has used a variety of techniques including single-locus probe and multi-locus probe RFLP methods and more recently PCR (polymerase chain reaction)-based assays. Numerous advances have been made in the last 15 years in terms of sample processing speed and sensitivity. Instead of requiring large blood stains with well-preserved DNA, tiny amounts of sample, as little as a single cell in some cases, can yield a useful DNA profile.

The gamut of DNA typing technologies used over the past 15 years for human identity testing is compared in Figure 1.1. The various DNA markers have been divided into four quadrants based on their power of discrimination, i.e., their ability to discern the difference between individuals, and the speed at which

Figure 1.1

Comparison of DNA typing technologies. Forensic DNA markers are arbitrarily plotted in relationship to four quadrants defined by the power of discrimination for the genetic system used and the speed at which the analysis for that marker may be performed. Note that this diagram does not reflect the usefulness of these markers in terms of forensic cases.

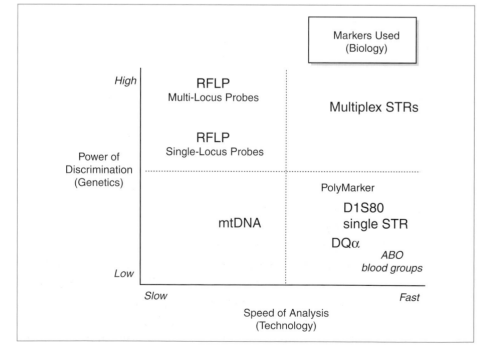

they can be analyzed. New and improved methods have developed over the years such that tests with a high degree of discrimination can now be performed in a few hours.

An ABO blood group determination, which was the first genetic tool used for distinguishing between individuals, can be performed in a few minutes but is not very informative. There are only four possible groups that are typed – A, B, AB, and O – and 40% of the population is type O. Thus, while the ABO blood groups are useful for excluding an individual from being the source of a crime scene sample, the test is not very useful when an inclusion has been made, especially if the sample is type O.

On the other extreme, multi-locus RFLP probes are highly variable between individuals but require a great deal of labor, time, and expertise to produce a DNA profile. Analysis of multi-locus probes (MLP) cannot be easily automated, a fact that makes them undesirable as the demand for processing large numbers of DNA samples has increased. Deciphering sample mixtures, which are common in forensic cases, is also a challenge with MLP RFLP methods, which is the primary reason that laboratories went to single-locus RFLP probes used in serial fashion.

The best solution including a high power of discrimination and a rapid analysis speed has been achieved with short tandem repeat (STR) DNA markers, shown in the upper right quadrant of Figure 1.1. Also because STRs by definition are short, they can be analyzed three or more at a time. Multiple STRs can be examined in the same DNA test, or 'multiplexed.' Multiplex STRs are valuable because they can produce highly discriminating results (Chapter 5) and can successfully measure sample mixtures and biological materials containing degraded DNA molecules (Chapter 7). In addition, the detection of multiplex STRs can be automated, which is an important benefit as demand for DNA testing increases.

It should be noted though that Figure 1.1 does not fully reflect the usefulness of these markers in terms of forensic cases. Mitochondrial DNA (mtDNA), which is shown in the quadrant with the lowest power of discrimination and longest sample processing time, can be very helpful in forensic cases involving severely degraded DNA samples or when associating maternally related individuals (Chapter 10). In many situations, multiple technologies may be used to help resolve an important case or identify victims of mass disasters, such as those from the World Trade Center collapse (Chapter 24).

Over the past 20 years, there has been a gradual evolution in adoption of the various DNA typing technologies shown in Figure 1.1. When early methods for DNA analysis are superseded by new technologies, there is usually some overlap as forensic laboratories implement the new technology. Validation of the new methods is crucial to maintaining high quality results (Chapter 16). Table 1.1 lists some of the major historical events in forensic DNA typing. The implementation of new methods by the FBI Laboratory has been listed in this historical timeline because the DNA casework protocols used by the FBI create an important trend within the United States and around the world.

STEPS IN DNA SAMPLE PROCESSING

This book contains a review of the steps involved in processing forensic DNA samples with STR markers. STRs are a smaller version of the VNTR sequences first described by Dr. Jeffreys. Samples obtained from crime scenes or paternity investigations are subjected to defined processes involving biology, technology, and genetics (Figure 1.2).

BIOLOGY

Following collection of biological material from a crime scene or paternity investigation, the DNA is first extracted from its biological source material and

Figure 1.2

Overview of biology, technology, and genetic components of DNA typing using short tandem repeat (STR) markers.

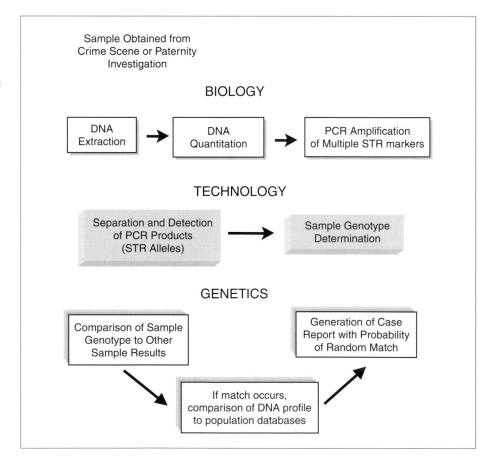

then measured to evaluate the quantity of DNA recovered (Chapter 3). After isolating the DNA from its cells, specific regions are copied with a technique known as the polymerase chain reaction, or PCR (Chapter 4). PCR produces millions of copies for each DNA segment of interest and thus permits very minute amounts of DNA to be examined. Multiple STR regions can be examined simultaneously to increase the informativeness of the DNA test (Chapter 5).

TECHNOLOGY

The resulting PCR products are then separated and detected in order to characterize the STR region being examined. The separation methods used today include slab gel and capillary electrophoresis (CE) (Chapter 12). Fluorescence detection methods have greatly aided the sensitivity and ease of measuring PCR-amplified STR alleles (Chapter 13). The primary instrument platform used in the United States for fluorescence detection of STR alleles is currently the ABI Prism 310 Genetic Analyzer (Chapter 14). After detecting the STR alleles, the number of repeats in a DNA sequence is determined, a process known as sample genotyping (Chapter 15).

The specific methods used for DNA typing are validated by individual laboratories to ensure that reliable results are obtained (Chapter 16) and before new technologies (see Chapter 17) are implemented. DNA databases, such as the one described earlier in this chapter to match Montaret Davis to his crime scene, are valuable tools and will continue to play an important role in law enforcement efforts (Chapter 18).

GENETICS

The resulting DNA profile for a sample, which is a combination of individual STR genotypes, is compared to other samples. In the case of a forensic investigation, these other samples would include known reference samples such as the victim or suspects that are compared to the crime scene evidence. With paternity investigations, a child's genotype would be compared to his or her mother's and the alleged father(s) under investigation (Chapter 23). If there is not a match between the questioned sample and the known sample, then the samples may be considered to have originated from different sources (see D.N.A. Box 1.2). The term used for failure to match between two DNA profiles is 'exclusion.'

If a match or 'inclusion' results, then a comparison of the DNA profile is made to a population database, which is a collection of DNA profiles obtained from unrelated individuals of a particular ethnic group (Chapter 20). For example, due to genetic variation between the groups, African-Americans and Caucasians have different population databases for comparison purposes.

Forensic DNA testing can play a role in protecting the innocent as well as implicating the guilty. In recent years, the use of DNA evidence to free people from prison has been highly publicized and has altered some perceptions of the criminal justice system. For example, capital punishment in Illinois was put on hold by the governor after learning of several inmates being exonerated by post-conviction DNA testing.

As of May 2004, a total of 143 people including some 'death row' inmates previously incarcerated for crimes they did not commit have been released from prison thanks to the power of modern forensic DNA typing technologies. Many of these wrongfully convicted individuals were found 'guilty' prior to the development of DNA typing methods in the mid-1980s based on faulty eyewitness accounts or circumstantial evidence. Fortunately for the 143 so far exonerated by post-conviction DNA testing some evidence was preserved in police lockers that after many years could be used for DNA testing. The results successfully excluded them as the perpetrator of the crimes for which they were falsely convicted and imprisoned.

Defense attorneys Barry Scheck and Peter Neufeld launched the Innocence Project in 1992 at the Benjamin N. Cardozo School of Law in New York City. This non-profit legal clinic promotes cases where evidence is available for post-conviction DNA testing and can help demonstrate innocence. The Innocence Project has grown to include an Innocence Network of more than 40 law schools and other organizations around the United States and Australia. Law students and staff carefully evaluate thousands of requests for DNA testing to prove prisoners' innocence. In spite of careful screening, when post-conviction testing is conducted, DNA test results more often than not further implicate the defendant. However, the fact that truly innocent people have been behind bars for a decade or more has promoted legislation in a number of states and also at the federal level to fund post-conviction DNA testing. The increased use of DNA analysis for this purpose will surely impact the future of the criminal justice system.

Source:
http://www.innocenceproject.org

Finally a case report or paternity test result is generated. This report typically includes the random match probability for the match in question (see example in D.N.A. Box 1.3). This random match probability is the chance that a randomly selected individual from a population will have an identical STR profile or combination of genotypes at the DNA markers tested (Chapter 21).

STR MULTIPLEX EXAMPLE

An example of DNA profiles obtained from two different individuals using STR markers is shown in Figure 1.3. In a single amplification reaction, unique sites

D.N.A. Box 1.3

DNA evidence and Monica Lewinsky's blue dress

Can a simple DNA test have the power to impact world events? In 1998, independent counsel Kenneth Starr was investigating allegations that U.S. President William Jefferson Clinton had a sexual relationship with a young White House intern, Monica Lewinsky. President Clinton had publicly denied the allegations quite emphatically and at that time there was no concrete evidence to the contrary.

During the course of the investigation, a dark blue dress belonging to Monica Lewinsky was brought to the FBI Laboratory for examination. Semen was identified on evidence item Q3243, as the dress was cataloged. The unknown semen stain was quickly examined with seven RFLP single locus probes. Late on the evening of 3 August 1998, a reference blood sample was drawn from President Clinton for comparison purposes (Woodward 1999).

As in the O.J. Simpson case (see D.N.A. Box 3.2), conventional RFLP markers were used to match the sample of President Clinton's blood to the semen stain on Monica Lewinski's dress. At the time these samples were run in the FBI Laboratory (early August 1998), STR typing methods were being validated but were not yet in routine use within the FBI's DNA Analysis Unit. High molecular weight DNA from the semen stain (FBI specimen Q3243-1) and President Clinton's blood (FBI specimen K39) was digested with the restriction enzyme *Hae*III. A seven-probe match was obtained at all seven RFLP loci examined.

This match was reported in the following manner: 'Based on the results of these seven genetic loci, specimen K39 (CLINTON) is the source of the DNA obtained from specimen Q3243-1, to a reasonable degree of scientific certainty.' The random match probability was calculated to be on the order of 1 in 7.8 trillion when compared to a Caucasian population database.

When faced with this indisputable DNA evidence, President Clinton found himself in a tight spot. Earlier statements that he had not had 'sexual relations' with Miss Lewinsky were now in doubt. The DNA results along with other evidence and testimony resulted in the impeachment of President Clinton on 19 December 1998 – only the second President in U.S. history to be impeached. This physical evidence played an important role in demonstrating that a sexual relationship had existed between Miss Lewinsky and President William Jefferson Clinton. Although during the Senate impeachment trial, it was determined that his deeds were not serious enough for him to be removed from office, President Clinton's career will always be tainted by the semen stain on the now famous blue dress.

Sources:

Woodward, B. (1999) *Shadow: Five Presidents and the Legacy of Watergate.* New York: Simon & Schuster.

Grunwald, L. and Adler, S.J. (eds) (1999) *Letters of the Century: America 1900–1999*, p. 673. New York: The Dial Press.

on ten different chromosomes were probed with this DNA test to provide a random match probability of approximately 1 in 3 trillion. Note that every single site tested produces a different result between these two DNA samples. For example, marker A has two peaks in the top panel and only one peak in the bottom panel. Likewise, marker J produces two peaks in both samples but they result in different patterns due to different sizes at the site measured in the two DNA samples. These results can be reliably obtained in as little as a few hours from a very small drop of blood or bloodstain.

Each STR allele is distinguished from the others in the amplification reaction by separating it based on its length and color. The color results from a fluorescent dye that is attached during the amplification reaction. In this example, DNA markers B, E, H, and J are labeled with a blue colored dye, markers A, D, and G are labeled with a yellow dye, and markers C, F, I, and the gender ID are labeled in green. The gender ID results in two peaks for a male sample (X,Y) and a single peak for a female sample (X,X). Chapter 5 will describe the identity of the DNA markers represented in Figure 1.3.

COMPARISONS TO COMPUTER TECHNOLOGY

In order to get a better feel for how rapidly forensic DNA analysis methods have progressed in the last two decades, a comparison to computer technology may be helpful. The use of computers at home and in the workplace has increased dramatically since personal computers became available in the mid-1980s.

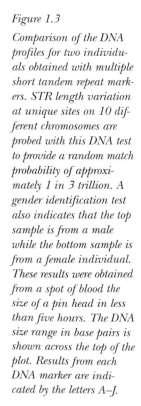

Figure 1.3

Comparison of the DNA profiles for two individuals obtained with multiple short tandem repeat markers. STR length variation at unique sites on 10 different chromosomes are probed with this DNA test to provide a random match probability of approximately 1 in 3 trillion. A gender identification test also indicates that the top sample is from a male while the bottom sample is from a female individual. These results were obtained from a spot of blood the size of a pin head in less than five hours. The DNA size range in base pairs is shown across the top of the plot. Results from each DNA marker are indicated by the letters A–J.

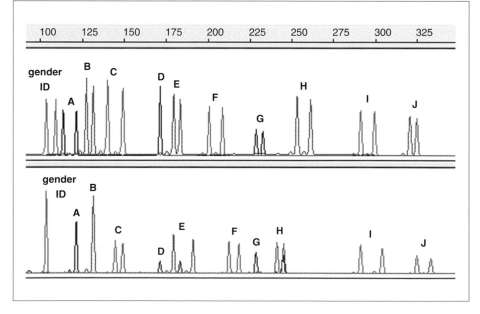

These computers get faster and more powerful every year. It is almost inconceivable that the Internet, which has such a large impact on our daily lives, was just an idea a few years ago.

When multi-locus RFLP probes were first reported in 1985, the average computer operating speed was less than 25 MHz. Almost 20 years later in the year 2004, computing speeds of 2500 MHz (2.5 GHz) are now common. Computer processing speeds and capabilities have increased rapidly every year. Likewise, the ability of laboratories to perform DNA typing methods has improved dramatically along a similar timeline due to rapid progress in the areas of biology, technology, and understanding of genetic theories. In addition, the power of discrimination for DNA tests has steadily increased in the late 1990s (see Table 5.3, Table 20.8).

Some interesting parallels can be drawn between the Microsoft Corporation, the company that has led the computer technology revolution, and the timing for advancements in the field of forensic DNA typing (Table 1.1). In 1985, the

Table 1.1

Major historical events in forensic DNA typing shown by year. The events relating to forensic DNA (first column) are described in context with parallel developments in biotechnology (second column) and key events relating to Microsoft Corporation, which have impacted the computer age (final column).

Year	Forensic DNA Science & Application	Parallel Developments in Biotechnology	Microsoft Corporation Chronology
1985	Alec Jeffreys develops multi-locus RFLP probes	PCR process first described	First version of Windows shipped
1986	DNA testing goes public with Cellmark and Lifecodes in United States	Automated DNA sequencing with 4-colors first described	Microsoft goes public
1988	FBI begins DNA casework with single-locus RFLP probes		
1989	TWGDAM established; NY *v.* Castro case raises issues over quality assurance of laboratories	DNA detection by gel silver-staining, slot blot, and reverse dot blots first described	
1990	Population statistics used with RFLP methods are questioned; PCR methods start with DQA1	Human Genome Project begins with goal to map all human genes	Windows 3.0 released (quality problems); exceeds $1 billion in sales
1991	Fluorescent STR markers first described; Chelex extraction		Windows 3.1 released
1992	NRC I Report; FBI starts casework with PCR-DQA1	Capillary arrays first described	
1993	First STR kit available; sex-typing (amelogenin) developed	First STR results with CE	
1994	Congress authorizes money for upgrading state forensic labs; 'DNA wars' declared over; FBI starts casework with PCR-PM	Hitachi FMBIO and Molecular Dynamics gel scanners; first DNA results on microchip CE	

Year	Forensic DNA Science & Application	Parallel Developments in Biotechnology	Microsoft Corporation Chronology
1995	O.J. Simpson saga makes public more aware of DNA; DNA Advisory Board setup; UK DNA Database established; FBI starts using D1S80/amelogenin	ABI 310 Genetic Analyzer and TaqGold DNA polymerase introduced	Windows 95 released
1996	NRC II Report; FBI starts mtDNA testing; first multiplex STR kits become available	STR results with MALDI-TOF and GeneChip mtDNA results demonstrated	
1997	13 core STR loci defined; Y-chromosome STRs described		Internet Explorer begins overtaking Netscape
1998	FBI launches national Combined DNA Index System; Thomas Jefferson and Bill Clinton implicated with DNA	2000 SNP hybridization chip described	Windows 98 released; anti-trust trial with U.S. Justice Department begins
1999	Multiplex STR kits are validated in numerous labs; FBI stops testing DQA1/PM/D1S80	ABI 3700 96-capillary array for high-throughput DNA analysis; chromosome 22 fully sequenced	
2000	FBI and other labs stop running RFLP cases and convert to multiplex STRs; PowerPlex 16 kit enables first single amplification of CODIS STRs	First copy of human genome completed	Bill Gates steps down as Microsoft CEO; Windows 2000 released
2001	Identifiler STR kit released with 5-dye chemistry; first Y-STR kit becomes available	ABI 3100 Genetic Analyzer introduced	Windows XP released
2002	FBI mtDNA population database released; Y-STR 20plex published		Windows XP Tablet PC Edition released
2003	U.S. DNA database (NDIS) exceeds 1 million convicted offender profiles; the UK National DNA Database passes the 2 million sample mark	Human Genome Project completed with the 'final' sequence coinciding with 50th anniversary of Watson–Crick DNA discovery	Windows Server 2003 released; 64-Bit Operating Systems expand capabilities of software

Table 1.1 (Continued)

year that Alec Jeffreys first published his work with multi-locus RFLP probes, Microsoft shipped its first version of Windows software to serve as a computer operating system. In 1986, as DNA testing began to 'go public' in the United States with Cellmark and Lifecodes performing multi-locus RFLP, Microsoft went public with a successful initial public offering.

In the late 1980s, single-locus RFLP probes began to be used by the FBI Laboratory in DNA casework. Due to issues over the use of statistics for population genetics and the quality of results obtained in forensic laboratories, RFLP methods were questioned by the legal community in 1989 and the early 1990s. At this same time, Microsoft had quality problems of their own with the

Windows 3.0 operating system. However, they 'turned the corner' with their product release of Windows 3.1 in 1991. In the same year, improved methods for DNA typing were introduced, namely fluorescent STR markers and Chelex extraction.

The popularity of Microsoft products improved in 1995 with the release of Windows 95. During this same year, forensic DNA methods gained public exposure and popularity due to the O.J. Simpson trial. The United Kingdom also launched a National DNA Database that has revolutionized the use of DNA as an investigative tool. The United States launched their national Combined DNA Index System (CODIS) in 1998, concurrent with the release of Windows 98.

To aid sample throughput and processing speed, the FBI Laboratory and many other forensic labs have stopped running RFLP cases as of the year 2000. On 13 January 2000, Bill Gates stepped down as the CEO of Microsoft in order to help his company move into new directions.

The development and release of Windows 2000 and Windows XP at the beginning of the 21st century continue to improve the capabilities of multitasking computer software. In like manner, the development and release of new DNA testing kits capable of single amplification reactions for examining 16 regions of the human genome furthers the capability of multiplexing DNA information (see Chapter 5).

We recognize that due to the rapid advances in the field of forensic DNA typing, some aspects of this book may be out of date by the time it is published, much like a computer is no longer the latest model by the time it is purchased. However, a reader should be able to gain a fundamental understanding of forensic DNA typing from the following pages. While we cannot predict the future with certainty, short tandem repeat DNA markers have had and will continue to have an important role to play in forensic DNA typing due to their use in DNA databases.

The match on Mr. Davis described at the beginning of this chapter was made with eight STR markers. These eight STRs are a subset of 13 STR markers described in detail throughout this book that will most likely be used in DNA databases around the world for many years to come. Perhaps with odds of getting caught becoming greater than ever before, violent criminals like Mr. Davis will think twice before carrying out such heinous actions.

BIOLOGY

CHAPTER 2

DNA BIOLOGY REVIEW

Today, we are learning the language in which God created life.

(President Bill Clinton, 26 June 2000
announcing the first draft sequence of the human genome)

BASIC DNA PRINCIPLES

The basic unit of life is the cell, which is a miniature factory producing the raw materials, energy, and waste removal capabilities necessary to sustain life. Thousands of different proteins, called enzymes, are required to keep these cellular factories operational. An average human being is composed of approximately 100 trillion cells, all of which originated from a single cell. Each cell contains the same genetic programming. Within the nucleus of our cells is a chemical substance known as DNA that contains the informational code for replicating the cell and constructing the needed enzymes. Because the DNA resides in the nucleus of the cell, it is often referred to as nuclear DNA. (As will be discussed in Chapter 10, some minor extranuclear DNA exists in human mitochondria, which are the cellular powerhouses.)

Deoxyribonucleic acid, or DNA, is sometimes referred to as our genetic blueprint because it stores the information necessary for passing down genetic attributes to future generations. Residing in every cell of our body (with the exception of red blood cells, which lack nuclei), DNA provides a 'computer program' that determines our physical features and many other attributes. The complete set of instructions for making an organism, i.e., the entire DNA in a cell, is referred to collectively as its genome.

DNA has two primary purposes: (1) to make copies of itself so cells can divide and carry on the same information; and (2) to carry instructions on how to make proteins so cells can build the machinery of life. Information encoded within the DNA structure itself is passed on from generation to generation with one-half of a person's DNA information coming from their mother and one-half coming from their father.

DNA STRUCTURE AND DEFINITIONS

Nucleic acids including DNA are composed of nucleotide units that are made up of three parts: a nucleobase, a sugar, and a phosphate (Figure 2.1). The nucleobase or 'base' imparts the variation in each nucleotide unit while the phosphate and sugar portions form the backbone structure of the DNA molecule.

The DNA alphabet is composed of only four characters representing the four nucleobases: A (adenine), T (thymine), C (cytosine), and G (guanine).

Figure 2.1

Basic components of nucleic acids: (a) phosphate sugar backbone with bases coming off the sugar molecules, (b) chemical structure of phosphates and sugar molecules illustrating numbering scheme on the sugar carbon atoms. DNA sequences are conventionally written from 5′ to 3′.

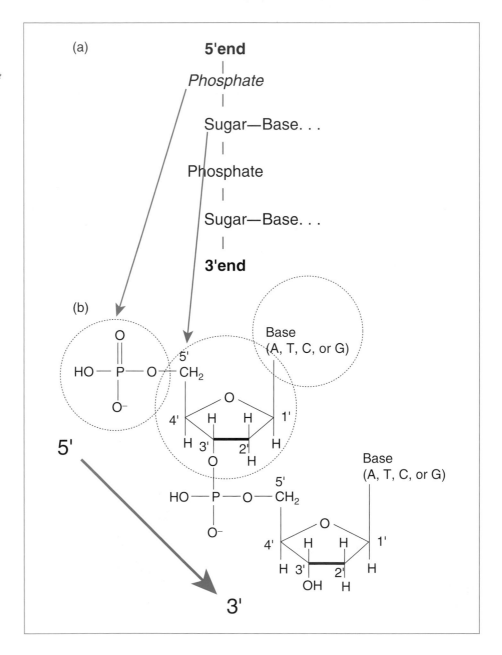

The various combinations of these four letters, known as nucleotides or bases, yield the diverse biological differences among human beings and all living creatures. Humans have approximately three billion nucleotide positions in their genomic DNA. Thus, with four possibilities (A, T, C, or G) at each position, literally trillions of combinations are possible. The informational content of DNA is encoded in the order (sequence) of the bases just as computers store binary information in a string of ones and zeros.

Directionality is provided when listing a DNA sequence by designating the '5′-end' and the '3′-end.' This numbering scheme comes from the chemical structure of DNA and refers to the position of carbon atoms in the sugar ring of the DNA backbone structure (Figure 2.1). A sequence is normally written (and read) from 5′ to 3′ unless otherwise stated. DNA polymerases, the enzymes that copy DNA, only 'write' DNA sequence information from 5′ to 3′, much like we read words and sentences from left to right.

BASE PAIRING AND HYBRIDIZATION OF DNA STRANDS

In its natural state in the cell, DNA is actually composed of two strands that are linked together through a process known as *hybridization*. Individual nucleotides pair up with their 'complementary base' through hydrogen bonds that form between the bases. The base pairing rules are such that adenine can only hybridize to thymine and cytosine can only hybridize to guanine (Figure 2.2).

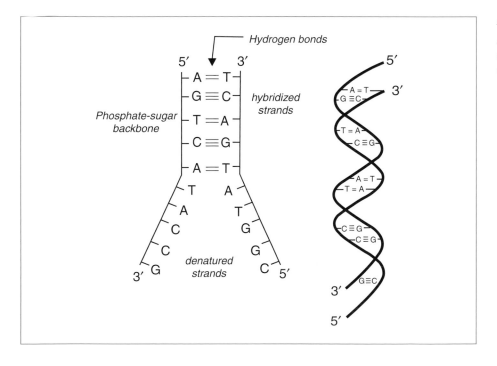

Figure 2.2

Base pairing of DNA strands to form double-helix structure.

There are two hydrogen bonds between the adenine–thymine base pair and three hydrogen bonds between the guanine–cytosine base pair. Thus, GC base pairs are stuck together a little stronger than AT base pairs. The two DNA strands form a double helix due to this 'base-pairing' phenomenon (Figure 2.2).

The two strands of DNA are 'anti-parallel', that is one strand is in the 5′ to 3′ orientation and the other strand lines up in the 3′ to 5′ direction relative to the first strand. By knowing the sequence of one DNA strand, its complementary sequence can easily be determined based on the base pairing rules of A with T and G with C. These combinations are sometimes referred to as Watson–Crick base pairs for James Watson and Francis Crick who discovered this structural relationship in 1953.

Hybridization of the two strands is a fundamental property of DNA. However, the hydrogen bonds holding the two strands of DNA together through base pairing may be broken by elevated temperature or by chemical treatment, a process known as *denaturation*. A common method for denaturing double-stranded DNA is to heat it to near boiling temperatures. The DNA double helix can also be denatured by placing it in a salt solution of low ionic strength or by exposing it to chemical denaturants such as urea or formamide, which destabilize DNA by forming hydrogen bonds with the nucleotides and preventing their association with a complementary DNA strand.

Denaturation is a reversible process. If a double-stranded piece of DNA is heated up, it will separate into its two single strands. As the DNA sample cools, the single DNA strands will find their complementary sequence and rehybridize or anneal to each other. The process of the two complementary DNA strands coming back together is referred to as *renaturation* or *reannealing*.

CHROMOSOMES, GENES, AND DNA MARKERS

There are approximately three billion base pairs in a single copy of the human genome. Obtaining a complete catalog of our genes was the focus of the Human Genome Project, which announced a final reference sequence for the human genome in April 2003 (D.N.A. Box 2.1). The information from the Human Genome Project will benefit medical science as well as forensic human identity testing and help us better understand our genetic makeup.

Within human cells, DNA found in the nucleus of the cell (nuclear DNA) is divided into chromosomes, which are dense packets of DNA and protection proteins called histones. The human genome consists of 22 matched pairs of autosomal chromosomes and two sex determining chromosomes (Figure 2.3). Thus, normal human cells contain 46 different chromosomes or 23 pairs of chromosomes. Males are designated XY because they contain a single copy of the X chromosome and a single copy of the Y chromosome while females

D.N.A. Box 2.1
The Human Genome Project

Molecular biology's equivalent to NASA's Apollo Space Program began in 1990 when a multi-billion, 15-year project was launched to decipher the DNA sequence contained inside a human cell. The Human Genome Project began under the leadership of James Watson, the scientist who with Francis Crick first determined the double-helix structure of DNA in 1953. With joint funding from the U.S. National Institutes of Health and the Department of Energy, efforts in the United States began with examining genetic and physical maps of human DNA and other model organisms such as yeast, *Drosophila* (fruit fly) and the mouse.

In 1992, Francis Collins took over the helm of the Human Genome Project. Amazingly over the years the project met or exceeded its milestones and stayed under budget. In 1999, a private sector enterprise named Celera under the leadership of Craig Venter challenged the public sector to a sequencing duel. The competition in large measure drove the Human Genome Project forward leading to the announcement of a draft sequence in June 2000, its publication in February 2001, and a 'final' sequence in April 2003.

The medical community will likely be the largest beneficiaries of the Human Genome Project as we come to better understand the genetic basis for various diseases. This information raises legal and ethical issues as scientists and policy makers struggle with genetic privacy concerns and intellectual property rights. Undertaking such an enormous project has accelerated technology development and will continue to aid in the understanding of our species. Now that a reference sequence is in place, efforts have turned to understanding normal variation that occurs among different individuals in the International Haplotype Mapping ('HapMap') Project.

Source:
http://www.genome.gov

contain two copies of the X chromosomes and are designated XX. Most human identity testing is performed using markers on the autosomal chromosomes, and gender determination is done with markers on the sex chromosomes. As will be discussed in Chapters 9 and 10, the Y chromosome and mitochondrial DNA, a small, multi-copy genome located in cell's mitochondria, can also be used in human identification applications.

Chromosomes in all body (somatic) cells are in a *diploid* state; they contain two sets of each chromosome. On the other hand, gametes (sperm or egg) are in a *haploid* state; they have only a single set of chromosomes. When an egg cell and a sperm cell combine during conception, the resulting zygote becomes diploid again. Thus, one chromosome in each chromosomal pair is derived from each parent at the time of conception.

Mitosis is the process of nuclear division in somatic cells that produces daughter cells, which are genetically identical to each other and to the parent cell.

Figure 2.3

The human genome contained in every cell consists of 23 pairs of chromosomes and a small circular genome known as mitochondrial DNA. Chromosomes 1–22 are numbered according to their relative size and occur in single copy pairs within a cell's nucleus with one copy being inherited from one's mother and the other copy coming from one's father. Sex-chromosomes are either X,Y for males or X,X for females. Mitochondrial DNA is inherited only from one's mother and is located in the mitochondria with hundreds of copies per cell. Together the nuclear DNA material amounts to over three billion base pairs (bp) while mitochondrial DNA is only about 16569 bp in length.

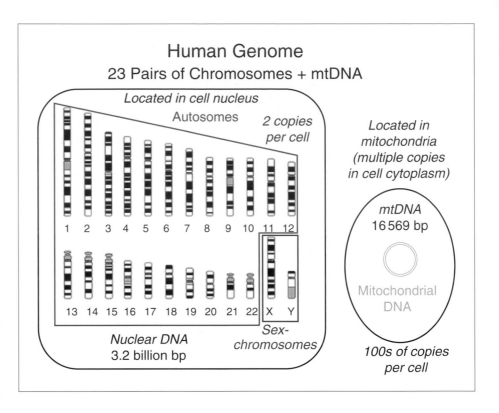

Meiosis is the process of cell division in sex cells or gametes. In meiosis, two consecutive cell divisions result in four rather than two daughter cells, each with a haploid set of chromosomes.

The DNA material in chromosomes is composed of 'coding' and 'non-coding' regions. The coding regions are known as *genes* and contain the information necessary for a cell to make proteins. A gene usually ranges from a few thousand to tens of thousands of base pairs in size. One of the big surprises to come out of the Human Genome Project is that humans have less than 30 000 protein-coding genes rather than the 50 000–100 000 previously thought.

Genes consist of *exons* (protein-coding portions) and *introns* (the intervening sequences). Genes only make up ~5% of human genomic DNA. Non-protein coding regions of DNA make up the rest of our chromosomal material. Because these regions are not related directly to making proteins they are sometimes referred to as 'junk' DNA. Markers used for human identity testing are found in the non-coding regions either between genes or within genes (i.e., introns) and thus do not code for genetic variation.

Polymorphic (variable) markers that differ among individuals can be found throughout the non-coding regions of the human genome. The chromosomal position or location of a gene or a DNA marker in a non-coding region is commonly referred to as a *locus* (plural: *loci*). Thousands of loci have been characterized and mapped to particular regions of human chromosomes through the worldwide efforts of the Human Genome Project.

Pairs of chromosomes are described as *homologous* because they are the same size and contain the same genetic structure. A copy of each gene resides at the same position (locus) on each chromosome of the homologous pair. One chromosome in each pair is inherited from an individual's mother and the other from his or her father. The DNA sequence for each chromosome in the homologous pair may or may not be identical since mutations may have occurred over time.

The alternative possibilities for a gene or genetic locus are termed *alleles*. If the two alleles at a genetic locus on homologous chromosomes are different they are termed *heterozygous* and if the alleles are identical at a particular locus, they are termed *homozygous*. Detectable differences in alleles at corresponding loci are essential to human identity testing.

A *genotype* is a characterization of the alleles present at a genetic locus. If there are two alleles at a locus, A and a, then there are three possible genotypes: AA, Aa, and aa. The AA and aa genotypes are homozygous while the Aa genotype is heterozygous. A *DNA profile* is the combination of genotypes obtained for multiple loci. DNA typing or DNA profiling is the process of determining the genotype present at specific locations along the DNA molecule. Multiple loci are typically examined in human identity testing to reduce the possibility of a random match between unrelated individuals.

NOMENCLATURE FOR DNA MARKERS

The nomenclature for DNA markers is fairly straightforward. If a marker is part of a gene or falls within a gene, the gene name is used in the designation. For example, the short tandem repeat (STR) marker TH01 is from the human **t**yrosine **h**ydroxylase gene located on chromosome 11. The '01' portion of TH01 comes from the fact that the repeat region in question is located within intron 1 of the tyrosine hydroxylase gene. Sometimes the prefix HUM- is included at the beginning of a locus name to indicate that it is from the human genome. Thus, the STR locus TH01 would be correctly listed as HUMTH01.

DNA markers that fall outside of gene regions may be designated by their chromosomal position. The STR loci D5S818 and DYS19 are examples of markers that are not found within gene regions. In these cases, the 'D' stands for DNA. The next character refers to the chromosome number, 5 for chromosome 5 and Y for the Y chromosome. The 'S' refers to the fact that the DNA marker is a single copy sequence. The final number indicates the order in which the marker was discovered and categorized for a particular chromosome. Sequential numbers are used to give uniqueness to each identified DNA marker. Thus, for the DNA marker D16S539:

D16S539
D: DNA
16:     chromosome 16
S: single copy sequence
539:    539th locus described on chromosome 16

DESIGNATING PHYSICAL CHROMOSOME LOCATIONS

The basic regions of a chromosome are illustrated in Figure 2.4. The center region of a chromosome, known as the *centromere*, controls the movement of the chromosome during cell division. On either side of the centromere are 'arms' that terminate with *telomeres* (Figure 2.4). The shorter arm is referred to as 'p' while the longer arm is designated 'q'.

Figure 2.4

Basic chromosome structure and nomenclature. The centromere is a distinctive feature of chromosomes and plays an important role during mitosis. On either side of the centromere are 'arms' that extend to terminal regions, known as telomeres. The short arm of a chromosome is designated as 'p' while the long arm is referred to as 'q'. The band nomenclature refers to physical staining with a Giemsa dye (G-banded). Band localization is determined by G-banding the image of a metaphase spread during cell division. Bands are numbered outward from the centromere with the largest values near the telomeres.

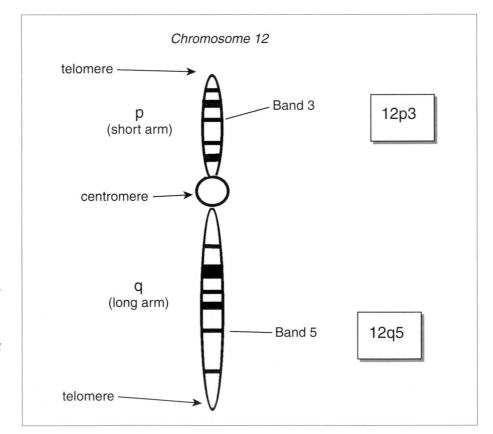

Human chromosomes are numbered based on their overall size with chromosome 1 being the largest and chromosome 22 the smallest. The complete sequence of chromosome 22 was reported in December 1999 to be over 33 million nucleotides in length. Since the Human Genome Project completed its monumental effort in April 2003, we now know the sequence and length of all 23 pairs of human chromosomes.

During most of a cell's life cycle, the chromosomes exist in an unraveled linear form. In this form, it can be transcribed to code for proteins. Regions of chromosomes that are transcriptionally active are known as *euchromatin*. The transcriptionally inactive portions of chromosomes, such as centromeres, are *heterochromatin* regions and are generally not sequenced due to complex repeat patterns found therein. Prior to cell division, during the metaphase step of mitosis, the chromosomes condense into a more compact form that can be observed under a microscope following chromosomal staining. Chromosomes are visualized under a light microscope as consisting of a continuous series of light and dark bands when stained with different dyes. The pattern of light and dark bands results because of different amounts of A and T versus G and C bases across the chromosomes.

A common method for staining chromosomes to obtain a banding pattern is the use of a Giemsa dye mixture that results in so-called 'G-bands' via the 'G-staining' method. These G-bands serve as signposts on the chromosome highway to help determine where a particular DNA sequence or gene is located compared to other DNA markers. The differences in chromosome size and banding patterns allow the 24 chromosomes (22 autosomes and X and Y) to be distinguished from one another, an analysis called a *karotype*.

A DNA or genetic marker is physically mapped to a chromosome location using banding patterns on the metaphase chromosomes. Bands are classified according to their relative positions on the short arm (p) or the long arm (q) of specific chromosomes (Figure 2.4). Thus, the chromosomal location 12p1 means band 1 on the short arm (p) of chromosome 12. The band numbers increase outward from the centromere to the telomere portion of the chromosome. Thus, band 3 is closer to the telomere than band 2. When a particular band is resolved further into multiple bands, its components are named p11, p12, etc. If additional sub-bands are seen as techniques are developed to improve resolution, then these are renamed p11.1, p11.11, etc. For DNA markers close to the terminal ends of the chromosome, the nomenclature 'ter' is often used as a suffix to the chromosome arm designation. The location of a DNA marker might therefore be listed as 15qter, meaning the terminus of the long arm of chromosome 15. Sometimes a DNA marker is not yet mapped with a high degree of accuracy in which case the chromosomal location would be listed as being in a particular range, i.e., 2p23-pter or somewhere between band 23 and the terminus of the short arm on chromosome 2.

POPULATION VARIATION

A vast majority of our DNA molecules (over 99.7%) is the same between people. Only a small fraction of our DNA (0.3% or ~10 million nucleotides) differs between people and makes us unique individuals. These variable regions of DNA provide the capability of using DNA information for human identity purposes. Methods have been developed to locate and characterize this genetic variation at specific sites in the human genome.

TYPES OF DNA POLYMORPHISMS

DNA variation is exhibited in the form of different alleles, or various possibilities at a particular locus. Two forms of variation are possible at the DNA level: sequence polymorphisms and length polymorphisms (Figure 2.5).

As discussed earlier, a genotype is an indication of a genetic type or allele state. A sample containing two alleles, one with 13 and the other with 18 repeat units, would be said to have a genotype of '13,18'. This shorthand method of designating the alleles present in a sample makes it easier to compare results from multiple samples.

In DNA typing, multiple markers or loci are examined. The more DNA markers examined and compared, the greater the chance that two unrelated individuals will have different genotypes. Alternatively, each piece of matching information adds to the confidence in connecting two matching DNA profiles from the same individual. If each locus is inherited independent of the other loci, then a calculation of a DNA profile frequency can be made by multiplying each individual genotype frequency together (see Chapter 21). This is known as the *product rule*.

Owing to the fact that it is currently not feasible in terms of time and expense to evaluate an individual's entire DNA sequence, multiple discrete locations are

Figure 2.5

Two forms of variation exist in DNA: (a) sequence polymorphisms and (b) length polymorphisms. The short tandem repeat DNA markers discussed in this book are length polymorphisms.

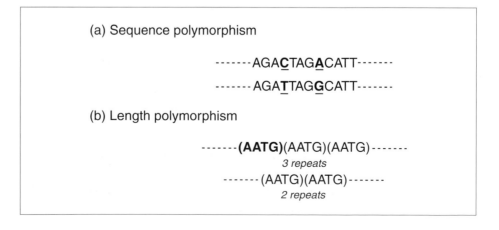

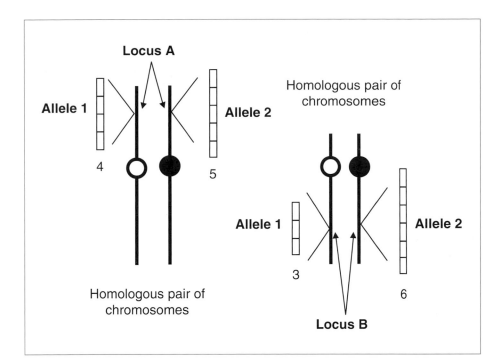

Figure 2.6

Schematic representation of two different STR loci on different pairs of homologous chromosomes. The chromosomes with the open circle centromeres are paternally inherited while the solid centromere chromosomes are maternally inherited. Thus, this individual received the four repeat allele at locus A and the three repeat allele at locus B from their father, and the five repeat allele at locus A and the six repeat allele at locus B from their mother.

evaluated (Figure 2.6). The variability that is observed at these locations is used to include or exclude samples, i.e., do they match or not. Because absolute certainty in DNA identification is not possible in practice, the next best thing is to claim virtual certainty due to the extreme small probabilities of a coincidental (random) match.

DNA searches can be narrowed down by comparing multiple data points in a manner analogous to how the U.S. Postal Service delivers mail. The entire United States has over 290 million individuals but by including the zip code, state, city, street, street number, and name on an envelope, a letter can be delivered to a single, unique individual. Likewise, the use of more and more information from DNA markers can be used to narrow a search down to a single individual. If marker 1, marker 2, marker 3, and so on match on a DNA profile between crime scene evidence and a suspect, one can become more confident that the two DNA types are from the same source. The likelihood increases with each marker match.

GENETIC VARIABILITY

Large amounts of genetic variability exist in the human population. This is evidenced by the fact that with the exception of identical twins, we all appear different from each other. Hair color, eye color, height, and shape all represent alleles in our genetic makeup. To gain a better appreciation for how the numbers of alleles present at a particular locus impacts the variability, let us consider the ABO blood group. Three alleles are possible: A, B, and O. These three alleles can be combined to form three possible homozygous genotypes (AA, BB, and OO) and three heterozygous genotypes (AO, BO, and AB). Thus, with three alleles there are six possible genotypes. However, with AA and AO appearing the same and BB and BO being phenotypically equivalent, there are only four phenotypically expressed blood types: A, B, AB, and O.

With larger numbers of alleles for a particular DNA marker, a greater number of genotypes result. In general, if there are n alleles, there are n homozygous genotypes and $n(n-1)/2$ heterozygous ones. Thus, a locus with ten possible alleles would exhibit 10 homozygous possibilities plus $[10 \times (10-1)]/2$ heterozygous possibilities or $10 + 45 = 55$ total genotypes. A locus with 20 possible alleles would exhibit $20 + (20 \times 19)/2 = 210$ genotypes. A combination of 10 loci with 10 alleles in each locus would have over 2.5×10^{17} possible genotypes ($55 \times 55 \times 55 \times \ldots$). Whereas the use of four loci with 30 alleles in each locus would have 465 genotypes each and 4.7×10^{10} possible genotypes ($465 \times 465 \times 465 \times 465$). The number of observed alleles per locus and the number of loci per DNA test both help produce a larger number of genetically possible genotypes.

RECOMBINATION: SHUFFLING OF GENETIC MATERIAL

Recombination is the process by which progeny derive a combination of genes different from that of either parent. During the process of meiosis or gamete cell production, each reproductive cell receives at random one representative of each pair of chromosomes, or 23 in all. Since there are two chromosomes in each pair, meiosis results in 2^{23}, or about 8.4 million, different possible combinations of chromosomes in human eggs or sperm cells. The union of egg and sperm cells therefore results in over 70 trillion ($2^{23} \times 2^{23}$) different possible combinations – each one representing half of the genetic material from the father and half from the mother. In this manner, human genetic material is effectively shuffled with each generation producing the diversity seen in the world today.

GENBANK: A DATABASE OF DNA SEQUENCES

Genetic variation from DNA sequence information around the world is cataloged in a large computer database known as GenBank. GenBank is maintained by the National Center for Biotechnology Information (NCBI), which is part of the National Library of Medicine with the U.S. National Institutes of Health. The NCBI was established in 1988 as a national resource for molecular biology information to improve understanding of molecular processes affecting human health and disease. As of December 2003, GenBank contained over 36 billion nucleotide bases from more than 30 million different records.

This repository of DNA sequence information is not from humans alone. Over 120 000 different species are represented in GenBank. GenBank DNA sequences may be viewed and retrieved over the Internet via the NCBI home page at http://www.ncbi.nlm.nih.gov.

METHODS FOR MEASURING DNA VARIATION

Techniques used by forensic DNA laboratories for human identity testing purposes are based on the same fundamental principles and methods used for medical diagnostics and gene mapping. A person's genetic makeup can be directly determined from very small amounts of DNA present in blood stains, saliva, bone, hair, semen, or other biological material. Because all the cells in the human body descend by successive divisions from a single fertilized egg, the DNA material is (barring mutations) identical between any cells collected from that individual and therefore provides the same forensic information.

Primary approaches for performing DNA typing can be classified into restriction fragment length polymorphism (RFLP) methods and polymerase chain reaction (PCR)-based methods. Some of the characteristics of these techniques are compared in Table 2.1. PCR-based methods have rapidly overtaken RFLP

Table 2.1

Comparison of RFLP and PCR-based DNA typing methods.

Characteristic	RFLP Methods	PCR Methods
Time required to obtain results	6–8 weeks with radioactive probes; ~1 week with chemiluminescent probes	1–2 days
Amount of DNA needed	50–500 ng	0.1–1 ng
Condition of DNA needed	High molecular weight, intact DNA	May be highly degraded
Capable of handling sample mixtures	Yes (single-locus probes)	Yes
Allele identification	Binning required since a distribution of sizes are observed	Discrete alleles obtained
Form used in analysis	DNA must be double-stranded for restriction enzymes to work	DNA can be either single-stranded or double-stranded
Power of discrimination	~1 in 1 billion with 6 loci	~1 in 1 billion with 8–13 loci (requires more loci)
Automatable and capable of high-volume sample processing	No	Yes
Commonly used DNA markers	D1S7, D2S44, D4S139, D5S110, D7S467, D10S28, D17S79	DQA1, D1S80, STR loci: TH01, VWA, FGA, TPOX, CSF1PO, D3S1358, D5S818, D7S820, D8S1179, D13S317, D16S539, D18S51, D21S11

methods due to the ability of PCR to handle forensic samples that are of low quantity and of poor quality. The desire for a rapid turnaround time and the capabilities for high volume sample processing have also driven the acceptance of PCR-based methods and markers. The most recent and probably most rapidly accepted forensic DNA markers are short tandem repeats (STRs) due to a number of advantages.

ADVANTAGES OF STR MARKERS

This book covers the use of short tandem repeat DNA markers for human identity testing (see D.N.A. Box 2.2). These markers have become popular for forensic DNA typing because they are PCR-based and work with low-quantity DNA templates or degraded DNA samples. STR typing methods are amenable to automation and involve sensitive fluorescent detection, which enables scientists to collect data quickly from these markers. When sites on multiple chromosomes are examined, STRs are highly discriminating between unrelated and even closely related individuals. Finally, discrete alleles make results easier to interpret and to compare through the use of computerized DNA databases than RFLP-based systems where similar DNA sizes were grouped together.

D.N.A. Box 2.2

Web sites with additional information on forensic DNA typing

Short Tandem Repeat Internet Database (STRBase) with details on STR typing
>http://www.cstl.nist.gov/biotech/strbase

Denver District Attorney's Office (court case summaries involving DNA testing)
>http://www.denverda.org

American Prosecutors Research Institute (information to help in prosecuting DNA cases)
>http://www.ndaa-apri.org/apri/index.html
>http://www.ndaa-apri.org/publications/apri/dna_pubs.html

Smith Alling Lane web site on policies impacting forensic DNA typing
>http://www.dnaresource.com

National Institute of Justice (NIJ) with funding opportunities
>http://www.ojp.usdoj.gov/nij

FBI's Combined DNA Index System (CODIS)
>http://www.fbi.gov/hq/lab/codis/index1.htm

Forensic Science Service
>http://www.forensic.gov.uk

ADDITIONAL READING

Cantor, C.R. and Smith, C.L. (1999) *Genomics: The Science and Technology Behind the Human Genome Project.* New York: John Wiley & Sons.

Kreeger, L.R. and Weiss, D.M. (2003) *Forensic DNA Fundamentals for the Prosecutor: Be Not Afraid.* Alexandria, Virginia: American Prosecutors Research Institute. Available online at: http://www.ndaa-apri.org/publications/apri/dna_pubs.html.

Lee, H.C., Ladd, C., Bourke, M.T., Pagliaro, E. and Tirnady, F. (1994) *American Journal of Forensic Medicine and Pathology*, 15, 269–282.

National Research Council (1996) *The Evaluation of Forensic DNA Evidence*. Washington, DC: National Academy Press.

Primrose, S.B. (1998) *Principles of Genome Analysis: A Guide to Mapping and Sequencing DNA from Different Organisms*, 2nd edn. Malden, MA: Blackwell Science.

Tagliaferro, L. and Bloom, M.V. (1999) *The Complete Idiot's Guide to Decoding Your Genes*. New York: Alpha Books.

CHAPTER 3

SAMPLE COLLECTION, DNA EXTRACTION, AND DNA QUANTITATION

... The blood or semen that [the perpetrator of a crime] deposits or collects – all these and more bear mute witness against him. This is evidence that does not forget... Physical evidence cannot be wrong; it cannot perjure itself; it cannot be wholly absent... Only human failure to find, study and understand it can diminish its value.

(Paul Kirk, *Crime Investigation*, 1953)

Before a DNA test can be performed on a sample, it must be collected and the DNA isolated and put in the proper format for further characterization. This chapter covers the important topics of sample collection and preservation, DNA extraction, and DNA quantitation. Each of these steps is vital to obtaining a successful result regardless of the DNA typing procedure used. If the samples are not handled properly in the initial stages of an investigation, then no amount of hard work in the final analytical or data interpretation steps can compensate.

SAMPLE COLLECTION

DNA SAMPLE SOURCES

DNA is present in every nucleated cell and is therefore present in biological materials left at crime scenes. DNA has been successfully isolated and analyzed from a variety of biological materials. Introduction of the polymerase chain reaction (PCR) has extended the range of possible DNA samples that can be successfully analyzed because many copies are made of the DNA markers to be examined (see Chapter 4). Some of the biological materials that have been tested with PCR-based DNA typing method are included in Table 3.1 (see also D.N.A. Box 3.1). The most common materials tested in forensic laboratories are blood and semen or bloodstains and semen stains.

BIOLOGICAL EVIDENCE AT CRIME SCENES

The different types of biological evidence discussed in the previous section can be used to associate or to exclude an individual from involvement with a crime. In particular, the direct transfer of DNA from one individual to another individual

Table 3.1

Sources of biological materials used for PCR-based DNA typing.

Material	Reference
Blood and blood stains	Budowle *et al.* (1995)
Semen and semen stains	Budowle *et al.* (1995)
Bones	Gill *et al.* (1994)
Teeth	Alvarez Garcia *et al.* (1996)
Hair with root	Higuchi *et al.* (1988)
Hair shaft	Wilson *et al.* (1995)
Saliva (with nucleated cells)	Sweet *et al.* (1997)
Urine	Benecke *et al.* (1996), Yasuda *et al.* (2003)
Feces	Hopwood *et al.* (1996)
Debris from fingernails	Wiegand *et al.* (1993)
Muscle tissue	Hochmeister (1998)
Cigarette butts	Hochmeister *et al.* (1991)
Postage stamps	Hopkins *et al.* (1994)
Envelope sealing flaps	Word and Gregory (1997)
Dandruff	Herber and Herold (1998)
Fingerprints	Van Oorschot and Jones (1997)
Personal items: razor blade, chewing gum, wrist watch, ear wax, toothbrush	Tahir *et al.* (1996)

or to an object can be used to link a suspect to a crime scene. As noted by Dr. Henry Lee (Lee *et al.* 1991, Lee 1996), this direct transfer could involve:

1. The suspect's DNA deposited on the victim's body or clothing;
2. The suspect's DNA deposited on an object;
3. The suspect's DNA deposited at a location;
4. The victim's DNA deposited on suspect's body or clothing;
5. The victim's DNA deposited on an object;
6. The victim's DNA deposited at a location;
7. The witness' DNA deposited on victim or suspect; or
8. The witness' DNA deposited on object or at location.

DNA evidence collection from a crime scene must be performed carefully and a chain of custody established in order to produce DNA profiles that are

D.N.A. Box 3.1
DNA recovery from irradiated samples

The U.S. Postal Service began using electron-beam irradiation of mail (for some ZIP postal codes in Washington, DC) as a protective measure against terrorism with biological agents following the anthrax attacks on the Senate Office Building in October 2001. The irradiation is performed at levels demonstrated to cleave microbial DNA and prevent passage of harmful materials such as anthrax.

However, recovering human DNA and developing a DNA profile from licked stamps and envelope flaps can sometimes be important in tracing the origin of threatening letters. Two studies have been published recently examining the effects of electron-beam irradiation on buccal-cell DNA (Castle *et al.* 2003, Withrow *et al.* 2003). Both studies concluded that while electron-beam irradiation reduces the yields and quality of DNA extracted from buccal-cell collections, the short tandem repeat DNA typing systems used in human identity testing could still be successfully amplified.

Sources:

Castle, P.E. *et al.* (2003) Effects of electron-beam irradiation on buccal-cell DNA. *American Journal of Human Genetics,* 73, 646–651.

Withrow, A.G. *et al.* (2003) Extraction and analysis of human nuclear and mitochondrial DNA from electron beam irradiated envelopes. *Journal of Forensic Sciences,* 48, 1302–1308.

meaningful and legally accepted in court. DNA testing techniques have become so sensitive that biological evidence too small to be easily seen with the naked eye can be used to link suspects to crime scenes. The evidence must be carefully collected, preserved, stored, and transported prior to any analysis conducted in a forensic DNA laboratory. The National Institute of Justice has produced a brochure entitled 'What Every Law Enforcement Officer Should Know About DNA Evidence' (see http://www.ojp.usdoj.gov/nij) that contains helpful hints for law enforcement personnel who are the first to arrive at a crime scene.

EVIDENCE COLLECTION AND PRESERVATION

The importance of proper DNA evidence collection cannot be overemphasized. If the DNA sample is contaminated from the start, obtaining unambiguous information becomes a challenge at best and an important investigation can be compromised (see D.N.A. Box 3.2). Samples for collection should be carefully chosen as well to prevent needless redundancy in the evidence for a case. The following suggestions may be helpful during evidence collection to preserve it properly (see page 38).

D.N.A. Box 3.2
Importance of carefully collecting DNA evidence: the O.J. Simpson case

On the night of 12 June 1994, Nicole Brown Simpson and Ronald Goldman were found brutally murdered at Nicole's home. A few days later Nicole's ex-husband, Orenthal James (O.J.) Simpson, was picked up by Los Angeles police officers and became the chief suspect in the murder investigation. Due to O.J. Simpson's successful football career and popularity, the case immediately drew the public's attention. Over 100 pieces of biological evidence were gathered from the crime scene consisting primarily of blood droplets and stains. DNA samples were sent to three laboratories for testing. Over the summer months of 1994, the Los Angeles Police Department (LAPD) DNA Laboratory, the California Department of Justice (CA DOJ) DNA Laboratory in Berkeley, and a private contract laboratory from Maryland named Cellmark Diagnostics performed the DNA testing using both RFLP and PCR techniques. A number of RFLP and PCR markers were examined in this high-profile case. However, no STRs were typed.

The so-called 'Trial of the Century', *People of the State of California v. Orenthal James Simpson*, began in the fall of 1994. O.J. Simpson hired a legal 'dream team', which worked hard to acquit their client. O.J.'s defense team knew that the DNA evidence was the most powerful thing going against the football star and vigorously attacked the collection of the biological material from the crime scene. Through accusations of improper sample collection and handling as well as police conspiracies and laboratory contamination, the defense team managed to introduce a degree of 'reasonable doubt'. After a lengthy and exhausting trial, the jury acquitted O.J. Simpson on 3 October 1995.

There were seven sets of bloodstains collected by the LAPD and analyzed by the three DNA laboratories mentioned above. These sets of samples are reviewed below along with the challenges put forward by the defense team. For each sample, the statistics for the odds of a random match ranged from 1 in 40 when only PCR testing with the DQ-alpha marker was evaluated to more than 1 in 40 billion when all RFLP markers were examined.

To gain a better understanding of the magnitude of the DNA testing conducted in the O.J. Simpson case, 61 items of evidence were received by CA DOJ from LAPD (Sims *et al.* 1995). From these evidence items, 108 samples were extracted in 22 sets and tested alongside 21 quality control samples that were co-extracted and 24 extraction reagent blanks. These extraction reagent blanks were performed to verify that no contamination was introduced in the CA DOJ laboratory.

From a scientific point of view, the results from the three testing laboratories agreed and more than a score of DNA markers were examined with no exclusions between the crime scene samples and Mr. Simpson. The acquittal verdict goes to show that DNA evidence is not always understood and can be quite complex to explain to the general public. Expert witnesses have the challenge of presenting the difficult subjects of DNA biology, technology, and genetics and jury members must make sense of concepts such as contamination and mixture analysis that can be fairly complex.

To their credit, the defense team focused on the evidence collection and preservation as the most important issues in the trial rather than attacking the validity of DNA testing. They implicated the LAPD in planting some of O.J. Simpson's liquid blood reference sample collected on 13 June – the day after the murders took place. Furthermore, the defense attacked the manner in which the evidence was handled in the LAPD DNA laboratory and alleged that contamination of the evidence samples by O.J.'s reference blood sample resulted from sloppy work and failure to maintain sterile conditions in the laboratory.

The contamination allegation became the focus of their arguments because much of the evidence had been handled, opened, and supposedly contaminated in the LAPD lab before it was packed up and sent to other laboratories for further testing. Thus, according to the defense, no matter how carefully the samples were handled by the California Department of Justice DNA Laboratory or Cellmark Diagnostics their testing results would not reflect the actual evidence from the crime scene. Since the samples were supposedly tainted by the LAPD laboratory, the defense argued that the evidence should not be considered conclusive. However, the sheer number of DNA samples that typed to O.J. makes it hard to believe that some random laboratory error made it possible to obtain such overwhelming incriminating results.

Since the conclusion of the O.J. Simpson trial in 1995, forensic DNA laboratories have improved their vigilance in conducting DNA evidence collection and performing the testing in a manner that is above reproach. Because PCR is an extremely sensitive technology, laboratories practicing the technique need to take extraordinary measures to prevent contamination in the laboratory. Hence, the value of laboratory accreditation and routine proficiency tests to verify that a laboratory is conducting its investigations in a proper and professional manner is clear (see Chapter 16).

The issuance of the DNA Advisory Board Quality Assurance Standards (see Appendix IV) has helped raise the professional status of forensic DNA testing. It is noteworthy that in a systematic analysis of circumstances normally encountered during casework, no PCR contamination was ever noted according to a recent study (Scherczinger *et al.* 1999). Significant contamination occurred only with gross deviations from basic preventative protocols, such as those outlined in the DAB Standards, and could not be generated by simple acts of carelessness. Arguably the most important outcome of the O.J. Simpson trial was the renewed emphasis placed on DNA evidence collection.

Samples/Location (Date collected)	Number of Samples Collected	DNA Match	Defense Challenge
Blood drops at Nicole Brown's home (13 June)	5 drops leading away from house	Simpson	Heavy degradation of the 'real' killer's DNA; tampering with evidence 'swatches'; sample contamination during laboratory investigation
Stains on rear gate at Brown's home (3 July)	3 stains	Simpson	Samples planted by rogue police officers prior to collection
Stains in O.J.'s Bronco (14 June)	5 stains around vehicle; bloody footprint; stain on center console	Simpson in 5 stains; Brown in footprint; Simpson/Goldman mixture on console	Simpson's DNA present for reasons unrelated to the crime; Detective Mark Fuhrman planted the blood footprint; Laboratory controls failed on console mixture analysis
Second collection of stains in O.J.'s Bronco (26 August)	3 stains	Mixture of Simpson, Brown, and Goldman	Blood planted in the vehicle between the crime and the collection
Stains at Simpson's home (13 June)	2 drops in driveway, 1 in foyer, 1 in master bedroom	Simpson	Simpson bled at these locations for reasons unrelated to the crime
Socks found in Simpson's bedroom (13 June)	Multiple stains	Simpson and Brown	Blood planted after the socks were collected
Bloody glove found on the grounds of Simpson's home (13 June)	15 stains identified	Goldman, Simpson, and Brown alone or as mixture	Glove was removed from murder scene and planted by Detective Mark Fuhrman; Simpson's DNA was present because of laboratory contamination

Sources:

Levy, H. (1996) O.J. Simpson: what the blood really showed. In Levy, H. (ed) *And the Blood Cried Out,* pp. 157–188. New York: Basic Books; Weir, B.S. (1995) *Nature Genetics,* 11, 365–368.

- Avoid contaminating the area where DNA might be present by not touching it with your bare hands, or sneezing and coughing over the evidence.
- Use clean latex gloves for collecting each item of evidence. Gloves should be changed between handling of different items of evidence.
- Each item of evidence must be packaged separately.
- Bloodstains, semen stains, and other types of stains must be thoroughly air-dried prior to sealing the package.
- Samples should be packaged in paper envelopes or paper bags after drying. Plastic bags should be avoided because water condenses in them, especially in areas of high humidity and water can speed the degradation of DNA molecules. Packages should be clearly marked with case number, item number, collection date, and initialed across the package seal in order to maintain a proper chain of custody.
- Stains on unmovable surfaces (such as a table or floor) may be transferred with sterile cotton swabs and distilled water. Rub the stained area with the moist swab until the stain is transferred to the swab. Allow the swab to air dry without touching any others. Store each swab in a separate paper envelope.

COLLECTION OF REFERENCE DNA SAMPLES

In order to perform comparative DNA testing with evidence collected from a crime scene, biological samples must also be obtained from suspects or convicted felons (see Chapter 18). In addition, family reference samples are used in missing persons investigations, paternity testing, and mass disaster victim identifications (see Chapter 24). It is advantageous to obtain these reference DNA samples as rapidly and painlessly as possible. Thus, many laboratories often use buccal cell collection rather than drawing blood. Buccal cell collection involves wiping a cotton swab similar to a Q-tip against the inside cheek of an individual's mouth to collect some skin cells. The swab is then dried or can be pressed against a treated collection card to transfer epithelial cells for storage purposes.

A simple Buccal DNA Collector (Fox *et al.* 2002, Schumm *et al.* 2004) may also be used for direct collection of buccal cell samples. A disposable toothbrush can be used for collecting buccal cells in a non-threatening manner (Burgoyne 1997, Tanaka *et al.* 2000). This method can be very helpful when samples need to be collected from children. After the buccal cells have been collected by gently rubbing a wet toothbrush across the inner cheek, the brush can be tapped onto the surface of treated collection paper for sample storage and preservation.

STORAGE AND TRANSPORT OF DNA EVIDENCE

Carelessness or ignorance of proper handling procedures during storage and transport of DNA from the crime scene to the laboratory can result in a specimen unfit for analysis. For example, bloodstains should be thoroughly dried prior to transport to prevent mold growth. A recovered bloodstain on a cotton swab should be air-dried in an open envelope before being sealed for transport. DNA can be stored as non-extracted tissue or as fully extracted DNA. DNA samples are not normally extracted though until they reach the laboratory.

Most biological evidence is best preserved when stored dry and cold. These conditions reduce the rate of bacterial growth and degradation of DNA. Samples should be packaged carefully and hand carried or shipped using overnight delivery to the forensic laboratory conducting the DNA testing. A nice evidence collection cardboard box was recently described for shipping and handling bloodstains and other crime scene evidence (Hochmeister *et al.* 1998). Inside the laboratory, DNA samples are either stored in a refrigerator at 4°C or a freezer at −20°C. For long periods of time, extracted DNA samples may even be stored at −70°C.

PRESUMPTIVE TESTS FOR BLOOD, SEMEN, AND SALIVA

Forensic evidence from crime scenes comes in many forms. For example, a bed sheet may be collected from the scene of a sexual assault. This sheet will have to be carefully examined in the forensic laboratory before selecting the area to sample for further testing. Prior to taking the effort to extract DNA from a sample, presumptive tests are often performed to indicate whether or not biological fluids such as blood or semen are present on an item of evidence (e.g., a pair of pants). Locating a blood or semen stain on a soiled undergarment can be a trying task. Three primary stains of forensic interest come from blood, semen, and saliva. Identification of vaginal secretions, urine, and feces can also be important to an investigation.

Serology is the term used to describe a broad range of laboratory tests that utilize antigen and serum antibody reactions. For example, the ABO blood group types are determined using anti-A and anti-B serums and examining agglutination when mixed with a blood sample. Serology still plays an important role in modern forensic biology but has taken a backseat to DNA in many respects since presumptive tests do not have the ability to individualize a sample like a DNA profile can.

DETECTION OF BLOOD STAINS

Blood is composed of liquid plasma and serum with solid components consisting of red blood cells (erythrocytes), white blood cells (leukocytes), and platelets (thrombocytes). Most presumptive tests for blood focus on detecting the presence of hemoglobin molecules, which are found in the red blood cells and used for transport of oxygen and carbon dioxide. A simple immunochromatographic test for identification of human blood is available from Abacus Diagnostics (West Hills, CA) as the ABAcard® HemaTrace® kit. This test has a limit of detection of 0.07 μg hemoglobin/mL and shows specificity for human blood along with higher primate and ferret blood (Johnston *et al.* 2003).

Luminol is another presumptive test for identification of blood that has been popularized by the TV series *CSI: Crime Scene Investigation*. The luminol reagent is prepared by mixing 0.1 g 3-amino-phthalhydrazide and 5.0 g sodium carbonate in 100 mL of distilled water. Before use, 0.7 g of sodium perborate is added to the solution (Saferstein 2001). Large areas can be rapidly evaluated for the presence of bloodstains by spraying the luminol reagent onto the item under investigation. Objects that have been sprayed need to be located in a darkened area so that the luminescence can be more easily viewed. Luminol can be used to locate traces of blood that have been diluted up to 10 million times (Saferstein 2001). The use of luminol has been shown to not inhibit DNA testing of STRs that may need to be performed on evidence recovered from a crime scene (Gross *et al.* 1999). Demonstration that presumptive tests do not interfere with subsequent DNA testing can be important when making decisions on how biological evidence is processed in a forensic laboratory (Hochmeister *et al.* 1991, Budowle *et al.* 2000).

DETECTION OF SEMEN STAINS

Almost two-thirds of cases pursued with forensic DNA testing involve sexual assault evidence. Hundreds of millions of sperm are typically ejaculated in several milliliters of seminal fluid. Semen stains can be characterized with visualization of sperm cells, acid phosphatase (AP) or prostate specific antigen (PSA or p30) tests.

A microscopic examination to look for the presence of spermatozoa is performed in some laboratories on sexual assault evidence. However, aspermic or oligospermic males have either no sperm or a low sperm count in their seminal fluid ejaculate. In addition, vasectomized males will not release sperm. Therefore tests that can identify semen-specific enzymes are helpful in verifying the presence of semen in sexual assault cases.

Acid phosphatase is an enzyme secreted by the prostate gland into seminal fluid and found in concentrations up to 400 times greater in semen than in other body fluids (Sensabaugh 1979, Saferstein 2001). A purple color with the addition of a few drops of sodium alpha naphthylphosphate and Fast Blue B solution or the fluorescence of 4-methyl umbelliferyl phosphate under a UV light indicates the presence of AP. Large areas of fabric can be screened by pressing the garment or bed sheet against an equal sized piece of moistened filter paper and then subjecting the filter paper to the presumptive tests.

Alternatively systematic searches can be done over sections of the fabric under examination to narrow the semen stain location with each successive test (Saferstein 2001).

Prostate specific antigen was discovered in the 1970s and shown to have forensic value with the identity of a protein named p30 due to its apparent 30 000 molecular weight (Sensabaugh 1978). p30 was initially thought to be unique to seminal fluid although it has been reported at lower levels in breast milk (Yu and Diamandis 1995) and other fluids (Diamandis and Yu 1995). PSA varies in concentration from approximately 300–4200 ng/mL in semen (Shaler 2002). Seratec (Goettingen, Germany) and Abacus Diagnostics (West Hills, CA) market PSA/p30 test kits that are similar to home-pregnancy tests and which may be used for the forensic identification of semen stains (Hochmeister *et al.* 1999, Simich *et al.* 1999).

DIRECT OBSERVATION OF SPERM

Most forensic laboratories like to observe spermatozoa as part of confirming the presence of semen in an evidentiary sample. A common method of doing this is to recover dried semen evidence from fabric or on human skin with a deionized water-moistened swab. A portion of the recovered cells are then placed onto a microscope slide and fixed to the slide with heat. The immobilized cells are stained with a 'Christmas Tree' stain consisting of aluminum sulfate, nuclear fast red, picric acid, and indigo carmine (Shaler 2002). The stained slide is then examined under a light microscope for sperm cells with their characteristic head and long tail. The Christmas Tree stain marks the anterior sperm heads light red or pink, the posterior heads dark red, the spermatozoa's mid-piece blue, and the tails stain yellowish green (Shaler 2002). John Herr at the University of Virginia is developing some 'sperm paints' to fluorescently label the head and tail portions of spermatozoa with antibodies specific to sperm and thus make it easier to observe sperm cells in the presence of excess female epithelial cells.

DETECTION OF SALIVA STAINS

A presumptive test for amylase is used for indicating the presence of saliva, which is especially difficult to see since saliva stains are nearly invisible to the naked eye. Two common methods for estimating amylase levels in forensic samples include the Phadebas test and the starch iodine radial diffusion test (Shaler 2002). Saliva stains may be found on bite-marks, cigarette butts, and drinking vessels (Abaz *et al.* 2002, Shaler 2002). A molecular biology approach using messenger RNA profiling is also being taken to develop sensitive and specific tests for various body fluids including saliva (Juusola and Ballantyne 2003). Such a molecular biology test should be able to assay blood, semen, and saliva simultaneously with great specificity and sensitivity.

DNA EXTRACTION

A biological sample obtained from a crime scene in the form of a blood or semen stain or a liquid blood sample from a suspect or a paternity case contains a number of substances besides DNA. DNA molecules must be separated from other cellular material before they can be examined. Cellular proteins that package and protect DNA in the environment of the cell can inhibit the ability to analyze the DNA. Therefore, DNA extraction methods have been developed to separate proteins and other cellular materials from the DNA molecules. In addition, the quantity and quality of DNA often need to be measured prior to proceeding further with analytical procedures to ensure optimal results.

There are three primary techniques for DNA extraction used in today's forensic DNA laboratory: organic extraction, Chelex extraction, and FTA paper (Figure 3.1). The exact extraction or DNA isolation procedure varies depending on the type of biological evidence being examined. For example, whole blood must be treated differently from a bloodstain or a bone fragment.

Organic extraction, sometimes referred to as phenol chloroform extraction, has been in use for the longest period of time and may be used for situations where either RFLP or PCR typing is performed. High molecular weight DNA, which is essential for RFLP methods, may be obtained most effectively with organic extraction.

The Chelex method of DNA extraction is more rapid than the organic extraction method. In addition, Chelex extraction involves fewer steps and thus fewer opportunities for sample-to-sample contamination. However, it produces single stranded DNA as a result of the extraction process and therefore is only useful for PCR-based testing procedures.

All samples must be carefully handled regardless of the DNA extraction method to avoid sample-to-sample contamination or introduction of extraneous DNA. The extraction process is probably where the DNA sample is more susceptible to contamination in the laboratory than at any other time in the forensic DNA analysis process. For this reason, laboratories usually process the evidence samples at separate times and sometimes even different locations from the reference samples.

A popular method for preparation of reference samples is to make a blood stain by applying a drop of blood on to a cotton cloth, referred to as a swatch, to produce a spot about 1 cm² in area. Ten microliters of whole blood, about the size of a drop, contains approximately 70 000–80 000 white blood cells and should yield approximately 500 ng of genomic DNA. The actual yield will vary

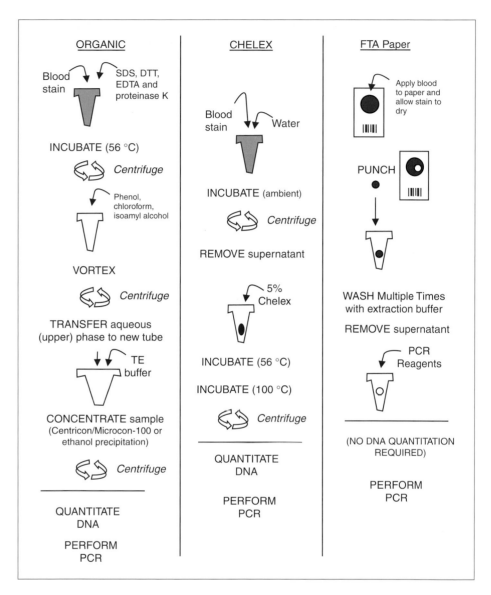

Figure 3.1
Schematic of commonly used DNA extraction processes.

with the number of white blood cells present in the sample and the efficiency of the DNA extraction process.

Extracted DNA is typically stored at −20°C, or even −80°C for long-term storage, to prevent nuclease activity. Nucleases are enzymes (proteins) found in cells that degrade DNA to allow for recycling of the nucleotide components. Nucleases need magnesium to work properly so one of the measures to prevent them from digesting DNA in blood is the use of purple-topped tubes containing a blood preservative known as EDTA. The EDTA chelates, or binds up, all of the free magnesium and thus prevents the nucleases from destroying the DNA in the collected blood sample.

ORGANIC (PHENOL-CHLOROFORM) EXTRACTION

Organic extraction involves the serial addition of several chemicals. First sodium dodecylsulfate (SDS) and proteinase K are added to break open the cell walls and to break down the proteins that protect the DNA molecules while they are in chromosomes. Next a phenol/chloroform mixture is added to separate the proteins from the DNA. The DNA is more soluble in the aqueous portion of the organic–aqueous mixture. When centrifuged, the unwanted proteins and cellular debris are separated away from the aqueous phase and double-stranded DNA molecules can be cleanly transferred for analysis. Some protocols involve a Centricon 100 (Millipore, Billerica, MA) dialysis and concentration step in place of the ethanol precipitation to remove heme inhibitors (Comey *et al.* 1994). While the organic extraction method works well for recovery of high molecular weight DNA, it is time-consuming, involves the use of hazardous chemicals, and requires the sample to be transferred between multiple tubes (a fact that increases the risk of error or contamination).

CHELEX EXTRACTION

An alternative procedure for DNA extraction that has become popular among forensic scientists is the use of a chelating-resin suspension that can be added directly to the sample (e.g., blood, bloodstain, or semen). Introduced in 1991 to the forensic DNA community, Chelex® 100 (Bio-Rad Laboratories, Hercules, CA) is an ion-exchange resin that is added as a suspension to the samples (Walsh *et al.* 1991). Chelex is composed of styrene divinylbenzene copolymers containing paired iminodiacetate ions that act as chelating groups in binding polyvalent metal ions such as magnesium. Like iron filings to a magnet, the magnesium ions are drawn in and bound up. By removing the magnesium from the reaction, DNA destroying enzymes known as nucleases are inactivated and the DNA molecules are thus protected.

In most protocols, biological samples such as bloodstains are added to a 5% Chelex suspension and boiled for several minutes to break open the cells and release the DNA. An initial, prior wash step is helpful to remove possible contaminants and inhibitors such as heme and other proteins (Willard *et al.* 1998). The exposure to 100°C temperatures denatures the DNA as well as disrupting the cell membranes and destroying the cell proteins. After a quick spin in a centrifuge to pull the Chelex resin and cellular debris to the bottom of the tube, the supernatant is removed and can be added directly to the PCR amplification reaction.

Chelex extraction procedures for recovering DNA from bloodstains or semen-containing stains are not effective for RFLP typing because Chelex denatures double-stranded DNA and yields single-stranded DNA from the extraction process.

Thus, it can only be followed by PCR-based analyses (see Table 2.1). However, Chelex extraction is an advantage for PCR-based typing methods because it removes inhibitors of PCR and uses only a single tube for the DNA extraction, which reduces the potential for laboratory-induced contamination.

The addition of too much whole blood or too large a bloodstain to the Chelex extraction solution can result in some PCR inhibition. The AmpF*l*STR kit manuals recommend 3 μL whole blood or a bloodstain approximately 3 mm × 3 mm (Applied Biosystems 1998).

FTA™ PAPER

Another approach to DNA extraction involves the use of FTA™ paper. In the late 1980s, FTA™ paper was developed by Lee Burgoyne at Flinders University in Australia as a method for storage of DNA (Burgoyne *et al.* 1994). FTA™ paper is an absorbent cellulose-based paper that contains four chemical substances to protect DNA molecules from nuclease degradation and preserve the paper from bacterial growth (Burgoyne 1996). As a result, DNA on FTA™ paper is stable at room temperature over a period of several years. However, a recent study evaluating FTA™ and three other commercial papers as DNA storage media found little difference in their ability to obtain typeable STR results after 19 months of storage (Kline *et al.* 2002).

Use of FTA paper simply involves adding a spot of blood to the paper and allowing the stain to dry. The cells are lysed upon contact with the paper and DNA from the white blood cells is immobilized within the matrix of the paper. A small punch of the paper is removed from the FTA card bloodstain and placed into a tube for washing. The bound DNA can then be purified by washing it with a solvent to remove heme and other inhibitors of the PCR reaction. This purification of the paper punch can be seen visually because as the paper is washed, the red color is removed with the supernatant. The clean punch is then added directly to the PCR reaction. Alternatively, some groups have performed a Chelex extraction on the FTA paper punch and used the supernant in the PCR reaction (Lorente *et al.* 1998, Kline *et al.* 2002).

A major advantage of FTA paper is that consistent results may be obtained without quantification. Furthermore the procedure may be automated on a robotic workstation (Belgrader and Marino 1997). For situations where multiple assays need to be run on the same sample, a bloodstained punch may be reused for sequential DNA amplifications and typing (Del Rio *et al.* 1996). Unfortunately, dry paper punches do not like to stay in their assigned tubes and due to static electricity can 'jump' between wells in a sample tray. Thus, this method is not as widely used today as was once envisioned. However, due to its preservation and storage capabilities, efforts have been made to use FTA cards for more widespread collection of crime scene evidence (Lorente *et al.* 2004).

DIFFERENTIAL EXTRACTION

Differential extraction is a modified version of the organic extraction method that separates epithelial and sperm cells (Figure 3.2). Differential extraction was first described in 1985 (Gill *et al.* 1985) and is commonly used today by the FBI Laboratory and other forensic crime laboratories to isolate the female and male fractions in sexual assault cases that contain a mixture of male and female DNA. By separating the male fraction away from the victim's DNA profile, it is much easier to interpret the perpetrator's DNA profile in a rape case.

The differential extraction procedure involves preferentially breaking open the female epithelial cells with incubation in a SDS/proteinase K mixture. Sperm nuclei are subsequently lysed by treatment with a SDS/proteinase K/dithiothreitol (DTT) mixture. The DTT breaks down the protein disulfide bridges that make up sperm nuclear membranes (Gill *et al.* 1985). Differential extraction works because sperm nuclei are impervious to digestion without DTT. The major difference between the regular version of organic extraction described earlier and differential extraction is the initial incubation in SDS/proteinase K without DTT present.

Differential extraction works well in most sexual assault cases to separate the female and male fractions from one another. Unfortunately, some perpetrators of sexual assaults have had a vasectomy in which case there is an absence of spermatozoa. Azoospermic semen, i.e., without sperm cells, cannot be separated from the female fraction with differential extraction. In the case of

Figure 3.2

Schematic of differential extraction process used to separate male sperm cells from female epithelial cells.

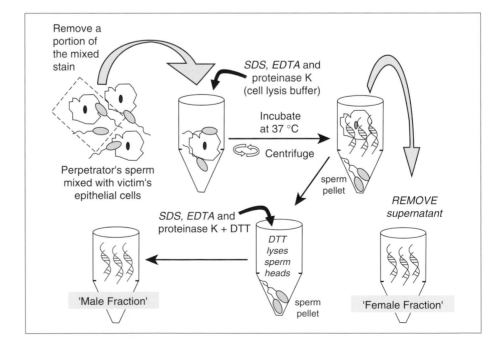

azoospermic perpetrators, the use of Y chromosome specific markers permit male DNA profiles to be deduced in the presence of excess female DNA (see Chapter 9). Failure to separate the male and female portions of a sexual assault sample results in a mixture of both the perpetrator's and the victim's DNA profiles (see Chapter 7).

DIRECT CAPTURE OF SPERM CELLS

Several approaches have been taken to directly sample sperm cells from sexual assault cases. One strategy is to physically separate the perpetrator's sperm cells from the victim's epithelial cells. Sperm cells can be collected on magnetic particles or beads that can be coated with antibodies specific for sperm proteins (Marshall 2002). The beads are then washed to remove the female epithelial cells. Finally, the purified sperm are placed into a PCR reaction to produce a DNA profile of the perpetrator. This approach depends on sperm being intact, which is not always the case with old sexual assault evidence.

Another exciting approach to selectively capturing sperm is the use of a clinical procedure known as laser-capture microdissection, which is commonly used to select tumor cells from surrounding tissue on microscope slides. Sperm cells from sexual assault evidence spread on microscope slides can be collected with laser-capture microdissection to perform reliable STR testing (Elliot *et al.* 2003). When sperm cells are observed in the field of view of the microscope, a tiny laser is activated and a thin plastic film placed over the slide melts at the specific point of laser light contact to capture or embalm the cell of interest. By moving the microscope slide around, dozens of sperm cells are collected onto this thin film that sits directly above the sample. The collection film is then transferred to a tube where DNA from the isolated sperm can be extracted and amplified using the polymerase chain reaction (see Chapter 4).

SOLID-PHASE EXTRACTION

Solid-phase extraction methods for DNA have been developed in recent years in formats that enable high-throughput DNA extractions. One of the most active efforts in this area is with silica-based extraction methods and products from QIAGEN, Inc. (Valencia, CA). QIAamp spin columns have also proven effective as a means of DNA isolation (Greenspoon *et al.* 1998). In this approach, nucleic acids selectively absorb to a silica support, such as small glass beads, in the presence of high concentrations of chaotropic salts such as guanidine hydrochloride, guanidine isothiocyanate, sodium iodide, and sodium perchlorate (Vogelstein and Gillespie 1979, Boom *et al.* 1990, Duncan *et al.* 2003). These chaotropic salts disrupt hydrogen-bonding networks in liquid water and thereby make denatured proteins and nucleic acids more thermodynamically stable than their correctly folded or structured counter-parts (Tereba *et al.* 2004).

If the solution pH is less than 7.5, DNA adsorption to the silica is typically around 95% and unwanted impurities can be washed away. Under alkaline conditions and low salt concentrations, the DNA will efficiently elute from the silica material. This solid-phase extraction approach can be performed with centrifugation or vacuum manifolds in single tube or 96-well plate formats (Hanselle *et al.* 2003, Yasuda *et al.* 2003) and is even being developed into formats that will work on a microchip (Wolfe *et al.* 2002).

Another solid-phase extraction approach is the DNA IQ™ system marketed by Promega Corporation. The DNA IQ™ system utilizes the same silica-based DNA binding and elution chemistries as QIAGEN kits but with silica-coated paramagnetic resin (Tereba *et al.* 2004). With this approach, DNA isolation can be performed in a single tube by simply adding and removing solutions. First, the DNA molecules are reversibly bound to the magnetic beads in solution with a solution pH of less than 7.5 (see above). A magnet is used to draw the silica-coated magnetic beads to the bottom or side of the tube leaving any impurities in solution. These solution impurities (proteins, cell debris, etc.,) can easily be removed by drawing the liquid off of the beads. The magnetic particles with DNA attached can be washed multiple times to more thoroughly clean the DNA. Finally, a defined amount of DNA can be released into solution via heating for a few minutes.

The quantity of DNA isolated with this approach is based on the number and capacity of the magnetic particles used. Since flow-through vacuum filtration or centrifugation steps are not used, magnetic bead procedures enable simple, rapid, and automated methods. This extraction method has been automated on the Beckman 2000 robot workstation and implemented into forensic casework by the Virginia Division of Forensic Science (Greenspoon *et al.* 2004).

OTHER METHODS FOR DNA EXTRACTION

Several other DNA extraction methods have been used to isolate DNA successfully prior to further sample processing (see Vandenberg *et al.* 1997). A simple closed tube DNA extraction method has been demonstrated with a thermal stable protease that looks very promising (Moss *et al.* 2003). Microwave extraction has been used to shorten the conventional organic extraction method by several hours and to yield genomic DNA that could be PCR-amplified (Lee *et al.* 1994). The addition of 6 M NaCl to a proteinase K-digested cell extract followed by vigorous shaking and centrifugation results in a simple precipitation of the proteins (Miller *et al.* 1988). The supernatant containing the DNA portion of cell extract can then be added to a PCR reaction. A simple alkaline lysis with 0.2 M NaOH for five minutes at room temperature has been shown to work as well (Rudbeck and Dissing 1998, Klintschar and Neuhuber 2000).

While each of these methods is effective for extracting most DNA samples, the forensic community has traditionally relied on the three main methods mentioned: organic extraction, Chelex extraction, and FTA paper. Higher-throughput approaches, though, are incorporating 96-well extraction formats with products such as the QIAamp® 96 DNA Blood Kit (Holland *et al.* 2003). The typical amounts of DNA that may be recovered from various biological materials are listed in Table 3.2.

PCR INHIBITORS AND DNA DEGRADATION

When extracting biological materials for the purpose of forensic DNA typing, it is important to try and avoid further degradation of the DNA template as well as to remove inhibitors of the polymerase chain reaction (PCR) where possible. The presence of inhibitors or degraded DNA can manifest themselves by complete PCR amplification failure or a reduced sensitivity of detection usually for the larger PCR products (see Chapter 7).

Two PCR inhibitors commonly found in forensic cases are hemoglobin and indigo dyes from denim. Melanin found in hair samples can be a source of PCR

Type of Sample	Amount of DNA
Liquid blood	20 000–40 000 ng/mL
Blood stain	250–500 ng/cm^2
Liquid semen	150 000–300 000 ng/mL
Post-coital vaginal swab	10–3000 ng/swab
Plucked hair (with root)	1–750 ng/root
Shed hair (with root)	1–10 ng/root
Liquid saliva	1000–10 000 ng/mL
Oral swab	100–1500 ng/swab
Urine	1–20 ng/mL
Bone	3–10 ng/mg
Tissue	50–500 ng/mg

Table 3.2

Typical DNA amounts that may be extracted from biological materials (Lee and Ladd 2001). Both quality and quantity of DNA recovered from evidentiary samples can be significantly affected by environmental factors.

inhibition when trying to amplify mitochondrial DNA (see Chapter 10). These inhibitors likely bind in the active site of the *Taq* DNA polymerase and prevent its proper functioning during PCR amplification.

DNA degrades through a variety of mechanisms including both enzymatic and chemical processes (Lindahl 1993). Once a cell (or organism) dies its DNA molecules face cellular nucleases followed by bacterial, fungal, and insect onslaughts depending on the environmental conditions (Poinar 2003). In addition, hydrolytic cleavage and oxidative base damage can limit successful retrieval and amplification of DNA. The main target for hydrolytic cleavage is the glycosidic base sugar bond. Breakage here leads to nucleobase loss and then a single stranded 'nick' at the abasic site. If a sufficient number of DNA molecules in the biological sample break in a region where primers anneal or between the forward and reverse primers, then PCR amplification efficiency will be reduced or the target region may fail to be amplified at all. Thus, heat and humidity, which speed up hydrolytic cleavage, are enemies of intact DNA molecules. Furthermore, UV irradiation (e.g., direct sunlight) can lead to cross-linking of adjacent thymine nucleotides on the DNA molecule, which will prevent passage of the polymerase during PCR.

DNA QUANTITATION

To ensure that DNA recovered from an extraction is human rather than from another source such as bacteria, DNA Advisory Board standard 9.3 requires human-specific DNA quantitation (see Appendix IV). Only after DNA in a sample has been isolated can its quantity and quality be reliably assessed. Determination of the amount of DNA in a sample is essential for most PCR-based assays because a narrow concentration range works best. For example, the Applied Biosystems' Profiler Plus™ and COfiler™ multiplex STR kits specify the addition of between 1–2.5 ng of template DNA for optimal results (Applied Biosystems 1998). Promega STR kits also work best in a similar DNA concentration range (Krenke *et al.* 2002). Too much DNA can result in split peaks or peaks that are off-scale for the measurement technique (see Chapter 6). Too little DNA template may result in allele 'drop-out' because the PCR reaction fails to amplify the DNA properly. This phenomenon is sometimes referred to as stochastic fluctuation (see Chapter 4).

Early methods for DNA quantitation typically involved either absorbance at a wavelength of 260 nm or fluorescence after staining a yield gel with ethidium bromide. Unfortunately, because these approaches are not very sensitive, they consume valuable forensic specimens that are irreplaceable. In addition, absorbance measurements are not specific for DNA and contaminating proteins or phenol left over from the extraction procedure can give falsely high signals.

To overcome these problems, several methods have been developed for DNA quantitation purposes. These include the slot blot procedure and fluorescence-based microtiter plate assays as well as so-called 'real-time or quantitative PCR' approaches. A nice review of various DNA quantification methods was recently published (Nicklas and Buel 2003b).

SLOT BLOT QUANTITATION

Probably the most commonly used method in forensic labs today for genomic DNA quantitation is the so-called 'slot blot' procedure. This test is specific for human and other primate DNA due to a 40 base pair (bp) probe that is complementary to a primate-specific alpha satellite DNA sequence D17Z1 located on chromosome 17 (Waye *et al.* 1989, Walsh *et al.* 1992). The slot blot assay was first described with radioactive probes (Waye *et al.* 1989) but has since been modified and commercialized with chemiluminescent or colorimetric detection formats (Walsh *et al.* 1992).

Slot blots involve the capture of genomic DNA on a nylon membrane followed by addition of a human-specific probe. Chemiluminescent or colorimetric signal intensities are compared between a set of standards and the samples (Figure 3.3). Slot blot quantitation is a relative measurement involving the comparison of unknown samples to a set of standards that are prepared usually via a serial dilution from a DNA sample of known concentration. While comparison of the

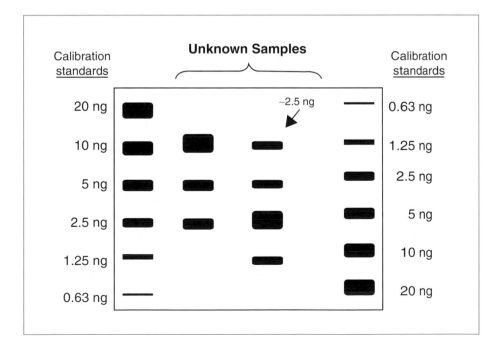

Figure 3.3

Illustration of a human DNA quantitation result with the slot blot procedure. A serial dilution of a human DNA standard is run on either side of the slot blot membrane for comparison purposes. The quantity of each of the unknown samples is estimated by visual comparison to the calibration standards. For example, the sample indicated by the arrow is closest in appearance to the 2.5 ng standard.

DNA standards and unknown samples on the slot blot membrane is often performed visually and therefore influenced by subjectivity of the analyst, digital capture and quantification of slot blot images has been demonstrated with a charged-coupled device (CCD) camera imaging system (Budowle *et al.* 2001).

Typically about 30 samples are tested on a slot blot membrane with 6–8 standard samples run on each side of the membrane for comparison purposes. For example, the standards might be a serial dilution of human DNA starting with 20 ng, 10 ng, 5 ng, 2.5 ng, etc. Typically only 5 μL of DNA extract is consumed for this quantification test.

The assay takes several hours to perform but is fairly sensitive as it can detect both single-stranded and double-stranded DNA down to levels of approximately 150 pg or about 50 copies of human genomic DNA. A 150 pg DNA standard can be detected after only a 15 minutes exposure to X-ray film (Walsh *et al.* 1992). With chemiluminescent detection, the sensitivity can be extended below 150 pg by performing longer exposures to the X-ray film and blotting additional low dilution DNA standards on the membrane. Levels of 10–20 pg have been reported with a three-hour exposure (Walsh *et al.* 1992). Even when no results are seen with this hybridization assay, some forensic scientists still go forward with DNA testing and often obtain a successful STR profile. Thus, the slot blot assay is not as sensitive as would be preferred.

As of early 2004, the slot blot assay is available as a commercial product from Applied Biosystems (Foster City, CA) called the QuantiBlot Human DNA Quantitation Kit. This kit uses a DNA probe from chromosome 17 and is thus useful for determining the level of human DNA present in a sample. It is important to note though that this assay, as with most 'human-specific' tests, works on primates such as chimpanzees and gorillas due to similarities in human and other primate DNA sequences.

PICOGREEN MICROTITER PLATE ASSAY

As higher throughput methods for DNA determination are being developed, more automated procedures are needed for rapid assessment of extracted DNA quantity prior to DNA amplification. To this end, the Forensic Science Service has developed a PicoGreen assay that is capable of detecting as little as 0.25 ng/mL of double-stranded DNA in a 96-well microtiter plate format (Hopwood *et al.* 1997). PicoGreen is a fluorescent interchelating dye whose fluorescence is greatly enhanced when bound to double-stranded DNA.

To perform this microtiter plate assay, 5 μL of sample are added to 195 μL of a solution containing the PicoGreen dye. Each sample is placed into an individual well on a 96-well plate and then examined with a fluorometer. A 96-well plate containing 80 individual samples and 16 calibration samples can be analyzed in under 30 minutes (Hopwood *et al.* 1997). The DNA samples are quantified through comparison to a standard curve. This assay has been demonstrated to be useful for the adjustment of input DNA into the amplification reaction of STR multiplexes (Hopwood *et al.* 1997). It has been automated on a robotic workstation as well. Unfortunately, this assay quantifies total DNA in a sample and is not specific for human DNA.

ALUQUANT™ HUMAN DNA QUANTITATION SYSTEM

The Promega Corporation has developed a human DNA quantitation system that enables sensitive detection of DNA (Mandrekar *et al.* 2001). The AluQuant™ assay probes *Alu* repeats that are in high abundance in the human genome (see Chapter 8). Probe-target hybridization initiates a series of enzymatic reactions that end in oxidation of luciferin with production of light. The light intensity is read by a luminometer and is proportional to the amount of DNA present in the sample. Sample quantities are determined by comparison to a standard curve. The AluQuant™ assay possesses a range of 0.1–50 ng for human DNA and can be automated on a robotic liquid handling workstation (Hayn *et al.* 2004).

REAL-TIME PCR APPROACHES TO DNA QUANTIFICATION

The primary purpose in performing a DNA quantification test is to determine the amount of 'amplifiable' DNA. A PCR amplification reaction may fail due to the presence of co-extracted inhibitors, highly degraded DNA, insufficient DNA quantity, or a combination of all of these factors. Thus, a test that can accurately reflect both the quality and the quantity of the DNA template present in an extracted sample is beneficial to making decisions about how to proceed. 'Real-time' PCR assays provide such an assessment. The theory behind real-time PCR and assay formats used will be discussed in more detail as part of Chapter 4.

Several real-time PCR assay have been published with the human identity testing market in mind (see Table 3.3). Commercial kits for detecting human DNA as well as a real-time PCR assay for determining the amount of human Y-chromosome DNA present in a sample are now available (Applied Biosystems 2003).

END-POINT PCR FOR DNA QUANTIFICATION

A less elegant (and less expensive) approach for testing the 'amplifiability' of a DNA sample is to perform an end-point PCR test. In this approach a single STR locus (Kihlgren *et al.* 1998, Fox *et al.* 2003) or other region of the human genome, such as an *Alu* repeat (Sifis *et al.* 2002, Nicklas and Buel 2003a), is amplified along with DNA samples of known concentrations. A standard curve

Kit or Assay	Principle Behind Detection	Limit of Detection	Volume of DNA Used	Human/Primate Specific	PCR Inhibition Detected	Reference
Quantiblot Kit	Hybridization	~150 pg	5 μL	Y	N	Walsh et al. (1992)
PicoGreen Assay	Intercalating Dye Fluorescence	250 pg	10 μL	N	N	Hopwood et al. (1997)
AluQuant Kit	Pyrophosphorylation and luciferase light production	100 pg	1–10 μL	Y	N	Mandrekar et al. (2001)
BodeQuant	End-point PCR	~100 pg	1–10 μL	Y	Y	Fox et al. (2003)
TQS-TH01	End-point PCR	~100 pg	1–10 μL	Y	Y	Nicklas and Buel (2003a)
Quantifiler Kit	Real-time PCR	20 pg	2 μL	Y	Y	Applied Biosystems (2003)
Alu Assay	Real-time PCR	1 pg	1–10 μL	Y	Y	Nicklas and Buel (2003c)
CFS TH01 Assay	Real-time PCR	20 pg	1–10 μL	Y	Y	Richard et al. (2003)
RB1 and mtDNA multiplex	Real-time PCR	20 pg	1–10 μL	Y	Y	Andreasson et al. (2002)

can be generated from the samples with known amounts to which samples of unknown concentration are compared. A fluorescent intercalating dye such as SYBR® Green (see Chapter 13) can be used to detect the generated PCR products. Based on the signal intensities resulting from amplification of the single STR marker or *Alu* repeat region, the level of DNA can be adjusted prior to amplifying the multiplex set of DNA markers in order to obtain the optimal results (see Chapters 5 and 6). This method is a functional test because it also monitors the level of PCR inhibitors present in the sample (see Chapter 7). In the end, each of the DNA quantitation methods described here has advantages and disadvantages and could be used depending on the equipment available and the needs of the laboratory.

Several inter-laboratory tests to evaluate DNA quantification methods have been conducted by the U.S. National Institute of Standards and Technology (NIST) to better understand the measurement variability seen with various techniques (Duewer *et al.* 2001, Kline *et al.* 2003). A ten-fold range of reported concentrations was observed in one study (Figure 3.4).

Most DNA quantitation measurements are precise to within a factor of two if performed properly (Kline *et al.* 2003). While this degree of imprecision may not seem reliable enough, quantitation results are usually sufficiently valid to estimate DNA template amounts that will enable PCR amplification.

Table 3.3 (Facing)
Summary of various DNA quantitation methods. Commercially available kits are listed in bold.

Figure 3.4
(a) Range of DNA concentrations reported for a 1 ng DNA sample supplied to 74 laboratories in an inter-laboratory study (Kline et al. 2003). Overall the median value was very close to the expected 1 ng level with 50% falling in the boxed region. However, laboratories returned values ranging from 0.1–3 ng. (b) A target plot examining concordance and apparent precision for the various laboratory methods used. Legend: A = ACES kit; q = Quantiblot with unreported visualization method; E = Quantiblot with chemiluminescent detection; T = Quantiblot with colorimetric detection; 1, 2, 3, 4, and 5 represent methods used by only one laboratory.

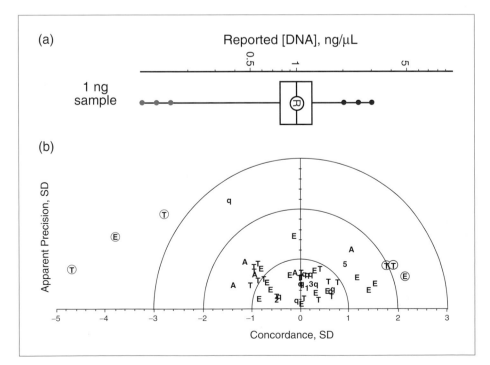

D.N.A. Box 3.3

Calculation of DNA quantities in genomic DNA

Important values for calculations:

1 bp = 618 g/mol   A: 313 g/mol; T: 304 g/mol; AT base pairs = 617 g/mol
G: 329 g/mol; C: 289 g/mol; GC base pairs = 618 g/mol

1 genome copy = ~3 × 10^9 bp = 23 chromosomes (one member of each pair)

1 mole = 6.02 × 10^{23} molecules

Standard DNA typing protocols with PCR amplification of STR markers typically ask for 1 ng of DNA template. How many actual copies of each STR locus exist in 1 ng?

1 genome copy = (~3 × 10^9 bp) x (618 g/mol/bp) = 1.85 × 10^{12} g/mol

= (1.85 × 10^{12} g/mol) × (1 mole/6.02 × 10^{23} molecules)

= 3.08 × 10^{-12} g = **3.08 pg**

Since a diploid human cell contains two copies of each chromosome, then *each diploid human cell contains ~6 pg genomic DNA*

Therefore 1 ng genomic DNA (1000 pg) = ~333 copies of each locus
(2 per 167 diploid genomes)

DNA QUANTITIES USED IN FLUORESCENCE-BASED STR TYPING

PCR amplification is dependent on the quantity of template DNA molecules added to the reaction. Based on the DNA quantitation results obtained using either the slot blot procedure or some other test, the extracted DNA for each sample is adjusted to a level that will work optimally in the PCR amplification reaction. Most commercial STR typing kits work best with an input DNA template of around 1 ng.

A quantity of 1 ng of human genomic DNA corresponds to approximately 333 copies of each locus that will be amplified (D.N.A. Box 3.3). There are approximately 6 pg (one million one millionth of a gram or 10^{-12} grams) of genomic DNA in each cell containing a single diploid copy of the human genome. Thus, a range of typical DNA quantities from 0.1–25 ng would involve approximately 30–8330 copies of every DNA sequence.

REFERENCES AND ADDITIONAL READING

Abaz, J., Walsh, S.J., Curran, J.M., Moss, D.S., Cullen, J., Bright, J.A., Crowe, G.A., Cockerton, S.L. and Power, T.E. (2002) *Forensic Science International*, 126, 233–240.

Ahn, S.J., Costa, J. and Emanuel, J.R. (1996) *Nucleic Acids Research*, 24, 2623–2625.

Alvarez Garcia, A., Munoz, I., Pestoni, C., Lareu, M.V., Rodriguez-Calvo, M.S. and Carracedo, A. (1996) *International Journal of Legal Medicine*, 109, 125–129.

Andersen, J. (1998) In Lincoln, P.J. and Thomson, J. (eds) *Methods in Molecular Biology, Vol. 98: Forensic DNA Profiling Protocols*, pp. 33–38. Totowa, NJ: Humana Press.

Andreasson, H., Gyllensten, U. and Allen, M. (2002) *Biotechniques*, 33, 402–411.

Applied Biosystems (1998) *AmpFlSTR® Profiler Plus™ PCR Amplification Kit User's Manual*. Foster City, California: Perkin Elmer Corporation.

Applied Biosystems (2003) *Quantifiler™ Human DNA Quantification Kit and Quantifiler™ Y Human Male DNA Quantification Kit User's Manual*. Foster City, California: Applied Biosystems.

Belgrader, P. and Marino, M.A. (1997) *Laboratory Robotics and Automation,* 9, 3–7.

Benecke, M., Schmitt, C. and Staak, M. (1996) *Proceedings of the First European Symposium on Human Identification,* p. 148. Madison, Wisconsin: Promega Corporation.

Boom, R., Sol, C.J., Salimans, M.M., Jansen, C.L., Wertheim-Van Dillen, P.M. and van der, N.J. (1990) *Journal of Clinical Microbiology*, 28, 495–503.

Budowle, B., Baechtel, F.S., Comey, C.T., Giusti, A.M. and Klevan, L. (1995) *Electrophoresis*, 16, 1559–1567.

Budowle, B., Leggitt, J.L., Defenbaugh, D.A., Keys, K.M. and Malkiewicz, S.F. (2000) *Journal of Forensic Sciences*, 45, 1090–1092.

Budowle, B., Hudlow, W.R., Lee, S.B. and Klevan, L. (2001) *Biotechniques*, 30, 680–685.

Burgoyne, L., Kijas, J., Hallsworth, P. and Turner, J. (1994) *Proceedings from the Fifth International Symposium on Human Identification,* p. 163. Madison, Wisconsin: Promega Corporation.

Burgoyne, L.A. (1996) Solid medium and method for DNA storage. U.S. Patent 5 496 562.

Burgoyne, L.A. (1997) *Proceedings from the Eighth International Symposium on Human Identification,* p. 153. Madison, Wisconsin: Promega Corporation.

Castle, P.E., Garcia-Closas, M., Franklin, T., Chanock, S., Puri, V., Welch, R., Rothman, N. and Vaught, J. (2003) *American Journal of Human Genetics*, 73, 646–651.

Comey, C.T., Koons, B.W., Presley, K.W., Smerick, J.B., Sobieralski, C.A., Stanley, D.M. and Baechtel, F.S. (1994) *Journal of Forensic Sciences*, 39, 1254–1269.

Del Rio, S.A., Marino, M.A. and Belgrader, P. (1996) *BioTechniques*, 20, 970–974.

Del Rio, S.A. (1997) *Proceedings from the Eighth International Symposium on Human Identification,* pp. 64–69. Madison, Wisconsin: Promega Corporation.

Diamandis, E.P. and Yu, H. (1995) *Lancet,* 345, 1186.

Duewer, D.L., Kline, M.C., Redman, J.W., Newall, P.J. and Reeder, D.J. (2001) *Journal of Forensic Sciences,* 46, 1199–1210.

Duncan, E., Setzke, E. and Lehmann, J. (2003) Isolation of genomic DNA. In Bowien, B. and Dürre, P. (eds) *Nucleic Acids Isolation Methods,* pp. 7–19. Stevenson Ranch, California: American Scientific Publishers.

Elliott, K., Hill, D.S., Lambert, C., Burroughes, T.R. and Gill, P. (2003) *Forensic Science International,* 137, 28–36.

Fox, J.C., Sangha, J., Douglas, E.K. and Schumm, J.W. (2002) New device and method for buccal cell collection and processing. *Proceedings of the Thirteenth International Symposium on Human Identification.* Available online at: http://www.promega.com/geneticidproc/ussymp13proc/contents/.

Fox, J.C., Cave, C.A. and Schumm, J.W. (2003) *Biotechniques,* 34, 314–322.

Gill, P., Jeffreys, A.J. and Werrett, D.J. (1985) *Nature,* 318, 577–579.

Gill, P., Ivanov, P.L., Kimpton, C., Piercy, R., Benson, N., Tully, G., Evett, I., Hagelberg, E. and Sullivan, K. (1994) *Nature Genetics,* 6, 130–135.

Greenspoon, S.A., Scarpetta, M.A., Drayton, M.L. and Turek, S.A. (1998) *Journal of Forensic Sciences,* 43, 1024–1030.

Greenspoon, S.A., Ban, J.D., Sykes, K., Ballard, E.J., Edler, S.S., Baisden, M. and Covington, B.L. (2004) *Journal of Forensic Sciences,* 49, 29–39.

Gross, A.M., Harris, K.A. and Kaldun, G.L. (1999) *Journal of Forensic Sciences,* 44, 837–840.

Hanselle, T., Otte, M., Schnibbe, T., Smythe, E. and Krieg-Schneider, F. (2003) *Legal Medicine (Tokyo),* 5, S145–S149.

Hayn, S., Wallace, M.M., Prinz, M. and Shaler, R.C. (2004) *Journal of Forensic Sciences,* 49, 87–91.

Herber, B. and Herold, K. (1998) *Journal of Forensic Sciences,* 43, 648–656.

Higuchi, R., von Beroldingen, C.H., Sensabaugh, G.F. and Erlich, H.A. (1988) *Nature,* 332, 543–546.

Hochmeister, M.N., Budowle, B. and Baechtel, F.S. (1991) *Journal of Forensic Sciences,* 36, 656–661.

Hochmeister, M.N. (1998) *Methods in Molecular Biology*, Vol. 98: *Forensic DNA Profiling Protocols,* pp. 19–26. Totowa, New Jersey: Humana Press.

Hochmeister, M., Rudin, O., Meier, R., Eisenberg, A., Nagy, R., Gehrig, C. and Dirnhofer, R. (1998) *Progress in Forensic Genetics*, 7, 24.

Hochmeister, M.N., Budowle, B., Rudin, O., Gehrig, C., Borer, U., Thali, M. and Dirnhofer, R. (1999) *Journal of Forensic Sciences*, 44, 1057–1060.

Holland, M.M., Cave, C.A., Holland, C.A. and Bille, T.W. (2003) *Croatian Medical Journal*, 44, 264–272.

Hopkins, B., Williams, N.J., Webb, M.B.T., Debenham, P.G. and Jeffreys, A.J. (1994) *Journal of Forensic Sciences*, 39, 526–532.

Hopwood, A.J., Mannucci, A. and Sullivan, K.M. (1996) *International Journal of Legal Medicine*, 108, 237–243.

Hopwood, A., Oldroyd, N., Fellows, S., Ward, R., Owen, S.-A. and Sullivan, K. (1997) *BioTechniques*, 23, 18–20.

Jobin, R.M. and De Gouffe, M. (2003) *Canadian Society of Forensic Sciences Journal*, 36(1), 1–10.

Johnston, S., Newman, J. and Frappier, R. (2003) *Canadian Society of Forensic Sciences Journal*, 36 (3), 173–183.

Juusola, J. and Ballantyne, J. (2003) *Forensic Science International*, 135, 85–96.

Khaldi, N., Miras, A., Botti, K., Benali, L. and Gromb, S. (2004) *Journal of Forensic Sciences*, 49(4), 749–753.

Kihlgren, A., Beckman, A. and Holgersson, S. (1998) *Progress in Forensic Genetics,* 7, 31–33.

Kline, M.C., Duewer, D.L., Redman, J.W., Butler, J.M. and Boyer, D.A. (2002) *Analytical Chemistry*, 74, 1863–1869.

Kline, M.C., Duewer, D.L., Redman, J.W. and Butler, J.M. (2003) *Analytical Chemistry*, 75, 2463–2469.

Klintschar, M. and Neuhuber, F. (2000) *Journal of Forensic Sciences*, 45, 669–673.

Kobilinsky, L. (1992) *Forensic Science Reviews*, 4, 67–87.

Krenke, B.E., Tereba, A., Anderson, S.J., Buel, E., Culhane, S., Finis, C.J., Tomsey, C.S., Zachetti, J.M., Masibay, A., Rabbach, D.R., Amiott, E.A. and Sprecher, C.J. (2002) *Journal of Forensic Sciences*, 47, 773–785.

Lee, H.C., Gaensslen, R.E., Bigbee, P.D. and Kearney, J.J. (1991) *Journal of Forensic Identification*, 41, 344–356.

Lee, H.C. (1996) *Proceedings of the Seventh International Symposium on Human Identification*, pp. 39–45. Madison, Wisconsin: Promega Corporation.

Lee, H.C. and Ladd, C. (2001) *Croatian Medical Journal*, 42, 225–228.

Lee, S.B., Ma, M., Worley, J.M., Sprecher, C., Lins, A.M., Schumm, J.W. and Mansfield, E.S. (1994) *Proceedings from the Fifth International Symposium on Human Identification*, pp. 137–145. Madison, Wisconsin: Promega Corporation.

Levy, H. (1996) O. J. Simpson: what the blood really showed. In Levy, H. *And the Blood Cried Out*, pp.157–188. New York: Basic Books.

Lindahl, T. (1993) *Nature*, 362, 709–715.

Lorente, J.A., Lorente, M., Lorente, M.J., Alvarez, J.C., Entrala, C., Lopez-Munoz, J. and Villanueva, E. (1998) *Progress in Forensic Genetics*, 7, 114–116.

Lorente, J.A., Martinez-Gonzales, L.J., Fernandez-Rosado, F., Martinez-Espin, E., Alvarez, J.C., Entrala, C., Lorente, M. and Villanueva, E. (2004) *Proceedings of the American Academy of Forensic Sciences meeting, Volume X,* p. 103. Dallas, TX.

Mandrekar, M.N., Erickson, A.M., Kopp, K., Krenke, B.E., Mandrekar, P.V., Nelson, R., Peterson, K., Shultz, J., Tereba, A. and Westphal, N. (2001) *Croatian Medical Journal*, 42, 336–339.

Marshall, P. (2002) Optimization of spermatozoa capture during the differential extraction process for STR typing with the potential for automation. Master's thesis. University of North Texas.

Miller, S.A., Dykes, D.D. and Polesky, H.F. (1988) *Nucleic Acids Research*, 16, 1215.

Moss, D., Harbison, S.A. and Saul, D.J. (2003) *International Journal of Legal Medicine*, 117, 340–349.

Nicklas, J.A. and Buel, E. (2003a) *Journal of Forensic Sciences*, 48, 282–291.

Nicklas, J.A. and Buel, E. (2003b) *Analytical Bioanalytical Chemistry*, 376, 1160–1167.

Nicklas, J.A. and Buel, E. (2003c) *Journal of Forensic Sciences*, 48, 936–944.

Pfeiffer, H., Huhne, J., Seitz, B. and Brinkmann, B. (1999) *International Journal of Legal Medicine*, 112, 142–144.

Poinar, H.N. (2003) The top 10 list: criteria of authenticity for DNA from ancient and forensic samples. In Brinkmann, B. and Carracedo, A. (eds) *Progress in Forensic Genetics 9,* pp. 575–579. New York: Elsevier Science.

Ravard-Goulvestre, C., Crainic, K., Guillon, F., Paraire, F., Durigon, M. and de Mazancourt, P. (2004) *Journal of Forensic Sciences*, 49, 60–63.

Richard, M.L., Frappier, R.H. and Newman, J.C. (2003) *Journal of Forensic Sciences*, 48, 1041–1046.

Rudbeck, L. and Dissing, J. (1998) *BioTechniques*, 25, 588–592.

Saferstein, R. (2001) *Criminalistics: An Introduction to Forensic Science, Seventh Edition*, Chapters 12 and 13, pp. 320–394. Upper Saddle River, New Jersey: Prentice Hall.

Scherczinger, C.A., Ladd, C., Bourke, M.T., Adamowicz, M.S., Johannes, P.M., Schercziger, R., Beesley, T. and Lee, H.C. (1999) *Journal of Forensic Sciences,* 44, 1042–1045.

Schmerer, W.M., Hummel, S. and Herrmann, B. (1999) *Electrophoresis*, 20, 1712–1716.

Schumm, J.W., Song, E.Y., Burger, M. and Sangha, J. (2004) *Progress in Forensic Genetics 10, International Congress Series*, 1261, 550–552.

Sensabaugh, G.F. (1978) *Journal of Forensic Sciences*, 23, 106.

Sensabaugh, G.F. (1979) *Journal of Forensic Sciences*, 30, 346–364.

Shaler, R.C. (2002) Chapter 10: Modern forensic biology. In Saferstein, R. (ed) *Forensic Science Handbook Volume I, 2nd Edition*, pp. 525–613. Upper Saddle River, New Jersey: Prentice Hall.

Sifis, M.E., Both, K. and Burgoyne, L.A. (2002) *Journal of Forensic Sciences*, 47, 589–592.

Simich, J.P., Morris, S.L., Klick, R.L. and Rittenhouse-Diakun, K. (1999) *Journal of Forensic Sciences*, 44, 1229–1231.

Sims, G., Montgomery, R., Myers, S. and Konzak, K. (1995) *Proceedings of the Sixth International Symposium on Human Identification,* pp. 116–117. Madison, Wisconsin: Promega Corporation.

Sweet, D., Lorente, M., Lorente, J.A., Valenzuela, A. and Villanueva, E. (1997) *Journal of Forensic Sciences*, 42, 320–322.

Tahir, M.A., Sovinski, S.M. and Novick, G.E. (1996) *Proceedings from the Seventh International Symposium on Human Identification,* p. 176. Madison, Wisconsin: Promega Corporation.

Tanaka, M., Yoshimoto, T., Nozawa, H., Ohtaki, H., Kato, Y., Sato, K., Yamamoto, T., Tamaki, K. and Katsumata, Y. (2000) *Journal of Forensic Sciences*, 45, 674–676.

Tereba, A.M., Bitner, R.M., Koller, S.C., Smith, C.E., Kephart, D.D. and Ekenberg, S.J. (2004) Simultaneous isolation and quantitation of DNA. U.S. Patent 6 673 631.

Vandenberg, N., van Oorschot, R.A.H. and Mitchell, R.J. (1997) *Electrophoresis,* 18, 1624–1626.

Van Oorschot, R.A.H. and Jones, M.K. (1997) *Nature*, 387, 767.

Vogelstein, B. and Gillespie, D. (1979) *Proceedings of the National Academy of Sciences U.S.A.*, 76, 615–619.

Von Wurmb-Schwark, N., Harbeck, M., Wiesbrock, U., Schroeder, I., Ritz-Timme S. and Oehmichen, M. (2003) *Legal Medicine*, 5, S169–S172.

Walsh, P.S., Metzger, D.A. and Higuchi, R. (1991) *BioTechniques*, 10, 506–513.

Walsh, P.S., Varlaro, J. and Reynolds, R. (1992) *Nucleic Acids Research*, 20, 5061–5065.

Waye, J.S., Presley, L.A., Budowle, B., Shutler, G.G. and Fourney, R.M. (1989) *BioTechniques,* 7, 852–855.

Wiegand, P., Bajanowski, T. and Brinkmann, B. (1993) *International Journal of Legal Medicine*, 106, 81–83.

Willard, J.M., Lee, D.A. and Holland, M.M. (1998) *Methods in Molecular Biology*, Vol. 98: *Forensic DNA Profiling Protocols*, pp. 9–18. Totowa, New Jersey: Humana Press.

Wilson, M.R., Polanskey, D., Butler, J., DiZinno, J.A., Replogle, J. and Budowle, B. (1995) *BioTechniques*, 18, 662–669.

Withrow, A.G., Sikorsky, J., Downs, J.C. and Fenger, T. (2003) *Journal of Forensic Sciences*, 48, 1302–1308.

Wolfe, K.A., Breadmore, M.C., Ferrance, J.P., Power, M.E., Conroy, J.F., Norris, P.M. and Landers, J.P. (2002) *Electrophoresis*, 23, 727–733.

Word, C.J. and Gregory, S. (1997) *Proceedings from the Eighth International Symposium on Human Identification,* p. 143. Madison, Wisconsin: Promega Corporation.

Yasuda, T., Iida, R., Takeshita, H., Ueki, M., Nakajima, T., Kaneko, Y., Mogi, K., Tsukahara, T. and Kishi, K. (2003) *Journal of Forensic Sciences*, 48, 108–110.

Yu, H. and Diamandis, E.P. (1995) *Clinical Chemistry*, 41, 54–60.

CHAPTER 4

THE POLYMERASE CHAIN REACTION (DNA AMPLIFICATION)

Out of a natural laziness, I always start with the easiest possible protocol and work from there. Better yet, I suggest that someone start from there, and I come back in a month to see how things worked out.

(Kary Mullis, inventor of PCR)

Forensic science and DNA typing laboratories have greatly benefited from the discovery of a technique known as the polymerase chain reaction or PCR. First described in 1985 by Kary Mullis and members of the Human Genetics group at the Cetus Corporation (now Roche Molecular Systems), PCR has revolutionized molecular biology with the ability to make millions of copies of a specific sequence of DNA in a matter of only a few hours. The impact of PCR has been such that its inventor, Kary Mullis, received the Nobel Prize in Chemistry in 1993 – less than 10 years after it was first described.

Without the ability to make copies of DNA samples, many forensic samples would be impossible to analyze. DNA from crime scenes is often limited in both quantity and quality and obtaining a cleaner, more concentrated sample is normally out of the question (most perpetrators of crimes are not surprisingly unwilling to donate more evidence material). The PCR DNA amplification technology is well suited to analysis of forensic DNA samples because it is sensitive, rapid, and not limited by the quality of the DNA as the restriction fragment length polymorphism (RFLP) methods are.

POLYMERASE CHAIN REACTION (PCR) PROCESS

PCR is an enzymatic process in which a specific region of DNA is replicated over and over again to yield many copies of a particular sequence (Saiki *et al.* 1988, Reynolds *et al.* 1991). This molecular 'xeroxing' process involves heating and cooling samples in a precise thermal cycling pattern over ~30 cycles (Figure 4.1). During each cycle, a copy of the target DNA sequence is generated for every molecule containing the target sequence (Figure 4.2). The boundaries of the amplified product are defined by oligonucleotide primers that are complementary to the 3′-ends of the sequence of interest.

Figure 4.1

*Thermal cycling tempera-
ture profile for PCR.
Thermal cycling typically
involves three different temperatures that are repeated
over and over again
25–35 times. At 94°C, the
DNA strands separate, or
'denature'. At 60°C,
primers bind or 'anneal'
to the DNA template and
target the region to be
amplified. At 72°C, the
DNA polymerase extends
the primers by copying the
target region using the
deoxynucleotide triphos-
phate building blocks. The
entire PCR process is about
3 hours in duration with
each cycle taking ~5 min-
utes on conventional thermal cyclers: 1 minute each
at 94°C, 60°C, and 72°C
and about 2 minutes
ramping between the three
temperatures.*

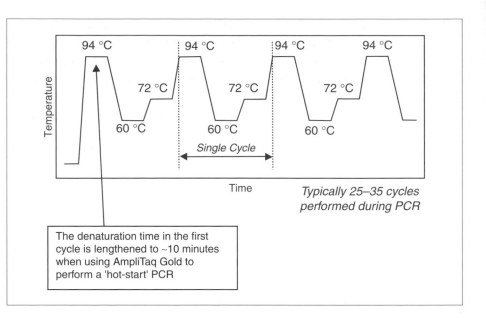

Figure 4.2

*DNA amplification
process with the
polymerase chain reaction.
In each cycle, the two
DNA template strands are
first separated (denatured)
by heat. The sample is
then cooled to an
appropriate temperature
to bind (anneal) the
oligonucleotide primers.
Finally the temperature of
the sample is raised to the
optimal temperature for
the DNA polymerase and
it extends the primers to
produce a copy of each
DNA template strand. For
each cycle, the number of
DNA molecules (with the
sequence between the two
PCR primers) doubles.*

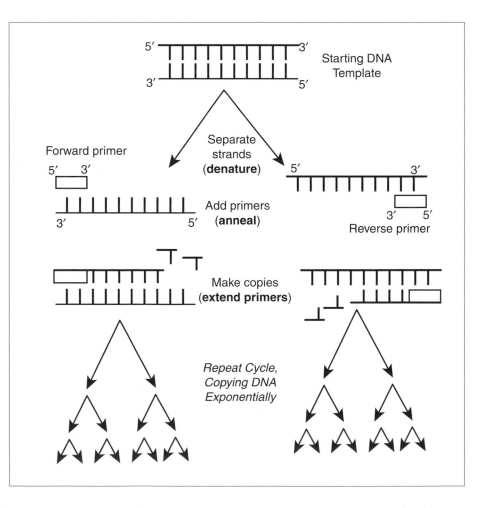

Theoretically after 30 cycles approximately a billion copies of the target region on the DNA template have been generated (Table 4.1). This PCR product, sometimes referred to as an 'amplicon', is then in sufficient quantity that it can be easily measured by a variety of techniques that will be discussed in more detail in the technology section.

PCR is commonly performed with a sample volume in the range of 5–100 μL. At such low volumes, evaporation can be a problem and accurate pipetting of the reaction components can become a challenge. On the other hand, larger solution volumes lead to thermal equilibrium issues for the reaction mixture because it takes longer for an external temperature change to be transmitted to the center of a larger solution than a smaller one. Therefore, longer hold times are needed at each temperature, which leads to longer overall thermal cycling times. Most molecular biology protocols for PCR are thus in the 20–50 μL range.

The sample is pipetted into a variety of reaction tubes designed for use in PCR thermal cyclers. The most common tube in use with 20–50 μL PCR reactions is a thin-walled 0.2 mL tube. These 0.2 mL tubes can be purchased as individual tubes with and without attached caps or as 'strip-tubes' with 8 or 12 tubes connected together in a row. In higher throughput labs, 96-well or 384-well plates are routinely used for PCR amplification.

PCR has been simplified in recent years by the availability of reagent kits that allow a forensic DNA laboratory to simply add a DNA template to a pre-made PCR mix containing all the necessary components for the amplification reaction. These kits are optimized through extensive research efforts on the part of commercial manufacturers. The kits are typically prepared so that a user adds an aliquot of the kit solution to a particular amount of genomic DNA. The best results with these commercial kits are obtained if the DNA template is added in an amount that corresponds to the concentration range designed for the kit.

PCR COMPONENTS

A PCR reaction is prepared by mixing several individual components and then adding deionized water to achieve the desired volume and concentration of each of the components. Commercial kits with pre-mixed components may also be used for PCR. These kits have greatly simplified the use of PCR in forensic DNA laboratories.

The most important components of a PCR reaction are the two primers, which are short DNA sequences that precede or 'flank' the region to be copied. A primer acts to identify or 'target' the portion of the DNA template to be copied. It is a chemically synthesized oligonucleotide that is added in a high concentration relative to the DNA template to drive the PCR reaction.

Table 4.1

Number of target DNA molecules created by PCR amplification (if reaction is working at 100% efficiency). The target PCR product is not completely defined by the forward and reverse primers until the third cycle.

Cycle Number	Number of Double-stranded Target Molecules (Specific PCR Product)
1	0
2	0
3	2
4	4
5	8
6	16
7	32
8	64
9	128
10	256
11	512
12	1024
13	2048
14	4096
15	8192
16	16 384
17	32 768
18	65 536
19	131 072
20	262 144
21	524 288
22	1 048 576
23	2 097 152
24	4 194 304
25	8 388 608
26	16 777 216
27	33 554 432
28	67 108 864
29	134 217 728
30	268 435 456
31	536 870 912
32	1 073 741 824

Some knowledge of the DNA sequence to be copied is required in order to select appropriate primer sequences.

The other components of a PCR reaction consist of template DNA that will be copied, building blocks made up of each of the four nucleotides, and a DNA polymerase that adds the building blocks in the proper order based on the template DNA sequence. The various components and their optimal concentrations are listed in Table 4.2. Thermal stable polymerases that do not fall apart during the near-boiling denaturation temperature steps have been important to the success of PCR (Saiki *et al.* 1988). The most commonly used thermal stable polymerase is *Taq*, which comes from a bacterium named *Thermus aquaticus* that inhabits hot springs.

When setting up a set of samples that contain the same primers and reaction components, it is common to prepare a 'master mix' that can then be dispensed in equal quantities to each PCR tube. This procedure helps to insure that there is more homogeneity between samples. Also by setting up a larger number of reactions at once, small pipetting volumes can be avoided, which improves the accuracy of adding each component (and thus the reproducibility of one's method). When performing a common test on a number of different samples, the goal should be to examine the variation in the DNA samples *not* variability in the reaction components used and the sample preparation method.

CONTROLS USED TO MONITOR PCR

Controls are used to monitor the effectiveness of the chosen experimental conditions and/or the technique of the experimenter. These controls typically

Reagent	Optimal Concentration
Tris-HCl, pH 8.3 (25°C)	10–50 mM
Magnesium chloride	1.2–2.5 mM
Potassium chloride	50 mM
Deoxynucleotide triphosphates (dNTPs)	200 μM each dATP, dTTP, dCTP, dGTP
DNA polymerase, thermal stable[a]	0.5–5 U
Bovine serum albumin (BSA)	100 μg/mL
Primers	0.1–1.0 μM
Template DNA	1–10 ng genomic DNA

[a]*Taq* and *Taq*Gold are the two most common thermal stable polymerase used for PCR.

Table 4.2

Typical components for PCR amplification.

include a 'negative control', which is the entire PCR reaction mixture without any DNA template. The negative control usually contains water or buffer of the same volume as the DNA template, and is useful to assess whether or not any of the PCR components have been contaminated by DNA (e.g., someone else in your lab). An extraction 'blank' is also useful to verify that the reagents used for DNA extraction are clean of any extraneous DNA templates (Presley and Budowle 1994).

A 'positive control' is a valuable indicator of whether or not any of the PCR components have failed or were not added during the reaction setup phase of experiments conducted. A standard DNA template of known sequence with good quality DNA should be used for the positive control. This DNA template should be amplified with the same PCR primers as used on the rest of the samples in the batch that is being amplified. The purpose of a positive control is to ensure confidence that the reaction components and thermal cycling parameters are working for amplifying a specific region of DNA.

STOCHASTIC EFFECTS FROM LOW LEVELS OF DNA TEMPLATE

Forensic DNA specimens often possess low levels of DNA. When amplifying very low levels of DNA template, a phenomenon known as *stochastic fluctuation* can occur. Stochastic effects, which are an unequal sampling of the two alleles present from a heterozygous individual, result when only a few DNA molecules are used to initiate PCR (Walsh *et al.* 1992). PCR reactions involving DNA template levels below approximately 100 pg of DNA, or about 17 diploid copies of genomic DNA, have been shown to exhibit allele dropout (Fregeau *et al.* 1993, Kimpton *et al.* 1994, Gill *et al.* 2000). False homozygosity results if one of the alleles fails to be detected. The problem of stochastic effects can be avoided by adjusting the cycle number of the PCR reaction such that approximately 20 or more copies of target DNA are required to yield a successful typing result (Walsh *et al.* 1992). However, efforts have been made to obtain results with low-copy number (LCN) DNA testing (Gill *et al.* 2000, Gill 2002). The challenges of LCN work and trying to interpret data obtained with less than 100 pg of DNA template will be addressed in Chapter 7. Whole genome amplification prior to PCR and locus-specific amplification also have the same issues in terms of stochastic fluctuations with low levels of DNA (Schneider *et al.* 2004; D.N.A. Box 4.1).

THERMAL CYCLING PARAMETERS

A wide range of PCR cycling protocols have been used for various molecular biology applications. To serve as an example of PCR cycling conditions

A common challenge with forensic casework is the recovery of limited quantities of DNA from evidentiary samples. Within the past few years a new DNA enrichment technology has been developed known as whole genome amplification (WGA). WGA involves a different DNA polymerase than the TaqGold enzyme commonly used in forensic DNA analysis. WGA amplifies the entire genome using random hexamers as priming points. The WGA enzymes work by multiple displacement amplification (MDA), which is sometimes referred to as rolling circle amplification. MDA is isothermal with an incubation temperature of 30°C and requires no heating and cooling like PCR.

Molecular Staging (New Haven, CT) and Amersham Biosciences (Piscataway, NJ) both offer phi29 DNA polymerase cocktails for performing WGA. The kit by Molecular Staging is called REPLI-g™ while Amersham's kit is GenomiPhi™. Yields of 4–7 µg of amplified genomic DNA are possible from as little as 1 ng of starting material. The phi29 enzyme has a high processivity and can amplify fragments of up to 100 kb because it displaces downstream product strands enabling multiple concurrent and overlapping rounds of amplification. In addition, phi29 has a higher replication fidelity compared to *Taq* polymerase due to 3′–5′ proofreading activity.

While all of these characteristics make WGA seem like a possible solution to the forensic problem of limited DNA starting material, studies have found that stochastic effects at low levels of DNA template prevent WGA from working reliably (Schneider *et al.* 2004). Allele dropouts from STR loci were observed at 50 pg and 5 pg levels of starting material (Schneider *et al.* 2004) just as are seen with current low-copy number DNA testing (see Chapter 7). While it is possible that WGA may play a limited role in enriching samples for archiving purposes that are in the low ng range, it will probably not be the end-all solution to low copy number DNA samples.

Sources:

Schneider, P.M. *et al.* (2004) Whole genome amplification – the solution for a common problem in forensic casework? *Progress in Forensic Genetics 10*, Elsevier Science, ICS 1261, 24–26.

Hosono, S. *et al.* (2003) Unbiased whole-genome amplification directly from clinical samples. *Genome Research*, 13, 954–964.

Dean *et al.* (2002) Comprehensive human genome amplification using multiple displacement amplification. *Proceedings of the National Academy of Sciences USA*, 99, 5261–5266.

Lizardi, P.M. (2000) Multiple displacement amplification, U.S. Patent 6 124 120.

See also http://www.rubicongenomics.com; http://www.the-scientist.com/yr2003/aug/tools2_030825.html; http://www.the-scientist.com/yr2003/jan/tools_030113.html

commonly used by forensic DNA laboratories, Table 4.3 contains the parameters used with the GenePrint® STR kits from Promega Corporation and the AmpF/STR® kits from Applied Biosystems. The primary reason that PCR protocols vary is that different primer sequences have different hybridization properties and thus anneal to the DNA template strands at different rates.

THERMAL CYCLERS

The instrument that heats and cools a DNA sample in order to perform the PCR reaction is known as a thermal cycler. Precise and accurate sample heating and cooling is crucial to PCR in order to guarantee consistent results. There are a wide variety of thermal cycler options available from multiple manufacturers. These instruments vary in the number of samples that can be handled at a time, the size of the sample tube and volume of reagents that can be handled, and the speed at which the temperature can be changed. Prices for thermal cycling devices range from a few thousand dollars to over 10 000 dollars.

Perhaps the most prevalent thermal cycler in forensic DNA laboratories is the GeneAmp® PCR System 9600 from Applied Biosystems (Foster City, CA). The '9600' can heat and cool 96 samples in an 8×12-well microplate format at a rate of approximately 1°C per second. The 9600 uses 0.2 mL tubes with tube caps. These tubes may be attached together in strips of 8 or 12, in which case they are referred to as 'strip-tubes.'

Table 4.3

Thermal cycling parameters used to amplify short tandem repeat DNA markers with commercially available PCR amplification kits. Cycling parameters differ because reaction components, in particular the primer concentrations and sequences, vary between the different manufacturers' kits.

Step in Protocol	AmpF/STR® kits (Applied Biosystems)	GenePrint® STR kits (Promega Corporation)
Initial Incubation	95°C for 11 minutes	95°C for 11 minutes
Thermal Cycling	28 cycles	30 cycles[a]
Denature	94°C for 1 minute	94°C for 30 seconds (cycle 1–10) 90°C for 30 seconds (cycle 11–30)
Anneal	59°C for 1 minute	60°C for 30 seconds
Extend	72°C for 1 minute	70°C for 45 seconds
Final Extension	60°C for 45 minutes	60°C for 30 minutes
Final Soak	25°C (until samples removed)	4°C (until samples removed)

[a]The first 10 cycles are run with a denaturation temperature of 94°C and the last 20 cycles are run at 90°C instead. The Promega PowerPlex 1.1, 2.1, and 16 kits also use specific ramp times between the different temperature steps that differ from the conventional 1°C/second.

Modern thermal cyclers use a heated lid to keep the PCR reagents from condensing at the top of the tube during the temperature cycling. However, there are some forensic laboratories that still use an older model thermal cycler called the DNA Thermal Cycler '480'. Samples amplified in the 480 require an overlay of a drop of mineral oil on top of the PCR reaction mix in order to prevent evaporation since there is no heated lid.

Thermal cyclers capable of amplifying 384 samples or more at one time are now available. The Dual 384-well GeneAmp® PCR System 9700 can run 768 reactions simultaneously on two 384-well sample blocks. Thermal cyclers capable of high sample volume processing are valuable in production settings but are not widely used in forensic DNA laboratories.

HOT START PCR

Regular DNA polymerases exhibit some activity below their optimal temperature, which for *Taq* polymerase is 72°C. Thus, primers can anneal non-specifically to the template DNA at room temperature when PCR reactions are being set-up and non-specific products may result. It is also possible at a low temperature for the primers to bind to each other creating products called 'primer dimers.' These are a particular problem because their small size relative to the PCR products means that they will be preferentially amplified.

Once low-temperature non-specific priming occurs, these undesirable products will be efficiently amplified throughout the remaining PCR cycles. Because the polymerase is busy amplifying these competing products, the target DNA region will be amplified less efficiently. If this happens, you will get less of what you are looking for and you may not have enough specific DNA to run your other tests.

Low-temperature mispriming can be avoided by initiating PCR at an elevated temperature, a process usually referred to as 'hot start' PCR. Hot start PCR may be performed by introducing a critical reaction component, such as the polymerase, after the temperature of the sample has been raised above the desired annealing temperature (e.g., 60°C). This minimizes the possibility of mispriming and misextension events by not having the polymerase present during reaction setup. However, this approach is cumbersome and time-consuming when working with large numbers of samples. Perhaps a more important disadvantage is the fact that the sample tubes must be opened at the thermal cycler to introduce the essential component, which gives rise to a greater opportunity for cross-contamination between samples. As will be discussed in the next section, a modified form of *Taq* DNA polymerase has been developed that requires thermal activation and thus enables a closed-tube hot start PCR.

This enzyme, named AmpliTaq Gold, has greatly benefited the specificity of PCR amplifications.

AMPLITAQ GOLD DNA POLYMERASE

AmpliTaq Gold™ DNA polymerase is a chemically modified enzyme that is rendered inactive until heated (Birch *et al.* 1996). An extended pre-incubation of 95°C, usually for 10 or 11 minutes, is used to activate the AmpliTaq Gold. The chemical modification involves a derivitization of the epsilon-amino groups of the lysine residues (Innis and Gelfand 1999). At a pH below 7.0 the chemical modification moieties fall off and the activity of the polymerase is restored. The Tris buffer in the PCR reaction is pH sensitive with temperature variation, and higher temperatures cause the solution pH to go down by approximately 0.02 pH units with every 1°C (Innis and Gelfand 1990). A Tris buffer with pH 8.3 at 25°C will go down to pH ~6.9 at 95°C. Thus, not only is the template DNA well denatured but the polymerase is activated just when it is needed, and not in a situation where primer dimers and mispriming can occur as easily.

It is important to note that AmpliTaq Gold is not compatible with the pH 9.0 buffers used for regular AmpliTaq DNA polymerases (Moretti *et al.* 1998). This fact is because the pH does not get low enough to remove the chemical modifications on TaqGold and thus the enzyme remains inactive. Tris buffers with a pH 8.0 or 8.3 at 25°C work the best with TaqGold.

PCR PRIMER DESIGN

Well-designed primers are probably the most important components of a good PCR reaction. The target region on the DNA template is defined by the position of the primers. PCR yield is directly affected by the annealing characteristics of the primers. For the PCR to work efficiently, the two primers must be specific to the target region, possess similar annealing temperatures, not interact significantly with each other or themselves to form 'primer dimers', and be structurally compatible. Likewise, the sequence region to which the primers bind must be fairly well conserved because if the sequence changes from one DNA template to the next then the primers will not bind appropriately. The general guidelines to optimal PCR primer design are listed in Table 4.4 (see also Dieffenbach *et al.* 1993).

A number of primer design software packages are commercially available including Primer Express (Applied Biosystems, Foster City, CA) and Oligo (Molecular Biology Insights, Cascade, CO). These programs use thermodynamic 'nearest neighbor' calculations to predict annealing temperatures and

Parameter	Optimal Values
Primer length	18–30 bases
Primer T_m (melting temperature)	55–72°C
Percentage GC content	40–60%
No self-complementarity (hairpin structure)	≤3 contiguous bases
No complementarity to other primer (primer dimer)	≤3 contiguous bases (especially at the 3′-ends)
Distance between two primers on target sequence	<2000 bases apart
Unique oligonucleotide sequence	Best match in BLAST[a] search
T_m difference between forward and reverse primers in pair	≤5°C
No long runs with the same base	<4 contiguous bases

Table 4.4

General guidelines for PCR primer design.

[a]BLAST search examines similarity of the primer to other known sequences that may result in multiple binding sites for the primer and thus reduce the efficiency of the PCR amplification reaction. BLAST searches may be conducted via the Internet: http://www.ncbi.nlm.nih.gov/BLAST.

primer interactions with themselves or other possible primers (Mitsuhashi 1996, SantaLucia 1998).

The Internet has become a valuable resource for tools that aid primer selection. For example, a primer design program called Primer 3 is available on the World Wide Web through the Whitehead Institute (http://www.genome.wi.mit.edu/cgi-bin/primer/primer3_www.cgi). With Primer 3, the user inputs a DNA sequence and specifies the target region within that sequence to be amplified. Parameters such as PCR product size, primer length and desired annealing temperature may also be specified by the user. The program then ranks the best PCR primer pairs and passes them back to the user over the Internet. Primer 3 works well for quickly designing singleplex primer pairs that amplify just one region of DNA at a time.

MULTIPLEX PCR

The polymerase chain reaction permits more than one region to be copied simultaneously by simply adding more than one primer set to the reaction mixture (Edwards and Gibbs 1994). The simultaneous amplification of two or more regions of DNA is commonly known as multiplexing or multiplex PCR (Figure 4.3). For a multiplex reaction to work properly the primer pairs need to be compatible. In other words, the primer annealing temperatures should be similar and excessive regions of complementarity should be avoided to prevent

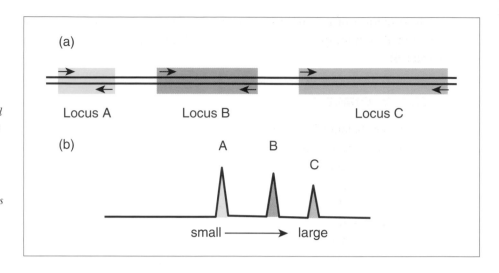

the formation of primer-dimers that will cause the primers to bind to one another instead of the template DNA. The addition of each new primer in a multiplex PCR reaction exponentially increases the complexity of possible primer interactions (Butler *et al.* 2001).

Each new PCR application is likely to require some degree of optimization in either the reagent components or thermal cycling conditions. Multiplex PCR is no exception. In fact, multiplex PCR optimization is more of a challenge than singleplex reactions because so many primer-annealing events must occur simultaneously without interfering with each other. Extensive optimization is normally required to obtain a good balance between the amplicons of the various loci being amplified (Kimpton *et al.* 1996, Markoulatos *et al.* 2002).

The variables that are examined when trying to obtain optimal results for a multiplex PCR amplification include many of the reagents listed in Table 4.2 as well as the thermal cycling temperature profile. Primer sequences and concentrations along with magnesium concentrations are usually the most crucial to multiplex PCR. Extension times during thermal cycling are often increased for multiplex reactions in order to give the polymerase time to fully copy all of the DNA targets. Obtaining successful co-amplification with well-balanced PCR product yields sometimes requires redesign of primers and tedious experiments with adjusting primer concentrations.

Primer design for the short tandem repeat (STR) DNA markers that are discussed in Chapter 5 include some additional challenges. Primers need to be adjusted on the STR markers to achieve good size separation between loci labeled with the same fluorescent dye. In addition, the primers must produce robust amplifications with good peak height balance between loci as well as specific amplification with no non-specific products that might interfere with proper interpretation of a sample's DNA profile. Finally, primers should produce a maximal non-template-dependent '+A' addition to all PCR products (see Chapter 6).

MULTIPLEX PCR OPTIMIZATION

Obtaining a nicely balanced multiplex PCR reaction with each PCR product having a similar yield is a challenging task. With the widespread availability of commercial kits, individual forensic laboratories rarely perform PCR optimization experiments any more. Rather, internal validation studies (see Chapter 16) focus on performance of the multiplex with varying conditions around the optimal parameters supplied with the kit protocol. For example, PCR product yields in the form of STR peak heights produced by a commercial kit might be evaluated at the optimal annealing temperature (e.g., 59°C) and 2°C and 4°C higher and lower (e.g., 55, 57, 61, and 63°C). Differences, if any, would then be noted relative to the optimal annealing temperature supplied in the kit protocol.

The development of an efficient multiplex PCR reaction requires careful planning and numerous tests and efforts in the area of primer design and balancing reaction components (Shuber *et al.* 1995, Henegariu *et al.* 1997, Markoulatos *et al.* 2002). A range of thermal cycling parameters including annealing temperatures and extension times are often examined in developing the final protocol. Primer concentrations are one of the largest factors in a multiplex PCR reaction determining the overall yield of each amplicon (Schoske *et al.* 2003). Repeated experiments and primer titrations are usually performed to achieve an optimal balance. Concentrations of magnesium chloride and deoxynucleotide triphosphates are typically increased slightly relative to singleplex reactions. A thorough evaluation of performance for a multiplex will also involve addition and removal of primer sets to see if overall balance in other amplification targets are affected.

Co-amplification of 25 PCR products followed by simultaneous detection of 35 Y chromosome single nucleotide polymorphism markers contained within the 25 amplicons has been performed in one of the most impressive demonstrations of multiplexing to date (Sanchez *et al.* 2003). The availability of 5-dye detection systems has enabled development of multiplexes capable of amplifying and analyzing in excess of 20 short tandem repeat loci (Butler *et al.* 2002, Hanson and Ballantyne 2004).

REAL-TIME (QUANTITATIVE) PCR

Instruments and assays are now available that can monitor the PCR process as it is happening enabling 'real-time' data collection. Real-time PCR, which was first described by Higuchi and co-workers at the Cetus Corporation in the early 1990s (Higuchi *et al.* 1992, Higuchi *et al.* 1993), is sometimes referred to as quantitative PCR or 'kinetic analysis' because it analyzes the cycle-to-cycle change in fluorescence signal resulting from amplification of a target sequence during PCR. This analysis is performed without opening the PCR tube and therefore can be referred as a closed-tube or homogeneous detection assay.

Several approaches to performing real-time PCR homogeneous detection assays have been published (see Foy and Parkes 2001, Nicklas and Buel 2003a). The most common approaches utilize either the fluorogenic 5′ nuclease assay – better known as TaqMan® – or use of an intercalating dye, such as SYBR® Green, that is highly specific for double-stranded DNA molecules (see Chapter 13). The TaqMan approach monitors change in fluorescence due to displacement of a dual dye-labeled probe from a specific sequence within the target region while the SYBR Green assay detects formation of any PCR product.

THE 5′ NUCLEASE ASSAY (TAQMAN)

TaqMan probes are labeled with two fluorescent dyes that emit at different wavelengths (Figure 4.4). The probe sequence is intended to hybridize specifically in the DNA target region of interest between the two PCR primers (Ong and Irvine 2002). Typically the probe is designed to have a slightly higher annealing temperature compared to the PCR primers so that the probe will be

Figure 4.4

Schematic of TaqMan (5′ nuclease) assay.

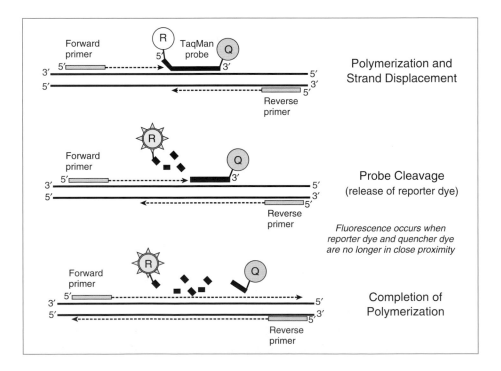

hybridized when extension (polymerization) of the primers begins. A minor groove binder is sometimes used near the 3′-end of TaqMan probes to enable the use of shorter sequences that still have high annealing temperatures (Applied Biosystems 2003). The 'reporter' (R) dye is attached at the 5′-end of the probe sequence while the 'quencher' (Q) dye is synthesized on the 3′-end. A popular combination of dyes is FAM or VIC for the reporter dye and TAMRA for the quencher dye (see Chapter 13 for more information on these dyes). When the probe is intact and the reporter dye is in close proximity to the quencher dye, little-to-no fluorescence will result because of suppression of the reporter fluorescence due to an energy transfer between the two dyes.

During polymerization, strand synthesis will begin to displace any TaqMan probes that have hybridized to the target sequence. The *Taq* DNA polymerase used has a 5′-exonuclease activity and therefore will begin to chew away at any sequences in its path (i.e., those probes that have annealed to the target sequence). When the reporter dye molecule is released from the probe and it is no longer in close proximity to the quencher dye, it can begin to fluoresce (Figure 4.4). Increase in fluorescent signal results if the target sequence is complementary to the TaqMan probe.

Some assays, such as the Quantifiler kit, include an internal PCR control (IPC) that enables verification that the polymerase, the assay, and the detection instrumentation are working correctly. In this case, the IPC is labeled with a VIC (green) reporter dye and hybridizes to a synthetic template added to each reaction. The TaqMan probe for detecting the target region of interest is labeled with a FAM (blue) reporter dye and therefore spectrally resolvable from the green VIC dye. Instruments such as the ABI Prism 7000 Sequence Detection System enable another dye like ROX (red dye) to be placed in each well to adjust for well-to-well differences across a plate through background subtraction.

REAL-TIME PCR ANALYSIS

There are three distinct phases that define the PCR process: geometric or exponential amplification, linear amplification, and the plateau region (Bloch 1991). These regions can be seen in a plot of fluorescence versus PCR cycle number (Figure 4.5). During exponential amplification, there is a high degree of precision surrounding the production of new PCR products. When the reaction is performing at close to 100% efficiency, then a doubling of amplicons occurs with each cycle (see Table 4.1). A plot of cycle number versus a log scale of the DNA concentration should result in a linear relationship during the exponential phase of PCR amplification.

A linear phase of amplification follows the exponential phase as one or more components fall below a critical concentration and amplification efficiency slows down to an arithmetic increase rather than the geometric one in the exponential phase. Since components such as dNTPs or primers may be used up at slightly different rates between reactions, the linear phase is not as precise from sample-to-sample and therefore is not as useful for comparison purposes.

The final phase of PCR is the plateau region where accumulation of PCR product slows to a halt as multiple components have reached the end of their effectiveness in the assay. The fluorescent signal observed in the plateau phase levels out. The accumulation of PCR product generally ceases when its concentration reaches approximately 10^{-7}M (Bloch 1991).

The optimal place to measure fluorescence versus cycle number is in the exponential phase of PCR where the relationship between the amount of product and input DNA is more likely to be consistent. Real-time PCR instruments use what is termed the cycle threshold (C_T) for calculations. The C_T value is the point in terms of PCR amplification cycles when the level of fluorescence exceeds some arbitrary threshold, such as 0.2, that is set by the real-time PCR software to be above the baseline noise observed in the early stages of PCR. The fewer cycles it takes to get to a detectable level of fluorescence (i.e., to cross the threshold set by the software), the greater the initial number of DNA molecules put into the PCR reaction. Thus a plot of the log of DNA concentrations versus the C_T value for each sample results in a linear relationship with a negative slope (Figure 4.5).

Figure 4.5

Real-time PCR output and example standard curve used to determine quantity of input DNA.

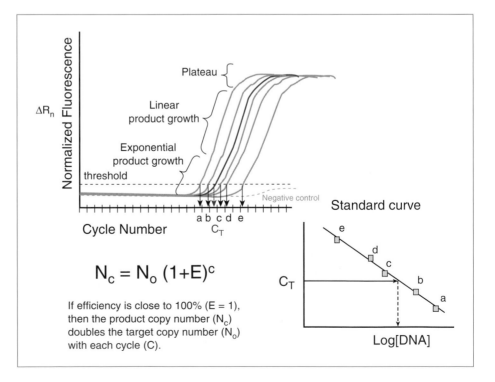

The cleavage of TaqMan probes or binding of SYBR Green intercalating dye to double-stranded DNA molecules results in an increase in fluorescence signal. This rise in fluorescence can be correlated to the initial DNA template amounts when compared with samples of known DNA concentration. For example in Figure 4.5, five samples (a,b,c,d,e) are used to generate a standard curve based on their measured C_T values. Provided that there is good sample-to-sample consistency and precision, a sample with an unknown DNA quantity can be compared to this standard curve to calculate its initial DNA template concentration. Several published real-time PCR assays and commercial kits, such as the Quantifiler Human DNA Quantification Kit, are briefly reviewed in Table 3.3.

PRECAUTIONS AGAINST CONTAMINATION

The sensitivity of PCR necessitates constant vigilance on the part of the laboratory staff to ensure that contamination does not affect DNA typing results. Contamination of PCR reactions is always a concern because the technique is very sensitive to low amounts of DNA. A scientist setting up the PCR reaction can inadvertently add his or her own DNA to the reaction if he or she is not careful. Likewise, the police officer or crime scene technician collecting the evidence can contaminate the sample if proper care is not taken. For this reason, each piece of evidence should be collected with clean tweezers or handled with disposable gloves that are changed frequently.

To aid discovery of laboratory contamination, everyone in a forensic DNA laboratory is typically genotyped in order to have a record of possible contaminating DNA profiles. This is often referred to as a staff elimination database (see Chapter 7). Laboratory personnel should be appropriately gowned during interactions with samples prior to PCR amplification (Rutty *et al.* 2003). The appropriate covering includes lab coats and gloves as well as facial masks and hairnets to prevent skin cells or hair from falling into the amplification tubes. These precautions are especially critical when working with miniscule amounts of sample or sample that has been degraded (see Chapter 7).

Some tips for avoiding contamination with PCR reactions in a laboratory setting include:

- Pre- and post-PCR sample processing areas should be physically separated. Usually a separate room or a containment cabinet is used for setting up the PCR amplification reactions.
- Equipment, such as pipettors, and reagents for setting up PCR should be kept separate from other laboratory supplies, especially those used for analysis of PCR products.

- Disposable gloves should be worn and changed frequently.
- Reactions may also be set up in a laminar flow hood, if available.
- Aerosol-resistant pipette tips should be used and changed on every new sample to prevent cross-contamination during liquid transfers.
- Reagents should be carefully prepared to avoid the presence of any contaminating DNA or nucleases.
- Ultraviolet irradiation of laboratory PCR setup space when the area is not in use and cleaning workspaces and instruments with isopropanol and/or 10% bleach solutions help to insure that extraneous DNA molecules are destroyed prior to DNA extraction or PCR setup (Kwok and Higuchi 1989, Prince and Andrus 1992).

PCR product carryover results from amplified DNA contaminating a sample that has not yet been amplified. Because the amplified DNA is many times more concentrated than the unamplified DNA template, it will be preferentially copied during PCR and the unamplified sample will be masked. The inadvertent transfer of even a very small volume of a completed PCR amplification to an unamplified DNA sample can result in the amplification and detection of the 'contaminating' sequence. For this reason, the evidence samples are typically processed through a forensic DNA laboratory prior to the suspect reference samples to avoid any possibility of contaminating the evidence with the suspect's amplified DNA.

Pipette tips should never be reused. Even a tiny droplet of PCR product left in a pipette tip contains as many as a billion copies of the amplifiable sequence. By comparison, a nanogram of human genomic DNA contains only about 300 copies of single-copy DNA markers (see D.N.A. Box 3.3).

ADVANTAGES AND DISADVANTAGES OF PCR WITH FORENSIC SPECIMENS

We conclude this chapter by reviewing the advantages and disadvantages of PCR amplification for forensic DNA analysis. The advantages of PCR amplification for biological evidence include the following:

- Very small amounts of DNA template may be used from as little as a single cell.
- DNA degraded to fragments only a few hundred base pairs in length can serve as effective templates for amplification.
- Large numbers of copies of specific DNA sequences can be amplified simultaneously with multiplex PCR reactions.
- Contaminant DNA, such as fungal and bacterial sources, will not amplify because human-specific primers are used.
- Commercial kits are now available for easy PCR reaction setup and amplification.

There are three potential pitfalls that could be considered disadvantages of PCR:

1. The target DNA template may not amplify due to the presence of PCR inhibitors in the extracted DNA (see Chapter 7 discussion on PCR inhibition).
2. Amplification may fail due to sequence changes in the primer-binding region of the genomic DNA template (see Chapter 6 discussion on null alleles).
3. Contamination from other human DNA sources besides the forensic evidence at hand or previously amplified DNA samples is possible without careful laboratory technique and validated protocols (see Chapter 7 discussion on PCR contamination and Chapter 16 on laboratory validation).

REFERENCES AND ADDITIONAL READING

Applied Biosystems (1998) *AmFlSTR® Profiler Plus™ PCR Amplification Kit User's Manual.* Foster City, California: Applied Biosystems.

Applied Biosystems (2003) *Quantifiler™ Human DNA Quantification Kit and Quantifiler™ Y Human Male DNA Quantification Kit User's Manual.* Foster City, California: Applied Biosystems.

Birch, D.E., Kolmodin, L., Wong, J., Zangenberg, G.A., Zoccoli, M.A., McKinney, N., Young, K.K.Y. and Laird, W.J. (1996) *Nature*, 381, 445–446.

Bloch, W. (1991) *Biochemistry*, 30, 2735–2747.

Butler, J.M., Ruitberg, C.M. and Vallone, P.M. (2001) *Fresenius Journal of Analytical Chemistry*, 369, 200–205.

Butler, J.M., Schoske, R., Vallone, P.M., Kline, M.C., Redd, A.J. and Hammer, M.F. (2002) *Forensic Science International,* 129, 10–24.

Dieffenbach, C.W., Lowe, T.M.J. and Dveksler, G.S. (1993) *PCR Methods and Applications*, 3, S30–S37.

Edwards, M.C. and Gibbs, R.A. (1994) *PCR Methods and Applications*, 3, S65–S75.

Foy, C.A. and Parkes, H.C. (2001) *Clinical Chemistry*, 47, 990–1000.

Fregeau, C.J. and Fourney, R.M. (1993) *BioTechniques*, 15, 100–119.

Gill, P., Whitaker, J., Flaxman, C., Brown, N. and Buckleton, J. (2000) *Forensic Science International*, 112, 17–40.

Gill, P. (2002) *Biotechniques*, 32, 366–372.

Hanson, E.K. and Ballantyne, J. (2004) *Journal of Forensic Sciences*, 49, 40–51.

Henegariu, O., Heerema, N.A., Dlouhy, S.R., Vance, G.H. and Vogt, P.H. (1997) *Biotechniques*, 23, 504–511.

Higuchi, R., Dollinger, G., Walsh, P.S. and Griffith, R. (1992) *BioTechnology*, 10, 413–417.

Higuchi, R., Fockler, C., Dollinger, G. and Watson, R. (1993) *BioTechnology*, 11, 1026–1030.

Innis, M.A. and Gelfand, D.H. (1990) In Innis, M.A. (ed) *PCR Protocols: A Guide to Methods and Applications*. San Diego: Academic Press.

Innis, M.A. and Gelfand, D.H. (1999) In Innis, M.A., Gelfand, D.H. and Sninsky, J.J. (eds) *PCR Applications: Protocols for Functional Genomics*. San Diego: Academic Press.

Kimpton, C., Fisher, D., Watson, S., Adams, M., Urquhart, A., Lygo, J. and Gill, P. (1994) *International Journal of Legal Medicine*, 106, 302–311.

Kimpton, C.P, Oldroyd, N.J., Watson, S.K., Frazier, R.R.E., Johnson, P.E., Millican, E.S., Urquhart, A., Sparkes, B.L. and Gill, P. (1996) *Electrophoresis*, 17, 1283–1293.

Kwok, S. and Higuchi, R. (1989) *Nature*, 339, 237–238.

Markoulatos, P., Siafakas, N. and Moncany, M. (2002) *Journal of Clinical Laboratory Analysis*, 16, 47–51.

Mitsuhashi, M. (1996) *Journal of Clinical Laboratory Analysis*, 10, 285–293.

Moretti, T., Koons, B. and Budowle, B. (1998) *BioTechniques*, 25, 716–722.

Nicklas, J.A. and Buel, E. (2003a) *Analytical Bioanalytical Chemistry*, 376, 1160–1167.

Nicklas, J.A. and Buel, E. (2003b) *Journal of Forensic Sciences*, 48, 936–944.

Ong, Y.-L. and Irvine, A. (2002) *Hematology*, 7, 59–67.

Presley, L.A. and Budowle, B. (1994) In Griffin, H.G. and Griffin, A.M. (eds) *PCR Technology: Current Innovations*, pp. 259–276. Boca Raton, Florida: CRC Press Inc.

Prince, A.M. and Andrus, L. (1992) *BioTechniques*, 12, 358.

Reynolds, R., Sensabaugh, G. and Blake, E. (1991) *Analytical Chemistry*, 63, 1–15.

Rutty, G. N., Hopwood, A. and Tucker, V. (2003) *International Journal of Legal Medicine*, 117, 170–174.

Saiki, R.K., Gelfand, D.H., Stoffel, S., Scharf, S.J., Higuchi, R., Horn, G.T., Mullis, K.B. and Erlich, H.A. (1988) *Science*, 239, 487–491.

Sanchez, J.J., Borsting, C., Hallenberg, C., Buchard, A., Hernandez, A. and Morling, N. (2003) *Forensic Science International*, 137, 74–84.

SantaLucia, J. (1998) *Proceedings of the National Academy of Sciences U.S.A.*, 95, 1460–1465.

Schoske, R., Vallone, P.M., Ruitberg, C.M. and Butler, J.M. (2003) *Analytical Bioanalytical Chemistry*, 375, 333–343.

Schneider, P.M., Balogh, K., Naveran, N., Bogus, M., Bender, K., Lareu, M. and Carracedo, A. (2004) *Progress in Forensic Genetics 10*, International Congress Series, 1261, 24–26.

Shuber, A.P., Grondin, V.J. and Klinger, K.W. (1995) *Genome Research*, 5, 488–493.

Walsh, P.S., Erlich, H.A. and Higuchi, R. (1992) *PCR Methods and Applications*, 1, 241–250.

CHAPTER 5

COMMONLY USED SHORT TANDEM REPEAT MARKERS AND COMMERCIAL KITS

Ever since their discovery in the early 1980s, the ubiquitous occurrence of microsatellites – also referred to as short tandem repeats (STRs) or simple sequence repeats (SSRs)– has puzzled geneticists... [Understanding STRs] is important if we wish to understand how genomes are organized and why most genomes are filled with sequences other than genes.

(Hans Ellegren 2004)

REPEATED DNA

Eukaryotic genomes are full of repeated DNA sequences (see Ellegren 2004). These repeated DNA sequences come in all types of sizes and are typically designated by the length of the core repeat unit and the number of contiguous repeat units or the overall length of the repeat region. Long repeat units may contain several hundred to several thousand bases in the core repeat.

These regions are often referred to as *satellite* DNA and may be found surrounding the chromosomal centromere. The term satellite arose due to the fact that frequently one or more minor satellite bands were seen in early experiments involving equilibrium density gradient centrifugation (Britten and Kohne 1968, Primrose 1998).

The core repeat unit for a medium length repeat, sometimes referred to as a *minisatellite* or a VNTR (variant number of tandem repeats), is in the range of approximately 10–100 bases in length (Tautz 1993, Chambers and MacAvoy 2000). The forensic DNA marker D1S80 is a minisatellite with a 16 bp repeat unit and contains alleles spanning the range of 16–41 repeat units (Kasai *et al.* 1990).

DNA regions with repeat units that are 2–6 bp in length are called *microsatellites*, simple sequence repeats (SSRs), or short tandem repeats (STRs). STRs have become popular DNA repeat markers because they are easily amplified by the polymerase chain reaction without the problems of differential amplification. This is due to the fact that both alleles from a heterozygous individual are similar in size since the repeat size is small. The number of repeats in STR markers can be highly variable among individuals, which make these STRs effective for human identification purposes.

Figure 5.1

Schematic of minisatellite and microsatellite (STR) DNA markers. PCR primers are designed to target invariant flanking sequence regions. The number of tandem repeat units in the repeat regions varies among individuals making them useful markers for human identification.

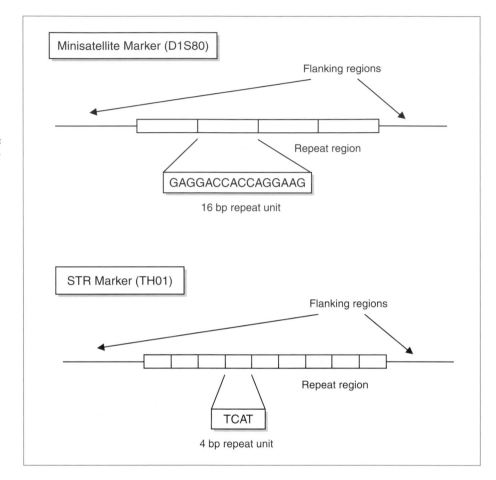

Literally thousands of polymorphic microsatellites have been characterized in human DNA and there may be more than a million microsatellite loci present depending on how they are counted (Ellegren 2004). Regardless, microsatellites account for approximately 3% of the total human genome (International Human Genome Sequencing Consortium 2001). STR markers are scattered throughout the genome and occur on average every 10 000 nucleotides (Edwards *et al.* 1991, Collins *et al.* 2003, Subramanian *et al.* 2003). Computer searches of the recently available human genome reference sequence have attempted to comprehensively catalog the number and nature of STR markers in the genome (see Collins *et al.* 2003, Subramanian *et al.* 2003). A large number of STR markers have been characterized by academic and commercial laboratories for use in disease gene location studies. For example, the Marshfield Medical Research Foundation in Marshfield, Wisconsin (http://research.marshfieldclinic.org/genetics/) has gathered genotype data on over 8000 STRs that are scattered across the 23 pairs of human chromosomes for the purpose of developing human genetic maps (Broman *et al.* 1998, Ghebranious *et al.* 2003).

ISOLATION AND TYPES OF STR MARKERS

In order to perform analysis on STR markers, the invariant flanking regions surrounding the repeats must be determined. Once the flanking sequences are known then PCR primers can be designed and the repeat region amplified for analysis. New STR markers are usually identified in one of two ways: (1) searching DNA sequence databases such as GenBank for regions with more than six or so contiguous repeat units (Weber and May 1989, Collins *et al.* 2003, Subramanian *et al.* 2003); or (2) performing molecular biology isolation methods (Edwards *et al.* 1991, Chambers and MacAvoy 2000).

STR repeat sequences are named by the length of the repeat unit. Dinucleotide repeats have two nucleotides repeated next to each other over and over again. Trinucleotides have three nucleotides in the repeat unit, tetranucleotides have four, pentanucleotides have five, and hexanucleotides have six repeat units in the core repeat. Theoretically, there are 4, 16, 64, 256, 1024, 4096 possible motifs for mono-, di-, tri-, tetra-, penta-, and hexanucleotide repeats, respectively (Jin *et al.* 1994). However, because microsatellites are tandemly repeated, some motifs are actually equivalent to others (D.N.A. Box 5.1). For reasons that will be discussed below, tetranucleotide repeats have become the most popular STR markers for human identification.

STR sequences not only vary in the length of the repeat unit and the number of repeats but also in the rigor with which they conform to an incremental repeat pattern. STRs are often divided into several categories based on the repeat pattern. *Simple repeats* contain units of identical length and sequence, *compound repeats* comprise two or more adjacent simple repeats, and *complex repeats* may contain several repeat blocks of variable unit length as well as variable intervening sequences (Urquhart *et al.* 1994). *Complex hypervariable repeats* also exist with numerous non-consensus alleles that differ in both size and sequence and are therefore challenging to genotype reproducibly (Urquhart *et al.* 1993, Gill *et al.* 1994). This last category of STR markers is not as commonly used in forensic DNA typing due to difficulties with allele nomenclature and measurement variability between laboratories, although two commercial kits now include the complex hypervariable STR locus SE33, sometimes called ACTBP2 (Urquhart *et al.* 1993, Promega Corporation 2002, Applied Biosystems 2002).

Not all alleles for a STR locus contain complete repeat units. Even simple repeats can contain non-consensus alleles that fall in between alleles with full repeat units. *Microvariants* are alleles that contain incomplete repeat units. Perhaps the most common example of a microvariant is the allele 9.3 at the TH01 locus, which contains nine tetranucleotide repeats and one incomplete

D.N.A. Box 5.1

List of possible microsatellite motifs

Theoretically, there are 4, 16, 64, 256, 1024, 4096 possible motifs for mono-, di-, tri-, tetra-, penta-, and hexanucleotide repeats, respectively. However, because microsatellites are tandemly repeated, some motifs are actually equivalent to others. Two rules can be used to identify whether motif A is equivalent to motif B. Motif A is considered equivalent to motif B when (1) motif A is inversely complementary to motif B, or (2) motif A is different from motif B or the inversely complementary sequence of motif B by frameshift. For example, $(GAAA)_n$ is equivalent to $(AGAA)_n$ or $(AAGA)_n$, to $(AAAG)_n$ or $(TTTC)_n$, to $(TTCT)_n$ or $(TCTT)_n$, or to $(CTTT)_n$. In other words, the eight motifs are equivalent. Note that $(AGAG)_n$ is considered a dinucleotide repeat instead of a tetranucleotide motif (Jin *et al.* 1994).

Because of this equivalence in repeat motif structure there are only 2, 4, 10, 33, 102, and 350 possible motifs for mono-, di-, tri-, tetra-, penta-, and hexanucleotide repeats, respectively (see below).

Mononucleotide repeats (2):

A   C

Dinucleotide repeats (4):

AC   AG   AT   CG

Trinucleotide repeats (10):

AAC   AAG   AAT   ACC   ACG   ACT   AGC   AGG   ATC   CCG

Tetranucleotide repeats (33):

AAAC AAAG AAAT AACC AACG AACT AAGC AAGG AAGT AATC
AATG AATT ACAG ACAT ACCC ACCG ACCT ACGC ACGG ACGT
ACTC ACTG **AGAT** AGCC AGCG AGCT AGGC AGGG ATCC ATCG
ATGC CCCG CCGG

AGAT or GATA motif is the most common for STR loci used by forensic scientists

Penta- (102) and hexanucleotide (350) repeats are not shown due to the sheer number of motifs possible.

Source:
Jin, L., Zhong, Y. and Chakraborty, R. (1994) The exact numbers of possible microsatellite motifs [letter]. *American Journal of Human Genetics*, 55, 582–583.

repeat of three nucleotides because the seventh repeat is missing a single adenine out of the normal AATG repeat unit (Puers *et al.* 1993).

DESIRABLE CHARACTERISTICS OF STRs USED IN FORENSIC DNA TYPING

For human identification purposes it is important to have DNA markers that exhibit the highest possible variation or a number of less polymorphic markers that can be combined in order to obtain the ability to discriminate between samples.

As will be discussed further in Chapter 7, forensic specimens are often challenging to PCR amplify because the DNA in the samples may be severely degraded (i.e., broken up into small pieces). Mixtures are prevalent as well in some forensic samples, such as those obtained from sexual assault cases containing biological material from both the perpetrator and victim.

The small size of STR alleles (~100–400 bp) compared to minisatellite VNTR alleles (~400–1000 bp) make the STR markers better candidates for use in forensic applications where degraded DNA is common. PCR amplification of degraded DNA samples can be better accomplished with smaller product sizes (see Chapter 7). Allelic dropout of larger alleles in minisatellite markers caused by preferential amplification of the smaller allele is also a significant problem with minisatellites (Tully *et al.* 1993). Furthermore, single base resolution of DNA fragments can be obtained more easily with sizes below 500 bp using denaturing polyacrylamide gel electrophoresis (see Chapter 12). Thus, for both biology and technology reasons the smaller STRs are advantageous compared to the larger minisatellite VNTRs.

Among the various types of STR systems, tetranucleotide repeats have become more popular than di- or trinucleotides. Penta- and hexanucleotide repeats are less common in the human genome but are being examined by some laboratories (Bacher *et al.* 1999). As will be discussed in Chapter 6, a biological phenomenon known as 'stutter' results when STR alleles are PCR amplified. Stutter products are amplicons that are typically one or more repeat units less in size than the true allele and arise during PCR because of strand slippage (Walsh *et al.* 1996). Depending on the STR locus, stutter products can be as large as 15% or more of the allele product quantity with tetranucleotide repeats. With di- and trinucleotides, the stutter percentage can be much higher (30% or more) making it difficult to interpret sample mixtures (see Chapter 7). In addition, the four base spread in alleles with tetranucleotides makes closely spaced heterozygotes easier to resolve with size-based electrophoretic separations (see Chapter 12) compared to alleles that could be two or three bases different in size with dinucleotides and trinucleotide markers, respectively.

Thus, to summarize, the advantages of using tetranucleotide STR loci in forensic DNA typing over VNTR minisatellites or di- and trinucleotide repeat STRs include:

- A narrow allele size range that permits multiplexing;
- A narrow allele size range that reduces allelic dropout from preferential amplification of smaller alleles;
- The capability of generating small PCR product sizes that benefit recovery of information from degraded DNA specimens; and
- Reduced stutter product formation compared to dinucleotide repeats that benefit the interpretation of sample mixtures.

In the past decade, a number of tetranucleotide STRs have been explored for application to human identification. The types of STR markers that have been sought out have included short STRs for typing degraded DNA materials, STRs with low stuttering characteristics for analyzing mixtures, and male-specific Y chromosome STRs for analyzing male-female mixtures from sexual crimes (Carracedo and Lareu 1998). The selection criteria for candidate STR loci in human identification applications include the following characteristics (Gill *et al.* 1996, Carracedo and Lareu 1998):

- High discriminating power, usually >0.9, with observed heterozygosity >70%;
- Separate chromosomal locations to ensure that closely linked loci are not chosen;
- Robustness and reproducibility of results when multiplexed with other markers;
- Low stutter characteristics;
- Low mutation rate; and
- Predicted length of alleles that fall in the range of 90–500 bp with smaller sizes better suited for analysis of degraded DNA samples.

In order to take advantage of the product rule, STR markers used in forensic DNA typing are typically chosen from separate chromosomes to avoid any problems with linkage between the markers (see Chapter 20).

A COMMON NOMENCLATURE FOR STR ALLELES

To aid in inter-laboratory reproducibility and comparisons of data, a common nomenclature has been developed in the forensic DNA community. DNA results cannot be effectively shared unless all parties are speaking the same language and referring to the same conditions. (It would do little good to describe the recipe for baking a cake in a language that is not understood by both the recipe giver and the chef. For example, if the recipe says to turn the oven on to 450 degrees Fahrenheit and the chef uses 450 Kelvin (~250°F), the results would be vastly different.) Likewise if one laboratory calls a sample 15 repeats at a particular STR locus and the same sample is designated 16 repeats by another laboratory, a match would not be considered, and the samples would be assumed to come from separate sources. As will be discussed in Chapter 18, the advent of national DNA databases with many laboratories worldwide contributing information has made it crucial to have internationally accepted nomenclature for designating STR alleles.

A repeat sequence is named by the structure (base composition) of the core repeat unit and the number of repeat units. However, because DNA has two strands, which may be used to designate the repeat unit for a particular STR marker, more than one choice is available and confusion can arise without a standard format. Also, where an individual starts counting the number of

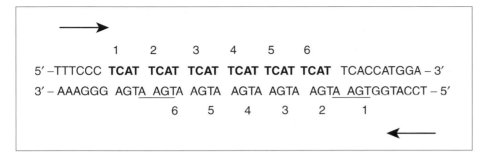

Figure 5.2

Example of the DNA sequence in a STR repeat region. Note that using the top strand versus the bottom strand results in different repeat motifs and starting positions. In this example, the top strand has six TCAT repeat units, while the bottom strand has six TGAA repeat units. Under ISFH recommendations (Bar et al. 1997), the top strand from GenBank should be used. Thus, this example would be described as having [TCAT] as the repeat motif. Repeat numbering, indicated above and below the sequence, proceeds in the 5' to 3' direction as illustrated by the arrows.

repeats can also make a difference. With double-stranded DNA sequences being read in the 5' to 3' direction (see Chapter 2), the choice of the strand impacts the sequence designation. For example, the 'top' strand for an STR marker may be 5'-...(GATA)$_n$...-3' while the 'bottom' strand for the same sequence would be 5'-...(TATC)$_n$...-3'. Depending on the sequence surrounding the repeat region, the core repeat could be shifted relative to the other strand (Figure 5.2).

Recognizing the need for standardization in STR repeat nomenclature, a committee of forensic DNA scientists, known as the DNA Commission of the International Society of Forensic Haemogenetics (ISFH), issued guidelines for designating STR alleles in 1994 (Bar *et al.* 1994) and again in 1997 (Bar *et al.* 1997). The ISFH is now known as the International Society of Forensic Genetics (ISFG; see http://www.isfg.org/). The ISFG 1994 recommendations focused on allelic ladders and designation of alleles that contain partial repeat sequences. The ISFG 1997 guidelines discuss the sequence and repeat designation of STRs.

When reviewing the STR literature prior to 1997, an individual should keep in mind that repeat nomenclatures often differ from the ISFG 1997 guidelines. This fact can lead to some confusion if one is not careful. For example, early descriptions of the STR locus TH01 by the Forensic Science Service label the repeat TCAT (Kimpton *et al.* 1993) while Caskey and co-workers described the TH01 repeat as AATG (Edwards *et al.* 1991).

The latest ISFG recommendations are reviewed below (Bar *et al.* 1997):

Choice of the Strand

- For STRs within protein coding regions (as well as in the intron of the genes), the coding strand should be used. This would apply to STRs such as VWA (GenBank: M25716), TPOX (GenBank: M68651), and CSF1PO (GenBank: X14720).

- For repetitive sequences without any connection to protein coding genes like many of the D#S### loci, the sequence originally described in the literature of the first public database entry shall become the standard reference (and strand) for nomenclature. Examples here include D18S51 (GenBank: L18333) and D21S11 (GenBank: M84567).

- If the nomenclature is already established in the forensic field but not in accordance with the aforementioned guideline, the nomenclature shall be maintained to avoid unnecessary confusion. This recommendation applies to the continued use by some laboratories of the 'AATG repeat' strand for the STR marker TH01. The GenBank sequence for TH01 uses the coding strand and therefore contains the complementary 'TCAT repeat' instead.

Choice of the Motif and Allele Designation

- The repeat sequence motif should be defined so that the first 5′-nucleotides that can define a repeat motif are used. For example, 5′-GG TCA TCA TCA TGG-3′ could be seen as having 3 × TCA repeats or 3 × CAT repeats. However, under the recommendations of the ISFH committee only the first one (3 × TCA) is correct because it defines the first possible repeat motif.

- Designation of incomplete repeat motifs should include the number of complete repeats and, separated by a decimal point, the number of base pairs in the incomplete repeat. Examples of 'microvariants' with incomplete repeat units include allele 9.3 at the TH01 locus. TH01 allele 9.3 contains nine tetranucleotide AATG repeats and one incomplete ATG repeat of three nucleotides (Puers *et al.* 1993). Another microvariant example is allele 22.2 at the FGA locus, which contains 22 tetranucleotide repeats and one incomplete repeat with two nucleotides (Barber *et al.* 1996).

- Allelic ladders containing sequenced alleles that are named according to the recommendations listed above should be used as a reference for allele designation in unknown samples. Allelic ladders may be commercially obtained or prepared in house and should contain all common alleles.

ALLELIC LADDERS

An allelic ladder is an artificial mixture of the common alleles present in the human population for a particular STR marker (Sajantila *et al.* 1992). They are generated with the same primers as tested samples and thus provide a reference DNA size for each allele included in the ladder. Allelic ladders have been shown to be important for accurate genotype determinations (Smith 1995). These allelic ladders serve as a standard like a measuring stick for each STR locus. They are necessary to adjust for different sizing measurements obtained from different instruments and conditions used by various laboratories (see Chapters 14 and 15).

Allelic ladders are constructed by combining genomic DNA or locus-specific PCR products from multiple individuals in a population, which possess alleles that are representative of the variation for the particular STR marker (Sajantila *et al.* 1992, Baechtel *et al.* 1993). The samples are then co-amplified to produce an artificial sample containing the common alleles for the STR marker (Figure 5.3). Allele quantities are balanced by adjusting the input amount of each component so that the alleles are fairly equally represented in the ladder.

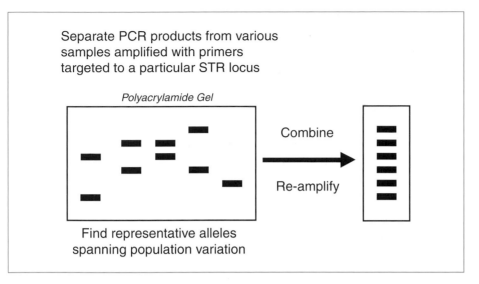

Separate PCR products from various samples amplified with primers targeted to a particular STR locus

Polyacrylamide Gel

Combine

Re-amplify

Find representative alleles spanning population variation

Figure 5.3

Principle of allelic ladder formation. STR alleles from a number of samples are separated on a polyacrylamide gel and compared to one another. Samples representing the common alleles for the locus are combined and re-amplified to generate an allelic ladder. Each allele in the allelic ladder is sequenced since it serves as the reference material for STR genotyping. Allelic ladders are included in commercially available STR kits.

For example, to produce a ladder containing five alleles with 6, 7, 8, 9, and 10 repeats, individual samples with genotypes of (6,8), (7,10), and (9,9) could be combined. Alternatively, the combination of genotypes could be (6,9), (7,8), and (10,10) or (6,6), (7,7), (8,8), (9,9), and (10,10).

Additional quantities of the same allelic ladder (second- and third-generation ladders) may be produced by simply diluting the original ladder 1/1000–1/1 000 000 parts with deionized water and then re-amplifying it using the same PCR primers (Baechtcl *et al.* 1993). It is imperative that allelic ladders be generated with the same PCR primers as used to amplify unknown samples so that the allele 'rungs' on the ladder will accurately line up with that of the repeat number of the unknown sample when the unknown is compared to the ladder. As will be seen in the next section, commercial manufacturers now provide allelic ladders in their STR typing kits so that individual laboratories do not have to produce their own allelic ladders.

CHOICE OF MARKERS USED BY THE FORENSIC DNA TYPING COMMUNITY

For DNA typing markers to be effective across a wide number of jurisdictions, a common set of standardized markers must be used. The STR loci that are commonly used today were initially characterized and developed either in the laboratory of Dr. Thomas Caskey at the Baylor College of Medicine (Edwards *et al.* 1991, Hammond *et al.* 1994) or at the Forensic Science Service in England (Kimpton *et al.* 1993, Urquhart *et al.* 1994). The Promega Corporation (Madison, Wisconsin) initially commercialized many of the Caskey markers while Applied Biosystems (Foster City, California) picked up on the Forensic Science Service (FSS) STR loci as well as developing some new markers.

Today both Applied Biosystems and the Promega Corporation have STR kits that address the needs of the DNA typing community and cover a common set of STR loci. The availability of STR kits that permit robust multiplex amplification of eight or more STR markers has truly revolutionized forensic DNA. Matching probabilities that exceed one in a billion are possible in a single amplification with 1 ng (or less) of DNA sample. Just as impressive is the fact that results can be obtained today in only a few hours compared to the weeks that restriction fragment length polymorphism (RFLP) methods took just a few years ago.

One of the first STR multiplexes to be developed was a quadruplex created by the Forensic Science Service that comprised the four loci TH01, FES/FPS, VWA, and F13A1 (Kimpton *et al.* 1994). This so-called 'first-generation multiplex' had a matching probability of approximately 1 in 10 000. The FSS followed with a second-generation multiplex (SGM) made up of six polymorphic STRs and a gender identification marker (Gill *et al.* 1996, Sparkes *et al.* 1996). The six STRs in SGM are TH01, VWA, FGA, D8S1179, D18S51, and D21S11 and provide a matching probability of approximately 1 in 50 million. The gender identification marker amelogenin will be described in more detail later in this chapter.

The first commercial STR kit capable of multiplex amplification became available from the Promega Corporation in 1994 for silver stain analysis. This kit consisted of the STR loci CSF1PO, TPOX, and TH01 and is often referred to as the 'CTT' triplex using the first letter in each locus. The CTT triplex only had a matching probability of ~1 in 500 but was still widely used in the United States in the mid-1990s as it was the first available STR multiplex kit and could be performed with a fairly low start-up cost.

THE 13 CODIS STR LOCI

In the United States, utilization of STRs initially lagged behind that of Europe, especially the efforts of the Forensic Science Service in the United Kingdom. However, beginning in 1996, the FBI Laboratory sponsored a community-wide forensic science effort to establish core STR loci for inclusion within the national DNA database known as CODIS (Combined DNA Index System). Chapter 18 covers CODIS and DNA databases in more detail. This STR Project beginning in April 1996 and concluding in November 1997 involved 22 DNA typing laboratories and the evaluation of 17 candidate STR loci. The evaluated STR loci were CSF1PO, F13A01, F13B, FES/FPS, FGA, LPL, TH01, TPOX, VWA, D3S1358, D5S818, D7S820, D8S1179, D13S317, D16S539, D18S51, and D21S11.

At the STR Project meeting on 13–14 November 1997, 13 core STR loci were chosen to be the basis of the future CODIS national DNA database (Budowle *et al.* 1998). The 13 CODIS core loci are CSF1PO, FGA, TH01, TPOX, VWA, D3S1358, D5S818, D7S820, D8S1179, D13S317, D16S539, D18S51, and D21S11. Table 5.1 lists the original references in the literature for these 13 STRs. When all 13 CODIS core loci are tested, the average random match probability is rarer than one in a trillion among unrelated individuals (Chakraborty *et al.* 1999). The genetics section of this book provides more information on the calculation of random match probability and evaluation of the 13 CODIS STRs in various populations (see Chapter 20).

The three most polymorphic markers are FGA, D18S51, and D21S11, while TPOX shows the least variation between individuals. A summary of information on the 13 STRs is contained in Table 5.2, which describes the chromosomal

Locus Name	Reference
CSF1PO	Hammond, H.A., Jin, L., Zhong, Y., Caskey, C.T. and Chakraborty, R. (1994) Evaluation of 13 short tandem repeat loci for use in personal identification applications. *American Journal of Human Genetics*, 55, 175–189.
FGA	Mills, K.A., Even, D. and Murray, J.C. (1992) Tetranucleotide repeat polymorphism at the human alpha fibrinogen locus (FGA). *Human Molecular Genetics*, 1, 779.
TH01	Polymeropoulos, M.H., Xiao, H., Rath, D.S. and Merril, C.R. (1991) Tetranucleotide repeat polymorphism at the human tyrosine hydroxylase gene (TH). *Nucleic Acids Research*, 19, 3753.
TPOX	Anker, R., Steinbrueck, T. and Donis-Keller, H. (1992) Tetranucleotide repeat polymorphism at the human thyroid peroxidase (hTPO) locus. *Human Molecular Genetics*, 1, 137.
VWA	Kimpton, C.P., Walton, A. and Gill, P. (1992) A further tetranucleotide repeat polymorphism in the vWF gene. *Human Molecular Genetics*, 1, 287.
D3S1358	Li, H., Schmidt, L., Wei, M.-H., Hustad, T., Lerman, M.I., Zbar, B. and Tory, K. (1993) Three tetranucleotide polymorphisms for loci: D3S1352, D3S1358, D3S1359. *Human Molecular Genetics*, 2, 1327.
D5S818	Cooperative Human Linkage Center GATA3F03.512
D7S820	Cooperative Human Linkage Center GATA3F01.511
D8S1179	Cooperative Human Linkage Center GATA7G07.37564
D13S317	Cooperative Human Linkage Center GATA7G10.415
D16S539	Cooperative Human Linkage Center GATA11C06.715
D18S51	Staub, R.E., Speer, M.C., Luo, Y., Rojas, K., Overhauser, J., Otto, L. and Gilliam, T.C. (1993) A microsatellite genetic linkage map of human chromosome 18. *Genomics*, 15, 48–56.
D21S11	Sharma, V. and Litt, M. (1992) Tetranucleotide repeat polymorphism at the D21S11 locus. *Human Molecular Genetics*, 1, 67.
Amelogenin	Sullivan, K.M., Mannucci, A., Kimpton, C.P. and Gill, P. (1993) A rapid and quantitative DNA sex test: fluorescence-based PCR analysis of X-Y homologous gene amelogenin. *BioTechniques*, 15, 637–641.

Table 5.1

Original reference describing each of the 13 CODIS STR loci and the gender identification marker amelogenin.

Cooperative Human Linkage Center information is available via the Internet: http://www.chlc.org

Locus Name	Chromosomal Location	Physical Position [a]	Repeat Motif ISFG Format [b]	GenBank Accession [c]	GenBank Allele	Allele Range [d]	Number of Alleles Seen [e]
CSF1PO	5q33.1 c-fms proto-oncogene, 6th intron	Chr 5 149.484 Mb	TAGA	X14720	12	5-16	20
FGA	4q31.3 alpha fibrinogen, 3rd intron	Chr 4 156.086 Mb	CTTT	M64982	21	12.2-51.2	80
TH01	11p15.5 tyrosine hydroxylase, 1st intron	Chr 11 2.156 Mb	TCAT	D00269	9	3-14	20
TPOX	2p25.3 thyroid peroxidase, 10th intron	Chr 2 1.436 Mb	GAAT	M68651	11	4-16	15
VWA	12p13.31 von Willebrand Factor, 40th intron	Chr 12 19.826 Mb	[TCTG][TCTA]	M25858	18	10-25	28
D3S1358	3p21.31	Chr 3 45.543 Mb	[TCTG][TCTA]	NT_005997	18	8-21	24
D5S818	5q23.2	Chr 5 123.187 Mb	AGAT	G08446	11	7-18	15
D7S820	7q21.11	Chr 7 83.401 Mb	GATA	G08616	12	5-16	30
D8S1179	8q24.13	Chr 8 125.863 Mb	[TCTA][TCTG]	G08710	12	7-20	17
D13S317	13q31.1	Chr 13 80.52 Mb	TATC	G09017	13	5-16	17
D16S539	16q24.1	Chr 16 86.168 Mb	GATA	G07925	11	5-16	19
D18S51	18q21.33	Chr 18 59.098 Mb	AGAA	L18333	13	7-39.2	51
D21S11	21q21.1	Chr 21 19.476 Mb	Complex [TCTA][TCTG]	AP000433	29	12-41.2	82

[a]Physical positions and chromosomal locations determined on July 2003 human genome reference sequence (NCBI build 34) using hgBLAT (http://genome.ucsc.edu).

[b]The DNA Commission of the International Society of Forensic Genetics (ISFG) has published several papers encouraging standardization in STR allele nomenclature (see Bar et al. 1994, 1997). STR repeats should be called on the strand sequence originally described in the first public database entry using the first 5'-nucleotides that can define a repeat motif.

[c]GenBank sequence information for a particular STR locus may be accessed at (http://www.ncbi.nlm.nih.gov/GenBank) by entering the accession number shown here. Reference sequences are also available at http://www.cstl.nist.gov/biotech/strbase/seq_ref.htm.

[d]Numbers in this column refer to the number of repeat units present in the alleles. More detail on alleles that have been observed and their PCR products with commercially available STR kits may be found in Appendix I.

[e]See Appendix I.

location, the repeat motif, allele range, and GenBank accession number where the DNA sequence for a reference allele may be found. The chromosomal locations for these STRs have been updated on the recently completed human genome reference sequence. We have included detailed allele sequence information and PCR product sizes with commercially available STR kits in Appendix I.

Table 5.2 (facing)

Summary information on the 13 CODIS core STR loci.

Using the previously described classification scheme for categorizing STR repeat motifs (Urquhart *et al.* 1994), the 13 CODIS core STR loci may be divided up into four categories:

1. Simple repeats consisting of one repeating sequence: TPOX, CSF1PO, D5S818, D13S317, D16S539;
2. Simple repeats with non-consensus alleles (e.g., 9.3): TH01, D18S51, D7S820;
3. Compound repeats with non-consensus alleles: VWA, FGA, D3S1358, D8S1179;
4. Complex repeats: D21S11.

COMMERCIALLY AVAILABLE STR KITS

A number of kits are available for single or multiplex PCR amplification of STR markers used in DNA typing. Two primary vendors for STR kits used by the forensic DNA community exist: the Promega Corporation located in Madison, Wisconsin, and Applied Biosystems located in Foster City, California. These companies have expended a great deal of effort over the past decade to bring STR markers to forensic scientists in kit form. More recently in Europe, companies such as Serac (Bad Homburg, Germany) and Biotype (Dresden, Germany) have begun offering commercial STR kits.

The technology has evolved quickly in the late 1990s for more sensitive, rapid, and accurate measurements of STR alleles. At the same time, the number of STRs that can be simultaneously amplified has increased from three or four with silver-stained systems to over 15 STRs using multiple-color fluorescent tags (see Chapter 13). A list of commercially available STR multiplexes and when they were released as products is shown in Table 5.3.

Table 5.3 (below)

Information on commercially available STR multiplexes (fluorescently-labeled).

Name	Source	Release Date	STR Loci Included
TH01, TPOX, CSF1PO monoplexes (silver stain)	Promega	Feb 1993	TH01, TPOX, CSF1PO
AmpF/STR® Blue	Applied Biosystems	Oct 1996	D3S1358, VWA, FGA
AmpF/STR® Green I	Applied Biosystems	Jan 1997	Amelogenin, TH01, TPOX, CSF1PO
CTTv	Promega	Jan 1997	CSF1PO, TPOX, TH01, VWA (vWF)

Name	Source	Release Date	STR Loci Included
FFFL	Promega	Jan 1997	F13A1, FES/FPS, F13B, LPL
GammaSTR	Promega	Jan 1997	D16S539, D13S317, D7S820, D5S818
PowerPlex™ (version 1.1 and 1.2 later)	Promega	Jan 1997 Sept 1998	CSF1PO, TPOX, TH01, VWA, D16S539, D13S317, D7S820, D5S818
AmpF/STR® Profiler™	Applied Biosystems	May 1997	D3S1358, VWA, FGA, Amelogenin, TH01, TPOX, CSF1PO, D5S818, D13S317, D7S820
AmpF/STR® Profiler Plus™	Applied Biosystems	Dec 1997	D3S1358, VWA, FGA, Amelogenin, D8S1179, D21S11, D18S51, D5S818, D13S317, D7S820
AmpF/STR® COfiler™	Applied Biosystems	May 1998	D3S1358, D16S539, Amelogenin, TH01, TPOX, CSF1PO, D7S820
AmpF/STR® SGM Plus™	Applied Biosystems	Feb 1999	D3S1358, VWA, D16S539, D2S1338, Amelogenin, D8S1179, D21S11, D18S51, D19S433, TH01, FGA
PowerPlex® 2.1 (for Hitachi FMBIO users)	Promega	June 1999	D3S1358, TH01, D21S11, D18S51, VWA, D8S1179, TPOX, FGA, Penta E
PowerPlex® 16	Promega	May 2000	CSF1PO, FGA, TPOX, TH01, VWA, D3S1358, D5S818, D7S820, D8S1179, D13S317, D16S539, D18S51, D21S11, Penta D, Penta E, amelogenin
PowerPlex® 16 BIO (for Hitachi FMBIO users)	Promega	May 2001	CSF1PO, FGA, TPOX, TH01, VWA, D3S1358, D5S818, D7S820, D8S1179, D13S317, D16S539, D18S51, D21S11, Penta D, Penta E, amelogenin
AmpF/STR® Identifiler™	Applied Biosystems	July 2001	CSF1PO, FGA, TPOX, TH01, VWA, D3S1358, D5S818, D7S820, D8S1179, D13S317, D16S539, D18S51, D21S11, D2S1338, D19S433, amelogenin
AmpF/STR® Profiler Plus™ ID (extra unlabeled D8-R primer)	Applied Biosystems	Sept 2001	D3S1358, VWA, FGA, Amelogenin, D8S1179, D21S11, D18S51, D5S818, D13S317, D7S820
PowerPlex® ES	Promega	Mar 2002	FGA, TH01, VWA, D3S1358, D8S1179, D18S51, D21S11, SE33, amelogenin
AmpF/STR® SEfiler™	Applied Biosystems	Sept 2002	FGA, TH01, VWA, D3S1358, D8S1179, D16S539, D18S51, D21S11, D2S1338, D19S433, SE33, amelogenin

Table 5.3
(Continued)

The adoption of the 13 core loci for CODIS in the United States has led to development of STR multiplexes that cover these markers. At the turn of the century, two PCR reactions were required to obtain information from all the 13 STRs: either PowerPlex® 1.1 and PowerPlex® 2.1 or Profiler Plus™ and COfiler™ (see Table 5.3). As an internal check to reduce the possibility of mixing up samples, both manufacturers included overlapping loci in their kits that should produce concordant data between samples amplified from the same biological material. The Profiler Plus™ and COfiler™ kits have the loci

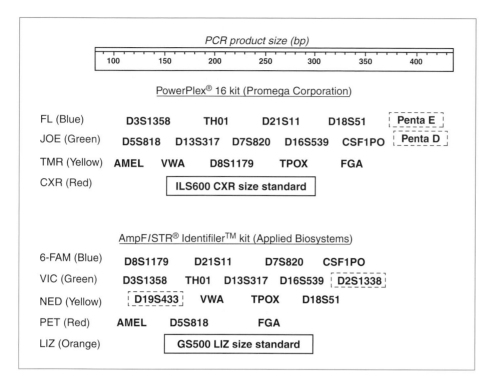

Figure 5.4

Commercially available STR kit solutions for a single amplification of the 13 CODIS core loci. General size ranges and dye-labeling strategies are indicated. The PowerPlex 16 kit uses four dyes while the Identifiler kit uses five dyes. Loci with dotted boxes are additional loci specific to each kit.

D3S1358 and D7S820 in common while the PowerPlex® 1.1 and PowerPlex® 2.1 have the loci TH01, TPOX, and VWA in common.

Since 2000, both Promega and Applied Biosystems have marketed multiplex PCR reactions that permit co-amplification of all 13 STRs in a single reaction along with the amelogenin sex-typing marker and two additional STR loci (Figure 5.4). Electropherograms with size separated PCR products for Promega's PowerPlex® 16 (Figure 5.5) can be viewed as color separated panels of loci or as an overlay of all colors. The allelic ladders for the Applied Biosystems' AmpF/STR® Identifiler™ kits are displayed in Figure 5.6.

As will be discussed in the Technology section, two primary methods are used in modern forensic DNA laboratories to separate and detect fluorescently labeled STR alleles. Some PowerPlex kits have been balanced to work with the Hitachi FMBIO II scanner while PowerPlex® 16, Identifiler™, Profiler Plus™, and COfiler™ reactions are typically analyzed on an ABI Prism 310 or 3100 Genetic Analyzer capillary electrophoresis system (see Chapter 14).

Commercial manufacturers of STR kits have spent a great deal of research effort defining which markers would be included in each kit as well as verifying if primer pairs are compatible and will work well in combination with each other during multiplex PCR conditions (Wallin *et al.* 2002, Krenke *et al.* 2002). Promega has published and patented their PCR primer sequences (Masibay *et al.* 2000, Krenke *et al.* 2002) whereas Applied Biosystems have kept their primer

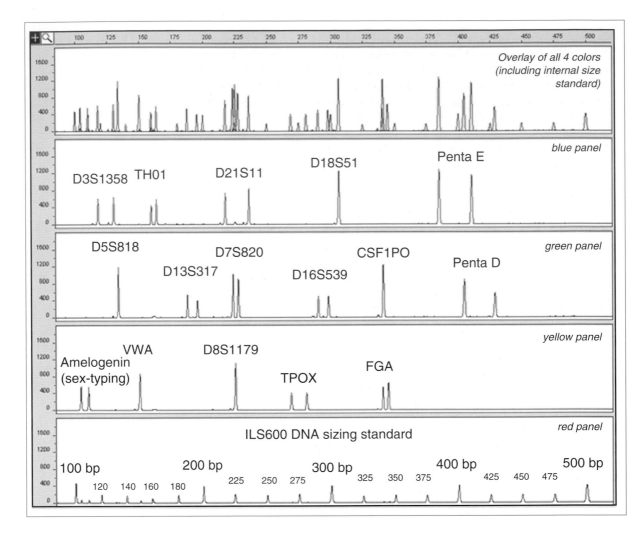

Figure 5.5

PowerPlex® 16 result from 1 ng genomic DNA.

sequences proprietary although some information has been revealed regarding the use of degenerate primers (see Chapter 6). The issue over failure to disclose kit primer sequences impacted several court cases early on in the legal acceptance of STR technology but appears to have been resolved now (D.N.A. Box 5.2).

Most laboratories do not have the time or resources to design primers, optimize PCR multiplexes, and quality control primer synthesis. The convenience of using ready-made kits is also augmented by the fact that widely used primer sets and conditions allow improved opportunities for sharing data between laboratories without fear of possible null alleles (see Chapter 6). Available STR multiplex sets vary based on which STR loci are included, the fluorescent dye combinations, the DNA strand that is labeled, allelic ladders present in kits, and most importantly, the primer sequences utilized for PCR amplification. It is important to keep in mind that commercially available kits quickly dictate which STRs will be used by the vast majority of forensic laboratories.

During the early adoption of STR typing technology in U.S. court systems, three cases ruled that DNA results would not be permissible as evidence because the commercial STR kit PCR primer sequences and developmental validation studies were not public information. These cases were People *v.* Bokin (San Francisco, California, May 1999), People *v.* Shreck (Boulder, Colorado, April 2000), and State *v.* Pfenning (Grand Isle, Vermont, Apr 2000).

Shortly after the Pfenning case, the Promega Corporation made the decision to publish their STR kit primer sequences (see news in *Nature* 27 July 2000 issue, volume 406, p. 336) and have done so since (Masibay *et al.* 2001, Krenke *et al.* 2002), along with obtaining several patents in the area of multiplex amplification of STR loci.

Applied Biosystems has repeatedly refused to release the primer sequences present in their STR kits claiming that this information is proprietary. The company is concerned that they would lose revenue if generic brand products were produced by other entities using the revealed primer information. However, in at least 16 cases, the primer sequences for the ProfilerPlus™ and COfiler™ kits have been supplied by Applied Biosystems under a protective court order. Numerous publications since 2000 have demonstrated the reliable use of Applied Biosystems STR kits including detailed validation studies (see Holt *et al.* 2002).

The arguments that not enough information exists to support the reliable use of commercial STR kits whose every component is not public knowledge have fallen by the wayside as millions of DNA profiles have been reliably generated with these kits in the past few years.

For further information:

Masibay, A. *et al.* (2000) Promega Corporation reveals primer sequences in its testing kits [letter]. *Journal of Forensic Sciences*, 45, 1360–1362.

Krenke, BE. *et al.* (2002) Validation of a 16-locus fluorescent multiplex system. *Journal of Forensic Science*, 47, 773–785.

Holt, CL. *et al.* (2002) TWGDAM validation of AmpFlSTR PCR amplification kits for forensic DNA casework. *Journal of Forensic Science*, 47, 66–96.

http://www.scientific.org/archive/archive.html

http://www.denverda.org/html_website/denver_da/DNA_resources.html

http://www.denverda.org/legalResource/AB Sequence case list.pdf

COMMERCIAL ALLELIC LADDERS

Each manufacturer of STR kits provides allelic ladders that may be used for accurate genotyping. It is important to note that kits from the Promega Corporation and Applied Biosystems for comparable STR markers often contain different alleles in their allelic ladders. For example, the PowerPlex® 1.1 kit from Promega contains alleles 7–15 in its D5S818 allelic ladder while the Profiler Plus™ kit from Applied Biosystems contains alleles 7–16 in its D5S818 allelic ladder. By having an allele present in the ladder, a laboratory can be more confident of a call from

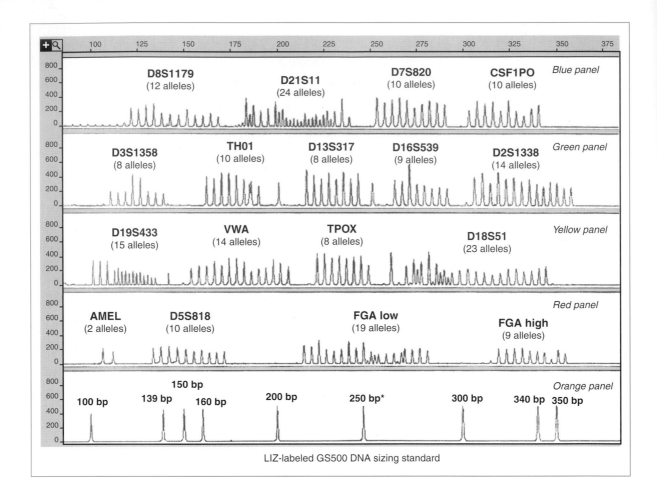

LIZ-labeled GS500 DNA sizing standard

Figure 5.6

AmpFlSTR® Identifiler™ allelic ladders (Applied Biosystems). A total of 205 alleles are included in this set of allelic ladders used for genotyping a multiplex PCR reaction involving 15 STR loci and the amelogenin sex-typing test.

an unknown sample that is being analyzed. In the D5S818 example listed here, one would be more confident typing an observed allele 16 when using the Applied Biosystems kit than the Promega kit because the D5S818 allelic ladder has an allele 16 in the ABI kit. The alleles present in the two sources of commercially available multiplex STR kits are reviewed and contrasted in Table 5.4.

Some of the more recent kits come with an amazing number of alleles in their ladders. For example, the Identifiler™ kit from Applied Biosystems contains 205 alleles (Table 5.4; Figure 5.6). Putting together and mass-producing such a large set of alleles is an impressive feat. The Promega PowerPlex® 16 kit has 209 alleles in its allelic ladders.

AMPFLSTR® IDENTIFILER™ KIT INNOVATIONS

Table 5.4 (facing)

Applied Biosystems introduced two new technologies with their AmpF*l*STR® Identifiler™ kit when it was released in 2001. The first, and most obvious,

Table 5.4

Comparison of represented alleles in commercially available STR allelic ladders

| | Promega Corporation STR Kits | | | | | | | | | | Applied Biosystems AmpFISTR Kits | | | | | | | | |
| | PP1.1 Alleles | # | PP2.1 Alleles | # | PP16 Alleles | # | PP ES Alleles | # | ProfTerPlus Alleles | # | COfiler Alleles | # | SGM Plus Alleles | # | Identifiler Alleles | # | SEfiler Alleles | # |
Loci/Kit																		
CSF1PO	6–15	10			6–15	10					6–15	10			6–15	10		
FGA			17–46.2	19	16–46.2	28	16–46.2	28	17–30	14			17–51.2	28	17–51.2	28	17–51.2	28
TH01	5–11	7	4–13.3	10	4–13.3	10	4–13.3	10			5–9.3,10	7	4–13.3	10	4–13.3	10	4–13.3	10
TPOX	6–13	8	6–13	8	6–13	8					6–13	8			6–13	8		
VWA	10–22	13	10–22	13	10–22	13	10–22	13	11–21	11			11–24	14	11–24	14	11–24	14
D3S1358			12–20	9	12–20	9	12–20	9	12–19	8	12–19	8	12–19	8	12–19	8	12–19	8
D5S818	7–15	9			7–16	10			7–16	10					7–16	10		
D7S820	6–14	9			6–14	9			6–15	10	6–15	10			6–15	10		
D8S1179			7–18	12	7–18	12	7–18	12	8–19	12			8–19	12	8–19	12	8–19	12
D13S317	7–15	9			7–15	9			8–15	8					8–15	8		
D16S539	5,8–15	9			5,8–15	9					5,8–15	9	5,8–15	9	5,8–15	9	5,8–15	9
D18S51			8–27	22	8–27	22	8–27	22	9–26	21			7,9–27	23	7,9–27	23	7,9–27	23
D21S11			24–38	24	24–38	24	24–38	24	24.2–38	22			24–38	24	24–38	24	24–38	24
D2S1338													15–28	14	15–28	14	15–28	14
D19S433													9–17.2	15	9–17.2	15	9–17.2	15
Penta D					2.2–17	14												
Penta E			5–24	20	5–24	20												
SE33							4.2–37	35									4.2–37	35
Amelogenin	X,Y	2			X,Y	2	X,Y	2	X,Y	2	X,Y	2	X,Y	2	X,Y	2	X,Y	2
Total Alleles		76		137		209		155		118		54		159		205		194

involves the use of 5-dye detection systems (see Chapter 13) where four different dyes (6FAM™, VIC™, NED™, and PET™) are used to label the PCR products rather than the traditional three dyes (5FAM, JOE, NED or FL, JOE, TMR) as used with the previous AmpF*l*STR or PowerPlex kits. A one dye detection channel is always used for an internal size standard to correlate electrophoretic mobilities to an apparent PCR product size (see Chapter 15). Thus, the fifth dye (LIZ™) in 5-dye detection and the fourth dye (ROX or CXR) in 4-dye detection are used for labeling the internal size standard. The extra dye channel for labeling PCR products enables smaller PCR products to be generated and placed in a separate dye channel rather than extending the size range for amplicons within the three previously available dye channels.

The second technology introduced with the Identifiler™ kit involves mobility modifying non-nucleotide linkers (Applied Biosystems 2001). The mobility modifier is composed of hexaethyleneoxide (HEO) that imparts a shift of approximately 2.5 nucleotides with each additional HEO unit (Grossman *et al.* 1994). This non-nucleotide linker is synthesized into the 5′-end of the PCR primer so that when the PCR product is created it contains these extra molecules on one end (Figure 5.7). By incorporating non-nucleotide linkers, mobilities for amplified alleles from one member of a pair of closely spaced STR loci can be shifted relative to the other. Thus, overlapping size ranges can be prevented (Figure 5.8).

The primary reason for introducing mobility modifiers is to permit continued use of the same PCR primers for amplifying STR loci and still have optimal inter-locus spacing within the various color channels. For example, if the loci D7S820 and CSF1PO, which are labeled with two different fluorophores in the COfiler kit and therefore do not interfere with one another, were labeled with the same colored fluorescent label (e.g., 6FAM) as they are in the Identifiler STR kit, the allelic ladder products would have overlapped by ~13 bp (Figure 5.8). To prevent this overlap in allele size ranges, either PCR primer binding sites must be altered to change the overall size of the PCR product or mobility

Figure 5.7

Illustration of mobility modifiers used in Applied Biosystems' Identifiler STR kit. Non-nucleotide linkers are synthesized into the primer between the fluorescent dye and 5′-end of the primer sequence. During PCR amplification, the dye and linker are incorporated into the amplicon. With the added non-nucleotide linker, the mobility of the generated STR allele will be shifted to a larger apparent size during electrophoresis. This shift of STR alleles for a particular locus then enables optimal inter-locus spacing for STR loci labeled with the same fluorescent dye without having to alter the PCR primer binding positions (see Figure 5.8).

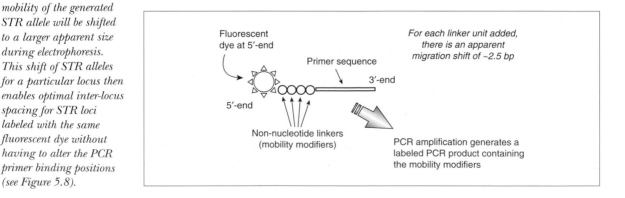

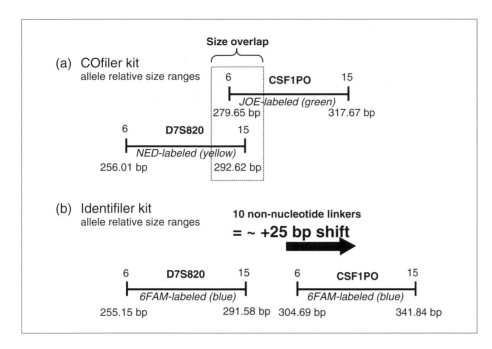

Figure 5.8

Illustration of how non-nucleotide linkers attached to CSF1PO PCR products in the Identifiler STR kit help with inter-locus spacing between D7S820 and CSF1PO. In the COfiler kit (a), CSF1PO and D7S820 are labeled with different colored fluorescent labels and thus do not interfere with one another. However, in the Identifiler kit (b), both D7S820 and CSF1PO are labeled with the same dye and would therefore have overlapping STR alleles unless primer positions were changed or mobility modifiers were used. A ~25 bp shift of the CSF1PO PCR products is accomplished by the addition of 10 non-nucleotide linkers. PCR product sizes for allelic ladder ranges displayed here are from the COfiler and Identifiler kit user's manuals. Note that sizes for D7S820 alleles do not match exactly because different dye labels are used with both the PCR products and the internal size standard thus impacting their relative mobilities.

modifiers can be introduced to shift the apparent molecular weight of the larger PCR product to an even larger size. In the case of the Identifiler™ kit, the locus CSF1PO was shifted by approximately 25 bp – most likely through the addition of 10 HEO non-nucleotide linkers to the 5′-end of the labeled PCR primer. Non-nucleotide linkers are also present on four other loci in the Identifiler™ kit: D2S1338, D13S317, D16S539, and TPOX.

Promega has changed primer sequences for a few of the loci between PowerPlex versions (see Masibay *et al.* 2000, Butler *et al.* 2001, Krenke *et al.* 2002). For example, between the PowerPlex® 1.1 and PowerPlex® 16 kits, the CSF1PO primer positions were drastically altered in order to achieve a 30 bp shift in PCR product size between the two kits (Figure 5.9). This primer change and subsequent PCR product shift was instituted so that CSF1PO and D16S539 loci could be labeled with the same dye in the PowerPlex® 16 kit. Note that if the original CSF1PO primers had been kept, there would have been a 13 bp overlap between D16S539 allele 15 (304 bp) and CSF1PO allele 6 (291 bp) making these systems incompatible in the same dye color without altering the PCR product size (i.e., primer positions) for one of them.

As will be discussed in Chapter 6, different primer positions have the potential to lead to allele dropout if a primer binding site mutation impacts one of the primer pairs. Hence concordance studies are needed between various STR kits to assess the level of potential allele dropout (Budowle *et al.* 2001). On the other hand, Applied Biosystems has maintained the same primers over time

Figure 5.9

Variation in CSF1PO primer positions between (a) PowerPlex 1.1 and (b) PowerPlex 16 STR kits. The base pair (bp) numbers in bold indicate the distance between the repeat region and 3'-end of the pertinent primer. The overall PCR product size for CSF1PO is shifted +30 bp with the primer changes from PowerPlex 1.1 to PowerPlex 16.

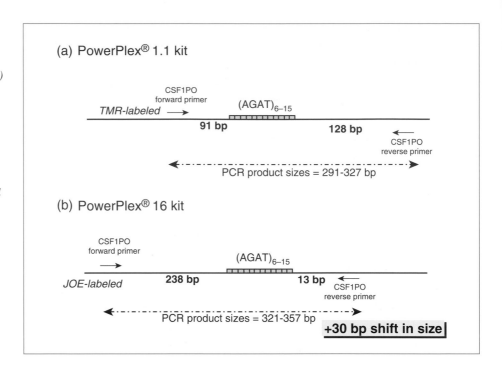

Figure 5.9

Variation in CSF1PO primer positions between (a) PowerPlex 1.1 and (b) PowerPlex 16 STR kits. The base pair (bp) numbers in bold indicate the distance between the repeat region and 3'-end of the pertinent primer. The overall PCR product size for CSF1PO is shifted +30 bp with the primer changes from PowerPlex 1.1 to PowerPlex 16.

and through their various AmpF*l*STR® kits (Holt *et al.* 2002) by introducing 5-dye chemistry and mobility modifiers for products that would normally overlap with one another (see Figure 5.8).

DETAILS ON ALLELES PRESENT IN THE 13 CODIS STR LOCI

Each of the 13 core STR loci has unique characteristics, either in terms of the number of alleles present, the type of repeat sequence, or the kinds of microvariants that have been observed. This section reviews some of the basic details on each of the 13 core STR loci. We have included in Appendix I a detailed summary of the alleles that have been reported as of June 2004 for the 13 core STR loci along with their expected sizes using various kits that are available from Promega or Applied Biosystems. The size difference in the PCR products produced by the different STR kits is important because a large difference is more likely to lead to null alleles when comparing results between two kits (see Chapter 6).

CSF1PO is a simple tetranucleotide repeat found in the sixth intron of the c-*fms* proto-oncogene for the CSF-1 receptor on the long arm of chromosome 5. Common alleles contain an AGAT core repeat and range in size from 6–15 repeats. An allele 16 has also been reported (Margolis-Nunno *et al.* 2001) as have several x.1 and x.3 variant alleles (see Appendix I). PCR products from Promega's PowerPlex® 1.1 STR kit are 11 bp larger than those generated with Applied Biosystems kits for equivalent alleles. Since PowerPlex 16 adds 30 bp to the size of CSF1PO relative to PowerPlex® 1.1 (see Figure 5.9), then PowerPlex® 16 PCR products for CSF1PO are 41 bp larger than those generated with AmpFℓSTR® kits.

FGA is a compound tetranucleotide repeat found in the third intron of the human alpha fibrinogen locus on the long arm of chromosome 4. FGA has also been referred to in the literature as FIBRA or HUMFIBRA. The locus contains a CTTT repeat flanked on either side by degenerate repeats. The spread in allele sizes is larger for FGA than any of the other core STR loci. Reported alleles range in size from 12.2 repeats to 51.2 repeats, spanning over 35 repeats! A 2 bp deletion, from the loss of a CT, in the region just prior to the core repeat motif is responsible for the x.2 microvariant alleles that are very prevalent in this STR system. PCR products from Promega's PowerPlex® 2.1 and 16 STR kits are 112 bp larger than those generated with Applied Biosystems AmpFℓSTR® kits for equivalent alleles. This size difference between these two primer sets is the largest of any of the 13 core loci. So far a total of 80 different FGA alleles have been reported (see Appendix I) making it one of the most polymorphic loci used in human identity testing.

TH01 is a simple tetranucleotide repeat found in intron 1 of the tyrosine hydroxylase gene on the short arm of chromosome 11. The locus name arises from the initials for tyrosine hydroxylase and intron 1 (i.e., 01). The locus is sometimes incorrectly referred to as 'THO1' with an 'O' instead of a 'zero.' In the literature, TH01 has also been referred to as TC11 and HUMTH01.

TH01 has a simple tetranucleotide sequence with a repeat motif of TCAT on the upper strand in the GenBank reference sequence. The repeat motif is commonly referenced as AATG, which is correct for the complementary (bottom) strand to the GenBank reference sequence. A common microvariant allele that exists in Caucasians contains a single base deletion from allele 10 and is designed allele 9.3. Other x.3 alleles have been reported such as 8.3, 10.3, and 13.3 (Griffiths et al. 1998). TH01 has probably been the most studied of the 13 core loci with over 1000 population studies reported in the literature using this DNA marker. PCR products from Promega's PowerPlex® 1.1 STR kit are 11 bp larger than those generated with Applied Biosystems AmpFℓSTR® kits for equivalent alleles. PowerPlex® 2.1 STR kits produce amplicons that are 19 bp smaller than PowerPlex® 1.1. The PowerPlex® 2.1 and PowerPlex® 16 STR kits contain the same PCR primers for TH01.

TPOX is a simple tetranucleotide repeat found in intron 10 of the human thyroid peroxidase gene near the very end of the short arm of chromosome 2. TPOX has also been referred to in the literature as hTPO. This STR locus possesses a simple AATG repeat and is the least polymorphic of the 13 core loci. PCR products from Promega's PowerPlex™ 1.1 STR kit are 7 bp larger than those generated with Applied Biosystems AmpF*l*STR® kits for equivalent alleles. PowerPlex® 2.1 STR kits produce amplicons that are 38 bp larger in size relative to PowerPlex® 1.1. The PowerPlex® 2.1 and PowerPlex® 16 STR kits contain the same PCR primers for TPOX. Tri-allelic (three banded) patterns are more prevalent in TPOX than any other forensic STR marker (see Chapter 6).

VWA is a compound tetranucleotide repeat found in intron 40 of the von Willebrand Factor gene on the short arm of chromosome 12. VWA has also been referred to in the literature as vWF and vWA. It possesses a TCTA repeat interspersed with a TCTG repeat. The VWA marker targeted by STR multiplex kits is only one of three repeats present in that region of the von Willebrand Factor. The other two have not been found to be as polymorphic (Kimpton *et al.* 1992). PCR products from Promega's PowerPlex® 1.1 STR kit are 29 bp smaller than those generated with Applied Biosystems AmpF*l*STR® kits for equivalent alleles. The PowerPlex® 1.1 and PowerPlex® 2.1 STR kits overlap at three STRs including VWA. Both kits produce amplicons that are equivalent in size for VWA alleles. The PowerPlex® 2.1 and PowerPlex® 16 STR kits contain the same PCR primers for VWA.

D3S1358 is a compound tetranucleotide repeat found on the short arm of chromosome 3. This locus possesses both AGAT and AGAC repeat units (Mornhinweg *et al.* 1998). The D3 marker is common to Applied Biosystems AmpF*l*STR® multiplexes Blue™, Profiler™, Profiler Plus™, COfiler™, SGM Plus™, SEfiler™, and Identifiler™. PCR products from Promega's PowerPlex® 2.1 STR kit are 2 bp larger than those generated with Applied Biosystems kits for equivalent alleles. The PowerPlex® 2.1 and PowerPlex® 16 STR kits contain the same PCR primers for D3S1358.

D5S818 is a simple tetranucleotide repeat found on the long arm of chromosome 5. The locus possesses AGAT repeat units with alleles ranging in size from 7–16 repeats. In both Promega and Applied Biosystems STR kits, D5S818 is one of the smaller sized loci and as such should appear more than some of the other loci in degraded DNA samples. Only a few rare microvariants have been reported at this STR marker. PCR products from Promega's PowerPlex® 1.1 STR kit are 15 bp smaller than those generated with Applied Biosystems kits for equivalent alleles and PowerPlex® 16 retains the original PowerPlex® 1.1 primers.

D7S820 is a simple tetranucleotide repeat found on the long arm of chromosome 7. The locus possesses primarily a GATA repeat. However, a number of new D7 microvariant alleles have been reported recently (see Appendix I). These *x*.1 and *x*.3 alleles likely result due to a variation in the number of T nucleotides found in a poly(T) stretch that occurs 13 bases downstream of the core GATA repeat. Sequencing has revealed that 'on-ladder' alleles contain nine tandem T's while x.3 alleles contain eight T's and x.1 alleles contain 10 T's (Egyed *et al.* 2000). PCR products from Promega's PowerPlex® 1.1 STR kit are 42 bp smaller than those generated with Applied Biosystems kits for equivalent alleles.

D8S1179 is a compound tetranucleotide repeat found on chromosome 8. In early publications by the Forensic Science Service, D8S1179 is listed as D6S502 because of a labeling error in the Cooperative Human Linkage Center database from which this STR was chosen (Oldroyd *et al.* 1995, Barber and Parkin 1996). The locus consists primarily of alleles containing TCTA although a TCTG repeat unit enters the motif for all alleles larger than 13 repeats, usually at the second or third position from the 5′-end of the repeat region (Barber and Parkin 1996). PCR products from Promega's PowerPlex® 2.1 and PowerPlex® 16 STR kits are 80 bp larger than those generated with Applied Biosystems kits for equivalent alleles. AmpF*l*STR® Identifiler™ and Profiler Plus™ ID kits possess an extra, unlabeled D8S1179 reverse primer to prevent allele dropout in Asian populations due to a mutation in the middle of the primer-binding site (Leibelt *et al.* 2003).

D13S317 is a simple TATC tetranucleotide repeat found on the long arm of chromosome 13. Common alleles contain between 7–15 repeat units although alleles 5, 6, and 16 have been reported (see Appendix I). PCR products from Promega's PowerPlex® 1.1 STR kit are 36 bp smaller than those generated with Applied Biosystems AmpF*l*STR® kits for equivalent alleles. A 4 bp deletion has been reported 24 bases downstream from the core TATC repeat that can impact allele calls with different primer sets (Butler *et al.* 2003, Drábek *et al.* 2004). PowerPlex® 16 primers, while generating the same size amplicons as the original PowerPlex® 1.1 primers, have been shifted to avoid this 4 bp deletion that is present in some African-American samples.

D16S539 is a simple tetranucleotide repeat found on the long arm of chromosome 16. Nine common alleles exist that possess a core repeat unit of GATA. These include an allele with five repeats and consecutive alleles ranging from 8–15 repeat units in length. PCR products from Promega STR kits are 31 bp larger than those generated with Applied Biosystems kits for equivalent alleles. A point mutation (T→A) 38 bp downstream of the STR repeat impacts the reverse primers for both Applied Biosystems and Promega primer sets. Applied Biosystems added an extra or 'degenerate' unlabeled primer in their COfiler™, SGM Plus™, and Identifiler™ kits so that both possible alleles could be amplified (Wallin *et al.* 2002). On the other hand, Promega altered their D16S539 reverse primer sequence between kits but kept the overall amplicon size the same (Butler *et al.* 2001). The 3′-end of the PowerPlex® 1.1 reverse primer was lengthened by five nucleotides to create the PowerPlex® 16 reverse primer and thus move the primer mismatch caused by this mutation further into the primer to prevent allele dropout (Nelson *et al.* 2001, Krenke *et al.* 2002).

D18S51 is a simple tetranucleotide repeat found on the long arm of chromosome 18. It has a repeat motif of AGAA. A number of *x*.2 allele variants exist due to a 2 bp deletion from a loss of AG in the 3′-flanking region (Barber and Parkin 1996). More than 50 alleles have been reported for D18S51 making it one of the more polymorphic of the 13 core loci. PCR products from Promega's PowerPlex® 2.1 STR kit are 22 bp larger than those generated with Applied Biosystems AmpF*l*STR® kits for equivalent alleles. The PowerPlex® 2.1 and PowerPlex® 16 STR kits contain the same PCR primers for D18S51.

D21S11 is a complex tetranucleotide repeat found on the long arm of chromosome 21. A variable number of TCTA and TCTG repeat blocks surround a constant 43 bp section made up of the sequence {[TCTA]$_3$ TA [TCTA]$_3$ TCA [TCTA]$_2$ TCCA TA}. The *x*.2 microvariant alleles arise primarily from a 2 bp (TA) insertion on the 3′-end of the repeat region (Brinkmann *et al.* 1996). PCR products from Promega's PowerPlex® 2.1 STR kit are 17 bp larger than those generated with Applied Biosystems AmpF*l*STR® kits for equivalent alleles. The PowerPlex® 2.1 and PowerPlex® 16 STR kits contain the same PCR primers for D21S11.

Early papers in the literature by the Forensic Science Service had alleles named based on the dinucleotide subunit CV, where the V represents either an A, T, or G (Urquhart *et al.* 1994), while other authors adopted a different allele naming scheme based on the primary tetranucleotide repeat (Moller *et al.* 1994). As outlined in the European DNA Profiling Group inter-laboratory study on D21S11 (Gill *et al.* 1997), a simple formula can be used to convert the Urquhart (U) designation into the Moller (M) equivalent:

$$M = \tfrac{1}{2} \times (U - 5) \tag{5.1}$$

Today most laboratories use the Moller allele notation since it fits the ISFG allele designation recommendation (Bar *et al.* 1997).

D21S11 is far more polymorphic than can be easily detected with sized-based length separations. A careful search of the literature has revealed more than 80 reported alleles, many of which are the same length (see Appendix I). Fine differences in the D21S11 allele structures can only be determined by DNA sequencing since so many of the alleles have the same length but different internal sequence structure because some of the repeat units are switched around. For example, there are four different alleles designated as 30 repeats, which are indistinguishable by size-based methods alone (Appendix I).

The three most polymorphic of the 13 loci are D21S11, FGA, and D18S51. These loci contain numerous microvariant alleles that are being uncovered as more and more samples are examined around the world.

ADDITIONAL STR LOCI COMMONLY USED

The 13 core loci used within the United States for CODIS are effective DNA markers for human identification and will most likely continue to be used for some time. However, these 13 markers are by no means the only STRs that have been evaluated or used by forensic labs around the world. Dozens of other markers have been used, some quite extensively (Table 5.5).

Table 5.5 (below) Some additional STR markers used in the forensic DNA community. STR markers in bold are part of commonly used multiplex kits. For information on Y chromosome STRs, see Tables 9.2 and 9.5.

Locus Name	Chromosomal Location	GenBank Accession	Repeat ISFG format	Allele Range	Amplicon Size Range	Reference
ARA	Xcen–q13	M21748	CAG	14–32	255–315 bp	Hammond *et al.* (1994)
APOAI1	11q23–qter	J00048	AAAG	Complex	263–291 bp	Dupuy and Olaisen (1997)
ACTBP2	6	V00481	AAAG	4.2–37	198–325 bp	Dupuy and Olaisen (1997)
CD4	12p12–pter	M86525	TTTTC	6–16	125–175 bp	Hammond *et al.* (1994)
CYAR04	15q21.1	M30798	AAAT	5–12	173–201 bp	Hammond *et al.* (1994)
F13A01	6p24.3–25.1	M21986	GAAA	3.2–16	281–331 bp	Hammond *et al.* (1994)
F13B	1q31–q32.1	M64554	TTTA	6–12	169–193 bp	Promega
FABP	4q28–31	M18079	ATT	10–15	199–220 bp	Hammond *et al.* (1994)
FES/FPS	15q25–qter	X06292	ATTT	7–14	222–250 bp	Hammond *et al.* (1994)
HPRTB	Xq26.1	M26434	TCTA	6–17	259–303 bp	Hammond *et al.* (1994)
LPL	8p22	D83550	TTTA	7–14	105–133 bp	Promega
Penta D	21q	AP001752	AAAGA	2.2–17	376–449 bp	**PowerPlex 16**
Penta E	15q	AC027004	AAAGA	5–24	379–474 bp	**PowerPlex 16**
PLA2A1	12q23–qter	M22970	AAT		118–139 bp	Hammond *et al.* (1994)
RENA4	1q32	M10151	ACAG		255–275 bp	Hammond *et al.* (1994)
D1S1656	1pter–qter	G07820	(TAGA) (TAGG)	9–19.3	125–168 bp	Wiegand *et al.* (1999)
D2S1242	2pter–qter	L17825	(GAAA) (GAAG)	10–18	141–175 bp	Reichenpfader *et al.* (1999)
D2S1338	2q35–37.1	G08202	(TGCC) (TTCC)	15–28	289–341 bp	**SGM Plus, Identifiler**
D3S1359	3p	AA306290	TCTA	11–25.3	196–255 bp	Poltl *et al.* (1998)
D3S1744	3q24	G08246	GATA	14–22	150–182 bp	Lifecodes

Locus Name	Chromosomal Location	GenBank Accession	Repeat ISFG format	Allele Range	Amplicon Size Range	Reference
D6S477	6pter–qter	G08543	TCTA	13.2–22	206–240 bp	Carracedo and Lareu (1998)
D7S809	7pter–qter	X73290	(AGGA) (AGGC)	9 alleles	241–289 bp	Tamaki *et al.* (1996)
D8S347	8q22.3–24.3	L12268	AGAT	16–28	340–388 bp	Poltl *et al.* (1997)
D8S639	8p21–p11	L24797	(AGAT) (AGGT)	20–33.3	316–371 bp	Seidl *et al.* (1999)
D9S302	9q31–33	G08746	ATCT	17 alleles	255–353 bp	Carracedo and Lareu (1998)
D10S2325	10pter–qter	G08790	TCTTA	6–17	113–168 bp	Wiegand *et al.* (1999)
D11S488	11q24.1–25	L04732	(AAAG) (GAAG)	26–41	242–302 bp	Seidl *et al.* (1999)
D11S554	11p11.2–12	M87277	AAAG	Complex	176–286 bp	Dupuy and Olaisen (1997)
D12S391	12	G08921	(AGAT) (AGAC)	15–26	209–253 bp	Lareu *et al.* (1996)
D12S1090	12q12	Not found	GATA	9–33	212–306 bp	Lifecodes
D18S535	18pter–qter	G07985	GATA	9–16	130–158 bp	Wiegand *et al.* (1999)
D18S849	18q12–q21	G07992	GATA	9–20	93–133 bp	Lifecodes
D19S433	19q12–13.1	G08036	AAGG	9–17.2	106–140 bp	**SGM Plus, Identifiler**
D20S161	20pter–qter	L16405	TAGA	14–22	156–187 bp	Hou *et al.* (1999)
D22S683	22pter–qter	G08086	(TA) (TATC)	12–21.2	168–206 bp	Carracedo and Lareu (1998)
DXS6807	Xpter–p22.2	G09662	GATA	11–17	251–275 bp	Edelmann and Szibor (1999)

Table 5.5
(Continued)

Applied Biosystems has created the AmpF/STR® SGM Plus™ kit that co-amplifies 10 STR loci including two new STRs: D19S433 and D2S1338. With the adoption of the SGM Plus kit by the Forensic Science Service and much of Europe, the amount of population data on the STR loci D19S433 and D2S1338 will continue to grow. These two loci are also part of the Identifiler 16plex STR kit. Likewise, the Promega Corporation has included two pentanucleotide STR loci, Penta E and Penta D, in their GenePrint® PowerPlex® 2.1 and PowerPlex® 16 kits. Because these markers are included in the STR multiplexes in conjunction with the 13 core loci for developing DNA databases, they will become more prevalent as the number of samples in the databases grows.

Owing to the fact that the German national DNA database requires analysis of the complex hypervariable STR locus SE33, Promega created the PowerPlex®

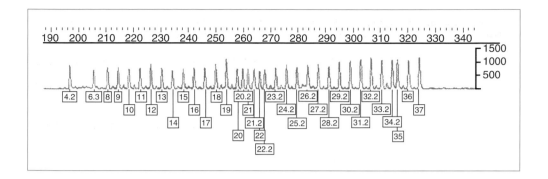

Figure 5.10

SE33 (ACTBP2) allelic ladder from PowerPlex ES kit produced by the Promega Corporation.

ES kit in 2001 and Applied Biosystems released the SEfiler™ kit in late 2002 to provide SE33 in a commercial kit form. The PowerPlex® ES allelic ladder for SE33 contains 35 alleles, which demonstrates this STR marker's variability (Figure 5.10).

Promega also has a multiplex commonly referred to as FFFL, which is used by many laboratories in South America to amplify the four STRs F13A01, F13B, FES/FPS, and LPL. Table 5.5 includes a listing of these markers as well as many others that have appeared in the literature along with useful information such as the GenBank accession number, references, and size ranges with a reported set of PCR primers. STR markers on the Y chromosome are described in Chapter 9. These Y-STRs are becoming increasingly popular due to their ability to aid sexual assault investigations through male-specific amplification.

GENDER IDENTIFICATION WITH AMELOGENIN

The ability to designate whether a sample originated from a male or a female source is useful in sexual assault cases, where distinguishing between the victim and the perpetrator's evidence is important. Likewise, missing persons and mass disaster investigations can benefit from gender identification of the remains. Over the years a number of gender identification assays have been demonstrated using PCR methods (Sullivan *et al.* 1993, Eng *et al.* 1994, Reynolds and Varlaro 1996). By far the most popular method for sex-typing today is the amelogenin system as it can be performed in conjunction with STR analysis.

Amelogenin is a gene that codes for proteins found in tooth enamel. The British Forensic Science Service was the first to describe the particular PCR primer sets that are used so prevalently in forensic DNA laboratories today (Sullivan *et al.* 1993). These primers flank a 6 bp deletion within intron 1 of the amelogenin gene on the X homologue (Figure 5.11). PCR amplification of this area with their primers results in 106 bp and 112 bp amplicons from the X and

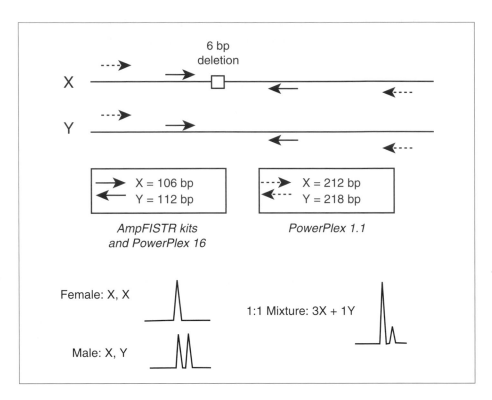

Figure 5.11

Schematic of the amelogenin sex-typing assay. The X and Y chromosomes contain a high degree of sequence homology at the amelogenin locus. The primer sets depicted here target a 6 bp deletion that is present only on the X chromosome. The presence of a single peak indicates that the sample comes from a female while two peaks identifies the sample's source as male. The primers to amplify the 106/112 bp fragments are used in the AmpFlSTR kits while the PowerPlex 1.1 kit uses the larger primer set.

Y chromosomes, respectively. Primers, which yield a 212 bp X-specific amplicon and a 218 bp Y-specific product by bracketing the same 6 bp deletion, were also described in the original amelogenin paper (Sullivan *et al.* 1993) and have been used in conjunction with the D1S80 VNTR system (Budowle *et al.* 1996).

An advantage with the above approach, i.e., using a single primer set to amplify both chromosomes, is that the X chromosome product itself plays a role as a positive control. This PCR-based assay is extremely sensitive. Mannucci and co-workers were able to detect as little as 20 pg (~3 diploid copies) as well as sample mixtures where female DNA was in 100-fold excess of male DNA (Mannucci *et al.* 1994).

Other regions of the amelogenin gene have size differences between the X and Y homologues and may be exploited for sex-typing purposes. For example, Eng and co-workers (1994) used a single set of primers that generated a 977 bp product for the X chromosome and a 788 bp fragment for the Y chromosome. In this case, a 189 bp deletion in the Y relative to the X chromosome was used to differentiate the two chromosomes.

A careful study found that 19 regions of absolute homology, ranging in size from 22–80 bp, exist between the human amelogenin X and Y genes that can be used to design a variety of primer sets (Haas-Rochholz and Weiler 1997). Thus, by spanning various deletions of the X and/or Y chromosome, it is possible to generate PCR products from the X and Y homologues that differ in size and contain size ranges that can be integrated into future multiplex STR amplifications.

ANOMALOUS AMELOGENIN RESULTS

While amelogenin is an effective method for sex-typing biological samples in most cases, the results are not full proof. A rare deletion of the amelogenin gene on the Y chromosome can cause the Y chromosome amplicon to be absent (Santos *et al.* 1998). In such a case, a male sample would falsely appear as a female. It appears that this deletion of the Y chromosome amelogenin region is more common in Indian populations (Thangaraj *et al.* 2002) than those of European or African origins. A study of almost 30 000 males in the Austrian National DNA database revealed that only six individuals lacked the amelogenin Y-amplicon (Steinlechner *et al.* 2002). These individuals were verified to be male with Y-STRs and amplification of the SRY region (see Chapter 9).

Amelogenin X allele dropout has also been observed in males. In this case only the amelogenin Y-amplicon is present (Shewale *et al.* 2000). This phenomenon was observed only three times out of almost 7000 males examined and likely results from a rare polymorphism in the primer binding sites for the amelogenin primers used in commercial STR kits. A different set of amelogenin primers targeting the same 6 bp deletion on the X chromosome amplified both the X and Y alleles of amelogenin (Shewale *et al.* 2000).

STRBASE: A DYNAMIC SOURCE OF INFORMATION ON STR MARKERS

The rapid growth of the human identification applications for STR loci insures that static written materials, such as this book, will quickly become out-of-date. New alleles are constantly being discovered (including 'off-ladder' microvariant alleles), additional STR markers are being developed, and population data increases with each month of published journals. Indeed, a growing list of publications describing the application of STR loci to forensic DNA typing has exceeded 2000 references.

The growth of the World Wide Web now permits dynamic sources of information to be widely available. Several years ago a web site was created to enable forensic scientists to keep abreast with the rapidly evolving field of DNA typing. In anticipation of the impact of STR markers on DNA typing and the need for a common source of information that could evolve as the process improved, an internet-accessible informational database was created in early 1997. STRBase was officially launched in July 1997 and is maintained by the DNA Technologies Group of the National Institute of Standards and Technology (Butler *et al.* 1997, Ruitberg *et al.* 2001). STRBase may be reached via the World Wide Web using

Figure 5.12

Homepage for STRBase, an internet-accessible database of information on STR markers used in forensic DNA typing. STRBase may be accessed via the URL: http:// www.cstl.nist.gov/biotech /strbase/ and contains among other things a comprehensive listing of all papers relating to STR typing for human identity testing purposes now numbering over 2000 references.

Short Tandem Repeat DNA Internet Database

These data are intended to benefit research and application of short tandem repeat DNA markers to aid human identity testing. The authors are solely responsible for the information herein. [Purpose of Database]

This database has been accessed `19519` *times since 10/02/97. (Counter courtesy* www.digits.com *- see* disclaimer.*)*

Created by John M. Butler and Dennis J. Reeder (NIST Biotechnology Division), with invaluable help from Christian Ruitberg and Michael Tung

Site creators' curriculum vitaes available using links above.

Partial support for the design and maintenance of this website is being provided by The National Institute of Justice through the NIST Office of Law Enforcement Standards.

- Background information on STRs
- Description of each STR system (STR Fact Sheets)
- Sequence Information
- Chromosomal Locations
- Non-published Variant Allele Reports **Updated**

- Allele Frequency Distribution Tables
- Sex-typing markers
- Technology for resolving STR alleles
- Y-chromosome STRs
- Population data

- Validation studies
- Multiplex STR sets
- PCR primers
- FBI Core STR Loci
- NIST Standard Reference Material for PCR-Based Testing
- DNA Advisory Board Quality Assurance Standards

- Reference List
- Original papers describing common STR systems
- Addresses for scientists working with STRs
- Links to other web sites
- Glossary of commonly used terms

the following URL: http://www.cstl.nist.gov/biotech/strbase. The home page for STRBase is shown in Figure 5.12.

STRBase contains a number of useful elements. Continually updated information includes the listing of references related to STRs and DNA typing (over 2000 references), addresses for scientists working in the field, and new microvariant or 'off-ladder' STR alleles. Other information that is updated less frequently includes STR fact sheets (with allele information similar to Appendix I), links to other web pages, a review of technology used for DNA typing as well as published primer sequence information, and population data for STR markers.

STR markers have become important tools for human identity testing. Commercially available STR kits are now widely used in forensic and paternity testing laboratories. The adoption of the 13 CODIS core loci for the U.S. national

DNA database ensures that these STR markers will be used for many years to come. However, as we will see in the next two chapters, results from STR markers require careful interpretation in order to be effective tools for law enforcement.

REFERENCES AND ADDITIONAL READING

Applied Biosystems (2001) AmpF*lSTR® Identifiler™ PCR Amplification Kit User's Manual. Foster City, CA.

Applied Biosystems (2002) AmpF*lSTR® SEfiler™ PCR Amplification Kit User's Manual. Foster City, CA.

Baechtel, F.S., Smerick, J.B., Presley, K.W. and Budowle, B. (1993) *Journal of Forensic Sciences*, 38, 1176–1182.

Bar, W., Brinkmann, B., Lincoln, P., Mayr, W.R. and Rossi, U. (1994) *International Journal of Legal Medicine*, 107, 159–160.

Bar, W., Brinkmann, B., Budowle, B., Carracedo, A., Gill, P., Lincoln, P., Mayr, W.R. and Olaisen, B. (1997) *International Journal of Legal Medicine*, 110, 175–176.

Barber, M.D., McKeown, B.J. and Parkin, B.H. (1996) *International Journal of Legal Medicine*, 108, 180–185.

Barber, M.D. and Parkin, B.H. (1996) *International Journal of Legal Medicine*, 109, 62–65.

Bacher, J.W., Hennes, L.F., Gu, T., Tereba, A., Micka, K.A., Sprecher, C.J., Lins, A.M., Amiott, E.A., Rabbach, D.R., Taylor, J.A., Helms, C., Donis-Keller, H. and Schumm, J.W. (1999) *Proceedings of the Ninth International Symposium on Human Identification*, pp. 24–37. Madison, Wisconsin: Promega Corporation.

Brinkmann, B., Meyer, E. and Junge, A. (1996) *Human Genetics*, 98, 60–64.

Britten, R.J. and Kohne, D.E. (1968) *Science*, 161, 529–540.

Broman, K.W., Murray, J.C., Sheffield, V.C., White, R.L. and Weber, J.L. (1998) *American Journal of Human Genetics*, 63, 861–869.

Budowle, B., Moretti, T.R., Niezgoda, S.J. and Brown, B.L. (1998) *Proceedings of the Second European Symposium on Human Identification*, pp. 73–88. Madison, Wisconsin: Promega Corporation.

Budowle, B., Masibay, A., Anderson, S.J., Barna, C., Biega, L., Brenneke, S., Brown, B.L., Cramer, J., DeGroot, G.A., Douglas, D., Duceman, B., Eastman, A., Giles, R., Hamill, J., Haase, D.J., Janssen, D.W., Kupferschmid, T.D., Lawton, T., Lemire, C., Llewellyn, B., Moretti, T., Neves, J., Palaski, C., Schueler, S., Sgueglia, J., Sprecher, C., Tomsey, C. and Yet, D. (2001) *Forensic Science International*, 124, 47–54.

Butler, J.M., Ruitberg, C.M. and Reeder, D.J. (1997) *Proceedings from the Eighth International Symposium on Human Identification*, pp. 38–47. Madison, Wisconsin: Promega Corporation.

Butler, J.M., Devaney, J.M., Mario, M.A. and Vallone, P.M. (2001) Comparison of primer sequences used in commercial STR kits. *Proceedings of the 53rd American Academy of Forensic Sciences* (Seattle, Washington); Presentation available at: http://www.cstl.nist.gov/biotech/strbase/NISTpub.htm.

Butler, J.M., Shen, Y. and McCord, B.R. (2003) *Journal of Forensic Sciences*, 48, 1054–1064.

Carracedo, A. and Lareu, M.V. (1998) *Proceedings from the Ninth International Symposium on Human Identification*, pp. 89–107. Madison, Wisconsin: Promega Corporation.

Chakraborty, R., Stivers, D.N., Su, Y. and Budowle, B. (1999) *Electrophoresis*, 20, 1682–1696.

Chambers, G.K. and MacAvoy, E.S. (2000) *Comparative Biochemistry and Physiology Part B*, 126, 455–476.

Collins, J.R., Stephens, R.M., Gold, B., Long, B., Dean, M. and Burt, S.K. (2003) *Genomics*, 82, 10–19.

Drábek, J., Chung, D., Butler, J.M. and McCord, B.R. (2004) *Journal of Forensic Sciences*, 49 (4), 859–860.

Dupuy, B.M. and Olaisen, B. (1997) *Forensic Science International*, 86, 207–227.

Edelmann, J. and Szibor, R. (1999) *Electrophoresis*, 20, 2844–2846.

Edwards, A., Civitello, A., Hammond, H.A. and Caskey, C.T. (1991) *American Journal of Human Genetics*, 49, 746–756.

Egyed, B., Furedi, S., Angyal, M., Boutrand, L., Vandenberghe, A., Woller, J. and Padar, Z. (2000) *Forensic Science International*, 113, 25–27.

Ellegren, H. (2004) *Nature Reviews Genetics*, 5, 435–445.

Ghebranious, N., Vaske, D., Yu, A., Zhao, C., Marth, G. and Weber, J.L. (2003) *BMC Genomics*, 4, 6.

Gill, P., Kimpton, C.P., d'Aloja, E., Andersen, J.F., Bar, W., Brinkmann, B., Holgersson, S., Johnsson, V., Kloosterman, A.D., Lareu, M.V., Nellemann, L., Pfitzinger, H., Phillips, C.P., Schmitter, H., Schneider, P.M. and Stenersen, M. (1994) *Forensic Science International*, 65, 51–59.

Griffiths, R.A.L., Barber, M.D., Johnson, P.E., Gillbard, S.M., Haywood, M.D., Smith, C.D., Arnold, J., Burke, T., Urquhart, A.J. and Gill, P. (1998) *International Journal of Legal Medicine*, 111, 267–272.

Grossman, P.D., Bloch, W., Brinson, E., Chang, C.C., Eggerding, F.A., Fung, S., Iovannisci, D.A., Woo, S. and Winn-Deen, E.S. (1994) *Nucleic Acids Research*, 22, 4527–4534.

Haas-Rochholz, H. and Weiler, G. (1997) *International Journal of Legal Medicine*, 110, 312–315.

Hammond, H.A., Jin, L., Zhong, Y., Caskey, C.T. and Chakraborty, R. (1994) *American Journal of Human Genetics*, 55, 175–189.

Holt, C.L., Buoncristiani, M., Wallin, J.M., Nguyen, T., Lazaruk, K.D. and Walsh, P.S. (2002) *Journal of Forensic Sciences*, 47, 66–96.

Hou, Y., Jin, Z.M., Li, Y.B., Wu, J., Walter, H., Kido, A. and Prinz, M. (1999) *International Journal of Legal Medicine*, 112, 400–402.

International Human Genome Sequencing Consortium. (2001) *Nature*, 409, 860–921.

Jin, L., Zhong, Y. and Chakraborty, R. (1994) *American Journal of Human Genetics*, 55, 582–583.

Kasai, K., Nakamura, Y. and White, R. (1990) *Journal of Forensic Sciences*, 35, 1196–1200.

Kimpton, C., Walton, A. and Gill, P. (1992) *Human Molecular Genetics*, 1, 287.

Kimpton, C.P., Gill, P., Walton, A., Urquhart, A., Millican, E.S. and Adams, M. (1993) *PCR Methods and Applications*, 3, 13–22.

Kimpton, C.P., Fisher, D., Watson, S., Adams, M., Urquhart, A., Lygo, J. and Gill, P. (1994) *International Journal of Legal Medicine*, 106, 302–311.

Krenke, B.E., Tereba, A., Anderson, S.J., Buel, E., Culhane, S., Finis, C.J., Tomsey, C.S., Zachetti, J.M., Masibay, A., Rabbach, D.R., Amiott, E.A. and Sprecher, C.J. (2002) *Journal of Forensic Sciences*, 47, 773–785.

Lareu, M.V., Pestoni, C., Schurenkamp, M., Rand, S., Brinkmann, B. and Carracedo, A. (1996) *International Journal of Legal Medicine*, 109, 134–138.

Leibelt, C., Budowle, B., Collins, P., Daoudi, Y., Moretti, T., Nunn, G., Reeder, D. and Roby, R. (2003) *Forensic Science International*, 133, 220–227.

Mannucci, A., Sullivan, K.M., Ivanov, P.L. and Gill, P. (1994) *International Journal of Legal Medicine*, 106, 190–193.

Margolis-Nunno, H., Brenner, L., Cascardi, J. and Kobilinsky, L. (2001) *Journal of Forensic Sciences*, 46, 1480–1483.

Masibay, A., Mozer, T.J. and Sprecher, C. (2000) *Journal of Forensic Sciences*, 45, 1360–1362.

Moller, A., Meyer, E. and Brinkmann, B. (1994) *International Journal of Legal Medicine*, 106, 319–323.

Mornhinweg, E., Luckenbach, C., Fimmers, R. and Ritter, H. (1998) *Forensic Science International*, 95, 173–178.

Nelson, M.S., Levedakou, E.N., Matthews, J.R., Early, B.E., Freeman, D.A., Kuhn, C.A., Sprecher, C.J., Amin, A.S., McElfresh, K.C. and Schumm, J.W. (2002) *Journal of Forensic Sciences*, 47, 345–349.

Oldroyd, N.J., Urquhart, A.J., Kimpton, C.P., Millican, E.S., Watson, S.K., Downes, T. and Gill, P.D. (1995) *Electrophoresis*, 16, 334–337.

Poltl, R., Luckenbach, C., Fimmers, R. and Ritter, H. (1997) *Electrophoresis*, 18, 2871–2873.

Poltl, R., Luckenbach, C. and Ritter, H. (1998) *Forensic Science International*, 95, 163–168.

Primrose, S.B. (1998) *Principles of Genome Analysis: A Guide to Mapping and Sequencing DNA from Different Organisms*, 2nd edn. Malden, MA: Blackwell Science.

Promega Corporation (2002) PowerPlex® ES System Technical Manual. Madison, WI.

Puers, C., Hammond, H.A., Jin, L., Caskey, C.T. and Schumm, J.W. (1993) *American Journal of Human Genetics*, 53, 953–958.

Reichenpfader, B., Zehner, R. and Klintschar, M. (1999) *Electrophoresis*, 20, 514–517.

Ruitberg, C.M., Reeder, D.J. and Butler, J.M. (2001) *Nucleic Acids Research*, 29, 320–322.

Sajantila, A., Puomilahti, S., Johnsson, V. and Ehnholm, C. (1992) *BioTechniques*, 12, 16–21.

Seidl, C., Muller, S., Jager, O. and Seifried, E. (1999) *International Journal of Legal Medicine*, 112, 355–359.

Shewale, J.G., Richey, S.L. and Sinha, S.K. (2000) Anomalous amplification of the amelogenin locus typed by AmpF/STR Profiler Plus amplification kit. *Forensic Science Communications* 2 (4). Available online at: http://www.fbi.gov/hq/lab/fsc/backissu/oct2000/shewale.htm.

Smith, R.N. (1995) *BioTechniques*, 18, 122–128.

Sparkes, R., Kimpton, C., Gibard, S., Carne, P., Andersen, J., Oldroyd, N., Thomas, D., Urquhart, A. and Gill, P. (1996) *International Journal of Legal Medicine*, 109, 195–204.

Steinlechner, M., Berger, B., Niederstatter, H. and Parson, W. (2002) *International Journal of Legal Medicine*, 116, 117–120.

Subramanian, S., Mishra, R.K. and Singh, L. (2003) *Genome Biology*, 4, R13. Available online at: http://genomebiology.com/2003/4/2/R13.

Sullivan, K.M., Mannucci, A., Kimpton, C.P. and Gill, P. (1993) *BioTechniques*, 15, 637–641.

Tamaki, K., Huang, X.-L., Nozawa, H., Yamamoto, T., Uchihi, R., Katsumata, Y. and Armour, J.A.L. (1996) *Forensic Science International*, 81, 133–140.

Tautz, D. (1993) Notes on definition and nomenclature of tandemly repetitive DNA sequences. In Pena, D.J., Chakraborty, R., Epplen, J.T. and Jeffreys, A.J. (eds), *DNA Fingerprinting: State of the Science*, pp. 21–28. Birkhauser Verlag: Basel.

Thangaraj, K., Reddy, A.G. and Singh, L. (2002) *International Journal of Legal Medicine*, 116, 121–123.

Tully, G., Sullivan, K.M. and Gill, P. (1993) *Human Genetics*, 92, 554–562.

Urquhart, A., Kimpton, C.P. and Gill, P. (1993) *Human Genetics*, 92, 637–638.

Urquhart, A., Kimpton, C.P., Downes, T.J. and Gill, P. (1994) *International Journal of Legal Medicine*, 107, 13–20.

Wallin, J.M., Holt, C.L., Lazaruk, K.D., Nguyen, T.H. and Walsh, P.S. (2002) *Journal of Forensic Sciences*, 47, 52–65.

Walsh, P.S., Erlich, H.A. and Higuchi, R. (1992) *PCR Methods and Applications*, 1, 241–250.

Walsh, P.S., Fildes, N.J. and Reynolds, R. (1996) *Nucleic Acids Research*, 24, 2807–2812.

Weber, J.L. and May, P.E. (1989) *American Journal of Human Genetics*, 44, 388–396.

Wiegand, P., Lareu, M.V., Schurenkamp, M., Kleiber, M. and Brinkmann, B. (1999) *International Journal of Legal Medicine*, 112, 360–363.

CHAPTER 6

BIOLOGY OF STRs: STUTTER PRODUCTS, NON-TEMPLATE ADDITION, MICROVARIANTS, NULL ALLELES AND MUTATION RATES

The most humbling aspect of the Human Genome Project so far has been the realization that we know remarkably little about what the vast majority of human genes do.

(James Watson, *DNA: The Secret of Life*, 2003, p. 217)

During polymerase chain reaction (PCR) amplification of short tandem repeat (STR) alleles, a number of artifacts can arise that may interfere with the clear interpretation and genotyping of the alleles present in the DNA template. In this chapter, we will focus on those PCR products that give rise to additional peaks besides the true, major allele peak(s). These artifacts include stutter products and non-template nucleotide addition. Other factors that impact STR typing, including microvariants, tri-allelic patterns, allele dropout, and mutations will also be covered.

STUTTER PRODUCTS

A close examination of electropherograms containing STR data typically reveals the presence of small peaks several bases shorter than each STR allele peak (Figure 6.1). These 'stutter product' peaks result from the PCR process when STR loci are copied by a DNA polymerase. In the literature, this stutter product has also been referred to as a shadow band or a DNA polymerase slippage product (Hauge and Litt 1993).

Sequence analysis of stutter products from the tetranucleotide repeat locus VWA has shown that they contain one repeat unit less than the corresponding main allele peak (Walsh *et al.* 1996). Stutter products that are larger in size by one repeat unit than the corresponding alleles are only rarely observed in commonly used tetranucleotide repeat STR loci.

Stutter products have been reported in the literature since STRs (microsatellites) were first described. The primary mechanism that has been proposed to explain the existence of stutter products is slipped-strand mispairing (Hauge and Litt 1993, Walsh *et al.* 1996). In the slipped-strand mispairing model, a region of primer-template complex becomes unpaired during primer extension allowing slippage of either primer or template strand such that one repeat forms a non-base-paired loop (Hauge and Litt 1993). The consequence of this

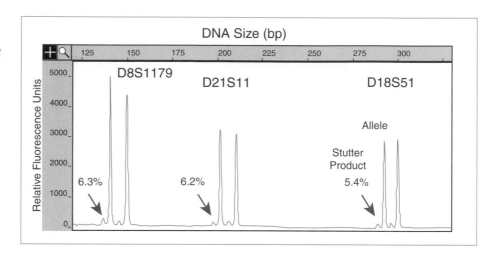

one repeat loop is a shortened PCR product that is less than the primary ampli-con (STR allele) by a single repeat unit (Figure 6.2).

IMPACT OF STUTTER PRODUCTS ON DATA INTERPRETATION

Stutter products impact interpretation of DNA profiles, especially in cases where two or more individuals may have contributed to the DNA sample (see Chapter 7).

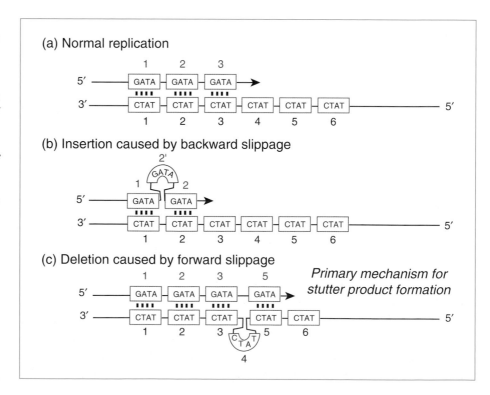

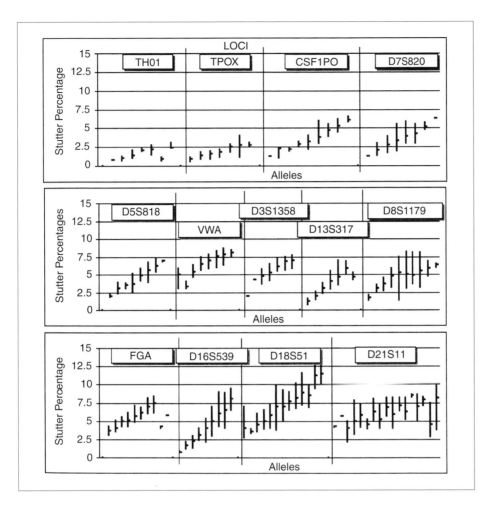

Figure 6.3

Stutter percentages for 13 CODIS STR loci (data adapted from AmpFlSTR manuals). Alleles for each STR locus are shown from smallest to largest. Each locus has a different average stutter percentage but all loci show the trend of increasing stutter with larger alleles (longer number of repeats).

Because stutter products are the same size as actual allele PCR products, it can be challenging to determine whether a small peak is a real allele from a minor contributor or a stutter product of an adjacent allele.

Mixture interpretation requires a good understanding of the behavior of stutter products in single source samples. Often a laboratory will quantify the percentage of stutter product peak heights compared to their corresponding allele peak heights. The percentage of stutter product formation for an allele is determined simply by dividing the stutter peak height by the corresponding allele peak height.

A plot of the alleles from each of the 13 standard STR loci reveals variation in the percentage of stutter for each locus as well as alleles from the same locus (Figure 6.3). Such a plot illustrates several important principles. First, each locus has a different amount of stutter product formation. Second, longer alleles for a STR locus exhibit a greater degree of stutter than smaller alleles for the same locus. Third, stutter percentage with the standard tetranucleotide repeats is generally less than 15% for all 13 CODIS core STR loci under standard amplification conditions.

REDUCED STUTTER PRODUCT FORMATION

The amount of stutter product formation may be reduced when using STR markers with longer repeat units, STR alleles with imperfect repeat units, and DNA polymerases with faster processivity. Several pentanucleotide repeat loci have been developed in an effort to produce STR markers that exhibit low amounts of stutter products to aid in mixture interpretation (Bacher and Schumm 1998). The first seven loci have been labeled Penta A through Penta G. Penta E has been incorporated in the GenePrint® PowerPlex™ 2.1 system and reportedly exhibits an average stutter percentage of less than 1% (Bacher *et al.* 1999). Both Penta D and Penta E are part of the PowerPlex® 16 kit (Krenke *et al.* 2002).

Alleles for a STR locus that contain variations on the common repeat motif exhibit a smaller amount of stutter product formation. For example, the common repeat motif for the STR marker TH01 is AATG. However, with allele 9.3, there is an ATG nucleotide sequence present in the middle of the repeat region (Puers *et al.* 1993). When the core repeat sequence has been interrupted, stutter product formation is reduced compared to alleles that are similar in length but possess uninterrupted core repeat sequences. This fact has been demonstrated with sequencing results from several VWA alleles (Walsh *et al.* 1996).

The amount of stutter may be related to the DNA polymerase processivity, or how rapidly it copies the template strand. Stutter products have been shown to increase relative to their corresponding alleles with a slower polymerase (Walsh *et al.* 1996). If thermal stable DNA polymerases become available in the future that are faster than the current 50–60 base processivity of *Taq* DNA polymerase, then it may be possible to reduce the stutter product formation. A faster polymerase would be able to copy the two DNA strands before they could come apart and re-anneal out of register during primer extension (see Figure 6.2).

A summary of stutter product formation is listed below:

- Primarily one repeat unit smaller than corresponding main allele peak;
- Typically less than 15% of corresponding allele peak height;
- Quantity of stutter depends on locus as well as PCR conditions and polymerase used;
- Propensity of stutter decreases with longer repeat units (pentanucleotide repeats<tetra-<tri-<dinucleotides);
- Quantity of stutter is greater for large alleles within a locus;
- Quantity of stutter is less if sequence of repeats is imperfect.

NON-TEMPLATE ADDITION

DNA polymerases, particularly the *Taq* polymerase used in PCR, often add an extra nucleotide to the 3′-end of a PCR product as they are copying the template strand (Clark 1988, Magnuson *et al.* 1996). This non-template addition is most often adenosine and is therefore sometimes referred to as 'adenylation' or the '+A' form of the amplicon. Non-template addition results in a PCR product that is one base pair longer than the actual target sequence.

Addition of the 3′ A nucleotide can be favored by adding a final incubation step at 60°C or 72°C after the temperature cycling steps in PCR (Clark 1988, Kimpton *et al.* 1993). However, the degree of adenylation is dependent on the sequence of the template strand, which in the case of PCR results from the 5′-end of the reverse primer (Figure 6.4). If the forward primer is labeled with

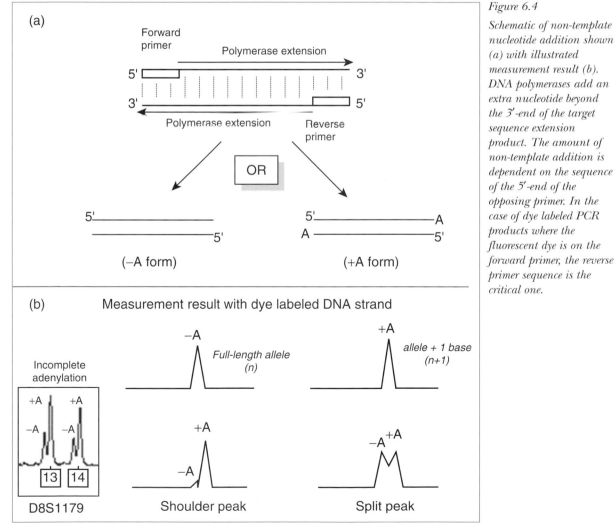

Figure 6.4

Schematic of non-template nucleotide addition shown (a) with illustrated measurement result (b). DNA polymerases add an extra nucleotide beyond the 3′-end of the target sequence extension product. The amount of non-template addition is dependent on the sequence of the 5′-end of the opposing primer. In the case of dye labeled PCR products where the fluorescent dye is on the forward primer, the reverse primer sequence is the critical one.

a fluorescent dye to amplify the STR allele, then only the top strand is detected by the fluorescent measurement. Since the sequence at the 3′-end of the top (labeled) strand serves as a template for polymerase extension, the terminal nucleotide of the labeled strand is determined by the 5′-end of the reverse primer used in generating the complementary unlabeled strand (Magnuson *et al.* 1996). One study found that if the 5′-terminus of the primer is a guanosine, then a complete addition is favored by the polymerase (Brownstein *et al.* 1996). Thus, every locus will have slightly different adenylation properties because the primer sequences differ.

Now why is all of this important? From a measurement standpoint, it is better to have all of the molecules as similar as possible for a particular allele. Partial adenylation, where some of the PCR products do not have the extra adenine (i.e., −A peaks) and some do (i.e., +A peaks), can contribute to peak broadness if the separation system's resolution is poor. Sharper peaks improve the likelihood that a system's genotyping software can make accurate calls. In addition, variation in the adenylation status of an allele across multiple samples can have an impact on accurate sizing and genotyping potential microvariants. For example, a non-adenylated TH01 10 allele would be the same size as a fully adenylated TH01 9.3 allele because they contain an identical number of base pairs. Therefore, it is beneficial if all PCR products for a particular amplification are either +A or −A rather than a mixture of +/−A products. Table 6.1 lists some of the methods that have been used to convert PCR products into either the −A or +A form.

Table 6.1

Ways to convert STR allele peaks to either −A or +A forms.

Method	Result	Reference
Conversion to fully adenylated products (+A form)		
Final extension at 60°C or 72°C for 30–45 minutes	Promotes full adenylation of all products	Kimpton *et al.* 1993, Applied Biosystems 1999
Addition of sequence GTTTCTT on the 5′-end of reverse primers ('PIG-tailing')	Promotes nearly 100% adenylation of the 3′ forward strand	Brownstein *et al.* 1996
Conversion to blunt-ended products (−A form)		
Restriction enzyme site built into reverse primer	Makes blunt end fragments following restriction enzyme digestion	Edwards *et al.* 1991
Enzymatic removal of one base overhang	Exonuclease activity of *Pfu* or T4 DNA polymerase removes +A	Ginot *et al.* 1996
Use of modified polymerase without terminal transferase activity	Polymerase does not add 3′ A nucleotide	Butler and Becker 2001

During PCR amplification most STR protocols include a final extension step to give the DNA polymerase extra time to completely adenylate all double-stranded PCR products. For example, the standard AmpF*l*STR® kit amplification parameters include a final extension at 60°C for 45 minutes at the end of thermal cycling (Applied Biosystems 1999). In order to make correct genotype calls, it is important that the allelic ladder and the sample have the same adenylation status for a particular STR locus. For all commercially available STR kits, this means that the STR alleles are all in the +A form.

Amplifying higher quantities of DNA than the optimal amount suggested by the manufacturer's protocols can result in incomplete 3′ A nucleotide addition and therefore split peaks. The addition of 10 ng of template DNA to a PCR reaction with AmpF*l*STR Profiler Plus results in split peaks compared to using only 2 ng of the same template DNA (Figure 6.5). Thus, quantifying the amount of DNA prior to PCR and adhering to the manufacturer's protocols will produce improved STR typing results with using commercial STR kits.

MICROVARIANTS AND 'OFF-LADDER' ALLELES

Rare alleles are encountered in the human population that may differ from common allele variants at tested DNA markers by one or more base pairs. Sequence variation between STR alleles can take the form of insertions, deletions, or nucleotide changes. Alleles containing some form of sequence variation compared to more commonly observed alleles are often referred to as *microvariants* because they are only slightly different from full repeat alleles. Because microvariant alleles often do not size the same as consensus alleles present in the reference allelic ladder, they can be referred to as 'off-ladder' alleles.

Figure 6.5

Incomplete non-template addition with high levels of DNA template. In the top panel, partial adenylation (both −A and +A forms of each allele) is seen because the polymerase is overwhelmed due to an abundance of DNA template. Note also that the peaks in the top panel are off-scale and flat-topped in the case of the smaller FGA allele. When the suggested level of DNA template is used, all alleles are fully adenylated (bottom panel).

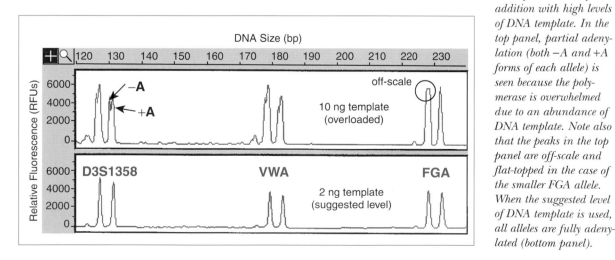

One example of a common microvariant is allele 9.3 at the STR locus TH01. The repeat region of TH01 allele 9.3 contains nine full repeats (AATG) and a partial repeat of three bases (ATG). The 9.3 allele differs from the 10 allele by a single base deletion of adenine in the seventh repeat (Puers *et al.* 1993).

Microvariants exist for most STR loci and are being identified in greater numbers as more samples are being examined around the world. In a recent study, 42 apparent microvariants were seen in over 10 000 samples examined at the CSF1PO, TPOX, and TH01 loci (Crouse *et al.* 1999). Microvariants are most commonly found in more polymorphic STR loci, such as FGA, D21S11, and D18S51, that possess the largest and most complex repeat structures compared to simple repeat loci, such as TPOX and CSF1PO (see Appendix I).

DETERMINING THE PRESENCE OF A MICROVARIANT ALLELE

Suspected microvariants can be fairly easily seen in heterozygous samples where one allele lines up with the fragment sizes in the allelic ladder and one does not (Figure 6.6). In the example shown here, the sample contains a peak that lines up with allele 25 from the FGA allelic ladder and a second peak that is labeled

Figure 6.6

Detection of a microvariant allele at the STR locus FGA. The sample in the bottom panel is compared to the allelic ladder shown in the top panel using Genotyper 2.5 software. Peaks are labeled with the allele category and the calculated fragment sizes using the internal sizing standard GS500-ROX.

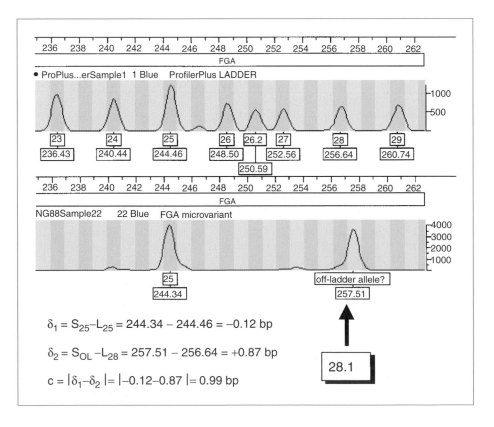

$$\delta_1 = S_{25} - L_{25} = 244.34 - 244.46 = -0.12 \text{ bp}$$

$$\delta_2 = S_{OL} - L_{28} = 257.51 - 256.64 = +0.87 \text{ bp}$$

$$c = |\delta_1 - \delta_2| = |-0.12 - 0.87| = 0.99 \text{ bp}$$

as an 'off-ladder allele' and lines up between the 28 and 28.2 shaded virtual bins created by the ladder. Each peak is labeled with its calculated size in base pairs determined by reference to the internal GS500 sizing standard (see Chapter 15). The relative size difference between the questioned sample and an allelic ladder marker run under the same electrophoretic conditions is then used to determine if the allele is truly a microvariant (Gill *et al.* 1996).

In Figure 6.6, the size difference between the sample allele 25 and the ladder allele 25 is −0.12 bp (δ_1) while the 'off-ladder allele' differs from the ladder allele 28 by +0.87 bp (δ_2). The relative peak shift between the two alleles in this heterozygous sample is 0.99 bp ($|\delta_1 - \delta_2|$) and therefore the 'off-ladder' allele is 1 bp larger than allele 28 making it a true 28.1 microvariant at the FGA locus.

The presence of a STR microvariant at a particular locus usually becomes evident following a comparison to an allelic ladder made up of characterized alleles for that locus. However, not all alleles (particularly rare microvariant alleles) can be incorporated into the standard allelic ladder used for genotyping STR markers. Therefore, interpolation of data from peaks that migrate between two characterized alleles or extrapolation of data from peaks that fall outside the expected allele range may be performed. Caution is in order though if 'off-ladder' alleles are more than a one or two repeats away from the nearest allele in the ladder since tetranucleotide repeats do not always size exactly 4.0 bp apart (see Gill *et al.* 1996, Applied Biosystems 1999, Butler *et al.* 2004).

If an allele peak falls in between the nominal alleles present in the allelic ladder, the sample may be designated by the allele number followed by a '.*x*' (Crouse *et al.* 1999). For example, the larger FGA allele shown in Figure 6.6 would be designated as a '28.*x*' allele. However, it is more common to label variant alleles by their calculated repeat content (e.g., 28.1). If an allele migrates above or below the defined allelic ladder, the allele is described as '>' or '<' than the nearest allele (Crouse *et al.* 1999).

SAME LENGTH BUT DIFFERENT SEQUENCE ALLELES

Complex repeat sequences, such as those found in D21S11, can contain variable repeat blocks in which the order is switched around for alleles that are the same length. For example, the STR locus D21S11 has four alleles that are all 210 bp when amplified with the Profiler Plus™ kit (Appendix I). While these alleles would be sized based on overall length to be 'allele 30', they contain repeat blocks of 4-6-CR-12, 5-6-CR-11, 6-5-CR-11, and 6-6-CR-10 for the pattern [TCTA]-[TCTG]-constant region (CR)-[TCTA]. In such cases, variant alleles would only be detectable with complete sequence analysis.

It is important to realize that from an operational point of view internal allele variation is not significant. In the end a match is being made against many loci not just one, such as D21S11, with possible internal sequence variation.

Most of the STR loci used in human identity testing have not exhibited internal sequence variation (see Appendix I), particularly the simple repeat loci TPOX, CSF1PO, D5S818, D16S539, TH01, D18S51, and D7S820. Remember that we are essentially binning alleles based on measured size anyway with STR typing since sequence analysis of individual alleles is too time consuming and would rarely reveal additional information because STR variation is primarily size-based.

PEAKS OUTSIDE THE ALLELIC LADDER RANGE AND THREE-BANDED PATTERNS

Occasionally new rare alleles may fall outside the allele range spanned by the locus allelic ladder. If these peaks fall between two STR loci in a multiplex set, they can be challenging to assign to a particular locus unless testing is performed with individual locus-specific primer sets or a different multiplex. These extreme 'off-ladder' alleles can be confirmed with singleplex amplification of the two loci in the multiplex bracketing the new allele. Alternatively the sample could be amplified again using a separate multiplex where the loci are present in a different order. For example, if a PCR product was observed between the typical VWA and D16S539 allele ranges when using the SGM Plus kit, then it could be either a large VWA allele or a small D16S539 allele, which is doubtful because allele 5 is the smallest run on the D16S539 ladder. A different kit, such as PowerPlex 16, could be used on this same sample to help address the source of the new allele since the loci are put together in a different combination. With PowerPlex 16 a large VWA allele would appear between the VWA and D8S1179 expected allele ranges, but a small D16S539 allele would fall between D7S820 and D16S539 (see Figure 5.4).

Three-banded or tri-allelic patterns are sometimes observed at a single locus in a multiplex STR profile. These extra peaks are not a result of a mixture but are reproducible artifacts of the sample. Extra chromosomal occurrences or primer point mutations have been known to happen and result in a three-banded pattern (Crouse *et al.* 1999). For example, three-banded patterns have been observed in the 9948 cell line with CSF1PO and in the K562 cell line with D21S11 (see D.N.A. Box 7.1).

The three peaks or bands seen at a particular locus may or may not be equal in intensity. While the TPOX three-banded patterns reported by Crouse and co-workers (1999) were approximately equal in intensity (similar to Figure 6.7a), there are also occasions when tri-allelic patterns occur with peaks of unequal intensity. In Figure 6.7b, a sum of two of the alleles is approximately equivalent in amount as the third allele as seen in Figure 6.7b for D18S51. The peak heights for alleles 14 and 15 when added together are similar to amount as the D18S51 allele 22. Thus, it is likely that the 14 and 15 alleles came from one parent while the 22 repeat allele came from the other. There is probably some

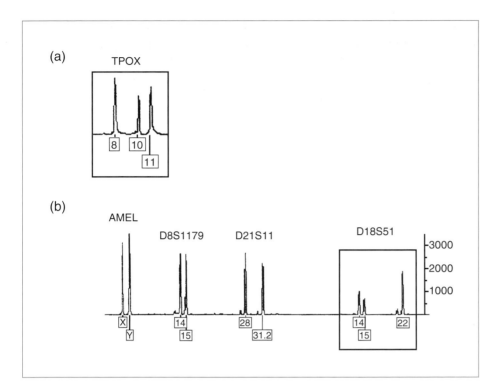

Figure 6.7

Tri-allelic patterns observed at (a) TPOX and (b) D18S51 from different samples. Allele calls are listed underneath each peak. (a) The TPOX result was obtained with the Identifiler STR kit and run on the ABI 3100. (b) This DNA sample was amplified with the Profiler Plus™ STR kit, separated on the ABI Prism 310 Genetic Analyzer and viewed with Genotyper software. Only the green dye-labeled PCR products are shown here for simplicity's sake.

kind of chromosome duplication for the region surrounding the D18S51 marker in this individual. Note that in this example the other STR loci besides D18S51 have two peaks of similar intensity, which suggests that a sample mixture is not likely (see Chapter 7).

More than 50 different tri-allelic patterns have been reported at all 13 CODIS STR loci with most of them being seen at TPOX and FGA. A frequently updated listing of tri-allelic patterns may be found on the STRBase web site: http://www.cstl.nist.gov/biotech/strbase/var_tab.htm.

ALLELE DROPOUT AND NULL ALLELES

When amplifying DNA fragments that contain STR repeat regions, it is possible to have a phenomenon known as *allele dropout*. Sequence polymorphisms are known to occur within or around STR repeat regions. These variations can occur in three locations (relative to the primer binding sites): within the repeat region, in the flanking region, or in the primer-binding region (Figure 6.8).

If a base pair change occurs in the DNA template at the PCR primer binding region, the hybridization of the primer can be disrupted resulting in a failure to amplify, and therefore failure to detect an allele that exists in the template DNA. More simply, the DNA template exists for a particular allele but fails to

Figure 6.8

Possible sequence variation in or around STR repeat regions and the impact on PCR amplification. The asterisk symbolizes a DNA difference (base change, insertion or deletion of a nucleotide) from a typical allele for a STR locus. In situation (a), the variation occurs within the repeat region and should have no impact on the primer binding and the subsequent PCR amplification (although the overall amplicon size may vary slightly). In situation (b), the sequence variation occurs just outside the repeat in the flanking region but interior to the primer annealing sites. Again, PCR should not be affected although the size of the PCR product may vary slightly. However, in situation (c) the PCR can fail due to a disruption in the annealing of a primer because the primer no longer perfectly matches the DNA template sequence.

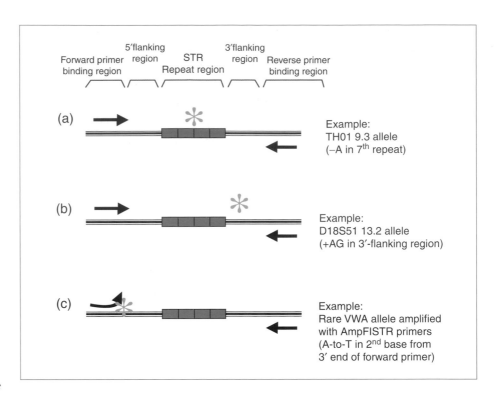

amplify during PCR due to primer hybridization problems. This phenomenon results in what is known as a *null allele*. Fortunately null alleles are rather rare because the flanking sequence around STR repeats is fairly stable and consistent between samples.

DISCOVERY OF NULL ALLELES

Null alleles have been 'discovered' by the observation of different typing results when utilizing independent STR primer sets. During a comparison of STR typing results on 600 population samples at the VWA locus, one sample typed 16,19 with Promega's PowerPlex kit and 16,16 with Applied Biosystem's AmpF*l*STR® Blue kit (Kline *et al.* 1998). In this case, VWA allele 19 dropped out with the AmpF*l*STR® VWA primer set due to a sequence polymorphism near the 3′-end of the forward primer (Walsh 1998).

Allele dropout may occur due to mutations (variants) at or near the 3′-end of a primer and thus produce little or no extension during PCR. In this case, the VWA allele 19 was present in the sample but failed to be amplified by one of the primer sets. It was later reported that the null allele resulted from a rare A–T nucleotide change in the DNA template at the second base from the 3′-end of the AmpF*l*STR® VWA forward primer (Walsh 1998).

Potential null alleles resulting from allele dropout can be predicted by statistical analysis of the STR typing data. The observed number of homozygotes can be compared to the expected number of homozygotes based on Hardy–Weinberg equilibrium (Chakraborty *et al.* 1992). An abnormally high level of homozygotes would indicate the possible presence of null alleles. Thus, each set of population data should be carefully examined when new STR markers are being tested in a forensic DNA laboratory (see Chapter 20).

A number of primer concordance studies have been conducted in the past few years as use of various STR kits has become more prevalent. An examination of over 2000 samples comparing the PowerPlex® 16 kit to the Profiler Plus™ and COfiler™ kit results found 22 examples of allele dropout due to a primer mismatch at seven of the 13 core STR loci: CSF1PO, D8S1179, D16S539, D21S11, FGA, TH01, and VWA (Budowle *et al.* 2001, Budowle and Sprecher 2001). Table 6.2 contains a summary of findings from some concordance studies that identified allele dropout with a particular primer pair.

SOLUTIONS TO NULL ALLELES

If a null allele is detected at a STR locus, there are several possible solutions. First, the problem PCR primer could be redesigned and moved away from the problematic site. This approach was taken early in the development of the D7S820 primers for the Promega PowerPlex® 1.1 kit (Schumm *et al.* 1996) and more recently with a D16S539 flanking region mutation (Nelson *et al.* 2002, see Chapter 5). However, this solution could result in the new primer interfering with another one in the multiplex set of primers or necessitate new PCR reaction optimization experiments. Clearly this solution is undesirable because it is time-consuming and labor intensive.

A second solution is to simply drop the STR locus from the multiplex mix rather than attempting to redesign the PCR primers to avoid the site. This approach is only desirable when early in the development cycle of a multiplex STR assay. The Forensic Science Service dropped the STR locus D19S253 from consideration in their prototype second-generation multiplex when a null allele was discovered (Urquhart *et al.* 1994).

A third, and more favorable, solution is to add a 'degenerate' primer that contains the known sequence polymorphism. This extra primer will then amplify alleles containing the problematic primer binding site sequence variant. This approach was taken with the AmpF/STR® kits for the D16S539 mutation mentioned previously (Holt *et al.* 2002, see Chapter 5). However, if the sequence variation at the primer binding site is extremely rare, it may not be worth the effort to add an additional primer to the multiplex primer mix.

Table 6.2

Summary of discordant results observed with STR kits as reported in the literature due to various concordance studies with different PCR primer sets. These discrepancies arise due to polymorphic nucleotides or insertions/ deletions that occur in the tested DNA templates near the 3'-end of a primer binding site that disrupt proper primer annealing and result in allele dropout upon PCR amplification. STR kits reported here include: PowerPlex 1.1 (PP1.1), PowerPlex 16 (PP16), Identifiler (ID), Profiler Plus (ProPlus), COfiler, and SGM Plus.

Locus	STR Kits/Assays Compared	Results	Reference
VWA	PP1.1 *vs* ProPlus	Loss of allele 19 with ProPlus; fine with PP1.1	Kline *et al.* (1998)
VWA	PP16 *vs* ProPlus	Loss of alleles 15 and 17 with ProPlus; fine with PP16	Budowle *et al.* (2001)
VWA	ID *vs* miniplexes	Loss of alleles 12, 13, and 14 with miniplex assay; fine with ID	Drabek *et al.* (2004)
VWA	SGM *vs* SGM Plus	Loss of allele 17 with SGM Plus; fine with SGM	Clayton *et al.* (2004)
D5S818	PP16 *vs* ProPlus	Loss of alleles 10 and 11 with PP16; fine with ProPlus	Alves *et al.* (2003)
D5S818	ID *vs* miniplexes	Loss of allele 12 with miniplex assay; fine with ID	Drabek *et al.* (2004)
D13S317	ID *vs* miniplexes	Shift of alleles 10 and 11 due to deletion outside of miniplex assay but internal to ID	Drabek *et al.* (2004)
D16S539	PP1.1 *vs* PP16 *vs* COfiler	Loss of alleles with PP1.1; fine with PP16 and COfiler	Nelson *et al.* (2002)
D16S539	PP16 *vs* COfiler	Loss of allele 12 with PP16; fine with COfiler	Budowle *et al.* (2001)
D8S1179	PP16 *vs* ProPlus	Loss of alleles 15, 16, 17, and 18 with ProPlus; fine with PP16	Budowle *et al.* (2001)
D8S1179	SGM *vs* SGM Plus	Loss of allele 16 with SGM Plus; fine with SGM	Clayton *et al.* (2004)
FGA	SGM *vs* SGM Plus	Loss of allele 26 with SGM Plus; weak amp of same allele with SGM	Cotton *et al.* (2000)
FGA	PP16 *vs* ProPlus	Loss of allele 22 with ProPlus; fine with PP16	Budowle and Sprecher (2001)
D18S51	SGM *vs* SGM Plus	Loss of alleles 17, 18, 19, and 20 with SGM Plus (in Kuwaiti individuals); fine with SGM	Clayton *et al.* (2004)
CSF1PO	PP16 *vs* COfiler	Loss of allele 14 with COfiler; fine with PP16	Budowle *et al.* (2001)
TH01	PP16 *vs* COfiler	Loss of allele 9 with COfiler; fine with PP16	Budowle *et al.* (2001)
TH01	SGM *vs* SGM Plus	Loss of allele 6 with SGM Plus; fine with SGM	Clayton *et al.* (2004)
D21S11	PP16 *vs* ProPlus	Loss of allele 32.2 with PP16; fine with ProPlus	Budowle *et al.* (2001)
D19S433	SGM *vs* SGM Plus	Loss of allele 11 with SGM Plus; fine with SGM	Clayton *et al.* (2004)

A fourth possible solution to correct for allele dropout that will work for some problematic primer binding sites is to re-amplify the sample with a lower annealing temperature and thereby reduce the stringency of the primer annealing. If the primer is only slightly destabilized, as detected by a peak height imbalance with a heterozygous sample (Figure 6.9, middle panel), then the peak height imbalance may be able to be corrected by lowering the annealing temperature during PCR.

No primer set is completely immune to the phenomenon of null alleles. However, when identical primer sets are used to amplify evidence samples and suspect reference samples, full concordance is expected from biological materials originating from a common source. If the DNA templates and PCR conditions are identical between two samples from the same individual, then identical DNA profiles should result regardless of how well or poorly the PCR primers amplify the DNA template.

The potential of null alleles is not a problem within a laboratory that uses the same primer set to amplify a particular STR marker. However, with the emergence of national and international DNA databases, which store only the genotype information for a sample, allele dropout could potentially result in a false negative or incorrect exclusion of two samples that come from a common source.

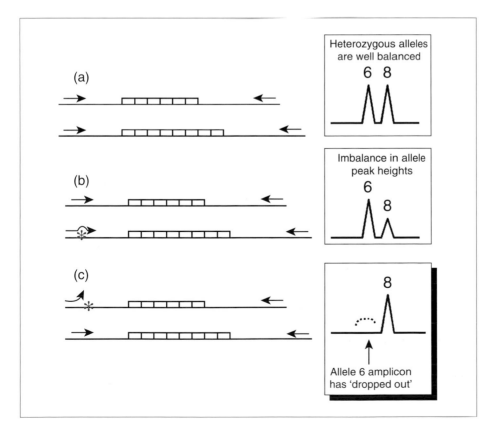

Figure 6.9

Impact of a sequence polymorphism in the primer binding site illustrated with a hypothetical heterozygous individual. Heterozygous allele peaks may be well-balanced (a), imbalanced (b), or exhibit allele dropout (c).

To overcome this potential problem, the matching criteria in database searches can be made less stringent when searching a crime stain sample against the DNA database of convicted offender profiles (see Chapter 18). That is, the database search might be programmed to return any profiles with a match at 25 out of 26 alleles instead of 26 out of 26.

When primers are selected for amplification of STR loci, candidate primers are evaluated carefully to avoid primer binding site mutations (Schumm *et al.* 1996, Wallin *et al.* 2002). Sequence analysis of multiple alleles is performed, family inheritance studies are conducted, within-locus peak signal ratios for heterozygous samples are examined, apparent homozygous samples are re-amplified with lower annealing temperatures, and statistical analysis of observed versus expected homozygosity is performed on population databases (Walsh 1998). It is truly a challenge to design multiplex STR primer sets in which primer binding sites are located in sequence regions that are as highly conserved as possible and yet do not interfere with primers amplifying other loci.

USE OF DEGENERATE PRIMERS IN COMMERCIAL KITS

The third solution mentioned above for solving potential allele dropout with primer binding site mutations is to add an additional PCR primer to the assay that can hybridize properly to the alternative allele when it exists in a sample. This has been the preferred solution for Applied Biosystems (e.g., Wallin *et al.* 2002) while Promega have moved their primers to overcome allele dropout problems (Nelson *et al.* 2002). According to their publications, Applied Biosystems has added an additional primer to correct for single point mutations in AmpF*l*STR® primer binding sites for D16S539 (Wallin *et al.* 2002), VWA (Lazaruk *et al.* 2001), and D8S1179 (Leibelt *et al.* 2003).

MUTATIONS AND MUTATION RATES

As with any region of DNA, mutations can and do occur at STR loci. By some not completely characterized mechanism, STR alleles can change over time (Ellegren 2004). Theoretically, all of the alleles that exist today for a particular STR locus have resulted from only a few 'founder' individuals by slowly changing over tens of thousands of years (Wiegand *et al.* 2000). The mutational event may be in the form of a single base change or in the length of the entire repeat. The molecular mechanisms by which STRs mutate are thought to involve replication slippage or defective DNA replication repair (Nadir *et al.* 1996, Ellegren 2004).

DISCOVERY OF STR ALLELE MUTATIONS

Estimation of mutational events at a DNA marker may be achieved by comparison of genotypes from offspring to those of their parents. Genotype data from

D.N.A. Box 6.1
Allele specific mutation rates

Until recently, only general information on STR mutation rates was reported – namely, how many mutations occurred relative to the number of meioses measured (see Table 6.3). The realization that certain alleles are more prone to mutation than others has prompted the American Association of Blood Banks (AABB) to carefully examine *which alleles* were mutating based on records from accredited parentage testing laboratories.

Appendix 5 in the AABB Annual Report Summary for Testing in 2002, prepared by the parentage testing program unit in November 2003, notes the number of paternal and maternal mutations by both locus and allele. For example, with the STR locus FGA an apparent change from allele 24 to 25 was observed 62 times (11.7%) out of 530 total paternal mutations seen in 2002, while an apparent change from allele 19 to 20 was seen only eight times (1.5%). In general longer alleles were seen to mutate more frequently.

The directionality of the mutation as either an expansion or a contraction of the repeat array can also vary significantly. For example, with paternal D16S539 mutations observed in 2002 there were 10 instances of allele 11 expanding to become allele 12 but only four examples of allele 11 contracting to allele 10. The process of expansion and contraction of the STR repeat regions probably occurs in a similar fashion as illustrated in Figure 6.2 for stutter product formation.

As this information continues to be collated in future studies, it should prove useful in refining mutation rates and aid in a better understanding of the process of STR origins and variability over human history.

Source:
American Association of Blood Banks Annual Report Summary for Testing in 2002 (see http://www.aabb.org/About_the_AABB/Stds_and_Accred/ptannrpt02.pdf)

paternity trios involving a father, a mother, and at least one child is examined. A discovery of an allele difference between the parents and the child is seen as evidence for a possible mutation (Figure 6.10). The search for mutations in STR loci involves examining many, many parent–child allele transfers because the mutation rate is rather low in most STRs.

The majority of STR mutations involve the gain or loss of a single repeat unit (see D.N.A. Box 6.1). Thus, a VWA allele with 14 repeats would show up as a 13

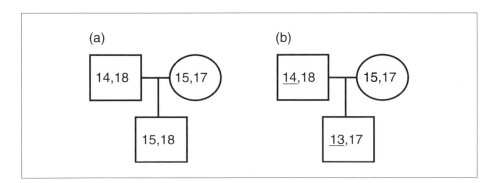

Figure 6.10

Mutational event observed in family trios. Normal transmission of alleles from a STR locus (a) is compared here to mutation of paternal allele 14 into the child's allele 13 (b).

or a 15 in the next generation following a mutational event (Figure 6.10). Paternal mutations appear to be more frequent than maternal ones for STR loci (Sajantila *et al.* 1999, Henke and Henke 1999). However, depending on the genotype combinations it can be difficult to ascertain from which parent the mutant allele was inherited.

MEASURING THE MUTATION RATE

Since the average mutation rate is below 0.1%, approximately 1000 parent–offspring allele transfers would have to be observed before one mutation would be seen in some STR markers (Weber and Wong 1993). Brinkman and co-workers (1998) examined 10 844 parent–child allele transfers at nine STR loci and observed 23 mutations. No mutations were observed at three of the loci (TH01, F13B, CD4). Sajantila *et al.* (1999) studied 29 640 parent–child allele transfers at five STRs and four minisatellites and observed only 18 mutational events (11 in three STR loci: D3S1359, VWA, and TH01). Two of the STRs, TPOX and FES/FPS, had no detectable mutations.

The mutation rates for the 13 core STR loci have been gathered from a number of studies in the literature and are summarized in Table 6.3. Most of these mutation rates are on the order of 1–5 mutations per 1000 allele transfers or generational events. The STR loci with the lowest observed mutation rates are CSF1PO, TH01, TPOX, D5S818, and D8S1179. Not surprisingly, the STR loci with the highest mutation rates – D21S11, FGA, D7S820, D16S539, and D18S51 – are among the most polymorphic and possess the highest number of observed alleles (Appendix I).

IMPACT OF MUTATION RATES ON PATERNITY TESTING

Low mutation rates are especially critical for paternity testing (see Chapter 23). This is because links are being made between the child and the alleged father based on the assumption that alleles remain the same when they are passed from one generation to the next. Parent–offspring allele transfer information tests for germ-line mutations. Additionally, genotypes from different kinds of tissues from the same individual are examined to demonstrate that no somatic mutations occur.

Mutations have practical consequences for paternity testing and mass disaster investigations (see Chapter 24) as well as population genetics where conclusions are being drawn from genetic data across one generation or many generations. In paternity testing situations, a high mutation rate for a STR marker could result in a false exclusion at that locus. With regards to population evolution studies, the mutation process must be subtracted from population demography and population history in order to accurately address the population

Table 6.3 (facing)

Observed mutation rates for the 13 core STR loci and other STR markers present in commercial kits. A total of 38 different paternity testing laboratories provided this STR mutation data, which is adapted from the American Association of Blood Banks (AABB) 2002 Annual Report issued in November 2003 that includes a compilation of multiple years (see http:// www.aabb.org/About_ the_AABB/Stds_and_ Accred/ptannrpt02.pdf, p. 15). The reported mutations are divided into maternal or paternal sources or from either when the source of the mutation observed in a child could not be determined.

STR System	Maternal Meioses (%)	Paternal Meioses (%)	Number from either	Total Number of Mutations	Mutation Rate
CSF1PO	70/179 353 (0.04)	727/504 342 (0.14)	303	1 100/683 695	0.16%
FGA	134/238 378 (0.06)	1 481/473 924 (0.31)	495	2 110/712 302	0.30%
TH01	23/189 478 (0.01)	29/346 518 (0.008)	23	75/535 996	0.01%
TPOX	16/299 186 (0.005)	43/328 067 (0.01)	24	83/627 253	0.01%
VWA	133/400 560 (0.03)	907/646 851 (0.14)	628	1 668/1 047 411	0.16%
D3S1358	37/244 484 (0.02)	429/336 208 (0.13)	266	732/580 692	0.13%
D5S818	84/316 102 (0.03)	537/468 366 (0.11)	303	924/784 468	0.12%
D7S820	43/334 886 (0.01)	550/461 457 (0.12)	218	811/796 343	0.10%
D8S1179	54/237 235 (0.02)	396/264 350 (0.15)	225	675/501 585	0.13%
D13S317	142/348 395 (0.04)	608/435 530 (0.14)	402	1 152/783 925	0.15%
D16S539	77/300 742 (0.03)	350/317 146 (0.11)	256	683/617 888	0.11%
D18S51	83/130 206 (0.06)	623/278 098 (0.22)	330	1 036/408 304	0.25%
D21S11	284/258 795 (0.11)	454/306 198 (0.15)	423	1 161/564 993	0.21%
Penta D	12/18 701 (0.06)	10/15 088 (0.07)	21	43/33 789	0.13%
Penta E	22/39 121 (0.06)	58/44 152 (0.13)	55	135/83 273	0.16%
D2S1338	2/25 271 (0.008)	61/81 960 (0.07)	31	94/107 231	0.09%
D19S433	22/28 027 (0.08)	16/38 983 (0.04)	37	75/67 010	0.11%
F13A01	1/10 474 (0.01)	37/65 347 (0.06)	3	41/75 821	0.05%
FES/FPS	3/18 918 (0.02)	79/149 028 (0.05)	None reported	82/167 946	0.05%
F13B	2/13 157 (0.02)	8/27 183 (0.03)	1	11/40 340	0.03%
LPL	0/8 821 (<0.01)	9/16 943 (0.05)	4	13/25 764	0.05%
SE33 (ACTBP2)	0/330 (<0.30)	330/51 610 (0.64)	None reported	330/51 940	0.64%

genetic questions being asked. Any time a family reference sample is used to try and match recovered remains during a mass disaster or missing persons investigation, mutations become an important issue because an exact match cannot be made when a mutation is present.

High mutation rates help keep STR markers polymorphic and therefore useful in human identity testing. It is important to keep in mind that while mutations can potentially impact kinship reference samples they will not affect direct matches between personal effects and victims or perpetrators and crime scene evidence since any mutation that occurs will be consistent over an individual's lifetime.

REFERENCES AND ADDITIONAL READING

Alves, C., Gusmao, L., Pereira, L. and Amorim, A. (2003) *Progress in Forensic Genetics, 9*, International Congress Series 1239, 131–135.

Applied Biosystems (1999) *AmpFlSTR® SGM Plus™ PCR Amplification Kit User's Manual*, P/N 4309589. Foster City, California: Applied Biosystems.

Bacher, J. and Schumm, J.W. (1998) *Profiles in DNA*, 2 (2), pp. 3–6. Madison, Wisconsin: Promega Corporation.

Bacher, J.W., Hennes, L.F., Gu, T., Tereba, A., Micka, K.A., Sprecher, C.J., Lins, A.M., Amiott, E.A., Rabbach, D.R., Taylor, J.A., Helms, C., Donis-Keller, H. and Schumm, J.W. (1999) *Proceedings of the Ninth International Symposium on Human Identification*, pp. 24–37. Madison, Wisconsin: Promega Corporation.

Brinkmann, B., Klintschar, M., Neuhuber, F., Huhne, J. and Rolf, B. (1998) *American Journal of Human Genetics*, 62, 1408–1415.

Brownstein, M.J., Carpten, J.D. and Smith, J.R. (1996) *BioTechniques*, 20, 1004–1010.

Budowle, B. and Sprecher, C.J. (2001) *Journal of Forensic Sciences*, 46, 637–641.

Budowle, B., Masibay, A., Anderson, S.J., Barna, C., Biega, L., Brenneke, S., Brown, B.L., Cramer, J., DeGroot, G.A., Douglas, D., Duceman, B., Eastman, A., Giles, R., Hamill, J., Haase, D.J., Janssen, D.W., Kupferschmid, T.D., Lawton, T., Lemire, C., Llewellyn, B., Moretti, T., Neves, J., Palaski, C., Schueler, S., Sgueglia, J., Sprecher, C., Tomsey, C. and Yet, D. (2001) *Forensic Science International*, 124, 47–54.

Butler, J.M. and Becker, C.H. (2001) *Improved analysis of DNA short tandem repeats with time-of-flight mass spectrometry*. Washington, DC: National Institute of Justice. Available online at: http://www.ojp.usdoj.gov/nij/pubs-sum/188292.htm.

Butler, J.M., Buel, E., Crivellente, F. and McCord, B.R. (2004) *Electrophoresis*, 25, 1397–1412.

Chakraborty, R., de Andrade, M., Daiger, S.P. and Budowle, B. (1992) *Annuals of Human Genetics*, 56, 45–57.

Clark, J.M. (1988) *Nucleic Acids Research*, 16, 9677–9686.

Clayton, T.M., Hill, S.M., Denton, L.A., Watson, S.K. and Urquhart, A.J. (2004) *Forensic Science International*, 139, 255–259.

Cotton, E.A., Allsop, R.F., Guest, J.L., Frazier, R.R., Koumi, P., Callow, I.P., Seager, A. and Sparkes, R.L. (2000) *Forensic Science International*, 112, 151–161.

Crouse, C., Rogers, S., Amiott, E., Gibson, S. and Masibay, A. (1999) *Journal of Forensic Sciences*, 44, 87–94.

Drábek, J., Chung, D.T., Butler, J.M. and McCord, B.R. (2004) *Journal of Forensic Sciences,* 49 (4), 859–860.

Edwards, A., Civitello, A., Hammond, H.A. and Caskey, C.T. (1991) *American Journal of Human Genetics*, 49, 746–756.

Ellegren, H. (2004) *Nature Reviews Genetics*, 5, 435–445.

Gill, P., Urquhart, A., Millican, E.S., Oldroyd, N.J., Watson, S., Sparkes, R. and Kimpton, C.P. (1996) *International Journal of Legal Medicine*, 109, 14–22.

Ginot, F., Bordelais, I., Nguyen, S. and Gyapay, G. (1996) *Nucleic Acids Research*, 24, 540–541.

Hauge, X.Y. and Litt, M. (1993) *Human Molecular Genetics*, 2, 411–415.

Henke, J. and Henke, L. (1999) *American Journal of Human Genetics*, 64, 1473.

Holt, C.L., Buoncristiani, M., Wallin, J.M., Nguyen, T., Lazaruk, K.D. and Walsh, P.S. (2002) *Journal of Forensic Sciences*, 47, 66–96.

Kimpton, C.P., Gill, P., Walton, A., Urquhart, A., Millican, E.S. and Adams, M. (1993) *PCR Methods and Applications*, 3, 13–22.

Kline, M.C., Jenkins, B. and Rogers, S. (1998) *Journal of Forensic Sciences*, 43, 250.

Krenke, B.E., Tereba, A., Anderson, S.J., Buel, E., Culhane, S., Finis, C.J., Tomsey, C.S., Zachetti, J.M., Masibay, A., Rabbach, D.R., Amiott, E.A. and Sprecher, C.J. (2002) *Journal of Forensic Sciences*, 47, 773–785.

Lazaruk, K., Wallin, J., Holt, C., Nguyen, T. and Walsh, P.S. (2001) *Forensic Science International*, 119, 1–10.

Leibelt, C., Budowle, B., Collins, P., Daoudi, Y., Moretti, T., Nunn, G., Reeder, D. and Roby, R. (2003) *Forensic Science International*, 133, 220–227.

Lins, A.M., Micka, K.A., Sprecher, C.J., Taylor, J.A., Bacher, J.A., Rabbach, D.R., Bever, R.A., Creacy, S.D. and Schumm, J.W. (1998) *Journal of Forensic Sciences*, 43, 1178–1190.

Magnuson, V.L., Ally, D., Nylund, S.J., Karanjawala, Z.E., Rayman, J.B., Knapp, J.I., Lowe, A.L., Ghosh, S. and Collins, F.S. (1996) *BioTechniques*, 21, 700–709.

Meldgaard, M. and Morling, N. (1997) *Electrophoresis*, 18, 1928–1935.

Mornhinweg, E., Luckenbach, C., Fimmers, R. and Ritter, H. (1998) *Forensic Science International*, 95, 173–178.

Nadir, E., Margalit, H., Gallily, T. and Ben-Sasson, S.A. (1996) *Proceedings of the National Academy of Sciences USA*, 93, 6470–6475.

Nelson, M.S., Levedakou, E.N., Matthews, J.R., Early, B.E., Freeman, D.A., Kuhn, C.A., Sprecher, C.J., Amin, A.S., McElfresh, K.C. and Schumm, J.W. (2002) *Journal of Forensic Sciences*, 47, 345–349.

Puers, C., Hammond, H.A., Jin, L., Caskey, C.T. and Schumm, J.W. (1993) *American Journal of Human Genetics*, 53, 953–958.

Sajantila, A., Lukka, M. and Syvanen, A.-C. (1999) *European Journal of Human Genetics*, 7, 263–266.

Schumm, J.W., Lins, A.M., Micka, K.A., Sprecher, C.J., Rabbach, D.R. and Bacher, J.W. (1996) *Proceeding of the Seventh International Symposium on Human Identification*, pp. 70–88. Madison, Wisconsin: Promega Corporation.

Szibor, R., Lautsch, S., Plate, I., Bender, K. and Krause, D. (1998) *International Journal of Legal Medicine*, 111, 160–161.

Thomson, J.A., Pilotti, V., Stevens, P., Ayres, K.L. and Debenham, P.G. (1999) *Forensic Science International*, 100, 1–16.

Urquhart, A., Chiu, C.T., Clayton, T., Downes, T., Frazier, R., Jones, S., Kimpton, C., Lareu, M.V., Millican, E., Oldroyd, N., Thompson, C., Watson, S., Whitaker, J. and Gill, P. (1994) *Proceedings of the Fifth International Symposium on Human Identification*, pp.73–83. Madison, Wisconsin: Promega Corporation.

Wallin, J.M., Holt, C.L., Lazaruk, K.D., Nguyen, T.H. and Walsh, P.S. (2002) *Journal of Forensic Sciences*, 47, 52–65.

Walsh, P.S., Fildes, N.J. and Reynolds, R. (1996) *Nucleic Acids Research*, 24, 2807–2812.

Walsh, P.S. (1998) *Journal of Forensic Sciences*, 43, 1103–1104.

Watson, J.D. and Berry, A. (2003) *DNA: The Secret of Life*. New York: Knopf.

Weber, J.L. and Wong, C. (1993) *Human Molecular Genetics*, 2, 1123–1128.

Wiegand, P., Meyer, E. and Brinkmann, B. (2000) *Electrophoresis*, 21, 889–895.

CHAPTER 7

FORENSIC ISSUES: DEGRADED DNA, PCR INHIBITION, CONTAMINATION, MIXED SAMPLES AND LOW COPY NUMBER

It is of the highest importance in the art of detection to be able to recognize out of a number of facts, which are incidental and which vital. Otherwise your energy and attention must be dissipated instead of being concentrated.

(Sherlock Holmes, *The Reigate Puzzle*)

UNIQUE NATURE OF FORENSIC SAMPLES

A forensic DNA laboratory often has to deal with DNA samples that are less than ideal. The biological material serving as evidence of a crime may have been left exposed to a harsh environment for days, months, or even years, such as in the case of an investigation into a missing person. The victims of homicides are typically taken to out of the way places where they remain until their bodies are discovered. Instead of being preserved in a freezer away from caustic chemicals that can break it down, the DNA molecules may have been left in direct sunlight or in damp woods.

Regardless of the situation, the DNA molecules from a crime scene come from a less than pristine environment that is normally found in molecular biology laboratories. Just as important is the fact that the retrieved biological sample may be limited in quantity. Thus, accurate sample analysis is critical since a forensic scientist may only obtain enough evidence for one attempt at analysis. In this chapter, we explore the forensic issues surrounding the analysis of short tandem repeats (STRs) including handling degraded DNA samples, avoiding contamination, overcoming polymerase chain reaction (PCR) inhibition, and interpreting mixtures, which are prevalent in forensic cases especially those involving sexual assault. In addition, we discuss the use of low-copy number DNA profiling to retrieve genetic information from samples with only a few cells available for testing.

DEGRADED DNA

Environmental exposure degrades DNA molecules by randomly breaking them into smaller pieces. Enemies to the survival of intact DNA molecules include water and enzymes called nucleases that chew up DNA. Both are ubiquitous in nature. With older technologies such as restriction fragment length polymorphism (RFLP), these severely degraded DNA samples would have been very difficult if not impossible to analyze. High molecular weight DNA molecules need to be present in the sample in order to detect large VNTR (variable number of tandem repeats) alleles (e.g., 20 000 bp) with RFLP techniques.

An ethidium-bromide stained agarose 'yield gel' may be run to evaluate the quality of a DNA sample. Typically high molecular weight, high quality genomic DNA runs as a relatively tight band of approximately 20 000 bp relative to an appropriate molecular weight marker. On the other hand, degraded DNA appears as a smear of DNA that is much less than 20 000 bp in size (Figure 7.1a).

Modern-day PCR methods, such as multiplex STR typing, are powerful because miniscule amounts of DNA can be measured by amplifying them to a level where they may be detected. Less than 1 ng of DNA can now be analyzed with multiplex PCR amplification of STR alleles compared to 100 ng or more that might have been required with RFLP only a few years ago. However, this sensitivity to low levels of DNA also brings the challenge of avoiding contamination from the police officer or crime scene technician who collects the biological evidence.

In order for PCR amplification to occur, the DNA template must be intact where the two primers bind as well as between the primers so that full extension can occur. Without an intact DNA strand that surrounds the STR repeat region to serve as a template strand, PCR will be unsuccessful because primer extension will halt at the break in the template. The more degraded a DNA sample becomes, the more breaks occur in the template and fewer and fewer DNA molecules contain the full length needed for PCR amplification.

BENEFITS OF STR MARKERS WITH DEGRADED DNA SAMPLES

Fortunately, because STR loci can be amplified with fairly small product sizes, there is a greater chance for the STR primers to find some intact DNA strands for amplification. In addition, the narrow size range of STR alleles benefits analysis of degraded DNA samples because allele dropout via preferential amplification of the smaller allele is less likely to occur since both alleles in a heterozygous individual are similar in size.

A number of experiments have shown that there is an inverse relationship between the size of the locus and successful PCR amplification from degraded DNA samples, such as those obtained from a crime scene or a mass disaster (Whitaker *et al.* 1995, Sparkes *et al.* 1996, Takahashi *et al.* 1997,

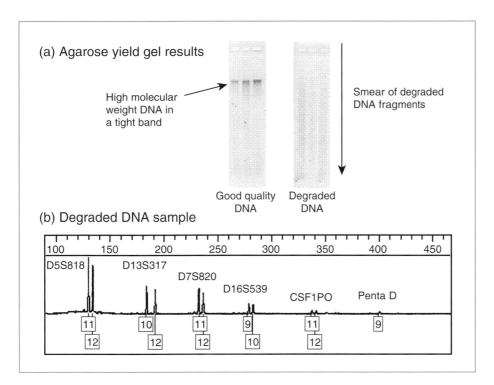

(a) Agarose yield gel results

High molecular weight DNA in a tight band

Smear of degraded DNA fragments

Good quality DNA Degraded DNA

(b) Degraded DNA sample

100   150   200   250   300   350   400   450

D5S818   D13S317
    D7S820
        D16S539
            CSF1PO   Penta D

11   10   11   9   11   9
12   12   12   10   12

Figure 7.1

Impact of degraded DNA on (a) agarose yield gel results and (b) STR typing. (a) Degraded DNA is broken up into small pieces that appear as a smear on a scanned yield gel compared to good quality DNA possessing intact high molecular weight DNA. (b) Signal strength is generally lost with larger size PCR products when STR typing is performed on degraded DNA, such as is shown from the green dye-labeled loci in the PowerPlex 16 kit. Thus, 180 bp D13S317 PCR products have a higher signal than 400 bp Penta D amplicons because more DNA molecules are intact in the 200 bp versus the 400 bp size range.

Schneider *et al.* 2004). The STR loci with larger sized amplicons in a multiplex amplification, such as D18S51 and FGA, are the first to drop out of the DNA profile when amplifying extremely degraded DNA samples (see Figure 7.1).

In one of the first studies demonstrating the value of multiplex STR analysis with degraded DNA samples, the Forensic Science Service was able to successfully type a majority of 73 duplicate pathological samples obtained from the Waco disaster with four STR markers (Whitaker *et al.* 1995). They observed no allele dropout and obtained concordant results on all samples where alleles were scored. A correlation was observed between successful typing at a locus and the average length of the alleles at that locus. The FES/FPS locus, which has alleles in the size range of 212–240 bp, only yielded 91 successful amplifications while the VWA locus with alleles ranging from 130–169 bp had 115 successful amplifications. Thus, loci with the larger alleles failed first. In addition, amelogenin amplicons (106 or 112 bp) were obtained on all 24 samples examined as part of the Waco identification program.

The potential for analysis of degraded DNA samples is an area where multiplex STR systems really shine over previously used DNA markers. STRs are more sensitive than single-locus probe RFLP methods, less prone to allelic dropout than VNTR systems (AmpFLPs) such as D1S80, and more discriminating than other PCR-based typing methods, such as HLA-DQA1 and AmpliType PolyMarker.

THE USE OF REDUCED SIZED PCR PRODUCTS (miniSTRs)

In an article entitled 'Less is more – length reduction of STR amplicons using redesigned primers,' Wiegand and Kleiber (2001) demonstrated that highly degraded DNA as well as very low amounts of DNA could be more successfully typed using some new redesigned PCR primers that were close to the STR repeat compared to the established sequences that generated longer amplicons for the same loci. STR loci used in commercially available kits can extend past 400 bp in size (see Table 5.5). Most of this length however comes from flanking sequence surrounding the STR repeat of interest. PCR primers for larger sized STR markers have been moved away from the repeat region that imparts variability to the locus in order to fit into a desired size range for a particular multiplex assay (e.g., Krenke *et al.* 2002).

For example, the two PCR primers used for the PowerPlex 16 locus Penta D anneal 71 bp upstream and 247 bp downstream of the core AAAGA repeat. Amplification with these PCR primers generates amplicons in the size range of 376–449 bp with alleles ranging from 2.2–17 repeats (Krenke *et al.* 2002). When primers are brought to within 11 bp upstream and 19 bp downstream of the repeat region, the overall PCR product sizes drop by 282 bp to a range of 94–167 bp for alleles 2.2–17 (Butler *et al.* 2003). Figure 7.2 illustrates this size reduction principle when creating reduced size STR amplicons or 'miniSTRs.' It is important to keep in mind that some loci can be reduced in size more than others (Table 7.1).

Several disadvantages do exist for miniSTRs. A major disadvantage is that only a few loci can be simultaneously amplified in a multiplex because the size aspect has been removed. Large multiplex assays like PowerPlex 16 pack four or more loci into a single dye color by shifting primers away from the repeat region to make larger PCR products. The 'miniplexes' created for amplifying miniSTRs have primers that are as close as possible to the repeat region and therefore typically only have one locus per dye color because all of the loci are about the same general size range of ~100 bp (Butler *et al.* 2003).

Due to the fact that different PCR primers are in use with miniSTRs compared to conventional STR megaplexes, it is important that concordance studies be performed to verify that allele dropout from primer binding site mutations is rare or non-existent. This is performed by examining the genotyping results to see if they are the same between the primer sets (see Chapter 6). Occasionally a point mutation or an insertion or deletion may occur in the flanking region *outside* of a miniSTR primer binding site which can lead to a

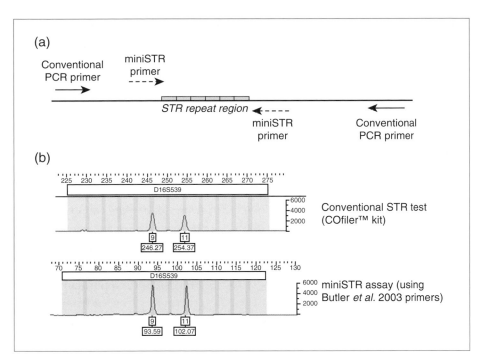

Figure 7.2

(a) MiniSTRs or reduced sized amplicons for STR typing are created by designing PCR primers that anneal closer to the repeat region than conventional STR kit primers. (b) PCR product sizes, such as demonstrated here with D16S539, can be reduced by over 150 bp relative to conventional tests. MiniSTR assays can produce the same typing result as those from larger STR amplicons produced by kits often with greater success on degraded DNA samples.

Table 7.1

PCR product size reduction obtained with new primers in several miniSTR studies.

Reference	Locus	miniSTR size (alleles range)	Size reduction over kit or previous primers*
Hellmann *et al.* (2001)	TH01	61–85 bp (alleles 5–10)	–103 bp
	TPOX	58–86 bp (alleles 6–13)	–157 bp
	FES/FPS	81–105 bp (alleles 8–14)	–132 bp
Tsukada *et al.* (2002)	TH01	74–98 bp (alleles 5–11)	–90 bp
	TPOX	107–135 bp (alleles 6–13)	–110 bp
	CSF1PO	90–122 (alleles 7–15)	–194 bp
	VWA	99–143 bp (alleles 10–21)	–53 bp
Butler *et al.* (2003)	TH01	51–98 bp (alleles 3–14)	–105 bp
	TPOX	65–101 bp (alleles 5–14)	–148 bp
	CSF1PO	89–129 bp (alleles 6–16)	–191 bp
	VWA	88–148 bp (alleles 10–15)	–64 bp
	FGA	125–281 bp (alleles 12.2–51.2)	–71 bp
	D3S1358	72–120 bp (alleles 8–20)	–25 bp
	D5S818	81–117 bp (alleles 7–16)	–53 bp
	D7S820	136–176 bp (alleles 5–15)	–117 bp
	D8S1179	86–134 bp (alleles 7–19)	–37 bp
	D13S317	88–132 bp (alleles 5–16)	–105 bp
	D16S539	81–121 bp (alleles 5–15)	–152 bp
	D18S51	113–193 bp (alleles 7–27)	–151 bp
	D21S11	153–211 bp (alleles 24–38.2)	–33 bp
	Penta D	94–167 bp (alleles 2.2–17)	–282 bp
	Penta E	80–175 bp (alleles 5–24)	–299 bp
	D2S1338	90–142 bp (alleles 15–28)	–198 bp

*Comparisons between various studies were adjusted to be against the same 'kit or previous primers' (usually the AmpF/STR kits).

problematic (and undetectable) difference in a heterozygous allele call (see Butler *et al.* 2003, Drabek *et al.* 2004).

Regardless of these disadvantages, it is likely that miniSTRs will play a role in the future of degraded DNA analysis probably to help recover information that has been lost with larger loci from conventional megaplex amplification. With DNaseI-digested DNA, miniSTR loci performed better than loci from a commercial STR kit (Chung *et al.* 2004). As will be described in Chapter 24, reduced size STR assays have helped make possible some of the World Trade Center victim identifications from burned and damaged bone samples (Schumm *et al.* 2004). Even telogen hair shafts, which contain very little nuclear DNA, have been successfully typed using reduced size STR amplicons (Hellman *et al.* 2001). New STR loci besides the CODIS markers and others that are currently used in forensic DNA typing are also being examined as potential miniSTR systems with a focus on loci possessing small alleles and narrow size ranges (Ohtaki *et al.* 2002, Coble and Butler 2005). Thus, a battery of additional assays should be available to aid researchers and forensic practitioners in the future when working with degraded DNA specimens.

PCR INHIBITION

Another important challenge to amplifying DNA samples from crime scenes is the fact that the PCR amplification process can be affected by inhibitors present in the samples themselves. Outdoor crimes may leave body fluids such as blood and semen on soil, sand, wood, or leaf litter that contain substances which may co-extract with the perpetrator's DNA and prevent PCR amplification. Textile dyes, leather, and wood from interior crime scenes may also contain DNA polymerase inhibitors.

Inhibitors can (1) interfere with the cell lysis necessary for DNA extraction, (2) interfere by nucleic acid degradation or capture, and (3) inhibit polymerase activity thus preventing enzymatic amplification of the target DNA (Wilson 1997). Occasionally substances such as textile dyes from clothing or hemoglobin from red blood cells can remain with the DNA throughout the sample preparation process and interfere with the polymerase to prevent successful PCR amplification (Akane *et al.* 1994, DeFanchis *et al.* 1988, Rådström *et al.* 2004).

The result of amplifying a DNA sample containing an inhibitor such as hematin is a loss of the alleles from the larger sized STR loci or even complete failure of all loci. Some example inhibitors that interfere with PCR amplification are listed in Table 7.2. Samples containing PCR inhibitors often produce partial profile results that look similar to a degraded DNA sample (see Applied Biosystems 1998). Thus, failure to amplify the larger STR loci for a sample can be either due to degraded DNA where there are not enough intact copies of the DNA template or due to the presence of a sufficient level of PCR inhibitor that

Possible Forensic Source	PCR Inhibitor	Reference
Blood	Heme (hematin)	Akane *et al.* (1994)
Tissue and Hair	Melanin	Eckhart *et al.* (2000)
Feces	Polysaccharides	Monteiro *et al.* (1997)
Feces	Bile salts	Lantz *et al.* (1997)
Soil	Humic compounds	Tsai and Olson (1992)
Urine	Urea	Mahony *et al.* (1998)
Blue jeans	Textile dyes (denim)	Shutler *et al.* (1999)

Table 7.2

Summary of some PCR inhibitors. Some of these inhibitors are removed during routine DNA extraction methods, such as Chelex (see Chapter 3), while others may need additional solutions (see Rådström et al. 2004). Most inhibitors bind to and interfere with polymerase activity.

reduces the activity of the polymerase. Reduced size STR amplicons can aid in recovery of information from a sample that is inhibited since smaller PCR products may be amplified more efficiently than larger ones.

SOLUTIONS TO PCR INHIBITION

A nice review of strategies to generate PCR-compatible samples was published recently (Rådström *et al.* 2004, see also Wilson 1997). PCR inhibitors may be removed or their effects reduced by one or more of the following solutions. The genomic DNA template may be diluted, which also dilutes the PCR inhibitor, and re-amplified in the presence of less inhibitor. Alternatively, more DNA polymerase can be added to overcome the inhibitor. With this approach some fraction of the *Taq* polymerase binds to the inhibiting molecule(s) and removes them from the reaction so that the rest of the *Taq* can do its job and amplify the DNA template. In addition, polymerases other than *Taq* have been shown to work well with blood and feces, which typically inhibit PCR when performed with *Taq* DNA polymerase (Al-Soud and Rådström 1998).

Additives to the PCR reaction, such as bovine serum albumin (BSA) (Comey *et al.* 1994) or betaine (Al-Soud and Rådström 2000), have been shown to prevent or minimize the inhibition of PCR. More recently a sodium hydroxide treatment of DNA has been shown to neutralize inhibitors of *Taq* polymerase (Bourke *et al.* 1999). The addition of aluminum ammonium sulfate proved helpful to prevent the co-purification of inhibitors with DNA from soil samples (Braid *et al.* 2003). Finally, a separation step may be performed prior to PCR to separate the extracted DNA from the inhibiting compound. Centricon-100 and Microcon-100 filters have been used for this purpose (Comey *et al.* 1994) as have low-melt agarose gel plugs (Moreira 1998).

CONTAMINATION ISSUES

The sensitivity of PCR with its ability to amplify low quantities of DNA can be a problem if proper care is not taken. Validated laboratory protocols must be adhered to so that contamination from higher concentrations of DNA, such as those of the DNA analyst, can be avoided. However, it is important to keep in mind that if contamination does occur, it will most likely result in an 'exclusion' or 'inconclusive' result and be in favor of the defendant (see Chapter 15).

Contamination implies the accidental transfer of DNA. There are three potential sources of contamination when performing PCR: sample contamination with genomic DNA from the environment, contamination between samples during preparation, and contamination of a sample with amplified DNA from a previous PCR reaction (Lygo *et al.* 1994). The first source of contamination is largely dependent on sample collection at the crime scene and the care taken there by the evidence collection team (see Chapter 3). Environment contamination can be monitored only in a limited sense by 'substrate controls' (Gill 1997). The latter two sources of contamination can be controlled and even eliminated by using appropriate laboratory procedures and designated work areas (see Chapter 4).

The possibility of laboratory contamination is addressed with 'negative controls' that test for contamination of PCR reagents and tubes. Basically, a negative control involves running a blank sample through the entire process in parallel with the forensic case evidence. The same volume of purified water as DNA template in the other samples is added to a negative control PCR reaction. If any detectable PCR products are observed in the negative control, then sources of contamination should be sought out and eliminated before proceeding further.

In a systematic examination of possible sources of PCR contamination, Henry Lee and co-workers found that under the circumstances normally encountered during casework analysis, PCR contamination was never noted (Scherczinger *et al.* 1999). Using reverse dot blot detection of AmpliType PM and DQA1 systems (the same PCR principles apply as with STR markers detected using fluorescence), they examined four general aspects of PCR during which contamination might occur: PCR amplification setup, handling the PCR products, aerosolization, and DNA storage. In these studies, detectable contamination occurred only when gross deviations from basic preventative protocols were employed. They concluded that contamination could not be generated by simple acts of carelessness (Scherczinger *et al.* 1999).

The genotypes of laboratory personnel are typically stored for comparison purposes so that any contamination by an individual in the lab can be picked up. This 'staff elimination database' is often searched prior to concluding that a generated DNA profile accurately reflects the evidence and is not due to contamination from laboratory personnel (Howitt *et al.* 2003). Only in the case of gross laboratory error, would a mixture of two DNA types result (i.e., that of the analyst and the sample being examined). These types of errors would most likely be sorted out by comparing the result with genotypes of laboratory personnel when the mixture analysis is performed (see discussion on mixtures below). Some expert system software has been developed to aid detection of potential cross-contamination within a batch of samples (see Chapter 17).

SPORADIC CONTAMINATION OF PCR TUBES

The Forensic Science Service recently noted that sporadic contamination of consumables used in DNA testing, such as the small tubes in which the PCR amplification is performed, can introduce extraneous DNA profiles (Howitt *et al.* 2003). The FSS observed 11 casework-contaminating DNA profiles in running over one million samples in a three-year period. These contaminating profiles were identified in negative controls and quality control testing of select tubes examined to indicate general batch quality. An analysis of anonymous samples provided by 300 employees of the tube manufacturer revealed complete matches to 10 of the 11 identified casework-contaminating DNA profiles (Howitt *et al.* 2003). Communication with the tube manufacturer resulted in implementation of additional anti-contamination measures in the production process, which has significantly lowered the number of contamination events attributable to their manufacturing staff.

The use of negative control log information and elimination databases for forensic laboratory personnel, manufacturing staff of the suppliers of DNA consumables, and police officers handling crime scene evidence help identify potential sources of contaminating DNA. In addition, increased used of robotics for process automation (see Chapter 17) help reduce the risk of contamination occurring since people have less direct contact with samples in the laboratory.

IMPACT OF CONTAMINATION ON CASEWORK

Issues surrounding the impact of contamination on casework reporting guidelines were explored by Peter Gill and Amanda Kirkham in a recent article entitled 'Development of a simulation model to assess the impact of contamination in casework using STRs' (Gill and Kirkham 2004). Negative controls were used to predict the level of overall contamination in an operational DNA unit. However, because PCR contamination can be tube-specific (see above), negative controls run with a batch of samples cannot provide complete confidence that the associated batch of extracted casework material is contaminant-free

(Gill and Kirkham 2004). This study concludes that the most likely outcome of a contamination event is false exclusion because contaminating DNA material can be preferentially amplified over extremely low levels of original material present from the casework sample or may mask the perpetrator's profile in a resulting mixture.

While this contamination possibility might only rarely impact a careful forensic DNA laboratory, it can have potential significance on old cases under review including the Innocence Project (see D.N.A. Box 1.2). For example, if biological evidence from a 20-year-old case was handled by ungloved police officers or evidence custodians (prior to knowledge regarding the sensitivity of modern DNA testing), then the true perpetrator's DNA might be masked by contamination from the collecting officer. Thus, when a DNA test is performed, the police officer's or evidence custodian's DNA would be detected rather than the true perpetrator. In the absence of other evidence, the individual in prison might then be falsely declared 'innocent' because his DNA profile was not found on the original crime scene evidence. This scenario emphasizes the importance of considering DNA evidence as an investigative tool within the context of a case rather than the sole absolute proof of guilt or innocence.

To reduce contamination problems during the laboratory examination of DNA samples, all pre-PCR and post-PCR amplification reactions should be kept physically separate (see Chapter 4). Laboratory contamination is probably impossible to avoid completely but can be proactively assessed with negative controls and staff elimination databases (Gill and Kirkham 2004). However, if the original DNA sample has been contaminated during the collection processes (see Chapter 3), then the DNA profile would be a mixture, and the results would need to be interpreted as will be described in the next section.

MIXTURES

Mixtures arise when two or more individuals contribute to the sample being tested. Mixtures can be challenging to detect and interpret without extensive experience and careful training. As detection technologies have become more sensitive with PCR sensitivity coupled with fluorescent measurements, the ability to see minor components in the DNA profile of mixed samples has improved dramatically over what was available with RFLP methods only a few years ago. Likewise, the theoretical aspects of statistical calculations for mixture interpretation have been examined more thoroughly (Curran *et al.* 1999). Some statistical approaches to reporting a mixture result will be discussed in Chapter 22.

VALUE OF HIGHLY POLYMORPHIC MARKERS IN DECIPHERING MIXTURES

The probability that a mixture will be detected improves with the use of more loci and genetic markers that have a high incidence of heterozygotes. The detectability of multiple DNA sources in a single sample relates to the ratio of DNA present from each source, the specific combinations of genotypes, and the total amount of DNA amplified. In other words, some mixtures will not be as easily detectable as other mixtures.

Using highly polymorphic STR markers with more possible alleles translates to a greater chance of seeing differences between the two components of a mixture. For example, D18S51 has 51 possible alleles while TPOX only has 15 known alleles (see Appendix I) making D18S51 a more useful marker for detecting mixtures. Likewise, the more markers examined (e.g., by using a multiplex STR amplification), the greater the chance to observe multiple components in a mixture.

The quantity of each component in a mixture makes a difference in the ability to detect all contributors to the mixed sample. For example, if the two DNA sources are in similar quantities they will be much easier to detect than if one is present at only a fraction of the other. The minor component of a mixture is usually not detectable for mixture ratios below the 5% level or 1:20 (Cotton 1995, Applied Biosystems 1998). The AmpF/STR kits recommend that the minor component be above 35 pg in quantity to obtain a reliable genotype result (Applied Biosystems 1998).

QUANTITATIVE INFORMATION FROM FLUORESCENCE MEASUREMENTS

The ability to obtain quantitative information from peaks in an electropherogram, either using the ABI 310, 377, or 3100 platform or a fluorescence scanner such as the FMBIO II (see Chapter 14), now permits relative peak heights or areas of STR alleles to be measured. This peak information can then be used to decipher the possible genotypes of the contributors to the mixed sample. Due to peak shape variation, peak areas have been advocated as being superior to peak heights when comparing allele peak information (Gill *et al.* 1998a). However, peak heights are successfully used in many laboratories for mixture interpretation.

Figure 7.3 illustrates how typical single-source samples differ from mixed samples in their STR profiles. STR allele peak patterns for heterozygous samples will generally have stutter products that are less than 15% of the associated allele peak height/area. In addition, the peak height ratio, as measured by dividing the height of the lower quantity peak in relative fluorescence units by the height of the higher quantity allele peak, should be greater than approximately

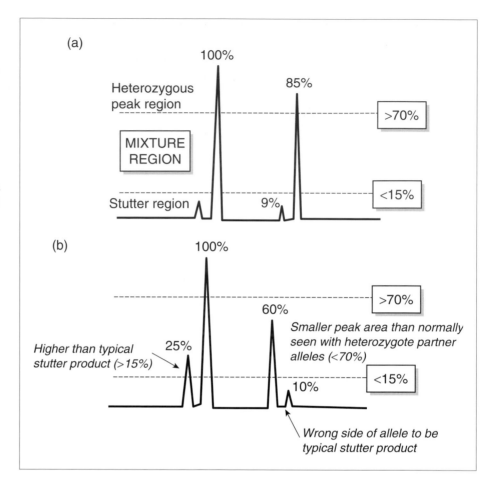

70% in a single-source sample (Gill *et al.* 1997, Applied Biosystems 1998). Thus, if peaks fall in the region between 15% and 70% of the highest peak at a particular STR locus, a mixed sample that has resulted from two or more contributors is probable. The observation of three or more alleles at multiple loci is also a strong indicator of the presence of a mixture.

DISTINGUISHING GENOTYPES IN A MIXED SAMPLE

Several clues exist to help determine that a mixture is present. Answers to the following questions can help ascertain the genotypes that make up the composite DNA profile of the mixture:

- Do any of the loci show more than two peaks in the expected allele size range?
- Is there a severe peak height imbalance between heterozygous alleles at a locus?
- Does the stutter product appear abnormally high (e.g., >15–20%)?

If the answer to any one of these three questions is yes, then the DNA profile may very well have resulted from a mixed sample. Mixture interpretation has been examined extensively by the Forensic Science Service (Clayton *et al.* 1998, Gill *et al.* 1998a, 1998b) and many of their strategies have been incorporated into this section's material.

Usually a mixture is first identified by the presence of three or more prominent peaks at one or more loci. At a single locus, a sample containing DNA from two sources can exhibit one, two, three, or four peaks due to the possible genotype combinations listed below.

Four peaks:
- heterozygote + heterozygote, no overlapping alleles (genotypes are unique).

Three peaks:
- heterozygote + heterozygote, one overlapping allele.
- heterozygote + homozygote, no overlapping alleles (genotypes are unique).

Two peaks:
- heterozygote + heterozygote, two overlapping alleles (genotypes are identical).
- heterozygote + homozygote, one overlapping allele.
- homozygote + homozygote, no overlapping alleles (genotypes are unique).

Single peak
- homozygote + homozygote, overlapping allele (genotypes are identical).

When two contributors to a mixed stain share one or more alleles, the alleles are 'masked' and the contributing genotypes may not be easily decipherable. For example, if two individuals at the FGA locus have genotypes 23,24 and 24,24, then a mixture ratio of 1:1 will produce a ratio of 1:3 for the 23:24 peak areas. In this particular case, the mixture could be interpreted as a homozygous allele with a large stutter product without further information. However, by examining the STR profiles at other loci that have unshared alleles, i.e., three or four peaks per locus, this sample may be able to be dissected properly into its components.

In an effort to see if it was possible for masking to occur at every locus in a multiplex, the Forensic Science Service conducted a simulated mixture study with 120 000 individual STR profiles in their Caucasian database (Gill *et al.* 1997). They found that the vast majority of these artificial mixtures showed 15–22 peaks across a six-plex STR marker multiplex. The maximum number in a mixture of two heterozygous individuals with no overlapping alleles at six STRs would be 24 peaks. Thus, in this example with unrelated individuals, simple mixtures can be identified by the presence of three or more alleles at several loci. Out of more than 212 000 pairwise comparisons, there were only four examples where one or two alleles were observed at each locus in the six-plex, and these could be designated mixtures because of peak imbalances (Gill *et al.* 1997).

MIXTURE INTERPRETATION

This next section will review the principles described by Gill *et al.* (1998b) and Clayton *et al.* (1998) for interpreting mixed forensic stains using STR typing results. Their six primary steps for interpreting mixtures are outlined in Figure 7.4. The interpretation steps will first be discussed and then an example mixture will be reviewed to put these steps into the context of a real sample.

An understanding of how non-mixtures behave is essential to being able to proceed with mixture interpretation. Mixed DNA profiles need to be interpreted against a background of biological and technological artifacts. Chapter 6 reviewed some of the prominent biological artifacts that exist for STR markers.

Figure 7.4

Steps in the interpretation of mixtures (Clayton et al. *1998).*

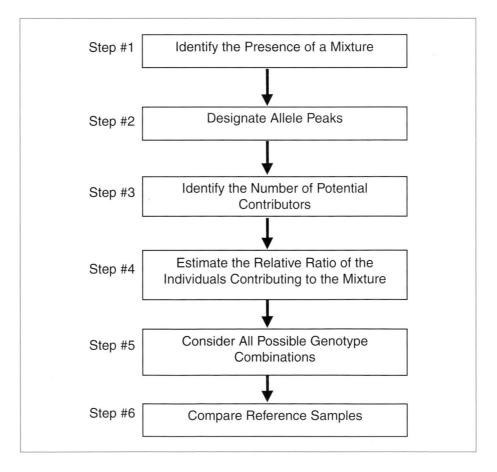

These include stutter products and null alleles. In addition, chromosomal abnormalities, such as tri-allelic (three-banded) patterns resulting from trisomy (the presence of three chromosomes instead of the normal two) or duplication of specific chromosomal regions can occur. In addition, non-specific amplification products can occasionally happen and must be considered prior to making an attempt to decipher a mixed profile. Issues surrounding technological artifacts from fluorescence detection will be covered in Chapters 13 and 15.

Stutter products represent the greatest challenge in confidently interpreting a mixture and designating the appropriate alleles. It is not always possible to exclude stutters since they are allelic products and differ from their associated allele by a single repeat unit. The general guideline for stutter identification of one repeat unit less than the corresponding allele and less than 15% of that allele's peak area is typically a useful one and can be used to mark suspected stutter products. The introduction of pentanucleotide repeat markers with stutter products of less than 1–2% has greatly simplified mixture interpretation (Bacher *et al.* 1999, Krenke *et al.* 2002).

After a mixture has been identified as such and all of the alleles have been called, the next step (Figure 7.4, step #3) is to identify the number of potential contributors. For a two-person mixture, the maximum number of alleles at any given locus is four if both individuals are heterozygous and there is no allele overlap. Thus, if more than four alleles are observed at a locus then a complex mixture consisting of more than two individuals is possible. Fortunately, the overwhelming majority of mixtures encountered in forensic casework involve two-person mixtures (Clayton *et al.* 1998).

Mixtures can range from equal proportions of each component to one component being greatly in excess. The varying proportions of a mixture are usually referred to in a ratio format (e.g., 1:1 or 1:5). Mixtures of known quantities of DNA templates have shown that the mixture ratio is approximately preserved during PCR amplification (Gill *et al.* 1998a, Perlin and Szabady 2001). Thus, the peak areas and heights observed in an electropherogram can in most cases be related back to the amount of DNA template components included in the mixed sample.

An approximate mixture ratio can be best determined by considering the profile as a whole and looking at all of the information from each locus. The ratio of mixture components is most easily determined when there are no shared alleles at a locus. Thus, it is best to first examine loci with four alleles as a starting point for estimating the relative ratio of the two individuals contributing to the mixture. Determining the ratio when there are shared alleles is more complex because there may be more than one possible combination of alleles that could explain the observed peak patterns (Clayton *et al.* 1998).

Table 7.3

Pairwise comparisons of two, three and four allele peak patterns. Reciprocal combinations are not shown.

Four Alleles (A,B,C,D)		Three Alleles (A,B,C)		Two Alleles (A,B)	
A,B	C,D	A,A	B,C	A,A	A,B
A,C	B,D	B,B	A,C	A,B	A,B
A,D	B,C	C,C	A,B	A,A	B,B
		A,B	A,C	A,B	B,B
		B,C	A,C		
		A,B	B,C		

The possible combinations of alleles for two, three, and four allele peak patterns are listed in Table 7.3. For four alleles at a locus, there are three possible pairwise comparisons that exist, if one does not worry about the reciprocal cases, i.e., which allele combinations belong to the minor contributor and which belong to the major contributor. For three alleles at a locus, there are six possible pairwise combinations and for two alleles at a locus there are four possible pairwise combinations (Table 7.3).

Amelogenin, the sex-typing marker, is an effective marker for deciphering the contributions of genetically normal male and female individuals. The predicted X and Y allele peak ratios for a number of possible male and female mixture ratios are listed in Table 7.4. The amelogenin X and Y peak areas are especially useful in determining whether the major contributor to the mixture is male or female.

The next step in examining a mixture is to consider all possible genotype combinations at each locus (Figure 7.4, step #5). Peaks representing the allele calls at each locus are labeled with the designations A, B, and so forth. The possible pairwise combinations from Table 7.3 are considered using the peak areas for each called allele. Each particular combination of alleles at the different loci is considered in light of the information determined previously regarding the mixture ratio for the sample under investigation (step #4). By stepping through each STR locus in this manner, the genotypes of the major and minor contributors to the mixture can be deciphered. In the example shown in Figure 7.5 and Table 7.5, some of these calculations are demonstrated.

The final step in the interpretation of a mixture is to compare the resultant genotype profiles for the possible components of the mixture with the genotypes of reference samples (Figure 7.4, step #6). In a sexual assault case, this reference sample could be the suspect and/or the victim. If the DNA profile from the suspect's reference sample matches the major or minor component of the mixture, then that person cannot be eliminated as a possible contributor to the mixed stain (Clayton *et al.* 1998).

Mixture Ratio		Allele Combination		Ratio of X:Y Peak Areas
Female (X,X)	Male (X,Y)	X	Y	X:Y
20	1	41	1	41:1
10	1	21	1	21:1
5	1	11	1	11:1
4	1	9	1	9:1
3	1	7	1	7:1
2	1	5	1	5:1
1	1	3	1	3:1
<u>1</u>	<u>2</u>	4	2	<u>2:1</u>
1	3	5	3	1.7:1
1	4	6	4	1.5:1
1	5	7	5	1.4:1
1	10	12	10	1.2:1
1	20	22	20	1.1:1

Table 7.4

Possible amelogenin X and Y allele peak ratios with varying quantities of DNA. The 2:1 peak area ratio for the X and Y alleles observed in Figure 7.6 and calculated in Table 7.5 matches a mixture ratio of one part female to two parts male as indicated by the numbers underlined. Thus, there is twice as much male DNA compared to female DNA in the mixed sample shown in Figure 7.5.

AN EXAMPLE MIXTURE

An example mixture will now be examined to demonstrate how the steps illustrated in Figure 7.4 may be used to interpret a mixture. Figure 7.5 shows a mixed sample that is a combination of male and female DNA, typical of what might be seen in a sexual assault investigation. The STR markers for the mixture are separated into three panels based on their dye label in order to visualize each STR locus more easily.

The first thing that is obvious in this example is the presence of more than two peaks at a majority of the loci. For example, D3S1358 contains four peaks and VWA has three peaks. There is also an imbalance in the X and Y alleles of the amelogenin sex-typing marker.

With the presence of a mixture established, the alleles are determined. Because there are a maximum of four peaks at any one locus, it is unlikely that there are more than two contributors to this example mixture. Each of the called alleles is labeled with a letter: 'A', 'B', 'C', or 'D'. These universal designations are used to track possible allele combinations through the rest of the mixture calculations.

Figure 7.5

Mixture of male and female DNA typical of what might be seen in a forensic case involving mixed samples. Profiler Plus™ multiplex STR data is displayed here with GeneScan® 3.1. Sample mixture indicators include an imbalance in the amelogenin X and Y alleles as well as greater than two peaks at multiple STR markers.

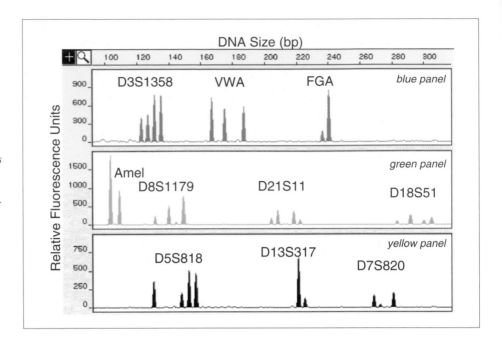

The relative ratio of the individuals contributing to this example mixture is then estimated by examining loci with four peaks. The green panel STR data from Figure 7.5 is shown in Figure 7.6 with labeled peak areas. There are four peaks present at both the D21S11 and D18S51 loci. Since the peak areas of the D21S11 A and the D alleles are similar and the peak areas of the B and the C alleles are similar to each other, we can assume that AD and BC represent the best possible combination of alleles to explain the data. Likewise,

Figure 7.6

Peak areas for green panel data from example mixture in Figure 7.5. Mixture ratio calculations for these STR markers are shown in Table 7.5. RFUs = relative fluorescence units.

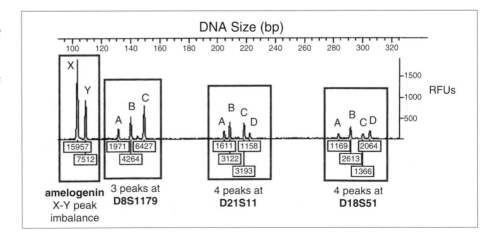

the D18S51 A and C alleles have similar peak areas and can be grouped together so that the best possible combination of alleles at this locus is AC and BD (Table 7.5).

The mixture ratio for the D21S11 alleles is calculated to be 2.3, or approximately 2:1, by dividing the sum of the larger alleles (B and C) by the sum of the smaller alleles (A and D). Thus, the major contributor to this mixture has about twice as much DNA present as the minor contributor. Using the same approach, the D18S51 mixture ratio is calculated to be 1.8 or approximately 2:1

Locus	Component	Allele Call	Peak Area	Possible Combinations	Mixture Ratio
D21S11	A	29	1611	AD BC	$\frac{B+C}{A+D} = \frac{3122+3193}{1611+1158} = 2.3$
	B	30	3122		
	C	32.2	3193		~2:1
	D	33.2	1158	29,33.2 30,32.2	
D18S51	A	12	1169	AC BD	$\frac{B+D}{A+C} = \frac{2613+2064}{1169+1366} = 1.8$
	B	14	2613		
	C	16	1366		~2:1
	D	17	2064	12,16 & 14,17	
D8S1179	A	10	1971	AA BC	
	B	12	4264	BB AC	
	C	14	6427	CC AB	
				AB AC	Based on peak profile
				BC AC	appearance for 2:1 mixture
				AB BC	(see Figures 7.6 and 7.7)
Amel	X	X	15957	2X:1Y	$\frac{X}{Y} \frac{15957}{7512} = 2.1$
	Y	Y	7512		

2 parts male to 1 part female

Table 7.5

Peak information for four loci from the data displayed in Figure 7.6. The mixture ratio is determined by comparing the peak areas for the appropriate peak combinations. The original source genotypes are also listed for the male and female components of the mixture, along with the expected number of peaks at each locus.

Original source genotypes for Figure 7.5 mixed sample example

	D3	VWA	FGA	Amel	D8	D21	D18	D5	D13	D7
Male	17,18	14,19	24,24	X,Y	12,14	30,32.2	14,17	12,13	12,12	9,12
Female	15,16	16,16	23,24	X,X	10,14	29,33.2	12,16	7,11	12,13	10,12
Mixture # Peaks	4	3	2	2	3	4	4	4	2	3
Major	CD	AC	BB	XY	BC	BC	BD	CD	AA	AC
Minor	AB	BB	AB	XX	AC	AD	AC	AB	AB	BC

for the major contributor (Table 7.5). These mixture ratios will not always be exact due to the influence of stutter products and imbalances in heterozygote peak areas.

Calculating mixture ratios is much easier at loci with four observed alleles than at loci where one or more of the alleles are shared. D8S1179 in this example represents such a situation. Three peaks are present at D8S1179 with one of these peaks representing an allele from both the major and the minor contributor. Each possible combination of alleles is therefore carefully considered to determine which one best fits the observed data.

Expected peak patterns for each of the possible 2:1 mixture combinations are displayed in Figure 7.7. The observed data for D8S1179 (Figure 7.6) fits the scenario of BC and AC allele combinations with BC belonging to the major contributor. Thus, in this case allele C, or 14, is shared. The major contributor's genotype is therefore 12,14 at D8S1179 while the minor contributor's genotype is 10,14.

The major contributor in this example mixture was the male individual at a ratio of two times that of the female DNA in the mixture (Table 7.5). This fact was determined by examining the peak areas for the X and Y alleles and comparing them to the information found in Table 7.4. Thus, in this example it was possible to decipher that the major contributor was a male and that the D21S11 alleles 30 and 32.2, D18S51 alleles 14 and 17, and D8S1179 alleles 12 and 14 belonged to him. By continuing through all of the loci in the manner indicated above, the genotype profile of the major and minor contributors can be distinguished.

Mixture ratios cannot always be calculated at every locus with complete confidence especially those with two or three peaks that have shared alleles between the contributors. Note that in Figure 7.7 when a homozygous individual is the minor component of a mixture all three scenarios have indistinguishable profiles of three fairly balanced peak signals (e.g., VWA in Figure 7.5). Of course, stutter products and imbalanced heterozygote peak signals make calculation of mixture ratios less accurate.

SOFTWARE FOR DECIPHERING MIXTURE COMPONENTS

Computer programs can be used to aid the process of deciphering mixture components and determining mixture ratios. A linear mixture analysis approach was reportedly able to derive estimated mixture ratios from quantitative STR peak information that were similar to known input mixture proportions (Perlin and Szabady 2001). Researchers at the University of Tennessee are also working on algorithms to decipher mixture components in an automated fashion (Wang *et al.* 2002).

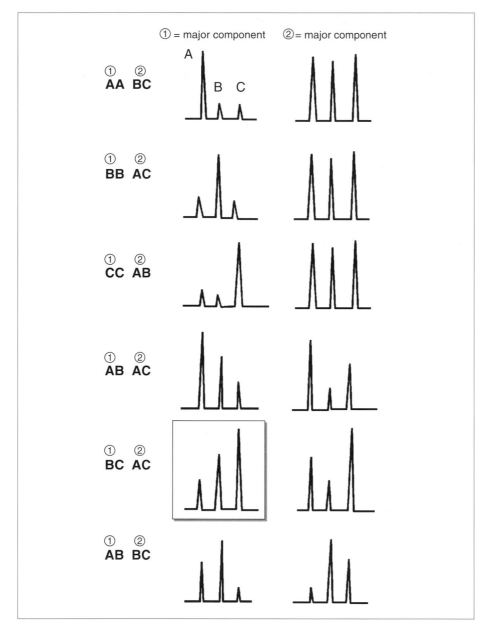

Figure 7.7

Expected peak profiles for 2:1 mixture combinations involving three peaks. These peak profiles assume no overlap with stutter products and homozygote alleles that possess twice the signal strength of each heterozygote allele.

CHROMOSOMAL ABNORMALITIES

Chromosomal abnormalities do exist and can give rise to extra allele peaks at a particular STR locus. Chromosomal translocations, somatic mutations and trisomies may occur in the cells of the donor of a forensic stain. However, the STR profile from the individual with the chromosomal abnormality would most likely show only a single extra peak and the same pattern would be present in both the forensic stain and the reference sample from the matching suspect (Clayton *et al.* 1998). The rare cases where a chromosomal abnormality is observed can even help strengthen the final conclusions.

An excellent example of a chromosomal abnormality is found in the standard cell line K562. Three peaks are obtained at the D21S11 locus and at least five other STR loci have heterozygous peak patterns that are not balanced (D.N.A. Box 7.1). At first glance, one might suspect this sample to have arisen from more than one source rather than a sample with an abnormal number of chromosomes.

As described in Chapter 6, tri-allelic patterns have been reported for a number of STR loci (see Figure 6.7). In fact, more than 56 different tri-allelic patterns have been reported spanning all 13 CODIS core loci (see http://www.cstl.nist.gov/biotech/strbase/tri_tab.htm).

CONCLUSIONS ON INTERPRETING MIXTURES

Mixed sample stains are present in many forensic investigations and STR typing procedures have been demonstrated to be an effective means of differentiating components of a mixed sample. However, a case may contain multiple stains and not all of these will be mixtures. In fact, the proportions of a mixture can vary across the forensic stain itself. Thus, if additional samples can be tested that are easier to interpret they should be sought after versus complicated mixtures (Gill *et al.* 1998b). As recommended by Peter Gill of the Forensic Science Service, the best advice is 'Don't do mixture interpretation unless you have to'.

As an example of the number of mixture samples encountered in typical casework, Torres *et al.* (2003) reviewed all of the mixture STR profiles seen in their laboratory over a four-year time period. From 1547 criminal cases worked which involved a total of 2424 samples, only 163 showed a mixed profile or 6.7%.

Some forensic DNA laboratories may decide not to go through the trouble of fully deciphering the genotype possibilities and assigning them to the major and minor contributors. An easier approach is to simply include or exclude a suspect's DNA profile from the crime scene mixture profile. If all of the alleles from a suspect's DNA profile are represented in the crime scene mixture, then the suspect cannot be excluded as contributing to the crime scene stain. Likewise, the alleles in a victim's DNA profile could be subtracted out of the mixture profile to simplify the alleles that need to be present in the perpetrator's DNA profile. Approaches to attaching a statistical value to mixture results are presented in Chapter 22.

D.N.A. Box 7.1

Abnormal STR peak heights in K562 cell line DNA profile

For many years, DNA extracted from the cell line K562 was supplied as a control sample from the Promega Corporation with their GenePrint® STR typing kits. However, as can be seen from the peak profiles below, some of the STR loci exhibit imbalanced heterozygous alleles and/or multiple peaks that would make the sample appear to come from more than one source of DNA. In this particular case, these extra peaks or peak imbalances are the result of an abnormal number of chromosomes present in the sample rather than a problem with the DNA typing system. K562 cells are derived from a female human subject with a diagnosis of chronic myelogeneous leukemia. Because mutant cells are present with chromosomes that possess somatic mutations that affect the number of repeats in various STR markers, the K562 cell line results do not possess the normal balance of chromosomal material seen in healthy individuals. Thus, balanced heterozygous allele peaks are not always seen. Shown below are six STR markers amplified from K562 genomic DNA that possess a significant variation in the balance of the STR allele peak heights.

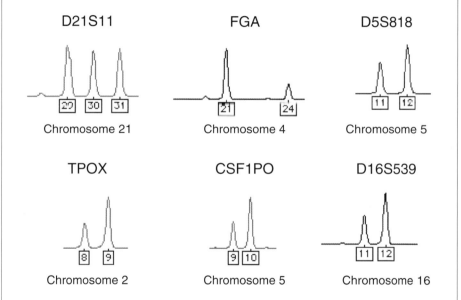

Genotyper data from AmpF/STR® Profiler Plus™ and COfiler™ PCR amplifications with ABI Prism 310 detection.

LOW-COPY NUMBER DNA TESTING

An exciting area of research that is on-going in many laboratories is the ability to obtain DNA profiles from very small amounts of sample. Low-copy number (LCN) DNA testing typically refers to examination of less than 100 pg of input DNA (Gill *et al.* 2000). In fact, fluorescent multiplexes have been used to obtain STR typing results from as little as a single buccal cell (Findlay *et al.* 1997).

STR ANALYSIS FROM EXTREMELY LOW LEVELS OF HUMAN DNA

The number of PCR cycles is often increased to improve the amplification yield from samples containing extremely low levels of DNA template. For example, by increasing the PCR cycle number from 28 to 34, STR typing results have been routinely demonstrated for samples containing less than 100 pg (Gill *et al.* 2000, Whitaker *et al.* 2001). However, application of LCN results should be approached with caution due to the possibilities of allele dropout, allele drop-in, and increased risks of collection-based and laboratory-based contamination.

Remarkably, DNA profiles may be obtained from fingerprint residues due to cells that are left on the objects that are touched (van Oorschot and Jones 1997, van Hoofstat *et al.* 1998, Abaz *et al.* 2002, Alessandrini *et al.* 2003). DNA technology may permit the handles of tools used in crimes, such as knives or guns, to be effectively evaluated and used to link a perpetrator to his crime. A nice review of the theory and application of trace DNA detection was published by Wickenheiser (2002).

The ability to obtain DNA profiles from small amounts of biological material has expanded the types of samples available for analysis. For example, crime scenes can contain insects that may be useful in linking a suspect to the crime in question. Several interesting studies have found that human DNA consumed by the insects may be extracted and successful identified. In particular, human crab louse feces have been reported as a source of human DNA (Replogle *et al.* 1994) as have mosquitoes (Kreike *et al.* 1999). As demonstrated by these examples, the capability of obtaining a useful DNA profile is often only limited by the ability of the forensic investigator to find and collect the appropriate evidence.

ISSUES WITH LCN WORK

Trying to generate a reliable STR profile with only a few cells from a biological sample is similar to looking for an object in the mud or trying to decipher the image in a fuzzy photograph. Since the sensitivity of the STR typing assay is turned up so high, it is often not immediately clear if you have a reliable result or even a probative one. Recovered DNA profiles may not be associated with the crime event itself but rather have been left innocently before the crime occurred (Gill 2001). Secondary transfer of skin cells due to casual contact such as hand shaking has been demonstrated to occur in controlled laboratory settings (Lowe *et al.* 2002). This phenomenon occurs to a variable degree depending on what kind of a 'shedder' the individuals are (Lowe *et al.* 2002).

When LCN testing is performed at least three artifacts typically arise: (1) additional alleles are often observed from sporadic contamination in what is referred to as allele 'drop-in', (2) allele 'dropout' is common where an allele fails to amplify due to stochastic effects (see Chapter 4), and (3) stutter product amounts are enhanced so that they are often higher than the typical 5–10% of the nominal allele (Whitaker *et al.* 2001). Heterozygote peak imbalance is typically exacerbated due to stochastic PCR amplification, where one of the alleles is amplified by chance during the early rounds of PCR in a preferential fashion. Allele dropout can be thought of as an extreme form of heterozygote peak imbalance.

RULES TO INSURE OPTIMAL RESULTS WITH LCN

The allele drop-in phenomenon is usually not reproducible and can be detected through testing the sample multiple times. As noted in an early paper describing reliable genotyping of samples with very low DNA quantities, the probability of obtaining a particular extra allele (i.e., allele drop-in) does not exceed 5%, and thus the probability of obtaining this same extra allele in two independent PCR reactions is <1% (Taberlet *et al.* 1996). Thus, the routine application of LCN involves at least two amplifications from the same DNA extract with a rule that an allele cannot be scored unless it is present at least twice in replicate samples (Taberlet *et al.* 1996, Whitaker *et al.* 2001, Gill 2001). The consensus or composite STR profile is then the one that is reported. This approach has been confirmed with consensus profiles from separate single cell PCR experiments matching the actual profile of the cell donor (Kloosterman and Kersbergen 2003).

It is essential that LCN be performed in a sterile environment to prevent contamination from laboratory personnel (Gill 2001, Rutty *et al.* 2003). DNA extractions and setting up PCR reactions should be performed in a dedicated laboratory similar to what is done in the 'ancient DNA' field (Capelli *et al.* 2003). Laboratory personnel should wear disposable lab coats, gloves, and face masks. In addition, benches and equipment should be frequently treated with bleach to destroy any extraneous DNA as previously described in Chapter 4. STR profile results should be compared against a staff elimination database as well as anyone who may have legitimately come in contact with the crime scene evidence prior to the DNA testing.

A WORD OF CAUTION REGARDING LCN

In practice many LCN profiles are mixtures and difficult to interpret reliably due to the issues of allele dropout and drop-in described above. The success rate for obtaining a clean profile can be poor due to the limited amount of sample available. Thus a lot of work can go into generating DNA profiles that may not be probative or useful in a particular forensic case. Although a DNA profile may be obtained, it is usually not possible to identify the type of cells from which the DNA originated or when the cells were deposited (Gill 2001). Furthermore, with such little starting material, it may not be possible to preserve LCN evidence to enable confirmatory testing by a second laboratory should that be required. Nevertheless, even with these caveats LCN results have enabled recovery of DNA profiles from burglaries and other situations where only a few cells from the perpetrator were present and have thereby extended the power of DNA testing.

In some cases, it may be possible to obtain reliable results from low levels of DNA without having to boost the PCR cycle number and push the sensitivity of the amplification. Budowle *et al.* (2001) list several alternatives to LCN to enable boosting a signal for a STR profile without increasing the PCR cycle number and the concomitant increased risk of contamination. These alternatives include: (1) reducing the PCR volume to get a more concentrated PCR product, (2) filtration of the PCR product to remove ions that compete with the STR amplicons when being injected into the capillary (see Chapter 14), (3) use of a formamide with lower conductivity, (4) adding more amplified product to the analysis tube, and (5) increasing the injection time on the capillary electrophoresis instrument.

Some important LCN reporting guidelines include: (1) multiple tube PCR amplifications with demonstrated duplication of every allele before reporting results, (2) if negative controls associated with a particular batch of samples show duplicated alleles that correspond to any of the samples, then the samples should not be reported and where possible samples should be retested, and (3) if there is one allele in a sample that does not match the suspect's STR profile, then further testing may be pursued (Gill *et al.* 2000). Generating STR data with an increased number of PCR cycles and invoking a LCN philosophy can provide a useful lead in many instances for an investigation but it is unlikely to provide definitive probative evidence of a crime in every instance.

OTHER USES FOR STR TYPING

In this chapter we have focused on forensic issues with STR typing. The final section will cover other uses for STR typing that involve the capability of these markers for mixture detection.

CHARACTERIZING CELL LINES

STR typing with the same core set of markers and commercial STR kits is being used for several other purposes besides forensic and parentage testing.

Human cell line authentication is now being carried out by the American Type Culture Collection (ATCC) along with other international suppliers of cell lines (Masters *et al.* 2001). STR typing enables rapid discovery of cross-contamination between cell lines and may serve as a universal reference standard for characterizing human cell lines. This type of analysis has been dubbed 'cell culture forensics' (O'Brien 2001). Over the past several years, the ATCC has created a database of over 500 human cell lines that have been run with eight STR loci present in the PowerPlex® 1.2 kit from the Promega Corporation (see http://www.atcc.org/Cultures/str.cfm). It is important to note that cell cultures, such as K562 (D.N.A. Box 7.1), may not always have a regular diploid complement of chromosomes and thus may possess tri-allelic patterns and severe peak imbalances due to the presence of additional copies of one allele (Masters *et al.* 2001).

MONITORING TRANSPLANTS

Monitoring the engraftment of donor cells after bone marrow transplants (Millison *et al.* 2000) or allogeneic blood stem cell transplantation (Thiede *et al.* 1999) is another important application of STR testing. Examination of STR profiles from transplant recipients can help diagnosis graft failure or a relapse of the disease. In these cases, mixtures are detected as mixed chimerism that exists within the recipient from their own cells and those of the donor.

DETECTING GENETIC CHIMERAS

Chimerism, which is the presence of two genetically distinct cell lines in an organism, can be acquired through blood stem cell transplantation, blood transfusion, or can be inherited (Yu *et al.* 2002). Approximately 8% of non-identical twins can have chimeric blood (van Dijk *et al.* 1996). Several years ago an interesting case was reported of a phenotypically normal woman who possessed different DNA types in different body tissues due to tetragametic chimerism (see D.N.A. Box 7.2).

A study with 203 matched related donor-recipient pairs ranked 27 different STRs, including the 13 CODIS core loci, in terms of their ability to detect chimeric mixtures (Thiede *et al.* 2004). Not surprisingly, the loci with the highest heterozygosities, namely Penta E, SE33, D2S1338, and D18S51, worked the best.

MONITORING NEEDLE SHARING

In yet another application of the capability to perform mixture detection with STRs, a laboratory method was described by Shrestha *et al.* (2000) using the CODIS STR marker D8S1179 (called D6S502 in the paper) to differentiate between single and multi-person use of syringes by intravenous drug users.

D.N.A. Box 7.2

Natural mixtures and chimeric individuals

In May 2002, the *New England Journal of Medicine* published a report of the genetic analysis of a phenotypically normal chimeric individual who was unexpected identified because histocompatibility testing of family members suggested that she was not the biological mother of two of her three children (Yu *et al.* 2002). The doctors examining this chimeric individual proposed that her condition had arisen because two fertilized eggs, destined originally to be fraternal twins, had fused to form a zygote that possessed DNA of two different types. Thus, from a genetic perspective she was both her children's mother and their aunt.

Among the various genetic tests performed on this chimeric individual was analysis of 22 STR loci. All of the CODIS core loci except CSF1PO were examined in this study. This unusual patient possessed some differences in her STR profiles among various tissues tested. While the buccal and blood samples tested matched exactly, a mixture containing another type was present as the minor component in the thyroid, hair, and skin cells of this chimeric patient.

While chimeric individuals such as the one described above are most likely extremely rare in the general population, it is possible in theory for DNA testing from different tissues of a chimeric individual to not match one another and thus lead to a false exclusion. This situation may increase in frequency with the rise of *in vitro* fertilization since multiple eggs are sometimes fertilized in order to increase the success rate of the procedure.

Sources:

Yu *et al.* (2002) *New England Journal of Medicine*, 346, 1545–1552.

David Baron, 'DNA tests shed light on hybrid human', National Public Radio-Morning Edition, August 11, 2003 (http://www.npr.org).

Pearson, H. (2002) Human genetics: dual identities (see http://www.nature.com/nsu/nsu_pf/020429/020429-13.html)

Monitoring needle sharing can help determine the source of spreading blood-borne pathogens among drug users.

DETECTING CANCER TUMORS

Loss of heterozygosity (LOH) is a method of monitoring genetic deletions common in tumors for many types of cancer. LOH is manifested by severe allelic imbalance at a locus in a single-source DNA sample so that a true heterozygote almost appears as a homozygote since some of the chromosomes have a deletion present in the region of the locus being PCR-amplified.

Probably the only time that LOH would have an impact on human identity testing is if an archived clinical specimen from a tissue biopsy was used as a reference sample to identify someone from a mass disaster (see Chapter 24). However, it is worth being aware of the fact that normal and cancerous tissue from an individual can vary fairly dramatically in some instances in terms of their STR allele peak heights (Vauhkonen *et al.* 2004). An examination of a cancer biopsy tissue specimen compared to normal tissue with the nine STR loci present in the AmpF*l*STR Profiler kit found that the D13S317 locus exhibited a severe peak imbalance consistent with that seen arising from LOH (Rubocki *et al.* 2000). The authors suggest that this LOH might be due to a deletion of 13q21–22 seen previously with prostate cancer (Hyytinen *et al.* 1999) that is near the physical location of D13S317 on chromosome 13.

MAPPING GENETIC DISEASES

Genome scans for disease gene mapping are routinely performed with around 400 STRs covering the human genome at 5–10 centiMorgan (cM) distances (Ghebranious *et al.* 2003). Marshfield Genetics (http://research.marshfieldclinic. org/genetics/) and the Center for Inherited Disease Research (CIDR; http:// www.cidr.jhmi.edu/) perform high volume genetic testing using hundreds of STRs per DNA sample. Studies of STR allele frequencies between normal and disease patient populations are used to help make associations with the genetic disease. When correlations are made between a STR locus and a disease gene through linkage analysis, then the known location of the STR marker can help pinpoint the previously unknown location of the disease gene of interest.

STR loci that are examined in these types of genetic mapping studies may be associated with disease gene locations. Thus, it is helpful to know which loci used in human identity testing are also widely used in genetic mapping studies. For example, the 408 loci included in Marshfield Set 12 incorporate five of the 13 CODIS core STR loci – namely TPOX, D7S820, D8S1179, D13S317, and D16S539. The locus D19S433 that is part of the SGM Plus and Identifiler kits (see Chapter 5) is also included in the commonly used genome scan studies for genetic disease associations. Likewise, the 405 markers used by CIDR as of June 2004 include D8S1179, D13S317, and D16S539 along with F13A1 and D19S433 that are present in various commercial STR kits used for human identity testing. A genetic disease called Meckel–Gruber syndrome has been reported to be located near D8S1179 (Morgan *et al.* 2002). While the determination of linkage to some genetic disease with a STR marker may cause concern for some users of these loci, it is important to keep in mind that it is likely that many or possibly most STRs will eventually be shown to be useful in following a genetic disease or other genetic trait within a family and therefore this possibility must be recognized at the outset of the use of such systems (Kimpton *et al.* 1995).

EXAMINING HUMAN POPULATION DIVERSITY

Whole genome scans with 377 autosomal STR loci (Marshfield set 10) were used to genotype 1056 individuals from 52 populations in order to study human population structure (Rosenberg *et al.* 2002). Studies with this same set of data have identified particular STR loci that are effective indicators of ancestral origin (Rosenberg *et al.* 2003). Analysis of Y chromosome STRs (see Chapter 9) and mitochondrial DNA (see Chapter 10) have also been used for genetic genealogy studies (Helgason *et al.* 2003). Both STR markers and single nucleotide polymorphisms (SNPs), which will be discussed in the next chapter, should continue to play an important role in understanding human diversity at the genetic level.

REFERENCES AND ADDITIONAL READING

Abaz, J., Walsh, S.J., Curran, J.M., Moss, D.S., Cullen, J., Bright, J.A., Crowe, G.A., Cockerton, S.L. and Power, T.E. (2002) *Forensic Science International*, 126, 233–240.

Akane, A., Matsubara, K., Nakamura, H., Takahashi, S. and Kimura, K. (1994) *Journal of Forensic Sciences*, 39, 362–372.

Alessandrini, F., Cecati, M., Pesaresi, M., Turchi, C., Carle, F. and Tagliabracci, A. (2003) *Journal of Forensic Sciences*, 48, 586–592.

Al-Soud, W.A. and Rådström, P. (1998) *Applied and Environmental Microbiology*, 64, 3748–3753.

Al-Soud, W.A. and Rådström, P. (2000) *Journal of Clinical Microbiology*, 38, 4463–4470.

Al-Soud, W.A. and Rådström, P. (2001) *Journal of Clinical Microbiology*, 39, 485–493.

Applied Biosystems (1998) *AmpFISTR Profiler Plus PCR Amplification Kit User's Manual*. Foster City, California: Applied Biosystems.

Bacher, J.W., Hennes, L.F., Gu, T., Tereba, A., Micka, K.A., Sprecher, C.J., Lins, A.M., Amiott, E.A., Rabbach, D.R., Taylor, J.A., Helms, C., Donis-Keller, H. and Schumm, J.W. (1999) *Proceedings of the Ninth International Symposium on Human Identification*, pp. 24–37. Madison, Wisconsin: Promega Corporation.

Banaschak, S., Moller, K. and Pfeiffer, H. (1998) *International Journal of Legal Medicine*, 111, 284–285.

Bourke, M.T., Scherczinger, C.A., Ladd, C. and Lee, H.C. (1999) *Journal of Forensic Sciences*, 44, 1046–1050.

Braid, M.D., Daniels, L.M. and Kitts, C.L. (2003) *Journal of Microbiological Methods*, 52, 389–393.

Budowle, B., Hobson, D.L., Smerick, J.B. and Smith, J.A.L. (2001) *Proceedings of the Twelfth International Symposium on Human Identification*. Madison, Wisconsin: Promega Corporation. Available at: http://www.promega.com/geneticidproc/ussymp12proc/contents/budowle.pdf.

Butler, J.M., Shen, Y. and McCord, B.R. (2003) *Journal of Forensic Sciences,* 48, 1054–1064.

Capelli, C., Tschentscher, F. and Pascali, V.L. (2003) *Forensic Science International,* 131, 59–64.

Chung, D.T., Drábek, J., Opel, K.L., Butler, J.M. and McCord, B.R. (2004) *Journal of Forensic Sciences,* 49 (4), 733–740.

Coble, M.D. and Butler, J.M. (2005) *Journal of Forensic Sciences*, (January), in press.

Comey, C.T., Koons, B.W., Presley, K.W., Smerick, J.B., Sobieralski, C.A., Stanley, D.M. and Baechtel, F.S. (1994) *Journal of Forensic Sciences*, 39, 1254–1269.

Cotton, R.W. (1995) *Proceedings from the Sixth International Symposium on Human Identification*, pp. 112–115. Madison, Wisconsin: Promega Corporation.

Curran, J.M., Triggs, C.M., Buckleton, J. and Weir, B.S. (1999) *Journal of Forensic Sciences*, 44, 987–995.

DeFranchis, R., Cross, N.C.P., Foulkes, N.S. and Cox, T.M. (1988) *Nucleic Acids Research*, 16, 10355.

Drábek, J., Chung, D.T., Butler, J.M. and McCord, B.R. (2004) *Journal of Forensic Sciences,* 49 (4), 859–860.

Duewer, D.L., Kline, M.C., Redman, J.W., Newall, P.J. and Reeder, D.J. (2001) *Journal of Forensic Sciences*, 46, 1199–1210.

Eckhart, L., Bach, J., Ban, J. and Tschachler, E. (2000) *Biochemical and Biophysical Research Communications*, 271, 726–730.

Findlay, I., Taylor, A., Quirke, P., Frazier, R. and Urquhart, A. (1997) *Nature,* 389, 555–556.

Ghebranious, N., Vaske, D., Yu, A., Zhao, C., Marth, G. and Weber, J.L. (2003) *BMC Genomics*, 4, 6. Available at: http://www.biomedcentral.com/1471-2164/4/6.

Gill, P. (1997) *Forensic Science International*, 85, 105–111.

Gill, P., Sparkes, R. and Kimpton, C. (1997) *Forensic Science International*, 89, 185–197.

Gill, P., Sparkes, R., Pinchin, R., Clayton, T., Whitaker, J. and Buckleton, J. (1998a) *Forensic Science International*, 91, 41–53.

Gill, P., Sparkes, B., Clayton, T.M., Whittaker, J., Urquhart, A. and Buckleton, J.S. (1998b) *Proceedings of the Ninth International Symposium on Human Identification*, pp. 7–18. Madison, Wisconsin: Promega Corporation.

Gill, P., Whitaker, J., Flaxman, C., Brown, N. and Buckleton, J. (2000) *Forensic Science International*, 112, 17–40.

Gill, P. (2001) *Croatian Medical Journal*, 42, 229–232.

Gill, P. (2002) *Biotechniques*, 32, 366–372.

Gill, P. and Kirkham, A. (2004) *Journal of Forensic Sciences*, 49 (3), 485–491.

Helgason, A., Hrafnkelsson, B., Gulcher, J. R., Ward, R. and Stefansson, K. (2003) *American Journal of Human Genetics*, 72, 1370–1388.

Hellmann, A., Rohleder, U., Schmitter, H. and Wittig, M. (2001) *International Journal of Legal Medicine*, 114, 269–273.

Howitt, T., Johnson, P., Cotton, L., Rowlands, D. and Sullivan, K. (2003) *Proceedings of the 14th International Symposium on Human Identification*. Madison, Wisconsin: Promega Corporation. Available at: http://www.promega.com/geneticidproc/ussymp14proc/oralpresentations/Howitt.pdf.

Hyytinen, E.R., Frierson, H.F., Jr., Boyd, J.C., Chung, L.W. and Dong, J.T. (1999) *Genes, Chromosomes and Cancer*, 25, 108–114.

Kimpton, C.P., Gill, P., d'Aloja, E., Andersen, J.F., Bar, W., Holgersson, S., Jacobsen, S., Johnsson, V., Kloosterman, A.D., Lareu, M.V., Nellemann, L., Pfitzinger, H., Phillips, C.P., Rand, S., Schmitter, H., Schneider, P.M., Sternersen, M. and Vide, M.C. (1995) *Forensic Science International*, 71, 137–152.

Kline, M.C., Duewer, D.L., Redman, J.W. and Butler, J.M. (2003) *Analytical Chemistry*, 75, 2463–2469.

Kloosterman, A.D. and Kersbergen, P. (2003) *Journal de la Societe de Biologie*, 197, 351–359.

Kreader, C.A. (1996) *Applied and Environmental Microbiology*, 62, 1102–1106.

Kreike, J. and Kampfer, S. (1999) *International Journal of Legal Medicine*, 112, 380–382.

Krenke, B.E., Tereba, A., Anderson, S.J., Buel, E., Culhane, S., Finis, C.J., Tomsey, C.S., Zachetti, J.M., Masibay, A., Rabbach, D.R., Amiott, E.A. and Sprecher, C.J. (2002) *Journal of Forensic Sciences*, 47, 773–785.

Lantz, P.-G., Matsson, M., Wadstrom, T. and Rädström, P. (1997) *Journal of Microbiological Methods*, 28, 159–167.

Lowe, A., Murray, C., Whitaker, J., Tully, G. and Gill, P. (2002) *Forensic Science International*, 129, 25–34.

Lygo, J.E., Johnson, P.E., Holdaway, D.J., Woodroffe, S., Whitaker, J.P., Clayton, T.M., Kimpton, C.P. and Gill, P. (1994) *International Journal of Legal Medicine*, 107, 77–89.

Mahony, J., Chong, S., Jang, D., Luinstra, K., Faught, M., Dalby, D., Sellors, J. and Chernesky, M. (1998) *Journal of Clinical Microbiology*, 36, 3122–3126.

Masters, J.R., Thomson, J.A., Daly-Burns, B., Reid, Y.A., Dirks, W.G., Packer, P., Toji, L.H., Ohno, T., Tanabe, H., Arlett, C.F., Kelland, L.R., Harrison, M., Virmani, A., Ward, T.H., Ayres, K.L. and Debenham, P.G. (2001) *Proceedings of the National Academy of Sciences U.S.A.*, 98, 8012–8017.

Miller, D.N., Bryant, J.E., Madsen, E.L. and Ghiorse, W.C. (1999) *Applied and Environmental Microbiology*, 65, 4715–4724.

Millson, A.S., Spangler, F.L., Wittwer, C.T. and Lyon, E. (2000) *Diagnostic Molecular Pathology*, 9, 91–97.

Monteiro, L., Bonnemaison, D., Vekris, A., Petry, K.G., Bonnet, J., Vidal, R., Cabrita, J. and Megraud, F. (1997) *Journal of Clinical Microbiology*, 35, 995–998.

Moreira, D. (1998) *Nucleic Acids Research*, 26, 3309–3310.

Morgan, N.V., Gissen, P., Sharif, S.M., Baumber, L., Sutherland, J., Kelly, D.A., Aminu, K., Bennett, C.P., Woods, C.G., Mueller, R.F., Trembath, R.C., Maher, E.R. and Johnson, C.A. (2002) *Human Genetics*, 111, 456–461.

O'Brien, S.J. (2001) *Proceedings of the National Academy of Sciences U.S.A.*, 98, 7656–7658.

Ohtaki, H., Yamamoto, I., Yoshimoto, I., Uchihi, R., Ooshima, C., Katsumata, Y. and Tokunaga, K. (2002) *Electrophoresis*, 23, 3332–3340.

Perlin, M.W. and Szabady, B. (2001) *Journal of Forensic Sciences*, 46, 1372–1378.

Rädström, P., Knutsson, R., Wolffs, P., Lovenklev, M. and Lofstrom, C. (2004) *Molecular Biotechnology*, 26, 133–146.

Replogle, J., Lord, W.D., Budowle, B., Meinking, T.L. and Taplin, D. (1994) *Journal of Medical Entomology*, 31, 686–690.

Rosenberg, N.A., Pritchard, J.K., Weber, J.L., Cann, H.M., Kidd, K.K., Zhivotovsky, L.A. and Feldman, M.W. (2002) *Science*, 298, 2381–2385.

Rosenberg, N.A., Li, L.M., Ward, R. and Pritchard, J.K. (2003) *American Journal of Human Genetics*, 73, 1402–1422.

Rubocki, R.J., Duffy, K.J., Shepard, K.L., McCue, B.J., Shepherd, S.J. and Wisecarver, J.L. (2000) *Journal of Forensic Sciences*, 45, 1087–1089.

Rubocki, R.J., McCue, B.J., Duffy, K.J., Shepard, K.L., Shepherd, S.J. and Wisecarver, J.L. (2001) *Journal of Forensic Sciences*, 46, 120–125.

Rutty, G.N., Hopwood, A. and Tucker, V. (2003) *International Journal of Legal Medicine*, 117, 170–174.

Scherczinger, C.A., Ladd, C., Bourke, M.T., Adamowicz, M.S., Johannes, P.M., Scherczinger, R., Beesley, T. and Lee, H.C. (1999) *Journal of Forensic Sciences*, 44, 1042–1045.

Schneider, P.M., Bender, K., Mayr, W.R., Parson, W., Hoste, B., Decorte, R., Cordonnier, J., Vanek, D., Morling, N., Karjalainen, M., Marie-Paule, C.C., Sabatier, M., Hohoff, C., Schmitter, H., Pflug, W., Wenzel, R., Patzelt, D., Lessig, R., Dobrowolski, P., O'Donnell, G., Garafano, L., Dobosz, M., de Knijff, P., Mevag, B., Pawlowski, R., Gusmao, L., Conceicao, V.M., Alonso, A.A., Garcia, F.O., Sanz, N.P., Kihlgreen, A., Bar, W., Meier, V., Teyssier, A., Coquoz, R., Brandt, C., Germann, U., Gill, P., Hallett, J. and Greenhalgh, M. (2004) *Forensic Science International*, 139, 123–134.

Schumm, J.W., Wingrove, R.S. and Douglas, E.K. (2004) *Progress in Forensic Genetics, 10*, ICS 1261, 547–549.

Shrestha, S., Strathdee, S.A., Brahmbhatt, H., Farzadegan, H., Vlahov, D. and Smith, M.W. (2000) *AIDS*, 14, 1507–1513.

Shutler, G.G., Gagnon, P., Verret, G., Kalyn, H., Korkosh, S., Johnston, E. and Halverson, J. (1999) *Journal of Forensic Sciences*, 44, 623–626.

Sparkes, R., Kimpton, C.P., Watson, S., Oldroyd, N.J., Clayton, T.M., Barnett, L., Arnold, J., Thompson, C., Hale, R., Chapman, J., Urquhart, A. and Gill, P. (1996) *International Journal of Legal Medicine*, 109, 186–194.

Taberlet, P., Griffin, S., Goossens, B., Questiau, S., Manceau, V., Escaravage, N., Waits, L.P. and Bouvet, J. (1996) *Nucleic Acids Research*, 24, 3189–3194.

Takahashi, M., Kato, Y., Mukoyama, H., Kanaya, H. and Kamiyama, S. (1997) *Forensic Science International*, 90, 1–9.

Thiede, C., Florek, M., Bornhauser, M., Ritter, M., Mohr, B., Brendel, C., Ehninger, G. and Neubauer, A. (1999) *Bone Marrow Transplantation*, 23, 1055–1060.

Thiede, C., Bornhauser, M. and Ehninger, G. (2004) *Leukemia*, 18, 248–254.

Torres, Y., Flores, I., Prieto, V., Lopez-Soto, M., Farfan, M.J., Carracedo, A. and Sanz, P. (2003) *Forensic Science International*, 134, 180–186.

Tsai, Y.L. and Olson, B.H. (1992) *Applied and Environmental Microbiology*, 58, 2292–2295.

Tsukada, K., Takayanagi, K., Asamura, H., Ota, M. and Fukushima, H. (2002) *Legal Medicine*, 4, 239–245.

van Dijk, B.A., Boomsma, D.I. and de Man, A.J. (1996) *American Journal of Medical Genetics*, 61, 264–268.

Van Hoofstat, D.E.O., Deforce, D.L.D., Brochez, V., De Pauw, I., Janssens, K., Mestdagh, M., Millecamps, R., Van Geldre, E. and Van den Eeckhout, E.G. (1998) *Proceedings of the Second European Symposium on Human Identification*, pp. 131–137. Madison, Wisconsin: Promega Corporation.

Van Oorschot, R.A.H. and Jones, M.K. (1997) *Nature*, 387, 767.

Vauhkonen, H., Hedman, M., Vauhkonen, M., Kataja, M., Sipponen, P. and Sajantila, A. (2004) *Forensic Science International*, 139, 159–167.

Wang, T., Xue, N. and Wickenheiser, R. (2002) Least-square deconvolution (LSD): a new way of resolving STR/DNA mixture samples. *Proceedings of the Thirteenth International Symposium on Human Identification.* Available at: http://www.promega.com/geneticidproc/ussymp13proc/contents/wang.pdf.

Whitaker, J.P., Clayton, T.M., Urquhart, A.J., Millican, E.S., Downes, T.J., Kimpton, C.P. and Gill, P. (1995) *BioTechniques*, 18, 670–677.

Wickenheiser, R.A. (2002) *Journal of Forensic Sciences*, 47, 442–450.

Wiegand, P. and Kleiber, M. (2001) *International Journal of Legal Medicine*, 114, 285–287.

Wilson, I.G. (1997) *Applied and Environmental Microbiology*, 63, 3741–3751.

Yu, N., Kruskall, M.S., Yunis, J.J., Knoll, J.H., Uhl, L., Alosco, S., Ohashi, M., Clavijo, O., Husain, Z., Yunis, E.J., Yunis, J.J. and Yunis, E.J. (2002) *New England Journal of Medicine*, 346, 1545–1552.

CHAPTER 8

SINGLE NUCLEOTIDE POLYMORPHISMS AND OTHER BI-ALLELIC MARKERS

Research into the identification and validation of more and better marker systems for forensic analysis should continue with a view of making each profile unique.

(NRC II Report, p. 7. Recommendation 5.3)

ROLE OF ADDITIONAL GENETIC MARKERS IN FORENSIC SCIENCE

In this chapter and the next two, we will examine other DNA markers that are being used or being developed for forensic DNA typing purposes. The 13 core loci described in Chapter 5 are being extensively used today and will probably continue to be used for many years in the future because they are part of the DNA databases that are growing around the world. Yet forensic DNA scientists often use additional markers as the need arises to obtain further information about a particular sample (see Gill *et al.* 2004).

Sex-typing is performed in conjunction with available short tandem repeat (STR) kits to provide the gender of the individual who is the source of the DNA sample in question. Additionally, in cases where samples may be extremely degraded and fail to result in useful information with conventional STR typing, mitochondrial DNA (see Chapter 10) may be used because it is in higher copy number per cell than nuclear DNA and thus more resistant to complete sample degradation.

Y chromosome systems are becoming more popular as a means to extract information from the male portion of a sample mixture (e.g., evidence in rape cases). In Chapter 9, we include a brief review of new Y chromosome STR markers that are being used to provide additional information in forensic DNA analysis. Information from non-human DNA sources (see Chapter 11) has already been used to solve forensic cases and will continue to grow in value as our knowledge of genomic DNA sequence diversity from human populations as well as other organisms improves.

It is conceivable that within a few years, a wide variety of validated marker sets and technologies will exist that will provide a forensic DNA laboratory with a smorgasbord of possibilities in their arsenal of weapons that may be used to solve crimes with biological evidence. In this chapter, we concentrate on a class of genetic markers known as single nucleotide polymorphisms (SNPs) that have received a lot of attention in recent years due to their abundance throughout the human genome.

BASICS OF SINGLE NUCLEOTIDE POLYMORPHISMS (SNPs)

A single base sequence variation between individuals at a particular point in the genome is often referred to as a single nucleotide polymorphism or SNP. SNPs are abundant in the human genome and as such are being used for linkage studies to track genetic diseases (Brookes 1999). Millions of SNPs exist per individual. The abundance of SNPs means that they will likely play a role in the future of differentiating individuals from one another. Table 8.1 compares and contrasts SNP and STR markers. A number of technologies are being developed to miniaturize and automate the procedure for SNP analysis. For example, a microchip-based SNP assay has been described where more than a thousand SNPs were examined simultaneously (Wang *et al.* 1998).

Table 8.1

Comparison of STR and SNP markers. SNPs are more common in the human genome than STRs but are not as polymorphic.

ADVANTAGES AND DISADVANTAGES OF SNPs

SNPs are appealing to the forensic DNA community for several reasons. First and foremost, the polymerase chain reaction (PCR) products from SNPs can be less than 100 bp in size, which means that these markers would be able to withstand degraded DNA samples better than STRs that have amplicons as large as

Characteristics	Short Tandem Repeats (STRs)	Single Nucleotide Polymorphisms (SNPs)
Occurrence in human genome	~1 in every 15 kb	~1 in every 1 kb
General informativeness	High	Low; only 20–30% as informative as STRs
Marker type	Di-, tri-, tetra-, pentanucleotide repeat markers with many alleles	Mostly bi-allelic markers with six possibilities: A/G, C/T, A/T, C/G, T/G, A/C
Number of alleles per marker	Typically >5	Typically 2
Detection methods	Gel/capillary electrophoresis	Sequence analysis; microchip hybridization
Multiplex capability	>10 markers with multiple fluorescent dyes	Potential of 1000s on microchip
Major advantage for forensic application	Many alleles enabling higher success rates for detecting and deciphering mixtures	PCR products can be made small potentially enabling higher success rates with degraded DNA samples

300–400 bp (see Chapter 7). Second, they can be potentially multiplexed to a higher level than STRs. Third, the sample processing and data analysis may be more fully automated because a size-based separation is not needed. Fourth, there is no stutter artifact associated with each allele, which should help simplify interpretation of the allele call. Finally, the ability to predict ethnic origin and certain physical traits may be possible with careful selection of SNP markers.

The vast majority of SNPs are bi-allelic meaning that they have two possible alleles and therefore three possible genotypes. For example, if the alleles for a SNP locus are A and B, then the three possible genotypes would be AA, BB, or AB. Mixture interpretation can present a challenge with SNPs because it may be difficult to tell the difference between a true heterozygote and a mixture containing two homozygotes or a heterozygote and a homozygote. The ability to obtain quantitative information from SNP allele calls is important when attempting to decipher mixtures (Gill 2001).

One of the biggest challenges at this time to using SNPs in forensic DNA typing applications is the inability to simultaneously amplify enough SNPs in robust multiplexes from low amounts of DNA. Because a single bi-allelic SNP by itself yields less information than a multi-allelic STR marker, it is necessary to analyze a larger number of SNPs in order to obtain a reasonable power of discrimination to define a unique profile. Progress is being made in the area of multiplex PCR amplification but as of January 2004 the best result so far is a 35plex with Y chromosome SNPs (Sanchez *et al.* 2003).

HOW MANY SNPs ARE NEEDED?

Since each SNP locus typically possesses only two possible alleles, more markers are needed to obtain a high discriminatory power than for STR loci that possess multiple alleles. Computational analyses have shown that on average 25–45 SNP loci are needed to yield equivalent random match probabilities as the 13 core STR loci (Chakraborty *et al.* 1999). Another study predicted that 50 SNPs possessing frequencies in the range of 20–50% for the minor allele can theoretically result in likelihood ratios similar to approximately 12 STR loci (Gill 2001). The number of SNPs needed may fluctuate in practice because some SNP loci have variable allele frequencies in different population groups. Most likely a battery of 50–100 SNPs will be required to match the same powers of discrimination and mixture resolution capabilities now achieved with 10–16 STR loci (Gill *et al.* 2004).

AVAILABLE HUMAN SNP MARKERS

Large national and international efforts have been underway over the past few years to catalog human variation found in the form of SNP markers. The SNP Consortium (TSC) was established in the spring of 1999 to create a high-density

SNP map of the human genome. The TSC is a collaboration between major pharmaceutical companies, the Wellcome Trust (the world's largest medical research charity), and five leading academic and genome sequencing centers (Holden 2002; see http://snp.cshl.org). The TSC effort has produced several million mapped and characterized human SNP markers that have been entered into public databases including dbSNP housed at the National Institute of Health's National Center for Biotechnology Information (Sherry *et al.* 2001; see http://www.ncbi.nlm.nih.gov/SNP). Additionally, the International HapMap project is a follow-on to the Human Genome Project (see Chapter 2) and includes a plan to type 270 individuals from African, European, and Asian populations with approximately one million SNPs (Gibbs *et al.* 2003; see http://www.hapmap.org/). With these large ventures on-going around the world, there will be no shortage of available SNP markers and accompanying population data.

Several members of the European forensic DNA typing community launched a project in 2003 known as SNP*for*ID that is developing SNP assays to directly aid forensic DNA analysis (Phillips *et al.* 2004). This group is endeavoring to develop highly multiplexed SNP assays using unlinked loci that are well spread throughout the human genome. Population data is also being gathered to measure SNP allele frequencies in various groups of interest.

SNP TYPING ASSAYS AND TECHNOLOGIES

Table 8.2

SNP analysis techniques.

A number of SNP typing methods are available, each with their own strengths and weaknesses. Several reviews of SNP typing technologies have been published and can be consulted for a more in-depth view of methodologies than will be presented here (see Gut 2001, Kwok 2001, Syvänen 2001). A summary and brief description of SNP analysis techniques are listed in Table 8.2.

Method	Description	References
Reverse dot blot or linear arrays	A series of allele-specific probes are attached to a nylon test strip at separate sites; biotinylated PCR products hybridize to their complementary probes and are then detected with a colorimetric reaction and evaluated visually	Saiki *et al.* (1989), Reynolds *et al.* (2000)
Genetic bit analysis	Primer extension with ddNTPs is detected with a colorimetric assay in a 96-well format	Nikiforov *et al.* (1994)
Direct sequencing	PCR products are sequenced and compared to reveal SNP sites	Kwok *et al.* (1994)
Denaturing HPLC	Two PCR products are mixed and injected on an ion-paired reversed-phase HPLC; single base differences in the two amplicons will be revealed by extra heteroduplex peaks	Hecker *et al.* (1999)

Continued

Method	Description	References
TaqMan 5′ nuclease assay	A fluorescent probe consisting of reporter and quencher dyes is added to a PCR reaction; amplification of a probe-specific product causes cleavage of the probe and generates an increase in fluorescence	Livak (1999)
Fluorescence polarization	Primer extension across the SNP site with dye-labeled ddNTPs; monitoring changes in fluorescence polarization reveals which dye is bound to the primer	Chen et al. (1999)
Mass spectrometry	Primer extension across the SNP site with ddNTPs; mass difference between the primer and extension product is measured to reveal nucleotide(s) present	Haff and Smirnov (1997), Li et al. (1999)
High-density arrays (Affymetrix chip)	Thousands of oligonucleotide probes are represented at specific locations on a microchip array; fluorescently labeled PCR products hybridize to complementary probes to reveal SNPs	Wang et al. (1998), Sapolsky et al. (1999)
Electronic dot blot (Nanogen chip)	Potential SNP alleles are placed at discrete locations on a microchip array; an electric field at each point in the array is used to control hybridization stringency	Sosnowski et al. (1997a), Gilles et al. (1999)
Molecular beacons	Hairpin stem on oligonucleotide probe keeps fluorophore and its quencher in contact until hybridization to DNA target, which results in fluorescence	Giesendorf et al. (1998)
Oligonucleotide ligation assay (OLA)	Colorimetric assay in microtiter 96-well format involving ligation of two probes if the complementary base is present	Delahunty et al. (1996)
T_m-shift genotyping	Allelic-specific PCR is performed with a GC-tail attached to one of the forward allele-specific primers; amplified allele with GC-tailed primer will exhibit a melting curve at a higher temperature	Germer and Higuchi (1999)
Pyrosequencing	Sequencing by synthesis of 20–30 nucleotides beyond primer site; dNTPs are added in a specific order and those incorporated result in release of pyrophosphate and light through an enzyme cascade	Anmadian et al. (2000), Andreasson et al. (2002)
Allele-specific hybridization (Luminex™ 100)	Dye-labeled PCR products hybridize to oligonucleotide probes (representing the various SNP types) attached to as many as 100 different colored beads; each bead is interrogated to determine its color and whether or not a PCR product is attached as the beads pass two lasers in a flow cytometer	Armstrong et al. (2000), Budowle et al. (2004)
Minisequencing (SNaPshot™ assay)	Allele-specific primer extension across the SNP site with fluorescently labeled ddNTPs; mobility modifying tails can be added to the 5′-end of each primer in order to spatially separate them during electrophoresis	Tully et al. (1996)
SNPstream® UHT	High-tech version of Genetic bit analysis with a 384-well tag array and 12plex PCR	Bell et al. (2002)

Some of the primary SNP typing methods that have received attention in the forensic community include pyrosequencing (Andreasson *et al.* 2002), TaqMan (Lareu *et al.* 2001), Luminex (Budowle *et al.* 2004), and minisequencing or SNaPshot™ (Tully *et al.* 1996, Sanchez *et al.* 2003). One of the important characteristics of a SNP assay is its ability to examine multiple markers simultaneously since SNPs are not as variable as STRs and typically a limited amount of DNA template is available in forensic casework. While pyrosequencing and TaqMan assays are limited in their multiplexing capabilities, Luminex and minisequencing assays enable multiplexed analysis of a dozen or more SNP markers simultaneously. Minisequencing is now a viable SNP typing option with the availability of the SNaPshot™ kit (Applied Biosystems, Foster City, CA) and multi-colored fluorescent detection electrophoresis instrumentation. In the next section, we go into detail about the SNaPshot assay.

SNaPshot: A PRIMER EXTENSION ASSAY CAPABLE OF MULTIPLEX ANALYSIS

Minisequencing, sometimes referred to as SNaPshot, involves allele-specific primer extension with fluorescent dye-labeled dideoxynucleotide triphosphates (ddNTPs) to help visualize the results. There are three primary steps in performing minisequencing: amplification, primer extension, and analysis (Figure 8.1). First, the region around each SNP locus is amplified using PCR. Amplicons can be pooled following singleplex PCR or simultaneously generated using multiplex PCR. The remaining dNTPs and primers following PCR are destroyed by simply adding two different enzymes to the initial reaction tube or well. Exonuclease (Exo) chews up the single-stranded primers while shrimp alkaline phosphatase (SAP) destroys the dNTP building blocks. These enzymes are often sold together as 'ExoSAP'. It is necessary to remove the primers and dNTPs so they do not interfere with the subsequent primer extension reaction.

Primer extension is performed by adding SNP extension primers, a mixture of the four possible ddNTPs each with a unique fluorescent dye label, and a polymerase to the ExoSAP-treated PCR products. The SNaPshot 'kit' from Applied Biosystems only supplies the fluorescently labeled ddNTPs, buffer, and polymerase making it generic to any primer set. The SNP extension primers are designed to anneal immediately adjacent to a SNP site so that the addition of a single ddNTP will interrogate the nucleotide present at the SNP site in the PCR product. The SNP extension reaction is heated and cooled, usually through 25 cycles on a thermal cycler, to permit a linear amplification of the fluorescent ddNTP addition to the SNP primer by the polymerase. If any dNTPs remain from the preceding PCR reaction, then extension can go beyond the single base. Likewise, the presence of remaining PCR primers could mean that

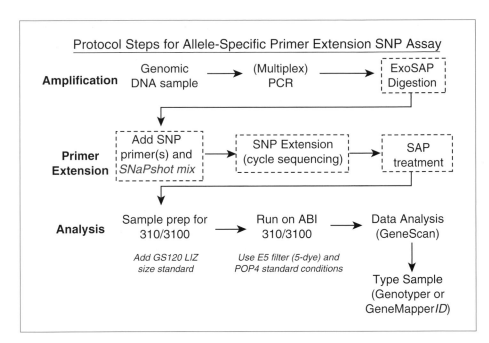

Figure 8.1

*Steps in allele-specific
primer extension SNP
detection (e.g., mini-
sequencing or SNaPshot)
assay. The boxed portions
illustrate additional steps
performed in this assay
relative to STR typing.*

competing side reactions occur and thus interfere with the desired primer extension of the SNP primer.

Following the SNP extension reaction, the products are treated with SAP to remove unincorporated fluorescent ddNTPs. If the SAP-treatment is incomplete, then dye artifacts (*a.k.a.* dye blobs) may occur in the electropherogram and obscure the SNP allele peaks being measured.

The availability of 5-dye detection with electrophoretic platforms (see Chapter 13) enables an internal size standard to be added in the fifth dye channel to correct for migration differences from run-to-run (see Chapter 15). Each of the four nucleotides has their own dye color: A (green), G (blue), C (yellow; usually displayed as black for better visual contrast), and T (red). Thus, the presence of a blue peak in the electropherogram would indicate that a G (ddGTP) had been incorporated by the polymerase at the SNP site.

Multiple primers can be analyzed simultaneously by linking a variable number of additional nucleotides to the 5′-end of the primers so that each primer differs by several nucleotides from its neighbor. Typically a poly(T) tail is used with a 3–5 base spread between primers (Tully *et al.* 1996, Vallone *et al.* 2004) although a mixed sequence that is not complementary to any human sequences has been used successfully (Sanchez *et al.* 2003). Thus, primer1 may contain a 5T 5′-tail, primer2 a 10T tail, primer3 a 15T tail, and primer4 a 20T tail in order to adequately resolve each locus during an electrophoretic separation (Figure 8.2).

Figure 8.2

*Allele-specific primer
extension results using
four autosomal SNP
markers on two different
samples (a). SNP loci are
from separate chromosomes
(1, 6, 14, and 20) and
therefore unlinked.
Electrophoretic resolution
of the SNP primer exten-
sion products occurs due
to poly-T tails that are five
nucleotides different from
one another (b).*

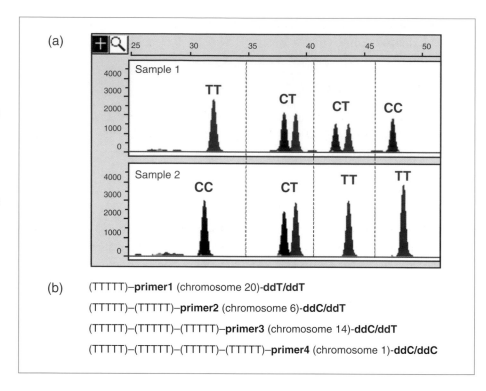

The color of a peak in a minisequencing assay conveys the nucleotide present at the SNP site of interest while the size position of a peak correlates back to its locus based on the 5′-tailing used to differentiate the SNP marker from its neighbors. Homozygous alleles appear as single peaks and heterozygous alleles as two adjacent peaks (Figure 8.2). Automated allele designation can be performed with computer programs such as Genotyper or GeneMapper designed to look for peaks in a particular color and size range.

A major advantage of allele-specific primer extension is that samples can be run on multi-color fluorescence detection electrophoresis instrumentation already available in most forensic DNA typing laboratories (e.g., ABI 310, 377, 3100; see Chapter 14). The technique is sensitive (Vallone *et al.* 2004), has the ability to be multiplexed (Tully *et al.* 1996, Sanchez *et al.* 2003), and has proven to be fairly robust with casework samples (Morley *et al.* 1999, Vallone *et al.* 2004).

POTENTIAL APPLICATIONS FOR SNPs IN HUMAN IDENTITY TESTING

Several potential applications of SNPs exist for human identity testing. These center around three areas: estimating ethnic origin of a sample, predicting physical traits of a perpetrator, and recovering more information from a degraded DNA sample.

ESTIMATING ETHNIC ORIGIN OF A SAMPLE

Forensic DNA typing currently focuses on STR markers that are highly polymorphic and are thus able to readily resolve unrelated individuals. While many more SNPs than STRs are required to obtain similar random match probabilities (Chakraborty *et al.* 1999), SNPs have the potential to be used in other ways to aid investigations, such as predicting a perpetrator's ancestral background (Frudakis *et al.* 2003a).

SNPs have a much lower mutation rate than STRs and therefore are more likely to become 'fixed' in a population. SNPs change on the order of once every 10^8 generations (Brookes 1999) while STR mutation rates are approximately one in a thousand (see Chapter 6). Because of their low mutation rate, SNPs and *Alu* insertions (discussed later in this chapter) are often found to be population-specific (Bamshad *et al.* 2003). These loci could thus be useful in predicting a perpetrator's ethnic origin to aid criminal investigations (see Cyranoski 2004).

The presence of rare STR or SNP alleles in particular population groups can be used to estimate the ethnic origin of a sample. Although efforts have been made with STRs (Lowe *et al.* 2001, Rowold and Herrera 2003), estimating ethnic origin is far from foolproof. Individuals with mixed ancestral background may not possess the expected phenotypic characteristics (e.g., dark colored skin for African-Americans). Thus, results from genetic tests attempting to predict ethnic origin or ancestry should always be interpreted with caution and only in the context of other reliable evidence.

A company in Florida named DNAPrint (Sarasota, Florida) is trying to predict an individual's ethnic/racial background with a panel of 56 SNPs (Frudakis *et al.* 2003a). DNAPrint has targeted pigmentation and xenobiotic metabolism genes in their search for ancestrally informative SNPs. Much of their work is based on the research efforts of Dr. Mark Shriver who is looking for ancestry informative markers (AIMs) that possess alleles with large frequency differences between populations (Shriver *et al.* 1997, Shriver *et al.* 2003). 'Population-specific alleles' (PSAs) have been found in both STR and SNP markers. While presently used AIMs are not 100% accurate for predicting ancestral background of samples (and perhaps never will be), the DNAPrint SNP typing approach was used to aid the investigation of an important serial rapist case in 2003 demonstrating the forensic value of this type of approach (see D.N.A. Box 8.1).

D.N.A. Box 8.1

Aiding a criminal investigation by predicting the ethnic origin of a biological sample

DNA typing tests with the standard 13 STRs linked five murders and rapes in the Baton Rouge, Louisiana area that occurred over an 18-month period. Based on an eyewitness report that a white male was seen leaving one of the crime scenes in a pickup truck, a police dragnet was initiated to collect DNA samples from more than 1000 white males in the area. However, the dragnet and seven long months of investigative work failed to find the culprit.

The police then turned to DNAPrint Genomics Inc. (Sarasota, Florida) to perform a genetic test with single nucleotide polymorphisms (SNPs) to predict the ethnic ancestry of the biological samples obtained from the crime scenes. The DNAPrint test revealed that the samples came from a person who had 85% African-American ancestry and 15% American-Indian ancestry. Authorities turned their attention to black males and within two months arrested Derrick Todd Lee, an African-American resident in the area with an extensive criminal record. Confirmatory testing with the standard 13 STRs matched Derrick Todd Lee's DNA profile with the ones found at the crime scenes.

In the future, this type of analysis to predict ethnicities and even phenotypic characteristics of perpetrators may be used in conjunction with DNA intelligence screens (see D.N.A. Box 18.3) to help narrow the list of possible suspects. Currently SNP tests, like the DNAPrint one, consume too much DNA material to be used routinely on precious crime scene samples. More validation studies will be needed in the future before such ethnicity tests become widely accepted.

Source:
http://www.dnaprint.com;
http://www.genomenewsnetwork.org/articles/06_03/serial.shtmlrl

PHYSICAL TRAITS IDENTIFICATION (PHENOTYPIC EVALUATION)

As more and more information becomes uncovered about the nature and content of the human genome, we will be able to identify the genetic variants that code for phenotypic characteristics (e.g., red hair or blue eyes). For example, the Forensic Science Service has developed an SNP typing assay involving mutations in the human melanocortin 1 receptor gene that are associated with red hair phenotype (Grimes *et al.* 2001). The company DNAPrint is also developing a genetic test for inference of eye color (Frudakis *et al.* 2003b).

Perhaps SNP sites can be identified in the future that will correlate to facial features thus aiding investigators with information about the possible appearance of a perpetrator. However, due to the complexity of multigenic traits and outside factors such as aging and environment, it is unlikely that a few carefully chosen SNPs will present a foolproof picture of a sample's source. Research will likely continue in this area though and hopefully provide beneficial information to investigations of the future.

TYPING WITH DEGRADED DNA SAMPLES

The primary selling point of SNPs in the forensic arena is the ability to make small PCR products that can overcome challenges of strong PCR inhibitors or samples possessing highly degraded material. Smaller PCR products should result in greater recovery of information from badly damaged samples. PCR products for SNP markers can be small because the target region is only a single nucleotide rather than an expandable array of 20–60 nucleotides as is present in tetra-nucleotide STRs with 5–15 repeats. While the promise of SNPs in this area has been promoted for many years, as of early 2004 no clear advantage of SNPs has been demonstrated scientifically in a quantitative fashion. Studies directly comparing SNP assays to STR assays on the same degraded DNA samples are needed.

SNPs have successfully been used on highly degraded samples from victims of the World Trade Center disaster of 11 September 2001 (see Chapter 24). Also valuable in those identification efforts was the use of reduced size STR markers (miniSTRs) (Butler *et al.* 2003). The PCR products from some STR markers can be reduced from over 300 bp to less than 100 bp in length by developing primers that anneal immediately adjacent to the repeat region (see Chapter 7). These reduced size STR assays have an advantage over SNPs in that more alleles exist to produce a higher power of discrimination. More importantly miniSTR assays offer compatibility with current DNA databases housing millions of convicted offender DNA profiles (see Chapter 18).

ADDITIONAL THOUGHTS ON SNPs

SNP typing is already in production use for human identity applications. A modified version of Genetic Bit Analysis (Nikiforov *et al.* 1994), called SNP-IT that runs on the SNPstream UHT platform (Bell *et al.* 2002), has been used by Orchid Cellmark (Dallas, TX) to genotype multiple SNP markers for paternity applications and examine highly degraded samples from mass disasters (see Chapter 24). The forensic DNA typing community will likely hear more about these SNP markers and assays in the future.

As the Human Genome Project and now the International HapMap effort, along with various commercial ventures, continue to improve SNP analysis the testing of multiple SNP sites will become less expensive and easier to perform. In the future, SNP detection platforms may be used in conjunction with fluorescent STR results to establish DNA profiles of forensic casework samples. Some technologies, such as capillary array electrophoresis, mass spectrometry and microchip array hybridization using electronic stringency (see Chapter 17), are capable of performing analysis on both STR and SNP markers.

For SNPs to become more widely used in the future of forensic DNA testing, it is important that the field settle on common loci. STR typing was propelled forward when the Second Generation Multiplex was introduced and used to launch the United Kingdom's National DNA Database (see Chapter 18). Likewise, the selection of 13 core loci for the Combined DNA Index System (CODIS) aided standardization of information in the United States and around the world (see Chapter 5). To aid in cataloging SNP loci of forensic interest and collating features of the markers in a common format, the U.S. National Institute of Standards and Technology has set up a forensic SNP web site: http://www.cstl.nist.gov/biotech/strbase/SNP.htm (Gill *et al.* 2004). This web site is intended to provide a resource to the community as further markers, assays, and technologies are developed for SNP analysis. With gathering this information many loci can be compared and examined for their forensic value to aid in the selection of a consistent set of SNP loci for the community to use as their standard (see Phillips *et al.* 2004).

It is important to keep in mind that SNPs will not replace STRs in the near or even medium term future as the primary source of information used in criminal investigations. The legacy data from national DNA databases means that STRs are here to stay. Replacing the millions of profiles that exist in large national DNA databases through re-typing convicted offender and casework samples with new SNP markers is neither financially practical nor prudent at this juncture.

JOINT ENFSI-SWGDAM ASSESSMENT ON THE FUTURE OF SNPs

In the spring of 2004, the DNA working group of the European Network of Forensic Science Institutes (ENFSI) and the U.S. based Scientific Working Group on DNA Analysis Methods (SWGDAM) jointly issued an assessment of whether SNPs will replace STRs in national DNA databases. The conclusion arrived at was that 'it is unlikely that SNPs will replace STRs as the preferred method of testing of forensic samples and database samples in the near to medium future' (Gill *et al.* 2004). The assessment goes on to praise the capabilities of SNPs for specific purposes including mass disaster and paternity analysis and comments on the need for standardization of SNP loci used for human identity testing applications.

OTHER BI-ALLELIC MARKERS

ALU INSERTION POLYMORPHISMS

Short, interspersed nuclear elements (SINEs) are another form of repeated DNA that has been investigated for population variation studies. SINEs consist of a short, identifiable sequence inserted at a location in the genome. The best studied SINEs are *Alu* insertion polymorphisms, which were named for an *Alu*I

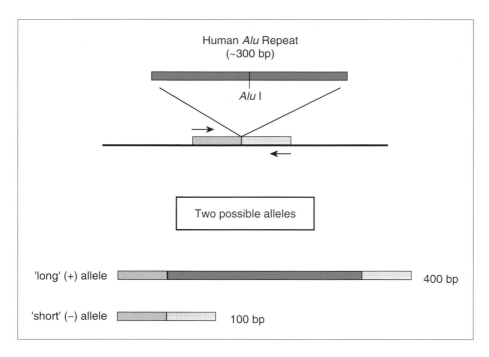

Figure 8.3

Schematic of the Alu *element insertion PCR assay. The* Alu *sequence, which is approximately 300 bp in size, may either be present or absent at a particular location in the human genome. When the flanking region of the* Alu *repeat is targeted with primers, PCR amplification will result in products that are 400 bp if the* Alu *element is inserted or 100 bp if it is absent. A simple ethidium bromide-stained agarose gel may be used to genotype individuals. Individuals that are homozygous for the insertion will amplify a 400 bp DNA fragment. Those who are heterozygous for the insertion will amplify both 400 bp and 100 bp fragments, and individuals that are homozygous for the lack of the* Alu *insertion element will exhibit only the 100 bp DNA fragment.*

restriction endonuclease site typical of the sequence. *Alu* units are found in nearly one million copies per haploid genome (5–10% of the human genome) and can be found flanking genes or clustered with other interspersed repeated sequences (Primrose 1998).

The insertion of an *Alu* element at a particular locus can be regarded as a unique event. Once inserted, *Alu* elements are stable genetic markers and do not appear to be subject to loss or rearrangement. Human-specific *Alu* insertions may be typed in a bi-allelic fashion by using PCR, agarose gel electrophoresis, and ethidium bromide staining (Batzer *et al.* 1993). The presence of the *Alu* insertion will be indicated by a 400 bp PCR product while the absence of the insertion will result in a 100 bp amplicon (Figure 8.3). Commonly used *Alu* insertion polymorphisms include APO, PV92, TPA25, FXIIIB, D1, ACE, A25, and B65 (Sajantila 1998).

Alu repeats have shown the potential to yield information about the geographic/ethnic origin of the sample being tested (Batzer *et al.* 1993, Bamshad *et al.* 2003). Since many *Alu* sequences are unique to humans (Batzer and Deininger 1991), it may be possible to design multiplex assays that are completely human-specific (i.e., no cross-reaction with even other primates). However, *Alu* repeats exhibit less variation than multiplex STR profiles and would therefore most likely be used to gain more information on an unknown sample rather than as an independent source of identification.

INDELS: INSERTION-DELETION POLYMORPHISMS

Another form of a bi-allelic (or di-allelic) polymorphism is an insertion-deletion or *indel*. An indel can be the insertion or deletion of a segment of DNA ranging from one nucleotide to hundreds of nucleotides (such as is seen with an *Alu* insertion). The two alleles for di-allelic indels can simply be classified as 'short' and 'long'. From a certain perspective, STR markers can be thought of as multi-allelic indels since the different alleles are typically insertions or deletions of a tandem repeat unit.

Most di-allelic indels exhibit allele-length differences of only a few nucleotides. James Weber and colleagues at the Marshfield Medical Research Foundation recently characterized over 2000 bi-allelic indels in the human genome (Weber *et al.* 2002). A total of 71% of these indels possessed 2-, 3-, or 4-nucleotide length differences with only 4% having greater than a 16-nucleotide length difference. Allele frequencies for the short and long alleles have been measured in African, European, Japanese, and Native American populations (Weber *et al.* 2002). These markers can be easily typed and may prove useful for future genetic studies including human identity testing.

POINTS FOR DISCUSSION

1. Why are SNPs being considered for use in human identity testing?
2. What are the advantages and disadvantages of SNPs compared to currently used STR markers?
3. Will SNPs replace STRs as a primary means of forensic DNA testing? Why or why not?
4. Are there ethical challenges with using SNPs to predict ethnicity and physical traits? If so, what are they and how should the law enforcement community use this type of information in the future?

REFERENCES AND ADDITIONAL READING

Ahmadian, A., Gharizadeh, B., Gustafsson, A.C., Sterky, F., Nyren, P., Uhlen, M. and Lundeberg, J. (2000) *Analytical Biochemistry*, 280, 103–110.

Andreasson, H., Asp, A., Alderborn, A., Gyllensten, U. and Allen, M. (2002) *Biotechniques*, 32, 124–133.

Armstrong, B., Stewart, M. and Mazumder, A. (2000) *Cytometry*, 40, 102–108.

Bamshad, M.J., Wooding, S., Watkins, W.S., Ostler, C.T., Batzer, M.A. and Jorde, L.B. (2003) *American Journal of Human Genetics*, 72, 578–589.

Batzer, M.A. and Deininger, P.L. (1991) *Genomics*, 9, 481–487.

Batzer, M., Alegria-Hartman, M., Bazan, H., Kass, D.H., Shaikh, T.H., Novick, G.E., Ioannou, P.A., Boudreau, D.A., Scheer, W.D., Herrera, R.J., Stoneking, M. and Deininger, P.L. (1993) *Fourth International Symposium on Human Identification*, pp. 49–57. Madison, Wisconsin: Promega Corporation.

Bell, P.A., Chaturvedi, S., Gelfand, C.A., Huang, C.Y., Kochersperger, M., Kopla, R., Modica, F., Pohl, M., Varde, S., Zhao, R., Zhao, X., Boyce-Jacino, M.T. and Yassen, A. (2002) *Biotechniques*, Suppl, 70–77.

Berger, B., Holzl, G., Oberacher, H., Niederstatter, H., Huber, C.G. and Parson, W. (2002) *Journal of Chromatography B Analytical Technology in Biomedical Life Sciences*, 782, 89–97.

Blondal, T., Waage, B.G., Smarason, S.V., Jonsson, F., Fjalldal, S.B., Stefansson, K., Gulcher, J. and Smith, A.V. (2003) *Nucleic Acids Research*, 31, e155.

Bray, M.S., Boerwinkle, E. and Doris, P.A. (2001) *Human Mutation*, 17, 296–304.

Brookes, A.J. (1999) *Gene*, 234, 177–186.

Broude, N.E., Driscoll, K. and Cantor, C.R. (2001) *Antisense Nucleic Acid Drug Development*, 11, 327–332.

Budowle, B., Koons, B.W. and Errera, J.D. (1996) *Journal of Forensic Sciences*, 41, 660–663.

Budowle, B., Planz, J.V., Campbell, R.S. and Eisenberg, A.J. (2004) *Forensic Science Review*, 16, 21–36.

Butler, J.M., Shen, Y. and McCord, B.R. (2003) *Journal of Forensic Sciences*, 48, 1054–1064.

Callinan, P.A., Hedges, D.J., Salem, A.H., Xing, J., Walker, J.A., Garber, R.K., Watkins, W.S., Bamshad, M.J., Jorde, L.B. and Batzer, M.A. (2003) *Gene*, 317, 103–110.

Chakraborty, R., Stivers, D.N., Su, B., Zhong, Y. and Budowle, B. (1999) *Electrophoresis*, 20, 1682–1696.

Chen, X., Levine, L. and Kwok, P.-Y. (1999) *Genome Research*, 9, 492–498.

Cyranoski, D. (2004) *Nature*, 427, 383.

Dean, F.B., Hosono, S., Fang, L., Wu, X., Faruqi, A.F., Bray-Ward, P., Sun, Z., Zong, Q., Du, Y., Du, J., Driscoll, M., Song, W., Kingsmore, S.F., Egholm, M. and Lasken, R.S. (2002) *Proceedings of the National Academy of Sciences USA*, 99, 5261–5266.

Delahunty, C., Ankener, W., Deng, Q., Eng, J. and Nickerson, D.A. (1996) *American Journal of Human Genetics*, 58, 1239–1246.

Egeland, T., Dalen, I. and Mostad, P.F. (2003) *International Journal of Legal Medicine*, 117, 271–275.

Eng, B., Ainsworth, P. and Waye, J.S. (1994) *Journal of Forensic Sciences*, 39, 1356–1359.

Frudakis, T., Venkateswarlu, K., Thomas, M.J., Gaskin, Z., Ginjupalli, S., Gunturi, S., Ponnuswamy, V., Natarajan, S. and Nachimuthu, P.K. (2003a) *Journal of Forensic Sciences*, 48, 771–782.

Frudakis, T., Thomas, M., Gaskin, Z., Venkateswarlu, K., Chandra, K.S., Ginjupalli, S., Gunturi, S., Natrajan, S., Ponnuswamy, V.K. and Ponnuswamy, K.N. (2003b) *Genetics*, 165, 2071–2083.

Germer, S. and Higuchi, R. (1999) *Genome Research*, 9, 72–78.

Gibbs, R.A., Belmont, J.W., Hardenbol, P., Willis, T.D., Yu, F., Yang, H., Ch'ang, L.Y., Huang, W., Liu, B., Shen, Y., Tam, P.K., Tsui, L.C., Waye, M.M., Wong, J.T., Zeng, C., Zhang, Q., Chee, M.S., Galver, L.M., Kruglyak, S., Murray, S.S., Oliphant, A.R., Montpetit, A., Hudson, T.J., Chagnon, F., Ferretti, V., Leboeuf, M., Phillips, M.S., Verner, A., Kwok, P.Y., Duan, S., Lind, D.L., Miller, R.D., Rice, J.P., Saccone, N.L., Taillon-Miller, P., Xiao, M., Nakamura, Y., Sekine, A., Sorimachi, K., Tanaka, T., Tanaka, Y., Tsunoda, T., Yoshino, E., Bentley, D.R., Deloukas, P., Hunt, S., Powell, D., Altshuler, D., Gabriel, S.B., Zhang, H., Matsuda, I., Fukushima, Y., Macer, D.R., Suda, E., Rotimi, C.N., Adebamowo, C.A., Aniagwu, T., Marshall, P.A., Matthew, O., Nkwodimmah, C., Royal, C.D., Leppert, M.F., Dixon, M., Stein, L.D., Cunningham, F., Kanani, A., Thorisson, G.A., Chakravarti, A., Chen, P.E., Cutler, D.J., Kashuk, C.S., Donnelly, P., Marchini, J., McVean, G.A., Myers, S.R., Cardon, L.R., Abecasis, G.R., Morris, A., Weir, B.S., Mullikin, J.C., Sherry, S.T., Feolo, M., Altshuler, D., Daly, M.J., Schaffner, S.F., Qiu, R., Kent, A., Dunston, G.M., Kato, K., Niikawa, N., Knoppers, B.M., Foster, M.W., Clayton, E.W., Wang, V.O., Watkin, J., Gibbs, R.A., Belmont, J.W., Sodergren, E., Weinstock, G.M., Wilson, R.K., Fulton, L.L., Rogers, J., Birren, B.W., Han, H., Wang, H., Godbout, M., Wallenburg, J.C., L'Archeveque, P., Bellemare, G., Todani, K., Fujita, T., Tanaka, S., Holden, A.L., Lai, E.H., Collins, F.S., Brooks, L.D., McEwen, J.E., Guyer, M.S., Jordan, E., Peterson, J.L., Spiegel, J., Sung, L.M., Zacharia, L.F., Kennedy, K., Dunn, M.G., Seabrook, R., Shillito, M., Skene, B., Stewart, J.G., Valle, D.L., Jorde, L.B., Belmont, J.W., Chakravarti, A., Cho, M.K., Duster, T., Foster, M.W., Jasperse, M., Knoppers, B.M., Kwok, P.Y., Licinio, J., Long, J.C., Marshall, P.A., Ossorio, P.N., Wang, V.O., Rotimi, C.N., Royal, C.D., Spallone, P., Terry, S.F., Lander, E.S., Lai, E.H., Nickerson, D.A., Altshuler, D., Bentley, D.R., Boehnke, M., Cardon, L.R., Daly, M.J., Deloukas, P., Douglas, J.A., Gabriel, S.B., Hudson, R.R., Hudson, T.J., Kruglyak, L., Kwok, P.Y., Nakamura, Y., Nussbaum, R.L., Royal, C.D., Schaffner, S.F., Sherry, S.T., Stein, L.D. and Tanaka, T. (2003) *Nature*, 426, 789–796.

Giesendorf, B.A.J., Vet, J.A.M., Tyagi, S., Mensink, E.J.M.G., Trijbels, F.J.M. and Blom, H.J. (1998) *Clinical Chemistry*, 44, 482–486.

Gill, P. (2001) *International Journal of Legal Medicine*, 114, 204–210.

Gill, P., Werrett, D.J., Budowle, B. and Guerreri, R. (2004) *Science and Justice*, 44, 51–53.

Gilles, P.N., Wu, D.J., Foster, C.B., Dillon, P.J. and Chanock, S.J. (1999) *Nature Biotechnology*, 17, 365–370.

Grimes, E.A., Noake, P.J., Dixon, L. and Urquhart, A. (2001) *Forensic Science International*, 122, 124–129.

Gut, I.G. (2001) *Human Mutation*, 17, 475–492.

Haff, L.A. and Smirnov, I.P. (1997) *Genome Research*, 7, 378–388.

Hecker, K.H., Taylor, P.D. and Gjerde, D.T. (1999) *Analytical Biochemistry*, 272, 156–164.

Hedges, D.J., Walker, J.A., Callinan, P.A., Shewale, J.G., Sinha, S.K. and Batzer, M.A. (2003) *Analytical Biochemistry*, 312, 77–79.

Heller, M.J., Forster, A.H. and Tu, E. (2000) *Electrophoresis*, 21, 157–164.

Holden, A.L. (2002) *Biotechniques*, Suppl, 22–26.

Hosono, S., Faruqi, A.F., Dean, F.B., Du, Y., Sun, Z., Wu, X., Du, J., Kingsmore, S.F., Egholm, M. and Lasken, R.S. (2003) *Genome Research*, 13, 954–964.

Jobling, M.A. (2001) *Forensic Science International*, 118, 158–162.

Jurinke, C., Van Den, B.D., Cantor, C.R. and Koster, H. (2002) *Advances in Biochemical and Engineering Biotechnology*, 77, 57–74.

Kwok, P.-K., Carlson, C., Yager, T.D., Ankener, W. and Nickerson, D.A. (1994) *Genomics*, 23, 138–144.

Kwok, P.Y. (2001) *Annual Reviews in Genomics and Human Genetics*, 2, 235–258.

Lareu, M., Puente, J., Sobrino, B., Quintans, B., Brion, M. and Carracedo, A. (2001) *Forensic Science International*, 118, 163–168.

Li, J., Butler, J.M., Tan, Y., Lin, H., Royer, S., Ohler, L., Shaler, T.A., Hunter, J.M., Pollart, D.J., Monforte, J.A. and Becker, C.H. (1999) *Electrophoresis*, 20, 1258–1265.

Livak, K.J. (1999) *Genetic Analysis*, 14, 143–149.

Lovmar, L., Fredriksson, M., Liljedahl, U., Sigurdsson, S. and Syvanen, A.C. (2003) *Nucleic Acids Research*, 31, e129.

Lowe, A.L., Urquhart, A., Foreman, L.A. and Evett, I.W. (2001) *Forensic Science International*, 119, 17–22.

Marth, G., Schuler, G., Yeh, R., Davenport, R., Agarwala, R., Church, D., Wheelan, S., Baker, J., Ward, M., Kholodov, M., Phan, L., Czabarka, E., Murvai, J., Cutler, D., Wooding, S., Rogers, A., Chakravarti, A., Harpending, H.C., Kwok, P.Y. and Sherry, S.T. (2003) *Proceedings of the National Academy of Sciences USA*, 100, 376–381.

Morley, J.M., Bark, J.E., Evans, C.E., Perry, J.G., Hewitt, C.A. and Tully, G. (1999) *International Journal of Legal Medicine*, 112, 241–248.

Nikiforov, T.T., Rendle, R.B., Goelet, P., Rogers, Y-H., Kotewicz, M.L., Anderson, S., Trainor, G.L. and Knapp, M.R. (1994) *Nucleic Acids Research*, 22, 4167–4175.

Oliphant, A., Barker, D.L., Stuelpnagel, J.R. and Chee, M.S. (2002) *Biotechniques*, Suppl, 56–64.

Paracchini, S., Arredi, B., Chalk, R. and Tyler-Smith, C. (2002) *Nucleic Acids Research*, 30, e27.

Pastinen, T., Partanen, J. and Syvanen, A.C. (1996) *Clinical Chemistry*, 42, 1391–1397.

Phillips, C., Lareu, M., Sanchez, J., Brion, M., Sobrino, B., Morling, N., Schneider, P., Syndercombe Court, D. and Carracedo, A. (2004) Selecting single nucleotide polymorphisms for forensic applications. *Progress in Forensic Genetics 10*, ICS 1261, 18–20.

Primrose, S.B. (1998) *Principles of Genome Analysis: A Guide to Mapping and Sequencing DNA from Different Organisms*. Malden, Massachusetts: Blackwell Science.

Reynolds, R., Walker, K., Varlaro, J., Allen, M., Clark, E., Alavaren, M. and Erlich, H. (2000) *Journal of Forensic Sciences*, 45, 1210–1231.

Romualdi, C., Balding, D., Nasidze, I.S., Risch, G., Robichaux, M., Sherry, S.T., Stoneking, M., Batzer, M.A. and Barbujani, G. (2002) *Genome Research*, 12, 602–612.

Rowold, D.J. and Herrera, R.J. (2003) *Forensic Science International*, 133, 260–265.

Saiki, R.K., Walsh, P.S., Levenson, C.H. and Erlich, H.A. (1989) *Proceedings of the National Academy of Sciences of the United States of America*, 86, 6230–6234.

Sajantila, A. (1998) *Second European Symposium on Human Identification*, pp. 1–5. Madison, Wisconsin: Promega Corporation.

Sanchez, J.J., Borsting, C., Hallenberg, C., Buchard, A., Hernandez, A. and Morling, N. (2003) *Forensic Science International*, 137, 74–84.

Sapolsky, R.J., Hsie, L., Berno, A., Ghandour, G., Mittmann, M. and Fan, J.-B. (1999) *Genetic Analysis*, 14, 187–192.

Sherry, S.T., Ward, M.H., Kholodov, M., Baker, J., Phan, L., Smigielski, E.M. and Sirotkin, K. (2001) *Nucleic Acids Research*, 29, 308–311.

Shriver, M.D., Smith, M.W., Jin, L., Marcini, A., Akey, J.M., Deka, R. and Ferrell, R.E. (1997) *American Journal of Human Genetics*, 60, 957–964.

Shriver, M.D., Parra, E.J., Dios, S., Bonilla, C., Norton, H., Jovel, C., Pfaff, C., Jones, C., Massac, A., Cameron, N., Baron, A., Jackson, T., Argyropoulos, G., Jin, L., Hoggart, C.J., McKeigue, P.M. and Kittles, R.A. (2003) *Human Genetics*, 112, 387–399.

Sosnowski, R.G., Tu, E., Butler, W.F., O'Connell, J.P. and Heller, M.J. (1997a) *Proceedings of the National Academy of Sciences USA*, 94, 1119–1123.

Sosnowski, R.G., Canter, D., Duhon, M., Feng, L., Muralihar, M., Radtkey, R., O'Connell, J., Heller, M. and Nerenberg, M. (1997b) *Proceedings of the Eighth International Symposium on Human Identification*, pp. 119–125. Madison, Wisconsin: Promega Corporation.

Syvänen, A.C., Sajantila, A. and Lukka, M. (1993) *American Journal of Human Genetics*, 52, 46–59.

Syvänen, A.C. (1999) *Human Mutation*, 13, 1–10.

Syvänen, A.C. (2001) *Nature Reviews Genetics*, 2, 930–942.

Tapp, I., Malmberg, L., Rennel, E., Wik, M. and Syvanen, A.C. (2000) *Biotechniques*, 28, 732–738.

Taylor, J.D., Briley, D., Nguyen, Q., Long, K., Iannone, M.A., Li, M.S., Ye, F., Afshari, A., Lai, E., Wagner, M., Chen, J. and Weiner, M.P. (2001) *Biotechniques*, 30, 661–669.

Tully, G., Sullivan, K.M., Nixon, P., Stones, R.E. and Gill, P. (1996) *Genomics*, 34, 107–113.

Vallone, P.M., Hamm, R.S., Coble, M.D., Butler, J.M. and Parsons, T.J. (2004) *International Journal of Legal Medicine*, 118, 147–157.

Walker, J.A., Kilroy, G.E., Xing, J., Shewale, J., Sinha, S.K. and Batzer, M.A. (2003a) *Analytical Biochemistry*, 315, 122–128.

Walker, J.A., Hughes, D.A., Anders, B.A., Shewale, J., Sinha, S.K. and Batzer, M.A. (2003b) *Analytical Biochemistry*, 316, 259–269.

Wang, D.G., Fan, J.-B., Siao, C.-J., Berno, A., Young, P., Sapolsky, R., Ghandour, G., Perkins, N., Winchester, E., Spencer, J., Kruglyak, L., Stein, L., Hsie, L., Topaloglou, T., Hubbell, E., Robinson, E., Mittmann, M., Morris, M.S., Shen, N., Kilburn, D., Rioux, J., Nusbaum, C., Rozen, S., Hudson, T.J., Lipshutz, R., Chee, M. and Lander, E.S. (1998) *Science*, 280, 1077–1082.

Weber, J.L., David, D., Heil, J., Fan, Y., Zhao, C. and Marth, G. (2002) *American Journal of Human Genetics*, 71, 854–862.

White, L.D., Shumaker, J.M., Tollett, J.J. and Staub, R.W. (1998) *Proceedings of the Ninth International Symposium on Human Identification*. Madison, Wisconsin: Promega Corporation. Available online at: http://www.promega.com/geneticidproc/.

Ye, J., Parra, E.J., Sosnoski, D.M., Hiester, K., Underhill, P.A. and Shriver, M.D. (2002) *Journal of Forensic Sciences*, 47, 593–600.

CHAPTER 9

Y CHROMOSOME DNA TESTING

Until recently, the Y chromosome seemed to fulfill the role of juvenile delinquent among human chromosomes – rich in junk, poor in useful attributes, reluctant to socialize with its neighbors and with an inescapable tendency to degenerate…

(Mark Jobling and Chris Tyler-Smith 2003)

LINEAGE MARKERS

Autosomal DNA markers, such as the 13 core short tandem repeat (STR) loci, are shuffled with each generation because half of an individual's genetic information comes from his/her father and half from his/her mother. However, the Y chromosome and mitochondrial DNA markers that will be discussed in this chapter and the next one represent 'lineage markers'. They are passed down from generation-to-generation without changing (except for mutational events). Maternal lineages can be traced with mitochondrial DNA sequence information while paternal lineages can be followed with Y chromosome markers (Figure 9.1).

With lineage markers, the genetic information from each marker is referred to as a haplotype rather than a genotype because there is only a single allele per individual. Because Y chromosome markers are linked on the same chromosome and are not shuffled with each generation, the statistical calculations for a random match probability cannot involve the product rule. Therefore, haplotypes obtained from lineage markers can never be as effective in differentiating between two individuals as genotypes from autosomal markers that are unlinked and segregate separately from generation to generation. However, as will be discussed in this chapter and the next one, Y chromosome and mitochondrial DNA markers do have an important role to play in forensic investigations.

VALUE OF Y CHROMOSOME ANALYSIS IN HUMAN IDENTITY TESTING

The value of the Y chromosome in forensic DNA testing is that it is found only in males. The SRY (sex-determining region of the Y) gene determines maleness. Since a vast majority of crimes where DNA evidence is helpful, particularly

Figure 9.1

Illustration of inheritance patterns from recombining autosomal genetic markers and the lineage markers from the Y chromosome and mitochondrial DNA.

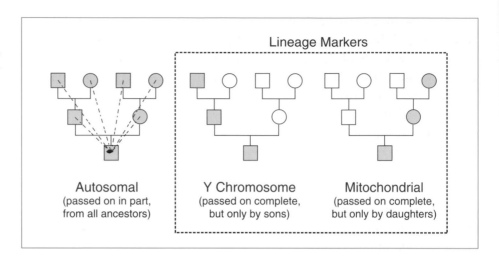

sexual assaults, involve males as the perpetrator, DNA tests designed to only examine the male portion can be valuable. With Y chromosome tests, interpretable results can be obtained in some cases where autosomal tests are limited by the evidence, such as high levels of female DNA in the presence of minor amounts of male DNA (Figure 9.2). These situations include sexual assault evidence from azospermic or vasectomized males and blood–blood or saliva–blood mixtures where the absence of sperm prevents a successful differential extraction for isolation of male DNA (Prinz and Sansone 2001). In addition, the number of individuals involved in a 'gang rape' may be easier to decipher

Figure 9.2

Schematic illustrating the types of autosomal or Y-STR profiles that might be observed with sexual assault evidence where mixtures of high amounts of female DNA may mask the STR profile of the perpetrator. Y-STR testing permits isolation of the male component without having to perform a differential lysis.

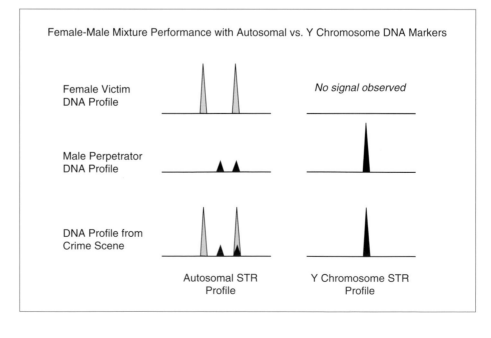

with Y chromosome results than with highly complicated autosomal STR mixtures. Using Y chromosome specific PCR primers can improve the chances of detecting low levels of the perpetrator's DNA in a high background of the female victim's DNA (Hall and Ballantyne 2003a). Y chromosome tests have also been used to verify amelogenin Y deficient males (Thangaraj *et al.* 2002), as mentioned at the end of Chapter 5.

The same feature of the Y chromosome that gives it an advantage in forensic testing, namely maleness, is also its biggest limitation. A majority of the Y chromosome is transferred directly from father to son (Figure 9.1) without recombination to shuffle its genes and provide greater variety to future generations. Random mutations are the only mechanisms for variation over time between paternally related males. Thus, while exclusions in Y chromosome DNA testing results can aid forensic investigations, a match between a suspect and evidence only means that the individual in question could have contributed the forensic stain – as could a brother, father, son, uncle, paternal cousin, or even a distant cousin from his paternal lineage (Figure 9.3)! Needless to say, inclusions with Y chromosome testing are not as meaningful as autosomal STR matches from a random match probability point of view (de Knjiff 2003).

On the other hand, the presence of relatives having the same Y chromosome (see Figure 9.3) expands the number of possible reference samples in missing persons investigations and mass disaster victim identification efforts. Deficient paternity tests where the father is dead or unavailable for testing are benefited if Y chromosome markers are used (Santos *et al.* 1993). However, an autosomal DNA test is always preferred if possible since it provides a higher power of discrimination.

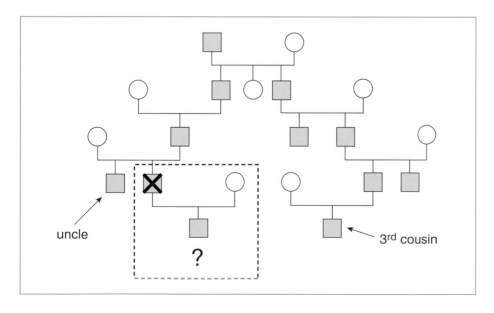

uncle

?

3rd cousin

Figure 9.3

An example pedigree showing patrilineal inheritance where all shaded males have the same Y chromosome barring any mutations. To help identify the person in question, any of the other males with the same patrilineage could provide a reference sample to assist in a missing persons investigation, mass disaster victim identification, or deficient paternity test (boxed region) where the father is deceased or not available for testing.

OTHER APPLICATIONS OF Y CHROMOSOME TESTING

Several uses of Y chromosome testing are listed in Table 9.1. The Y chromosome has become a popular tool for tracing historical human migration patterns through male lineages (Jobling and Tyler-Smith 1995, Jobling and Tyler-Smith 2003). Anthropological, historical, and genealogical questions can be answered through Y chromosome results. For example, as will be discussed later in the chapter, Y chromosome results in 1998 linked modern day descendants of Thomas Jefferson and Eston Hemings leading to the controversial conclusion that Jefferson fathered the slave.

Y CHROMOSOME STRUCTURE

A detailed analysis of the 'finished' reference Y chromosome sequence was described in the 19 June 2003 issue of *Nature* by researchers from the Whitehead Institute and Washington University. Although it is stated as being a 'finished' sequence, Skaletsky *et al.* (2003) report on only 23 Mb of the roughly 50 Mb present in a typical human Y chromosome. The unreported and as yet unsequenced ~30 Mb portion is a heterochromatin region located on the long arm of the Y chromosome (Figure 9.4) that is not transcribed and is composed of highly repetitive sequences, which are impossible to sequence reliably with current technology. At 50 Mb, the Y chromosome is the third smallest human chromosome – only slightly larger than chromosome 21 (47 Mb) and chromosome 22 (49 Mb).

The non-recombining region of the human Y chromosome (NRY) comprises approximately 95% of the chromosome (Figure 9.4a). The two tips of the

Table 9.1

Areas of use in Y chromosome testing (adapted from Butler 2003).

Use	Advantage
Forensic casework on sexual assault evidence	Male-specific amplification (can avoid differential extraction to separate sperm and epithelial cells)
Verification of amelogenin Y deficient males	Analysis of multiple regions along the Y chromosome that should not be affected by deletion of the amelogenin region (see Thangaraj *et al.* 2002, Steinlechner *et al.* 2002)
Paternity testing	Male children can be tied to fathers in motherless paternity cases or testing of male relatives if father is unavailable
Missing persons investigations	Patrilineal male relatives may be used for reference samples
Human migration and evolutionary studies	Lack of recombination enables comparison of male individuals separated by large periods of time
Historical and genealogical research	Surnames usually retained by males; can make links where paper trail is limited

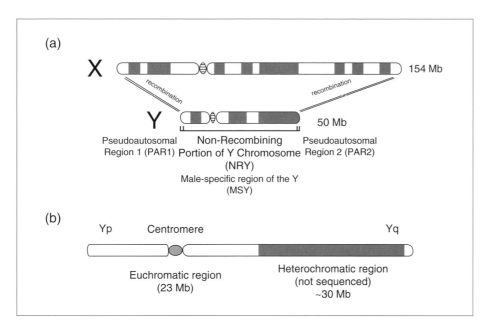

Figure 9.4

(a) Schematic of X and Y sex chromosomes. The two tips of the Y chromosome known as the pseudo-autosomal region 1 (PAR1) and 2 (PAR2) recombine with the tips of the X chromosome. The remaining 95% of the Y chromosome is referred to as the non-recombining portion of the Y chromosome (NRY) or male-specific region of the Y (MSY). (b) The Y chromosome is composed of both euchromatic and heterochromatic regions of which only the 23 Mb of euchromatin has been sequenced.

Y chromosome, known as pseudo-autosomal regions (PAR), recombine with X chromosome homologous regions. PAR1 located at the tip of the short arm (Yp) of the Y chromosome is approximately 2.5 Mb in length while PAR2 at the tip of the long arm (Yq) is less than 1 Mb in size (Graves *et al.* 1998). Skaletsky *et al.* (2003) renamed the NRY the male-specific region (MSY) because of evidence of frequent gene conversion or intrachromosomal recombination. A total of 156 known transcription units including 78 protein-coding genes are present on MSY.

The Y chromosome is highly duplicated either with itself or with the X chromosome. Three classes of sequences have been characterized in the Y chromosome: X-transposed, X-degenerate, and ampliconic (Skaletsky *et al.* 2003). Two blocks on the short arm of the Y chromosome with a combined length of 3.4 Mb make up the X-transposed sequences. These sequences are 99% identical to sequences found in Xq21, contain two coding genes, and do not participate in X–Y crossing over during male meiosis. X-degenerate segments of MSY occur in eight blocks on both the short arm and the long arm of the Y chromosome with an aggregate length of 8.6 Mb. These X-degenerate segments possess up to 96% nucleotide sequence identity to their X-linked homologues. These X-homologous regions can make it challenging to design Y chromosome assays that generate male-specific DNA results. If portions of an X-homologous region of the Y chromosome are examined inadvertently, then female DNA, which possesses two X chromosomes, will be detected. Thus, when testing Y chromosome-specific assays it is important to examine them in the presence of female DNA (high levels) to verify that there is little-to-no cross talk with X-homologous regions of the Y chromosome (Hall and Ballantyne 2003a).

The ampliconic segments are composed of seven large blocks scattered across both the short arm and the long arm and covering about 10.2 Mb of the Y chromosome (Skaletsky *et al.* 2003). Some 60% of these ampliconic sequences have intrachromosomal identities of 99.9% or greater. In other words, it is very difficult to tell these sequences apart from one another. Another interesting feature of these ampliconic segments is that many of them are palindromes – that is, the almost exact duplicate sequences are inverted with respect to each other's sequence essentially as mirror images. Eight large palindromes collectively comprise 5.7 Mb of Yq with at least six of these palindromes containing testis genes. Genetic markers within these palindromic regions will exist as multi-copy PCR products from single primer sets. For example, the DAZ (<u>d</u>eleted in <u>az</u>ospermic) gene occurs in four copies at ~24 Mb along the reference sequence (Saxena *et al.* 1996, Skaletsky *et al.* 2003).

DIFFERENT CLASSES OF Y CHROMOSOME GENETIC MARKERS

Two broad categories of DNA markers have been used to examine Y chromosome diversity: bi-allelic loci, which exhibit two possible alleles, and multi-allelic loci. Results from typing the lower resolution bi-allelic markers are classified into haplogroups while multi-allelic results are characterized as haplotypes (de Knijff 2000).

Bi-allelic markers include single nucleotide polymorphisms (Y-SNPs) and an *Alu* element insertion (see Chapter 8). The Y-*Alu* polymorphism (YAP) was the first discovered Y chromosome bi-allelic marker (Hammer 1994). Bi-allelic markers are sometimes referred to as unique event polymorphisms (UEPs) because of their low mutation rates (~10^{-8} per generation). Approximately 250 bi-allelic Y chromosome markers have been characterized (Y Chromosome Consortium 2002, Butler 2003).

Y chromosome multi-allelic markers include two minisatellites and over 200 short tandem repeat (Y-STR) markers (Butler 2003, Kayser *et al.* 2004). These multi-allelic loci can be used to differentiate Y chromosome haplotypes with fairly high resolution due to their higher mutation rates. Minisatellite loci have mutation rates as high as 6–11% per generation (Jobling *et al.* 1999) while the average mutation rate for Y-STRs is ~0.2% per generation (Kayer *et al.* 2000, Dupuy *et al.* 2004).

Y-STR MARKERS

MINIMAL HAPLOTYPE LOCI

The number of Y chromosome short tandem repeat (Y-STR) loci available for use in human identity testing has increased dramatically since the turn of the century and the availability of the human genome sequence. In the 1990s only a handful of Y-STR markers were characterized and available for use. At the beginning of 2002, only about 30 Y-STRs were available for researchers (Butler 2003). Since that time more than 200 new Y-STRs have been deposited in the Genome Database (GDB; http://www.gdb.org; Kayser *et al.* 2004).

Yet even with a limited number of loci available at the time, a core set was selected in 1997 that continue to serve as 'minimal haplotype' loci (Kayser *et al.* 1997, Pascali *et al.* 1998). The minimal haplotype is defined by the single copy Y-STR loci DYS19, DYS389I, DYS389II, DYS390, DYS391, DYS392, DYS393, and the highly polymorphic multi-copy locus DYS385 a/b (Schneider *et al.* 1998). By means of a multicenter study, more than 4000 male DNA samples from 48 different subpopulation groups were studied with the single copy loci in the minimal haplotype set (de Knijff *et al.* 1997). This work formed the basis for what is now the online Y-STR Haplotype Reference Database (http://www.ystr.org; http://www.yhrd.org) that will be described in more detail below.

In January 2003, the U.S. Scientific Working Group on DNA Analysis Methods (SWGDAM) recommended use of the minimal haplotype loci plus two additional single-copy Y-STRs: DYS438 and DYS439 (Ayub *et al.* 2000). Information regarding these core loci may be found in Table 9.2. Although other Y-STRs will be added to databases as their value is demonstrated and they become part of commercially available kits, the original minimal haplotype loci and SWGDAM recommended Y-STRs are likely to dominate human identity applications in the coming years.

STR Marker	Position (Mb)	Repeat Motif	Allele Range	PCR Product Size	STR Diversity
DYS393	3.04	AGAT	8–16	104–136 bp	0.363
DYS19	9.44	TAGA	10–19	232–268 bp	0.498
DYS391	13.41	TCTA	6–13	90–118 bp	0.552
DYS437	13.78	TCTA	13–17	183–199 bp	0.583
DYS439	13.83	AGAT	8–15	203–231 bp	0.639
DYS389I/II	13.92	TCTG TCTA	10–15/24–34	148–168 bp/256–296 bp	0.538 /0.675
DYS438	14.25	TTTTC	8–12	101–121 bp	0.594
DYS390	16.52	TCTA TCTG	18–27	191–227 bp	0.701
DYS385 a/b	20.00, 20.04	GAAA	7–25	243–315 bp	0.838
DYS392	21.78	TAT	7–18	294–327 bp	0.596

Table 9.2

Characteristics of minimal haplotype and SWGDAM-recommended Y-STR loci. Positions in megabases (Mb) along the Y chromosome were determined with the human genome reference sequence in January 2003 (see Butler 2003). Allele ranges are calculated from kit allelic ladders and do not represent the full range of alleles observed in world populations. PCR product sizes are those present in the PowerPlex® Y kit for the alleles listed. The STR diversity values are calculated from 244 U.S. Caucasian males (Schoske et al. 2004) and are useful to rank the relative informativeness of the loci. Note that DYS437 is not one of the recommended loci but is present in the PowerPlex® Y kit.

SINGLE COPY VS. MULTI-COPY MARKERS

Due to the duplicated, palindromic regions of the Y chromosome mentioned above, some Y-STR loci occur more than once and when amplified with a locus-specific set of primers produce more than one PCR product. This fact can lead to some confusion in terms of counting the number of loci present in a haplotype. A single set of primers can produce two amplicons, which may be thought of as 'two loci' for a Y chromosome haplotype.

For example, the Y-STR locus DYS385 is present in two regions along the long arm of the Y chromosome. These duplicated regions are located about 40 000 bp apart and can generate two different alleles when amplified with a single set of primers. The two alleles are typically labeled 'a' and 'b' with the 'a' designation going to the smaller sized allele. It is also possible to have both 'a' and 'b' alleles be the same size in which case only a single peak would appear in an electropherogram (see Figure 9.5a). Due to the presence of two alleles, this duplicated locus is usually referred to as DYS385 a/b. It has been recently demonstrated that with a nested PCR approach, the 'a' and 'b' alleles for DYS385 can be amplified separately (Kittler *et al.* 2003, Seo *et al.* 2003). Other multi-copy Y-STRs besides DYS385 a/b that have been used in human identity testing include YCAII a/b and DYS464 a/b/c/d (Redd *et al.* 2002, Butler 2003, Schoske *et al.* 2004).

Two PCR products can also be generated at the DYS389 locus with a single set of primers. However in this case the DYS389I PCR product is a subset of the

Figure 9.5

Schematic illustration of how multiple PCR primer binding sites give rise to multi-copy PCR products for (a) DYS385 a/b and (b) DYS389I/II. Arrows represent either forward 'F' or reverse 'R' primers. In the case of DYS385 a/b, the entire region around the STR repeat is duplicated and spaced about 40,775 bp apart on the long arm of the Y chromosome. Thus, amplification with a single set of primers gives rise to one peak if the 'a' repeat region is equal in size to the 'b' repeat region or separate peaks if 'a' and 'b' differ in length. DYS389 possesses two primary repeat regions that are flanked on one side by a similar sequence. Widely used forward primers bind adjacent to both repeats generating amplicons that differ in size by ~120 bp. Note that DYS389II is inclusive of the DYS389I repeat region and therefore some analyses subtract DYS389II–DYS389I repeats.

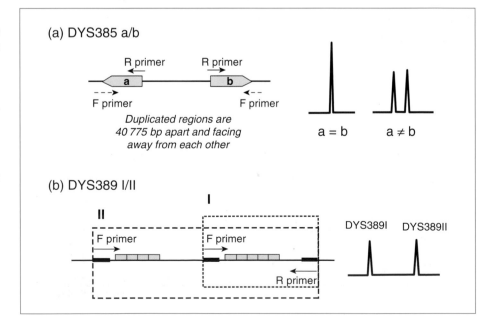

DYS389II amplicon because the forward PCR primer binds to the flanking region of two different repeat regions that are approximately 120 bp apart (Figure 9.5b). Some analyses with DYS389I/II treat the larger PCR product as DYS389II–DYS389I to get a handle on the variation occurring in the two regions independent of one another (e.g., Redd *et al.* 2002).

COMMERCIAL Y-STR KITS AVAILABLE

Forensic scientists rely heavily on commercially available kits to perform DNA testing (see Chapter 5). Thus, many laboratories especially in the U.S. have been reluctant to move into Y-STR typing until Y-STR kits were offered. The first kit was Y-PLEX™ 6 from ReliaGene Technologies (New Orleans, LA), which co-amplifies DYS19, DYS389II, DYS390, DYS391, DYS393, and DYS385 a/b (Figure 9.6). Table 9.3 lists the kits that were available as of June 2004 and the loci that they amplify. Kits that enable a single amplification of all the minimal haplotype and SWGDAM recommended loci are now available.

ReliaGene Technologies sells the Y-PLEX™ 12 kit, which amplifies the SWGDAM recommended loci plus the amelogenin marker (see Chapter 5). Inclusion of amelogenin enables confirmation that the PCR reaction has not failed on female DNA samples since a single X amplicon will result. In addition, mixture levels of male and female DNA can be confirmed with the amelogenin X and Y peak height ratios (see Chapter 7). While the amelogenin primers provide a measure of quality control on PCR amplifications, they have the disadvantage of possibly tying up and consuming PCR reagents when high levels of female DNA are present in a mixture. Shewale *et al.* (2003b) found that male–female mixtures down to 1:400 still resulted in the male component amplifying at the Y-STR loci.

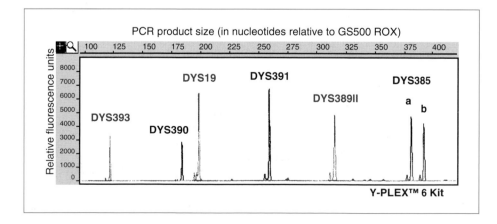

Figure 9.6

Result from the commercial Y-STR kit Y-PLEX™ 6.

Kit Name (Source)	Release Date	Dye Color	Loci Amplified Arranged by Size
Y-PLEX™ 6 (ReliaGene Technologies)	January 2001	B Y	DYS393, DYS19, DYS389II DYS390, DYS391, DYS385 a/b
Y-PLEX™ 5 (ReliaGene Technologies)	July 2002	B G Y	DYS389I, DYS389II DYS439 DYS438, DYS392
genRES® DYSplex-1 (Serac, GERMANY)	2002		DYS390, DYS39I, DYS385 a/b Amelogenin, DYS5389 I/II
genRES® DYSplex-2 (Serac, GERMANY)	2002		DYS392, DYS393 DYS19, DYS389 I/II
Y-PLEX™ 12 (ReliaGene Technologies)	September 2003	B G Y	DYS392, DYS390, DYS385 a/b DYS393, DYS389I, DYS391, DYS389II Amelogenin, DYS19, DYS439, DYS438
PowerPlex® Y (Promega Corporation)	October 2003	B G Y	DYS391, DYS389I, DYS439, DYS389II DYS438, DYS437, DYS19, DYS392 DYS393, DYS390, DYS385 a/b
MenPlex® Argus Y-MH (Biotype, GERMANY)	January 2004	B Y	DYS393, DYS390, DYS385 a/b DYS391, DYS19, DYS389I, DYS392, DYS389II
Yfiler™ (Applied Biosystems)	Fall 2004	B G Y R	DYS456, DYS389I, DYS390, DYS389II DYS458, DYS19, DYS385 a/b DYS393, DYS391, DYS439, C4, DYS392 H4, DYS437, DYS438, DYS448

Table 9.3

Commercially available Y-STR kits. Characteristics of each locus may be found in Table 9.2. An internal size standard is typically run in the fourth or fifth dye position on multi-color fluorescence detection systems for allele sizing purposes. Dye colors = Blue (B), Green (G), Yellow (Y), or Red (R). The underlined loci, such as DYS437 present in the PowerPlex Y kit, are not part of the minimal haplotype or SWGDAM recommended loci.

A major advantage in having commercial Y-STR kits is the availability of common allelic ladders. These allelic ladders provide a consistent currency that aids in quality assurance of results as well as compatibility of results going into DNA databases. The allelic ladders for the DYS385 locus with the Y-PLEX 12 and PowerPlex Y kits are shown in Figure 9.7. As was noted in Chapter 5 with autosomal STR markers, various kits differ with the alleles present in their ladders. Hence the ability to reliably call a DYS385 allele 23 is greater with the PowerPlex Y ladder since this rare allele is present in PowerPlex Y but outside the range of the Y-PLEX 12 ladder rungs.

Y-STR HAPLOTYPE DATABASES

The largest and most widely used Y-STR database was created by Lutz Roewer and colleagues at Humbolt University in Berlin, Germany and has been available online since 2000. The information in this database comes from 89 collaborating institutions located in 36 different countries (Roewer 2003). As of May 2004, more than 24 000 samples from greater than 224 populations

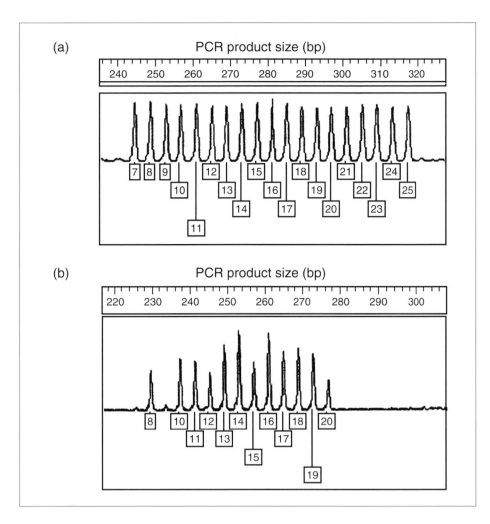

Figure 9.7

Alleles present in (a) PowerPlex® Y and (b) Y-PLEX™ 12 allelic ladders for the DYS385 a/b locus.

around the world can be searched via the Internet at the following web sites: http://www.ystr.org or http://www.yhrd.org. It is important to keep in mind that the samples entered into this database have been typed only at the minimal haplotype loci (DYS19, DYS389I/II, DYS390, DYS391, DYS392, DYS393, and DYS385 a/b). At present, these samples do not include the additional SWGDAM recommended loci DY438 and DYS439 that are now available in commercial Y-STR kits.

Other Internet-accessible Y-STR databases are also available as summarized in Table 9.4. These databases contain information from the minimal haplotype loci, a subset of the minimal haplotype loci, or additional Y-STRs and therefore cannot always be searched across all loci of interest. Hopefully a comprehensive Y-STR database will be developed and made available in the future that will include a larger number of loci typed and a greater number of samples. This database will need to go to efforts to insure that no duplicates are

Population Group	No. of Samples (as of June 2004)	Markers Tested	Reference
224 worldwide populations	24 474	Minimal haplotype loci	www.yhrd.org
94 European populations	13 892	Minimal haplotype loci	Roewer *et al.* (2001); www.ystr.org
30 regional U.S. populations	1705 total 599 African-Americans 628 European-Americans 478 Hispanic-Americans	Minimal haplotype loci	Kayser *et al.* (2002); Kayser *et al.* (2003); www.ystr.org/usa
22 Asian populations	2576	Minimal haplotype loci	Lessig *et al.* (2003); www.ystr.org/asia
U.S. groups	4623 total 2239 African-Americans 1826 Caucasians 454 Hispanics 104 Native Americans	Y-PLEX 6 loci	Sinha *et al.* (2003) www.reliagene.com
U.S. groups	3406 total 1605 African-Americans 1243 Caucasians 454 Hispanics 104 Native Americans	SWGDAM-recommended loci (with Y-PLEX 6 and Y-PLEX 5 kits)	www.reliagene.com
U.S. groups	2443 total 577 African-Americans 595 Caucasians 630 Hispanics 357 Native Americans 284 Asian-Americans	SWGDAM-recommended loci + DYS437 (with PowerPlex Y kit)	www.promega.com
Genealogists from around the world	2300	Up to 49 Y-STRs (run by genetic genealogy companies)	www.ybase.org
Genealogists from around the world	5300 records 4169 unique haplotypes	Up to 37 Y-STRs (run by FamilyTree DNA)	www.ysearch.org
Genealogists from around the world	8735 haplotypes associated with 296 424 individual ancestors	Allele frequencies for 37 Y-STRs (run by Sorenson Genomics/ Relative Genetics)	www.smgf.org

Table 9.4

Summary of available online Y-STR databases.

present so that reliable estimates of the rarity of a profile can be performed. Demonstration that samples are unique may be performed with autosomal STR typing of those specimens that possess similar Y-STR profiles (Butler *et al.* 2002).

DETERMINING THE RARITY OF A Y-STR PROFILE

As with mitochondrial DNA information that will be discussed in the next chapter, the size of the database matters in trying to estimate the rarity of a Y-STR profile. The lack of recombination between Y chromosome markers means that

Y-STR results have to be combined into a haplotype for searching available data-bases as well as estimating the rarity of a particular haplotype.

Generally speaking there are three possible interpretations resulting from a Y-STR test: (1) *exclusion* because the Y-STR profiles are different and could not have originated from the same source, (2) *inconclusive* where there are insufficient data to render an interpretation or ambiguous results were obtained, or (3) *inclusion* or failure to exclude as the Y-STR results from two samples are sufficiently similar and could have originated from the same source.

When the evidence and suspect do not match, then Y-STR typing is helpful in demonstrating the exclusion. However, estimating the strength of a match when a suspect's Y-STR haplotype cannot be excluded is more problematic (de Knijff 2003). Yet it is common practice to place some significance on the likelihood of a random match. Thus, statistics derived from population data would usually be applied.

Three approaches to evaluating the rarity of a coincidental match using Y-STR markers are reviewed by Budowle *et al.* (2003). These include (1) the counting method, (2) a Bayesian approach (Roewer *et al.* 2001, Krawczak 2001), and (3) the use of a 'mismatch' distribution of haplotypes present in a reference database to evaluate how often two randomly selected haplotypes would be at a molecular distance as close as the two matched haplotypes found in the case analysis (Pereira *et al.* 2002).

Within the European Y-STR Haplotype Database (http://www.ystr.org), the most frequent minimal haplotype occurs approximately 3% of the time (Roewer *et al.* 2001). This most common type has the following alleles: DYS19 (14), DYS389I (13), DYS389II (29), DYS390 (24), DYS391 (11), DYS392 (13), DYS393 (13), and DYS385 a/b (11,14). As will be discussed in more detail in a later section, the use of additional markers beyond the minimal haplotype has been shown to help differentiate individuals that have this most common type (Redd *et al.* 2002, Schoske *et al.* 2004).

A query of the Y-STR Haplotype Database (http://www.yhrd.org) on 12 June 2004 with a Y-STR minimal haplotype of DYS19 (14), DYS389I (13), DYS389II (29), DYS390 (24), DYS391 (11), DYS392 (14), DYS393 (13), and DYS385 a/b (11,15) revealed six matches in a worldwide sample of 23 597 haplotypes from a set of 211 populations. These matches can be broken down into the following population statistics in various subpopulations: 1/147 (Bogota, Colombia), 1/185 (Central Portugal), 1/135 (Cologne, Germany), 1/661 (Leipzig, Germany), 1/81 (Liguria, Italy), and 1/247 (London, England).

THE MEANING OF A Y CHROMOSOME MATCH

Due to the fact that the Y chromosome is passed down unchanged (except for mutations) from father to son, the observation of a match with Y-STRs does not carry the power of discrimination and weight into court as an autosomal STR match. Peter de Knijff (2003) discusses some of the challenges of presenting Y-STR results in court. He concludes that information from the Y-STR Haplotype Reference Database (http://www.yhrd.org) should be seen as qualitative rather than quantitative because this database cannot provide a reliable frequency of the population at large. Thus, the fact that the Y-STR profile listed in the previous paragraph is found six times out of 23 597 haplotypes queried in the worldwide database does not necessarily mean that this minimal haplotype profile is expected to be seen 0.025% (6/23 597) times from a random selection of unrelated males. Of course, this same Y-STR profile would also be seen in all brothers, male children, father, uncles, paternal grandfather, paternal cousins, etc., barring any mutations.

The following statement is an example of a conservative conclusion for a matching Y-STR profile as it might be reported to the court (de Knijff 2003): 'The Y-STR profile of the crime sample matches the Y-STR profile of the suspect. Therefore *we cannot exclude the suspect* as being the donor of the crime sample. In addition, we cannot exclude all patrilineal related male relatives and an unknown number of unrelated males as being the donor of the crime sample.' In spite of accuracy of the above conservative statement, courts are likely to require some kind of statistic to give meaning to a match (i.e., more than just a simple 'failure to exclude' statement). Budowle and co-workers (2003) advocate the counting method for this purpose because of its operational simplicity. Confidence intervals may be used to reflect the uncertainty involved in population database samplings of unrelated individuals, particularly since many possible rare haplotypes will not be observed with typical database sizes of hundreds to thousands of individuals. Thus, Y-STR profile frequency estimates with confidence intervals can be calculated in a similar fashion as mitochondrial DNA (see D.N.A. Box 10.3).

Using the counting method with an upper bound confidence limit and following the formulas laid out in D.N.A. Box 10.3, calculation for the above example profile that matched six times in a database of 23 597 yields $0.00025 + 1.96[\{(0.00025)(1-0.00025)\}/23\,597]^{1/2}$ which equals 0.00045 (0.045% or 1 in 2000). Of course, the relevance of using an entire world survey of Y chromosomes rather than a specific ethnic or geographic group for a population comparison must also be considered within the scope of the case. Most likely a number of different population databases would be utilized in a case report for comparison purposes. Frequency estimates calculated with the counting method while not as powerful as those produced with unlinked autosomal STRs may nevertheless be informative in many forensic casework scenarios and provide another piece of evidence in the overall framework of a case.

COMBINING Y-STR INFORMATION AND AUTOSOMAL DNA RESULTS

In some cases, such as a fingernail scrapping, a missing persons investigation, or a mass disaster reconstruction scenario, results from both Y-STR loci and a limited number of autosomal loci may be obtained. The question might then be asked can this information be combined to increase the rarity of a match since the autosomal data by itself may not be satisfactory? While this is a relatively new area and has not been investigated in detail yet, Sinha *et al.* (2004) reason that multiplication of the autosomal STR locus profile frequency obtained following the NRC II recommendations (see Appendix VI) and the Y-STR haplotype frequency obtained with a minimal frequency threshold and correction for sampling (as demonstrated in the previous example) is still conservative based on lack of dependence between Y-STR loci and biological independence of chromosomes. Those interested in a detailed discussion of the theoretical issues between joint match probabilities for Y-STRs and autosomal markers should consult a manuscript by Bruce Walsh from the University of Arizona (see http://nitro.biosci.arizona.edu/zdownload/current_ms/Y-autosomal.pdf).

STUDIES WITH ADDITIONAL Y-STR MARKERS

Within the last several years, a number of new Y chromosome STR markers have been characterized and new multiplex assays developed (Redd *et al.* 1997, Prinz *et al.* 1997, Gusmao *et al.* 1999, Bosch *et al.* 2002, Butler *et al.* 2002, Redd *et al.* 2002, Schoske 2003, Hanson and Ballantyne 2004). Information on additional Y-STR loci and assays is available on the NIST STRBase web site at http://www.cstl.nist.gov/biotech/strbase/y_str.htm.

A few population studies have been conducted that go beyond the minimal haplotype loci in order to assess the power of additional markers in resolving most common types. For example, Berger *et al.* (2003) found that addition of the multi-copy marker DYS464 to the minimal haplotype loci increased the number of different haplotypes in a set of 135 Austrian males from 110 to 122. Schoske and co-workers (2004) demonstrated that 25 samples, which possessed an indistinguishable most common minimal haplotype could be subdivided into 24 different groups (only one pair could not be resolved) with the addition of DYS438, DYS439, DYS464, DYS458, DYS460, and DYS437. Statistical studies have also been performed to compare various combinations of Y-STR loci to the minimal haplotype in order to determine the best order in which to apply the markers (Alves *et al.* 2003). Thus, other loci beyond the core minimal haplotype or SWGDAM-recommended loci are likely to play a valuable role with future forensic DNA analysis involving the Y chromosome.

Table 9.5

Information on additional Y-STR loci (adapted from Butler 2003).

Marker Name	Allele Range	Repeat Motif	GenBank Accession	Reference Allele
YCAII a/b	11–25	CA	AC015978	23
DYS388	10–18	ATT	AC004810	12
DYS425	10–14	TGT	AC095380	10
DYS426	10–12	GTT	AC007034	12
DYS434	9–12	TAAT (CTAT)	AC002992	10
DYS435	9–13	TGGA	AC002992	9
DYS436	9–15	GTT	AC005820	12
DYS441	12–18	CCTT	AC004474	14
DYS442	10–14	TATC	AC004810	12
DYS443	12–17	TTCC	AC007274	13
DYS444	11–15	TAGA	AC007043	14
DYS445	10–13	TTTA	AC009233	12
DYS446	10–18	TCTCT	AC006152	14
DYS447	22–29	TAAWA compound	AC005820	23
DYS448	20–26	AGAGAT	AC025227	22
DYS449	26–36	TTTC	AC051663	29
DYS450	8–11	TTTTA	AC051663	9
DYS452	27–33	YATAC compound	AC010137	31
DYS453	9–13	AAAT	AC006157	11
DYS454	10–12	AAAT	AC025731	11
DYS455	8–12	AAAT	AC012068	11
DYS456	13–18	AGAT	AC010106	15
DYS458	13–20	GAAA	AC010902	16
DYS459 a/b	7–10	TAAA	AC010682	9
DYS460 (A7.1)	7–12	ATAG	AC009235 (r&c)	10
DYS461 (A7.2)	8–14	(TAGA) CAGA	AC009235 (r&c)	12
DYS462	8–14	TATG	AC007244	11
DYS463	18–27	AARGG compound	AC007275	24
DYS464 a/b/c/d	9–20	CCTT	X17354	13
Y-GATA-H4	8–13 (25–30)	TAGA	AC011751 (r&c)	12
DYS635 (C4)	19–25	TSTA compound	G42673	21
Y-GATA-A10	13–18	TAGA	AC011751	13

Y-STR ALLELE NOMENCLATURE

The DNA Commission of the International Society of Forensic Genetics (ISFG) has made a series of recommendations on the use of Y-STR markers (Gill *et al.* 2001). Their recommendations address allele nomenclature, use of allelic ladders, population genetics, and reporting methods.

The ISFG recommendations for Y-STR allelic ladders include the following: (a) the alleles should span the distance of known allelic variants for a particular locus, (b) the rungs of the ladder should be one repeat unit apart wherever possible, (c) the alleles present in the ladder should be sequenced, and (d) the ladders should be widely available to enable reliable inter-laboratory comparisons. The existence of commercially available Y-STR kits has now facilitated the widespread use of consistent allelic ladders (see above).

Unfortunately, because various researchers in the field have taken different approaches to naming Y-STR alleles there are instances of multiple designations for the same allele. An example of this phenomenon that illustrates the importance of standardization is DYS439, which has been called three different ways in the literature (Figure 9.8).

In an effort to provide a unified nomenclature for STR loci, a comparative analysis of the repeat and sequence structure of Y chromosome markers in humans and chimpanzees has been proposed and 11 human Y-STRs have been

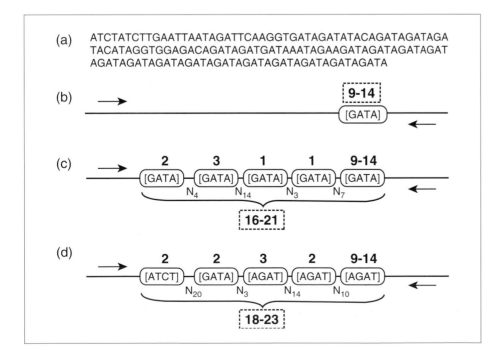

Figure 9.8

Various published allele nomenclatures for DYS439. (a) Sequence from DYS439 spanning the repeat regions used in the various nomenclatures; (b) schematic of allele designation by Ayub et al. 2000 – repeat range is 9–14; (c) schematic of allele designation by Grignani et al. 2000 – repeat range is 16–21; (d) schematic of allele designation by Gonzalez-Neira et al. 2001 – repeat range is 18–23. The most widely accepted designation and what is used in commercial Y-STR kits is (b) – that of Ayub et al. 2000.

studied (Gusmao *et al.* 2002b). Since the chimpanzees examined in their study did not vary in the other regions outside of the variable core GATA repeat for DYS439, Gusmao and co-workers (2002b) proposed a [GATA]$_n$ repeat structure for humans (see Figure 9.8, panel b). This nomenclature has now been adopted for all commercial STR kits typing DYS439.

REFERENCE SAMPLES-NIST SRM 2395

The use of reference samples will aid in obtaining calibrated and consistent results among laboratories performing Y-STR testing. The U.S. National Institute of Standards and Technology (NIST) released Standard Reference Material (SRM) 2395 in July 2003 that serves as a human Y chromosome DNA profiling standard. Five male DNA samples are provided with SRM 2395, each of which has been sequenced at over 20 different Y-STR loci including the common core loci present in all commercially available kits.

MUTATION RATES WITH Y-STR MARKERS

Several studies have been conducted to examine mutation rates among the commonly used Y-STR loci. Most studies have focused on the minimal haplotype loci. Two different approaches have been used: deep-rooted pedigrees (Heyer *et al.* 1997, Bonne-Tamir *et al.* 2003) and male germ-line transmissions from confirmed father/son pairs (Bianchi *et al.* 1998, Kayser *et al.* 2000, Dupuy *et al.* 2001, Dupuy *et al.* 2004, Kurihara *et al.* 2004). The pedigree approach has the advantage of not having to run as many samples but when differences are seen it can be hard to attribute the mutation to the proper generation (see Bonne-Tamir *et al.* 2003) or to potential illegitimacy (Heyer *et al.* 1997). Heyer *et al.* (1997) tested only 42 males but were able to infer information from 213 generations or meioses (once three illegitimate lines had been removed) while Bonne-Tamir *et al.* (2003) examined 74 male samples that spanned 139 generations. Of course the pedigree approach requires detailed genealogical records and no breakdown in the paternal lineages through illegitimacy.

The mutation rates for Y-STRs are in the same range as autosomal STRs, namely about 1–4 per thousand generational events (0.1–0.4%) (see Chapter 6). Table 9.6 includes a summary of mutation rate studies on the minimal haplotype loci. A compilation of the various studies reveals that compound repeat locus DYS390 is the most likely to mutate with DYS392 being the least likely to change. As with autosomal STRs, single repeat changes are favored over multiple repeat jumps. Allele gains are more common than allele losses as the mutations occur with not only locus-specific but also allele-specific differences in mutation rate (Dupuy *et al.* 2004). Mutations typically only occur when 11 or more homogeneous repeats are immediately adjacent to one another (Kayser *et al.* 2000).

Locus	No. of Mutations/ No. of Meioses	Reference	Summed Mutation Rate for Locus
DYS390	8/1766	Dupuy et al. (2004)	14/3144 =
	0/161	Kurihara et al. (2004)	4.45×10^{-3}
	2/150	Dupuy et al. (2001)	0.45%
	4/466	Kayser et al. (2000)	
	0/249	Bianchi et al. (1998)	
	0/139	Bonne-Tamir et al. (2003)	
	0/213	Heyer et al. (1997)	
DYS391	8/1766	Dupuy et al. (2004)	13/3093 =
	1/161	Kurihara et al. (2004)	4.20×10^{-3}
	1/150	Dupuy et al. (2001)	0.42%
	2/415	Kayser et al. (2000)	
	0/249	Bianchi et al. (1998)	
	1/139	Bonne-Tamir et al. (2003)	
	0/213	Heyer et al. (1997)	
DYS385 a/b	7/1766	Dupuy et al. (2004)	12/3381 =
	0/161	Kurihara et al. (2004)	3.55×10^{-3}
	1/150	Dupuy et al. (2001)	0.35%
	2/952	Kayser et al. (2000)	
	1/139	Bonne-Tamir et al. (2003)	
	1/213	Heyer et al. (1997)	
DYS389II	4/1766	Dupuy et al. (2004)	11/3103 =
	1/161	Kurihara et al. (2004)	3.54×10^{-3}
	2/150	Dupuy et al. (2001)	0.35%
	2/425	Kayser et al. (2000)	
	0/249	Bianchi et al. (1998)	
	0/139	Bonne-Tamir et al. (2003)	
	2/213	Heyer et al. (1997)	
DYS389I	4/1766	Dupuy et al. (2004)	8/2890 =
	1/161	Kurihara et al. (2004)	2.77×10^{-3}
	0/150	Dupuy et al. (2001)	0.28%
	1/425	Kayser et al. (2000)	
	0/249	Bianchi et al. (1998)	
	2/139	Bonne-Tamir et al. (2003)	
DYS19	3/1766	Dupuy et al. (2004)	7/3674 =
	0/161	Kurihara et al. (2004)	1.90×10^{-3}
	0/150	Dupuy et al. (2001)	0.19%
	2/996	Kayser et al. (2000)	
	0/249	Bianchi et al. (1998)	
	2/139	Bonne-Tamir et al. (2003)	
	0/213	Heyer et al. (1997)	
DYS393	1/1766	Dupuy et al. (2004)	1/3093 =
	0/161	Kurihara et al. (2004)	3.23×10^{-4}
	0/150	Dupuy et al. (2001)	0.032%
	0/415	Kayser et al. (2000)	
	0/249	Bianchi et al. (1998)	
	0/139	Bonne-Tamir et al. (2003)	
	0/213	Heyer et al. (1997)	

Table 9.6

Summary of mutation rates observed in Y-STRs studies.

Table 9.6
(Continued)

Locus	No. of Mutations/ No. of Meioses	Reference	Summed Mutation Rate for Locus
DYS392	0/1766	Dupuy *et al.* (2004)	1/3093 =
	0/161	Kurihara *et al.* (2004)	3.23 × 10⁻⁴
	0/150	Dupuy *et al.* (2001)	0.032%
	0/415	Kayser *et al.* (2000)	
	0/249	Bianchi *et al.* (1998)	
	0/139	Bonne-Tamir *et al.* (2003)	
	1/213	Heyer *et al.* (1997)	

Kayser and Sajantila (2001) discuss the implications of mutations for paternity testing and forensic analysis. They observed mutations at two Y-STRs within the same father/son pair suggesting that differences at three or more Y-STRs are needed before an 'exclusion' can be declared with paternity testing or kinship analysis, which is typically the same criteria used for paternity testing with autosomal loci (see Chapter 23).

Occasionally duplications or even triplications of a Y-STR locus have been reported, particularly for DYS19. It is important to keep this fact in mind so that two peaks at the DYS19 locus are not automatically interpreted as coming from a mixture of two males. Both of these issues, namely mutations impacting paternity analysis and duplications of loci potentially confusing mixture interpretation, suggest that analysis of additional Y-STR loci can be helpful in these situations.

ISSUES WITH USE OF Y-STRs IN FORENSIC CASEWORK

Y-STR assays have been used for several years on a limited basis to aid forensic casework. Their use has been much more widespread in Europe than the United States. Early work in the U.S. with Y-STRs was performed in the late 1990s by the New York City Office of the Chief Medical Examiner (OCME). ReliaGene Technologies, Inc. (New Orleans, LA) developed the first Y-STR kit and started doing Y chromosome testing in late 2000.

The New York City Office of the Chief Medical Examiner primarily uses Y-STR testing when any one of the four scenarios is met (Prinz 2003): (1) evidence is positive for semen but no DNA foreign to the victim can be detected, or potential male alleles are below the detection threshold with autosomal STR tests; (2) the evidence in question is amylase positive and a male/female mixture is expected; (3) a large number of semen stains need to be screened; and (4) the number of semen donors need to be determined (e.g., suspected gang rape).

There have been several published reports describing the use and value of Y-STR testing in forensic casework. Some of these published results are summarized in Table 9.7.

Kit/Loci Used	Reference	Comments
In-house assay with DYS19, DYS390, DYS389I/II	Prinz et al. (2001)	In one year at the New York City Office of the Chief Medical Examiner, Y-STR testing was performed in more than 500 cases with over 1000 evidence and reference samples examined. A full or partial profile was obtained on 81% of all tested evidence samples (740 worked/915 samples tested). Mixtures of at least two males were observed in 97 instances. In male/female mixtures of up to 1:4000, the male component could be cleanly detected.
In-house assay with 9 Y-STR loci amplified in 3 PCR reactions	Dekairelle and Hoste (2001)	Y-STR typing was attempted on 166 semen traces from 89 cases that failed to yield a detectable male autosomal profile following differential extraction. About half of the cases had sufficient DNA to produce a Y-STR profile.
In-house assay with DYS393, DYS389I/II	Sibille et al. (2002)	Y-STR results could still be obtained more than 48 hours after the sexual assault in 30% of the cases examined. In 104 swabs collected with no evidence of sperm, Y-STRs could be detected in ~29% of the samples tested.
In-house assay with DYS19, DYS390, DYS389I/II	Prinz (2003)	Six case studies are reviewed along with advantages and disadvantages of Y-STR testing in each case: (1) different semen donors on vaginal swab and underwear; (2) possible oligospermic perpetrator gave a nice Y-STR profile but failed to have a 'male' fraction with differential extraction; (3) oral intercourse with no autosomal results – not possible to enrich male cell fraction with differential extraction in cases involving saliva; (4) presence of multiple semen donors created a complex autosomal mixture that could be sorted out with Y-STR results; (5) sperm cell fraction lacked amelogenin Y-specific peak due to known deletion – Y-STR results confirmed that the sperm cell fraction DNA was of male origin; and (6) Y-STR testing was used to rapidly screen 18 semen stains for comparison to five suspects and thus save the time of performing the differential extraction
Y-PLEX 6 and Y-PLEX 5 kits	Sinha (2003)	Five cases are reviewed: (1) criminal paternity case with a male fetus where the alleged father could not be excluded as the biological father; (2) autosomal STR test resulted in an uninterpretable mixture – suspect was excluded at three of the seven Y-STR loci tested; (3) Y-PLEX 6 STR profile matched suspect with sweat stains on cloth found at crime scene; (4) fingernail cuttings from a victim matched a suspect at 11 Y-STR loci while another suspect was excluded at two loci; (5) semen positive stain with no sperm cells produced a Y-PLEX 6 profile consistent with the male suspect
Y-PLEX 6 and Y-PLEX 5 kits	Sinha et al. (2004)	Seven cases are reviewed (some are the same as Sinha 2003) and a list of cases where Y-STR results have been accepted in U.S. courts is provided.

ReliaGene reported use of Y-STRs on 188 forensic samples from 2000–2003 with their Y-PLEX™ 6 and Y-PLEX™ 5 kits (Sinha et al. 2004). Samples were from epithelial cells including azospermic seminal fluid, sweat or saliva, sperm, fingernails, blood, and other tissues. Y-STR testing has been accepted in several jurisdictions throughout the United States (Sinha et al. 2004).

Determining the amount of male DNA present rather than the total amount of male and female DNA is important to getting on-scale results with Y-STR testing.

Table 9.7

Some published reports describing use of Y-STRs in forensic casework.

One approach is to estimate the general level of male DNA present by assessing the strength of the p30 antibody signal (see Chapter 3) (Prinz 2003). The recent availability of real-time PCR assays specific to the male DNA component of a forensic mixture (e.g., Quantifiler Y kit) provides a more high-tech approach.

Duplications or triplications of several Y-STRs have been reported for DYS19, DYS390, and DYS391. For example, one study found nine duplications for DYS19 in 7772 individuals (Kayser *et al.* 2000). Triplicated DYS385 alleles have also been reported (Kayser *et al.* 2000, Butler *et al.* 2002, Kurihara *et al.* 2004). These possible multi-allelic patterns need to be kept in mind so that a mixture is not expected when encountering multiple alleles at a single locus that could legitimately come from a single-source sample (see Chapter 7).

Y-SNP AND BI-ALLELIC MARKERS

Single nucleotide polymorphisms, *Alu* insertions, and insertion/deletion markers exist on the Y chromosome just as they do throughout the rest of the human genome (see Chapter 8). Most of the focus to date in forensic DNA typing applications has been on Y-STRs rather than Y-SNPs due to the higher power of discrimination with the multi-allelic Y-STRs. Y-SNPs play an important role in human migration studies though because they enable effective evaluation of major differences between population groups.

Y-SNP alleles are typically designated as either 'ancestral' or 'derived' and can be recorded in a simple binary format of 0 or 1 for ancestral and derived, respectively. The ancestral state of a Y-SNP marker is usually determined by comparison to a chimpanzee DNA sequence for the same marker (Underhill *et al.* 2000).

THE Y CHROMOSOME CONSORTIUM UNIFIED TREE FOR Y HAPLOGROUPS

The Rosenta Stone for interpreting the plethora of Y chromosome haplogroups listed in the literature was published by the Y Chromosome Consortium (YCC) in the February 2002 issue of *Genome Research* (YCC 2002). The YCC is an international group of scientists lead by Michael Hammer from the University of Arizona, Peter Underhill from Stanford University, Mark Jobling from Leicester University, and Chris Tyler-Smith who at that time was at Oxford University. Their paper entitled '*A nomenclature system for the tree of human Y chromosomal binary haplogroups*' opened the way for an easier understanding of seven previously published methods for describing information from the same SNP markers. The 'YCC tree' as it is commonly called describes the position of almost 250 bi-allelic markers in differentiating 153 different haplogroups (YCC 2002).

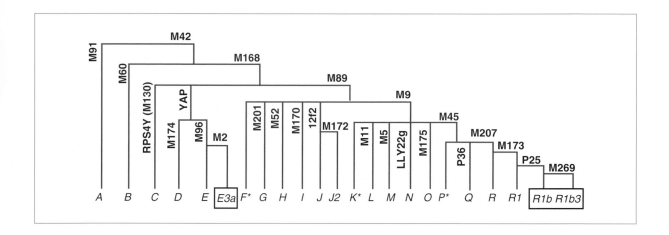

Figure 9.9

Y Chromosome Consortium tree with 18 major haplogroups (A–R). Representative Y-SNP markers that define each haplogroup are listed next to the branch point. The most common African-American haplogroup is E3a. The most common Caucasian (European) haplogroup is R1b/R1b3.

A slightly modified and updated YCC tree was published in August 2003 (Jobling and Tyler-Smith 2003).

Figure 9.9 highlights the major branches of the YCC tree along with some of the Y-SNP markers that help define the various branches. For example, observation of the derived allele for M2 in a sample classifies it into the E3a haplogroup. Y chromosome haplogroup designation and characterization has greatly benefited from the YCC tree.

Before 2002, if a 'G' (derived state) was observed in a sample when typing the M2 (sY81 or DYS271) marker, then the sample could be reported as belonging to Haplogroup (Hg) 8 by Jobling's nomenclature (Jobling and Tyler-Smith 2000), Hg III by Underhill's naming procedure (Underhill *et al.* 2000), or Hg 5 by Hammer's description (Hammer *et al.* 2001). On the YCC tree, M2 derived alleles define the Hg E3a. Needless to say, the unified and universal nomenclature is much easier to understand and permits comparisons of results across laboratories.

A number of the SNP typing technologies reviewed in Table 8.2 have also been used for Y-SNP typing (Butler 2003). Two of the more popular SNP typing methodologies have been allele-specific primer extension (ASPE) and allele-specific hybridization (ASH). Vallone and Butler (2004) observed complete concordance comparing ASPE and ASH in almost 4000 Y-SNP allele calls.

HISTORICAL AND GENEALOGICAL STUDIES WITH THE Y CHROMOSOME

Y chromosome testing is beginning to play a role in addressing some interesting historical questions (see D.N.A. Box 9.1) as well as aiding efforts in genealogical family history research (see D.N.A. Box 9.2). The use of Y chromosome

D.N.A. Box 9.1
Genetic legacy of Genghis Khan

In a study of more than 2100 males from Central Asia, a team of scientists lead by Chris Tyler-Smith from the University of Oxford found that approximately 8% of those studied had a unique Y chromosome lineage. These samples formed a central star-cluster that possessed the following Y-STR profile (repeat numbers in parentheses after each marker): DYS389I (10), DYS389II (26), DYS390 (25), DYS391 (10), DYS392 (11), DYS393 (13), DYS388 (14), DYS425 (12), DYS426 (11), DYS434 (11), DYS435 (11), DYS436 (12), DYS437 (8), DYS438 (10), and DYS439 (10). An analysis of at least 16 additional markers in the form of Y-SNPs placed all of these samples in haplogroup C*(xC3c), which is common in Asia (see Jobling and Tyler-Smith 2003). By making some assumptions regarding mutation rates and a generational time of 30 years, these researchers were able to calculate a time to the most recent common ancestor for these particular Y-lineages of ~1000 years ago. The highest frequency was in Mongolia leading to the assumption that it was the source of these particular male lineages, but it was spread throughout 16 different populations in Asia. Interestingly the geographical distribution of these populations closely matches the area of Genghis Khan's former Mongol Empire. The evidence that this Y-lineage was from Genghis Khan (circa 1162–1227) and his close male-line relatives was strengthened by a match to a group in Pakistan who by oral tradition consider themselves direct male-line descendants of Genghis Khan. Thus DNA testing can reveal some interesting historical clues into our past as a human race.

Sources:
Zerjal, T., *et al.* (2003) The genetic legacy of the Mongols. *American Journal of Human Genetics*, 72, 717–721.
Jobling, M.A. and Tyler-Smith, C. (2003) The human Y chromosome: an evolutionary marker comes of age. *Nature Reviews Genetics*, 4, 598–612.

information in addressing the issue of whether or not Thomas Jefferson fathered some of his slaves is presented below as an example of the power and the pitfalls of answering historical questions with DNA information from modern-day individuals. As in forensic casework, DNA information is only part of the evidence available in most investigations and should be considered carefully in the context of the 'case' without overstepping the bounds of conclusions that can be drawn.

THE THOMAS JEFFERSON-SALLY HEMINGS AFFAIR

In 1802, a year after becoming President of the United States, Thomas Jefferson was publicly accused by a Richmond, Virginia newspaper of fathering a child by his slave, Sally Hemings. While it is uncertain how this accusation arose, the connection between Thomas Jefferson and his slave Sally Hemings has been a source of controversy for almost 200 years.

Then, in November 1998, the prestigious scientific journal *Nature* published a report that introduced DNA evidence into this historical controversy (Foster *et al.* 1998). The report entitled 'Jefferson fathered slave's last child' used Y chromosome DNA markers to trace the Jefferson male line to a descendant of Sally Hemings's youngest son, Eston Hemings. The study involved 19 samples collected from living individuals who represented the Jefferson and Hemings line as well as other people who potentially could have been Jefferson's offspring or the father of Eston Hemings. These samples were tested at 19 different sites on the Y chromosome.

TRACKING DOWN LIVING RELATIVES

The study began in 1996 when Dr. Eugene Foster, a retired pathology professor, began tracking down living male-line relatives of President Thomas Jefferson. In order to show whether or not President Jefferson had fathered a child with Sally Hemings, direct male descendants were needed from both the Jefferson and the Hemings lines. Unfortunately, Jefferson's only legitimate son died in infancy. His two daughters who lived to adulthood obviously did not carry his Y chromosome and therefore their descendants were not useful in this study. There were two other possibilities for direct male-line descendants, Thomas Jefferson's brother Randolph and his father's brother Field.

The last of the direct male descendants of Jefferson's brother Randolph died in the 1920s or 1930s so Dr. Foster turned to the relatives of President Jefferson's paternal uncle, Field Jefferson. Seven living descendants of Field Jefferson were located. Five of them agreed to cooperate in the study and had their blood drawn for Y chromosome marker testing purposes.

On the Hemings side of the equation, it was even more difficult to come up with an abundance of living male relatives. Sally Hemings had at least six and possibly seven children: Harriet (1795–1797), Beverly (1798–post 1822), Harriet (1801–post 1822), an unnamed daughter (1799–1800), Madison (1805–1877), and Eston (1808–1856). According to the oral history of the descendants of Thomas Woodson (1790–1879), he was Sally Hemings's first child. Sally's son, Beverly and daughter, Harriet are listed as dying post 1822 because they disappeared into white society in the Washington, D.C. area in the year 1822.

Of the three known male sons from Sally Hemings, only descendants of Madison and Eston could possibly be located since Beverly's fate is unknown. Madison's Y chromosome line ended in the mid-1800s when one of his three sons vanished into white society and the other two had no children. Thus, Eston Hemings's descendants remained the last chance to find a male-line descendant of the man who fathered Sally Hemings's children.

Eston Hemings was born on 21 May 1808, at Monticello where he lived until President Jefferson's death in 1826, at which time he was freed. Eventually he married and moved to Ohio and finally to Madison, Wisconsin where he died and was buried in 1856. Eston assumed the surname of Jefferson when he left Virginia and gave everyone the impression that he was white because of his light skin color.

Eston Hemings Jefferson had two sons and a daughter. His youngest son, Beverly Jefferson, lived from 1838–1908 and had one son. This son, Carl Smith Jefferson, lived from 1876–1941 and had two sons, William Magill Jefferson (1907–1956) and Carl Smith Jefferson, Jr. (1910–1948). Only William had a son. This son, John Weeks Jefferson was born in 1946. As the *only* living male descendant of Eston Hemings, John Weeks Jefferson's blood was drawn to help answer the question of whether or not President Thomas Jefferson was Eston Hemings's father (Murray and Duffy 1998).

ADDITIONAL SAMPLES GATHERED FOR THIS STUDY

Several additional samples were gathered to serve as controls in this study and to address potential paternity questions. Thomas Woodson, who was mentioned earlier as possibly the first child born to Sally Hemings, was an African-American whose first known appearance in the documentary record is from a deed issued in 1807. He moved from Virginia to Ohio where he lived as a successful farmer until his death in 1879. His descendants now number over 1400 and are scattered across the United States (Murray and Duffy 1998). According to Woodson family tradition, he was the oldest child of Thomas Jefferson and Sally Hemings, born in 1790 shortly after Sally returned to Monticello from France (Monticello 2000). While there were no supporting documents for the claim of Thomas Woodson's family, Dr. Foster collected blood samples from five of his living descendants to help confirm or disprove this family tradition.

Another important set of samples for testing was gathered from direct male line descendants of Samuel and Peter Carr, who were Thomas Jefferson's nephews, the sons of his sister. According to Thomas Jefferson's grandchildren Thomas Jefferson Randolph and Ellen Coolidge, Samuel and Peter Carr were the fathers of the children of Sally Hemings and her sister (Monticello 2000). Dr. Foster collected three blood samples from living descendants of John Carr, the grandfather of Samuel and Peter Carr.

Finally, five male descendants from several old-line Virginia families around Charlottesville were sampled to serve as control samples. These controls were tested to provide a 'background' signal with the idea that potential similarities in the Y chromosome tests due to geographic proximity needed to be eliminated (Murray and Duffy 1998).

THE Y CHROMOSOME MARKERS EXAMINED

DNA samples from each of the 19 blood specimens gathered by Dr. Foster were carefully extracted by a pathologist at the University of Virginia (Murray and Duffy 1998). The DNA samples were coded by Dr. Foster and then taken to England where researchers at Oxford University examined them. Eventually the team of scientists involved expanded to include researchers from the University of Leicester in England and Leiden University in the Netherlands. A variety of tests were run independently at these three locations (Foster *et al.* 1998).

The Y chromosome markers used in this study are listed in Table 9.8. In all, there were 19 Y chromosome markers examined in this study. These included 11 STRs, seven SNPs, and one minisatellite MSY1, which proved to be the most polymorphic marker.

DNA Marker Tested	Field Jefferson Male-Line	Eston Hemings Male-Line	John Carr Male-Line	Thomas Woodson Male-Line
Number of individuals typed	5	1	3	5
Y STR Loci				
DYS19	15	15	14 ←	14 ←
DYS388	12	12	12	12
DYS389A	4	4	5 ←	5 ←
DYS389B	11	11	12 ←	11
DYS389C	3	3	3	3
DYS389D	9	9	10 ←	10 ←
DYS390	11	11	11	11
DYS391	10	10	10	13 ←
DYS392	15	15	13 ←	13 ←
DYS393	13	13	13	13
DXYS156Y	7	7	7	7
Y SNP Loci	(0 = ancestral state; 1 = derived state)			
DYS287 (YAP)	0	0	0	0
SRYm8299	0	0	0	0
DYS271 (SY81)	0	0	0	0
LLY22g	0	0	0	0
Tat	0	0	0	0
92R7	0	0	1 ←	1 ←
SRYm1532	1	1	1	1
Minisatellite Locus				
MSY1	(3)–5	(3)–5		
	(1)–14	(1)–14	(1)–17 ←	(1)–16 ←
	(3)–32	(3)–32	(3)–36 ←	(3)–27 ←
	(4)–16	(4)–16	(4)–21 ←	(4)–21 ←

Table 9.8

Y chromosome markers and results used to trace Thomas Jefferson's male-line ancestry (Foster et al. 1998). The Field Jefferson (uncle of President Thomas Jefferson) male-line matches the Eston Hemings (youngest son of Sally Hemings, one of President Jefferson's slaves) male-line exactly. The numbers shown in the table represent the number of repeats observed for each Y chromosome marker. Arrows have been placed next to the alleles in the Y haplotypes for John Carr and Thomas Woodson male-lines that differ from the Jefferson males.

All 19 regions of the Y chromosome examined in this study matched between the Jefferson and Hemings descendants. These DNA results were viewed by Dr. Foster and his co-authors as evidence for President Thomas Jefferson fathering the last child of Sally Hemings (Foster *et al.* 1998). Interestingly the John Carr and Thomas Woodson's male lines differed significantly from the Jefferson-Hemings results (Table 9.8). At least seven of the 19 tested DNA markers gave different results. Thus, Thomas Jefferson could not be linked as the father of Thomas Woodson nor was Samuel Carr or Peter Carr the father of Eston Hemings. The results of the Virginia old-line families were not reported, presumably because these samples served their purpose as effective controls and revealed no unusual Y chromosome patterns.

In this study, Y chromosome markers demonstrated their usefulness in monitoring paternal transmission of genetic information by tracing the male lineage of Thomas Jefferson across 15 generations (Figure 9.10). The ability to connect Y chromosome DNA information across the generation gaps meant that living relatives could be used in this investigation rather than disturbing the almost 200-year-old burial site of President Jefferson.

ALTERNATIVE SCENARIOS

Shortly after the results of Dr. Foster's study were announced, an alternative scenario was proposed. Could some other male Jefferson have fathered

Figure 9.10

Ancestry of Thomas Jefferson and Eston Hemings male lines. The shaded boxes represent the samples tested by Foster et al. *(1998) in their Jefferson Y chromosome study. A male descendant of Eston Hemings, son of Thomas Jefferson's slave Sally Hemings, was found to have a Y chromosome haplotype that matched male descendants of Field Jefferson, President Thomas Jefferson's uncle. A male descendant of Thomas Woodson, claimed by some to be descended from Jefferson, had a different Y haplotype and therefore could not have been a Jefferson.*

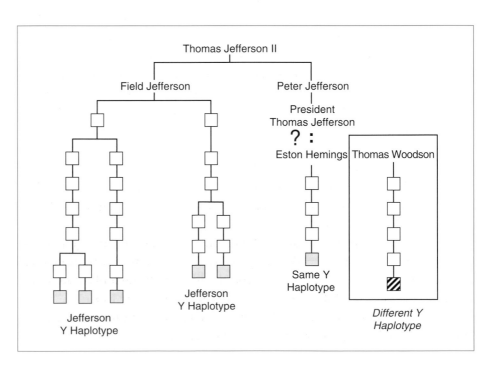

Eston Hemings? All the results in this study conclusively show is that there is a genetic match between descendants of Eston Hemings and Thomas Jefferson's uncle, Field Jefferson. Was it historically possible for another male Jefferson to have fathered Sally Hemings's children? The Thomas Jefferson Memorial Foundation, a private, non-profit organization established in 1923 that owns and operates Monticello with the goal of preservation and education, conducted a yearlong investigation into the historical record.

According to this careful historical investigation, 25 adult male descendants of Thomas Jefferson's father Peter and his uncle Field lived in Virginia during the 1794–1807 period of Sally Hemings's pregnancies (Monticello 2000). Most of them lived over 100 miles from Monticello and make no appearance in Thomas Jefferson's correspondence documents. Several male Jeffersons including President Jefferson's brother Randolph and his sons did live in the area of Monticello and visited occasionally. However, the historical records fail to indicate that any of these individuals were present at Monticello nine months before the births of Sally Hemings's children. This information combined with the fact that Thomas Jefferson was present at Monticello during the time of conception of each of Sally Hemings's six children led to the 26 January 2000 Thomas Jefferson Memorial Foundation report that he was the father of all of Sally Hemings's children (Monticello 2000).

A more recent study by a 13-member Scholars Commission of the Thomas Jefferson Heritage Society unanimously agreed that the allegations of a relationship are 'by no means proven.' The findings of this group are reported in a 565-page report available at the Heritage Society's web site: http://www.tjheritage.org. This report notes that the original DNA study indicated only that a Jefferson male had fathered one of Sally Hemings's children and that the available DNA evidence could not specify Thomas Jefferson as the father *to the exclusion of all other possibilities.* Thomas Jefferson's younger brother Randolph, who was known to fraternize with the Monticello slaves, is considered a likely possibility by many members of the Scholars Commission. Randolph and other family members would have visited Monticello when President Jefferson was home and therefore the circumstantial evidence of Thomas Jefferson being present on the plantation when Sally Hemings conceived might not be as strong as originally presented.

This study of Jefferson lineage DNA demonstrates one of the major disadvantages of Y chromosome DNA testing, namely that results only indicate connection to a male lineage and are not specific to an individual like autosomal STR profiles can be. While a Jefferson Y chromosome match exists between his descendants and those of Sally Hemings, the matter can probably never be definitely solved by Y chromosome information alone.

SURNAME TESTING AND GENETIC GENEALOGY

Genealogists in large numbers are beginning to turn to Y chromosome DNA testing to extend their research efforts (Brown 2002). Tens of thousands of genetic genealogy tests, primarily Y-STR typing of one to two dozen loci, are being conducted by several commercial enterprises (D.N.A. Box 9.2). Oxford Ancestors (Oxford England), FamilyTree DNA (Houston, TX), Relative Genetics (Salt Lake City, UT), and DNA Heritage (Dorset, England) offer Y-STR testing specifically for surname testing.

D.N.A. Box 9.2

The emerging field of genetic genealogy

A number of so-called 'genetic genealogy' companies have begun offering DNA testing services to avid family historians in order to help establish links between related individuals when the paper documentary trail runs cold (Brown 2002). The major assumption behind these efforts is that surnames, which generally are passed on from father to son, can be correlated to Y chromosome haplotype results. An early study with four Y-STR markers found a common core haplotype in 21 out of 48 men with the Sykes surname (Sykes and Irven 2000). Unfortunately, illegitimacy, adoption, and Y-STR mutations introduce a level of ambiguity into results (see Jobling 2001b). Nevertheless, this field is taking off quickly with demands for higher numbers of tested markers than are currently used in the forensic DNA typing community.

Below is a list of the companies offering Y-STR 'surname' (paternal lineage) testing and mitochondrial DNA 'maternal lineage' testing for genealogical purposes. Each company has its own database to enable comparisons to those with similar DNA results. The number of Y-STR markers offered in the various company tests as of January 2004 is also listed. Most mitochondrial DNA sequence results are for hypervariable region I only.

Family Tree DNA
(Houston, TX and Tucson, AZ)
http://www.familytreedna.com
12, 25, or 37 Y-STRs

Relative Genetics/GeneTree/ Ancestry.com
(Salt Lake City, UT and San Jose, CA)
Associated with Sorenson Genomics
(BYU Molecular Genealogy Group)
http://www.relativegenetics.com
http://www.genetree.com
26 Y-STRs

Oxford Ancestors
(Oxford, England)
http://www.oxfordancestors.com
10 Y-STRs

Trace Genetics
(Davis, CA)
http://www.tracegenetics.com
Focuses on Native American lineages

African Ancestry
(Washington, DC)
http://www.africanancestry.com
Focuses on African-American lineages
9 Y-STRs

DNA Heritage
(Dorset, England)
http://www.dnaheritage.com
http://www.ybase.org
21 Y-STRs

DNA Print Genomics
(Sarasota, FL)
Use SNPs to estimate ancestry
http://www.dnaprint.com
http://www.ancestrybydna.com
71 or 175 autosomal SNPs

Genetic genealogy using Y chromosome STR markers in conjunction with surname studies originated with a study published by Bryan Sykes in 2000 (Sykes and Irven 2000). Using four Y-STRs, DYS19, DYS390, DYS391, and DYS393, Sykes tested 48 men bearing the surname Sykes sampled from several regions of England. Of the 48 tested, 21 of them exhibited the core 'Sykes' haplotype and several others were only one mutational step away from the core haplotype. Sykes interpreted these results to reflect a common origin coming from an ancestor that lived about 700 years ago. While some interesting connections are being made with DNA to aid genealogical research, the field is still in its infancy and results should be interpreted with caution.

THE FUTURE OF Y CHROMOSOME TESTING

The field of Y chromosome analysis and its application to human identity testing has undergone rapid improvements in recent years. New markers and population groups are being characterized (Kayser *et al.* 2004). Commercial kits are now available to enable forensic practitioners to use core loci in male-specific amplifications. Validation studies and inter-laboratory studies have demonstrated that Y-STR typing is reliable. Internet-accessible databases house thousands of Y-STR haplotypes. Multiple examples exist of the value of Y-STR testing in forensic DNA casework (see Table 9.7).

The field appears poised to apply the power of paternal lineage testing yet the true potential of Y chromosome testing is still largely untapped. Only time will tell if Y chromosome analysis will routinely be performed in forensic DNA casework rather than as a specialized technique only referred to in unique situations. Databases will need to be continually enlarged to strengthen statistical estimates of a match. In addition, the more than 200 Y-STR and 250 Y-SNP markers now available need to be further characterized through evaluation of variation in different population groups to understand their value relative to the core loci widely used today.

POINTS FOR DISCUSSION

1. Will the availability and use of Y chromosome tests change the strategy for examining forensic DNA evidence in the future? If so, how?
2. What are some of the challenges with trying to determine a reliable Y-STR profile frequency estimate when a suspect cannot be excluded from contributing to a crime scene sample?

REFERENCES AND ADDITIONAL READING

Alves, C., Gusmao, L., Barbosa, J. and Amorim, A. (2003) *Forensic Science International*, 134, 126–133.

Ayub, Q., Mohyuddin, A., Qamar, R., Mazhar, K., Zerjal, T., Mehdi, S.Q. and Tyler-Smith, C. (2000) *Nucleic Acids Research*, 28, e8.

Berger, B., Niederstatter, H., Brandstatter, A. and Parson, W. (2003) *Forensic Science International*, 137, 221–230.

Bonne-Tamir, B., Korostishevsky, M., Redd, A.J., Pel-Or, Y., Kaplan, M.E. and Hammer, M.F. (2003) *Annals of Human Genetics*, 67, 153–164.

Bosch, E., Lee, A.C., Calafell, F., Arroyo, E., Henneman, P., de Knijff, P. and Jobling, M.A. (2002) *Forensic Science International*, 125, 42–51.

Brown, K. (2002) *Science*, 295, 1634–1635.

Budowle, B., Sinha, S.K., Lee, H.S. and Chakraborty, R. (2003) *Forensic Science Review*, 15, 153–164.

Butler, J.M., Schoske, R., Vallone, P.M., Kline, M.C., Redd, A.J. and Hammer, M.F. (2002) *Forensic Science International*, 129, 10–24.

Butler, J.M. (2003) *Forensic Science Review*, 15, 91–111.

Cali, F., Forster, P., Kersting, C., Mirisola, M.G., D'Anna, R., De Leo, G. and Romano, V. (2002) *International Journal of Legal Medicine*, 116, 133–138.

Carracedo, A., Beckmann, A., Bengs, A., Brinkmann, B., Caglia, A., Capelli, C., Gill, P., Gusmao, L., Hagelberg, C., Hohoff, C., Hoste, B., Kihlgren, A., Kloosterman, A., Myhre, D.B., Morling, N., O'Donnell, G., Parson, W., Phillips, C., Pouwels, M., Scheithauer, R., Schmitter, H., Schneider, P.M., Schumm, J., Skitsa, I., Stradmann-Bellinghausen, B., Stuart, M., Syndercombe, C.D. and Vide, C. (2001) *Forensic Science International*, 119, 28–41.

Cerri, N., Ricci, U., Sani, I., Verzeletti, A. and De Ferrari, F. (2003) *Croatian Medical Journal*, 44, 289–292.

Coyle, H.M., Budowle, B., Bourke, M.T., Carita, E., Hintz, J.L., Ladd, C., Roy, C., Yang, N.C., Palmbach, T. and Lee, H.C. (2003) *Journal of Forensic Sciences*, 48, 435–437.

de Knijff, P., Kayser, M., Caglia, A., Corach, D., Fretwell, N., Gehrig, C., Graziosi, G., Heidorn, F., Herrmann, S., Herzog, B., Hidding, M., Honda, K., Jobling, M., Krawczak, M., Leim, K., Meuser, S., Meyer, E., Oesterreich, W., Pandya, A., Parson, W., Penacino, G., Perez-Lezaun, A., Piccinini, A., Prinz, M., Schmitt, C., Schneider, P.M., Szibor, R., Teifel-Greding, J., Weichhold, G.M. and Roewer, L. (1997) *International Journal of Legal Medicine*, 110, 134–140.

de Knijff, P. (2000) *American Journal of Human Genetics*, 67, 1055–1061.

de Knijff, P. (2003) *Profiles in DNA*, 7, 3–5. Available online at: http://www.promega.com/profiles.

Dupuy, B.M., Stenersen, M., Egeland, T. and Olaisen, B. (2004) *Human Mutation*, 23, 117–124.

Foster, E.A., Jobling, M.A., Taylor, P.G., Donnelly, P., de Knijff, P., Mieremet, R., Zerjal, T. and Tyler-Smith, C. (1998) *Nature*, 396, 27–28.

Gill, P., Brenner, C., Brinkmann, B., Budowle, B., Carracedo, A., Jobling, M.A., de Knijff, P., Kayser, M., Krawczak, M., Mayr, W.R., Morling, N., Olaisen, B., Pascali, V., Prinz, M., Roewer, L., Schneider, P.M., Sajantila, A. and Tyler-Smith, C. (2001) *International Journal of Legal Medicine*, 114, 305–309.

Graves, J.A., Wakefield, M.J. and Toder, R. (1998) *Human Molecular Genetics*, 7, 1991–1996.

Gusmao, L., Gonzalez-Neira, A., Pestoni, C., Brion, M., Lareu, M.V. and Carracedo, A. (1999) *Forensic Science International*, 106, 163–172.

Gusmao, L., Gonzalez-Neira, A., Sanchez-Diz, P., Lareu, M.V., Amorim, A. and Carracedo, A. (2000) *Forensic Science International*, 112, 49–57.

Gusmao, L., Alves, C., Beleza, S. and Amorim, A. (2002a) *International Journal of Legal Medicine*, 116, 139–147.

Gusmao, L., Gonzalez-Neira, A., Alves, C., Lareu, M., Costa, S., Amorim, A. and Carracedo, A. (2002b) *Forensic Science International*, 126, 129–136.

Gusmao, L., Alves, C., Costa, S., Amorim, A., Brion, M., Gonzalez-Neira, A., Sanchez-Diz, P. and Carracedo, A. (2002c) *International Journal of Legal Medicine*, 116, 322–326.

Gusmao, L., Sanchez-Diz, P., Alves, C., Beleza, S., Lopes, A., Carracedo, A. and Amorim, A. (2003a) *Forensic Science International*, 134, 172–179.

Gusmao, L., Sanchez-Diz, P., Alves, C., Quintans, B., Garci, Poveda, E., Geada, H., Raimondi, E., Mari, Silva, d.l.F., Vide, M.C., Whittle, M.R., Zarrabeitia, M.T., Carvalho, M., Negreiros, V., Prieto, S.L., Riancho, J.A., Campos-Sanchez, R., Vieira-Silva, C., Toscanini, U., Amorim, A. and Carracedo, A. (2003b) *Forensic Science International*, 135, 150–157.

Hall, A. and Ballantyne, J. (2003a) *Forensic Science Review*, 15, 137–149.

Hall, A. and Ballantyne, J. (2003b) *Analytical and Bioanalytical Chemistry*, 376, 1234–1246.

Hall, A. and Ballantyne, J. (2003c) *Forensic Science International*, 136, 58–72.

Hammer, M.F. (1994) *Molecular Biology and Evolution*, 11, 749–761.

Hammer, M.F., Karafet, T.M., Redd, A.J., Jarjanazi, H., Santachiara-Benerecetti, S., Soodyall, H. and Zegura, S.L. (2001) *Molecular Biology and Evolution*, 18, 1189–1203.

Hammer, M.F. and Zegura, S.L. (2003) *Annual Reviews in Anthropology*, 31, 303–321.

Hanson, E.K. and Ballantyne, J. (2004) *Journal of Forensic Sciences*, 49, 40–51.

Henke, J., Henke, L., Chatthopadhyay, P., Kayser, M., Dulmer, M., Cleef, S., Poche, H. and Felske-Zech, H. (2001) *Croatian Medical Journal*, 42, 292–297.

Hurles, M.E. and Jobling, M.A. (2003) *Nature Genetics*, 34, 246–247.

Jobling, M.A. (2001a) *Forensic Science International*, 118, 158–162.

Jobling, M.A. (2001b) *Trends in Genetics*, 17, 353–357.

Jobling, M.A. and Tyler-Smith, C. (1995) *Trends in Genetics*, 11, 449–456.

Jobling, M.A., Pandya, A. and Tyler-Smith, C. (1997) *International Journal of Legal Medicine*, 110, 118–124.

Jobling, M.A. and Tyler-Smith, C. (2000) *Trends in Genetics*, 16, 356–362.

Jobling, M.A. and Tyler-Smith, C. (2003) *Nature Reviews Genetics*, 4, 598–612.

Johnson, C.L., Warren, J.H., Giles, R.C. and Staub, R.W. (2003) *Journal of Forensic Sciences*, 48, 1260–1268.

Kayser, M., Caglia, A., Corach, D., Fretwell, N., Gehrig, C., Graziosi, G., Heidorn, F., Herrmann, S., Herzog, B., Hidding, M., Honda, K., Jobling, M., Krawczak, M. , Leim, K., Meuser, S., Meyer, E., Oesterreich, W., Pandya, A., Parson, W., Penacino, G., Perez-Lezaun, A., Piccinini, A., Prinz, M., Schmitt, C., Schneider, P.M., Szibor, R., Teifel-Greding, J., Weichhold, G.M., de Knijff, P. and Roewer, L. (1997) *International Journal of Legal Medicine*, 110, 125–133, Appendix 141–149.

Kayser, M., Roewer, L., Hedman, M., Henke, L., Henke, J., Brauer, S., Kruger, C., Krawczak, M., Nagy, M., Dobosz, T., Szibor, R., de Knijff, P., Stoneking, M. and Sajantila, A. (2000) *American Journal of Human Genetics*, 66, 1580–1588.

Kayser, M. and Sajantila, A. (2001) *Forensic Science International*, 118, 116–121.

Kayser, M., Krawczak, M., Excoffier, L., Dieltjes, P., Corach, D., Pascali, V., Gehrig, C., Bernini, L.F., Jespersen, J., Bakker, E., Roewer, L. and de Knijff, P. (2001) *American Journal of Human Genetics*, 68, 990–1018.

Kayser, M., Brauer, S., Willuweit, S., Schadlich, H., Batzer, M.A., Zawacki, J., Prinz, M., Roewer, L. and Stoneking, M. (2002) *Journal of Forensic Sciences*, 47, 513–519.

Kayser, M., Brauer, S., Schadlich, H., Prinz, M., Batzer, M.A., Zimmerman, P.A., Boatin, B.A. and Stoneking, M. (2003) *Genome Research*, 13, 624–634.

Kayser, M. (2003) *Forensic Science Review*, 15, 77–89.

Kayser, M., Kittler, R., Erler, A., Hedman, M., Lee, A.C., Mohyuddin, A., Mehdi, S.Q., Rosser, Z., Stoneking, M., Jobling, M.A., Sajantila, A. and Tyler-Smith, C. (2004) *American Journal of Human Genetics*, 74, 1183–1197.

Kittler, R., Erler, A., Brauer, S., Stoneking, M. and Kayser, M. (2003) *European Journal of Human Genetics*, 11, 304–314.

Klintschar, M., Schwaiger, P., Regauer, S., Mannweiler, S. and Klieber, M. (2004) *Forensic Science International*, 139, 151–154.

Krawczak, M. (2001) *Forensic Science International*, 118, 114–115.

Lahn, B.T., Pearson, N.M. and Jegalian, K. (2001) *Nature Reviews Genetics*, 2, 207–216.

Lareu, M., Puente, J., Sobrino, B., Quintans, B., Brion, M. and Carracedo, A. (2001) *Forensic Science International*, 118, 163–168.

Lessiq, R. (2003) *Forensic Science Review*, 15, 181–188.

Lessig, R., Willuweit, S., Krawczak, M., Wu, F.C., Pu, C.E., Kim, W., Henke, L., Henke, J., Miranda, J., Hidding, M., Benecke, M., Schmitt, C., Magno, M., Calacal, G., Delfin, F.C., de Ungria, M.C., Elias, S., Augustin, C., Tun, Z., Honda, K., Kayser, M., Gusmao, L., Amorim, A., Alves, C., Hou, Y., Keyser, C., Ludes, B., Klintschar, M., Immel, U.D., Reichenpfader, B., Zaharova, B. and Roewer, L. (2003) *Legal Medicine (Toyko)*, 5 Suppl 1, S160–S163.

Monticello (2000) *Thomas Jefferson Memorial Foundation Research Committee Report on Thomas Jefferson and Sally Hemings*, 26 January 2000; available at: http://www.monticello.org.

Murray, B. and Duffy, B. (1998) *US News and World Report*, 9 November 1998, pp. 59–63.

Paracchini, S., Arredi, B., Chalk, R. and Tyler-Smith, C. (2002) *Nucleic Acids Research*, 30, e27.

Parson, W., Niederstatter, H., Kochl, S., Steinlechner, M. and Berger, B. (2001) *Croatian Medical Journal*, 42, 285–287.

Parson, W., Niederstatter, H., Brandstatter, A. and Berger, B. (2003) *International Journal of Legal Medicine*, 117, 109–114.

Pascali, V.L., Dobosz, M. and Brinkmann, B. (1998) *International Journal of Legal Medicine*, 112, 1.

Pereira, L., Prata, M.J. and Amorim, A. (2002) *Forensic Science International*, 130, 147–155.

Prinz, M., Boll, K., Baum, H. and Shaler, B. (1997) *Forensic Science International*, 85, 209–218.

Prinz, M., Ishii, A., Coleman, A., Baum, H.J. and Shaler, R.C. (2001) *Forensic Science International*, 120, 177–188.

Prinz, M. and Sansone, M. (2001) *Croatian Medical Journal*, 42, 288–291.

Prinz, M. (2003) *Forensic Science Review*, 15, 189–196.

Redd, A.J., Clifford, S.L. and Stoneking, M. (1997) *Biological Chemistry*, 378, 923–927.

Redd, A.J., Agellon, A.B., Kearney, V.A., Contreras, V.A., Karafet, T., Park, H., de Knijff, P., Butler, J.M. and Hammer, M.F. (2002) *Forensic Science International*, 130, 97–111.

Roewer, L. and Epplen, J.T. (1992) *Forensic Science International*, 53, 163–171.

Roewer, L., Kayser, M., de Knijff, P., Anslinger, K., Betz, A., Caglia, A., Corach, D., Furedi, S., Henke, L., Hidding, M., Kargel, H.J., Lessig, R., Nagy, M., Pascali, V.L., Parson, W., Rolf, B., Schmitt, C., Szibor, R., Teifel-Greding, J. and Krawczak, M. (2000) *Forensic Science International*, 114, 31–43.

Roewer, L., Krawczak, M., Willuweit, S., Nagy, M., Alves, C., Amorim, A., Anslinger, K., Augustin, C., Betz, A., Bosch, E., Caglia, A., Carracedo, A., Corach, D., Dekairelle, A., Dobosz, T., Dupuy, B.M., Furedi, S., Gehrig, C., Gusmao, L., Henke, J., Henke, L., Hidding, M., Hohoff, C., Hoste, B., Jobling, M.A., Kargel, H.J., de Knijff, P., Lessig, R., Liebeherr, E., Lorente, M., Martinez-Jarreta, B., Nievas, P., Nowak, M., Parson, W., Pascali, V.L., Penacino, G., Ploski, R., Rolf, B., Sala, A., Schmidt, U., Schmitt, C., Schneider, P.M., Szibor, R., Teifel-Greding, J. and Kayser, M. (2001) *Forensic Science International*, 118, 106–113.

Roewer, L. (2003) *Forensic Science Review*, 15, 163–170.

Rolf, B., Keil, W., Brinkmann, B., Roewer, L. and Fimmers, R. (2001) *International Journal of Legal Medicine*, 115, 12–15.

Rozen, S., Skaletsky, H., Marszalek, J.D., Minx, P.J., Cordum, H.S., Waterston, R.H., Wilson, R.K. and Page, D.C. (2003) *Nature*, 423, 873–876.

Ruiz, L.A., Nayar, K., Goldstein, D.B., Hebert, J.M., Seielstad, M.T., Underhill, P.A., Lin, A.A., Feldman, M.W. and Cavalli Sforza, L.L. (1996) *Annals of Human Genetics*, 60 (Pt 5), 401–408.

Sanchez, J.J., Borsting, C., Hallenberg, C., Buchard, A., Hernandez, A. and Morling, N. (2003) *Forensic Science International*, 137, 74–84.

Santos, F.R., Epplen, J.T. and Pena, S.D. (1993) *EXS*, 67, 261–265.

Saxena, R., Brown, L.G., Hawkins, T., Alagappan, R.K., Skaletsky, H., Reeve, M.P., Reijo, R., Rozen, S., Dinulos, M.B., Disteche, C.M. and Page, D.C. (1996) *Nature Genetics*, 14, 292–299.

Schneider, P.M., D'Aloja, E., Dupuy, B.M., Eriksen, B., Jangblad, A., Kloosterman, A.D., Kratzer, A., Lareu, M.V., Pfitzinger, H., Rand, S., Scheithauer, R., Schmitter, H., Skitsa, I., Syndercombe-Court, D. and Vide, M.C. (1999) *Forensic Science International*, 102, 159–165.

Schoske, R. (2003) PhD dissertation: the design, optimization and testing of Y chromosome short tandem repeat megaplexes. Available online at: http://www.cstl.nist.gov/biotech/strbase/NISTpub.htm.

Schoske, R., Vallone, P.M., Ruitberg, C.M. and Butler, J.M. (2003) *Analytical and Bioanalytical Chemistry*, 375, 333–343.

Schoske, R., Vallone, P.M., Kline, M.C., Redman, J.W. and Butler, J.M. (2004) *Forensic Science International*, 139, 107–121.

Seo, Y., Takami, Y., Takahama, K., Yoshizawa, M., Nakayama, T. and Yukawa, N. (2003) *Legal Medicine (Toyko)*, 5, 228–232.

Shewale, J.G. and Sinha, S.K. (2003) *Forensic Science Review*, 15, 115–136.

Shewale, J.G., Sikka, S.C., Schneida, E. and Sinha, S.K. (2003a) *Journal of Forensic Sciences*, 48, 127–129.

Shewale, J.G., Nasir, H., Schneida, E. and Sinha, S.K. (2003b) *Proceedings of the Fourteenth International Symposium on Human Identification*. Available online at: http://www.promega.com/geneticidproc/.

Sibille, I., Duverneuil, C., Lorin, d.l.G., Guerrouache, K., Teissiere, F., Durigon, M. and de Mazancourt, P. (2002) *Forensic Science International*, 125, 212–216.

Sinha, S.K. (2003a) *Forensic Science Review*, 15, 179–180.

Sinha, S.K. (2003b) *Forensic Science Review*, 15, 197–201.

Sinha, S.K., Budowle, B., Arcot, S.S., Richey, S.L., Chakrabor, R., Jones, M.D., Wojtkiewicz, P.W., Schoenbauer, D.A., Gross, A.M., Sinha, S.K. and Shewale, J.G. (2003a) *Journal of Forensic Sciences*, 48, 93–103.

Sinha, S.K., Nasir, H., Gross, A.M., Budowle, B. and Shewale, J.G. (2003b) *Journal of Forensic Sciences*, 48, 985–1000.

Sinha, S.K., Budowle, B., Chakraborty, R., Paunovic, A., Guidry, R.D., Larsen, C., Lal, A., Shaffer, M., Pineda, G., Sinha, S.K., Schneida, E., Nasir, H. and Shewale, J. (2004) *Journal of Forensic Sciences*, 49 (4), 691–700.

Skaletsky, H., Kuroda-Kawaguchi, T., Minx, P.J., Cordum, H.S., Hillier, L., Brown, L.G., Repping, S., Pyntikova, T., Ali, J., Bieri, T., Chinwalla, A., Delehaunty, A., Delehaunty, K., Du, H., Fewell, G., Fulton, L., Fulton, R., Graves, T., Hou, S.F., Latrielle, P., Leonard, S., Mardis, E., Maupin, R., McPherson, J., Miner, T., Nash, W., Nguyen, C., Ozersky, P., Pepin, K., Rock, S., Rohlfing, T., Scott, K., Schultz, B., Strong, C., Tin-Wollam, A., Yang, S.P., Waterston, R.H., Wilson, R.K., Rozen, S. and Page, D.C. (2003) *Nature*, 423, 825–837.

Steinlechner, M., Berger, B., Niederstatter, H. and Parson, W. (2002) *International Journal of Legal Medicine*, 116, 117–120.

Sykes, B. and Irven, C. (2000) *American Journal of Human Genetics*, 66, 1417–1419.

Thangaraj, K., Reddy, A.G. and Singh, L. (2002) *International Journal of Legal Medicine*, 116, 121–123.

Tilford, C.A., Kuroda-Kawaguchi, T., Skaletsky, H., Rozen, S., Brown, L.G., Rosenberg, M., McPherson, J.D., Wylie, K., Sekhon, M., Kucaba, T.A., Waterston, R.H. and Page, D.C. (2001) *Nature*, 409, 943–945.

Underhill, P.A., Jin, L., Lin, A.A., Mehdi, Q., Jenkins, T., Vollrath, D., Davis, R.W., Cavalli-Sforza, L.L. and Oefner, P.J. (1997) *Genome Research*, 7, 996–1005.

Underhill, P.A., Shen, P., Lin, A.A., Jin, L., Passarino, G., Yang, W.H., Kauffman, E., Bonne-Tamir, B., Bertranpetit, J., Francalacci, P., Ibrahim, M., Jenkins, T., Kidd, J.R., Mehdi, S.Q., Seielstad, M.T., Wells, R.S., Piazza, A., Davis, R.W., Feldman, M.W., Cavalli-Sforza, L.L. and Oefner, P.J. (2000) *Nature Genetics*, 26, 358–361.

Vallone, P.M. and Butler, J.M. (2004), *Journal of Forensic Sciences*, 49 (4), 723–732.

White, P.S., Tatum, O.L., Deaven, L.L. and Longmire, J.L. (1999) *Genomics*, 57, 433–437.

Yamamoto, T. (2003) *Forensic Science Review*, 15, 171–178.

Ye, J., Parra, E.J., Sosnoski, D.M., Hiester, K., Underhill, P.A. and Shriver, M.D. (2002) *Journal of Forensic Sciences*, 47, 593–600.

Y Chromosome Consortium (2002) *Genome Research*, 12, 339–348; see also http://ycc.biosci.arizona.edu.

Zerjal, T., Xue, Y., Bertorelle, G., Wells, R.S., Bao, W., Zhu, S., Qamar, R., Ayub, Q., Mohyuddin, A., Fu, S., Li, P., Yuldasheva, N., Ruzibakiev, R., Xu, J., Shu, Q., Du, R., Yang, H., Hurles, M.E., Robinson, E., Gerelsaikhan, T., Dashnyam, B., Mehdi, S.Q. and Tyler-Smith, C. (2003) *American Journal of Human Genetics*, 72, 717–721.

Zhivotovsky, L.A., Underhill, P.A., Cinnioglu, C., Kayser, M., Morar, B., Kivisild, T., Scozzari, R., Cruciani, F., Destro-Bisol, G., Spedini, G., Chambers, G.K., Herrera, R.J., Yong, K.K., Gresham, D., Tournev, I., Feldman, M.W. and Kalaydjieva, L. (2004) *American Journal of Human Genetics*, 74, 50–61.

CHAPTER 10

MITOCHONDRIAL DNA ANALYSIS

DNA is the messenger, which illuminates (our connection to the past), handed down from generation to generation, carried, literally, in the bodies of (our) ancestors. Each message traces a journey through time and space, a journey made by the long lines that spring from the ancestral mothers.

(Bryan Sykes, *The Seven Daughters of Eve*)

Conventional short tandem repeat (STR) typing systems do not work in every instance. Ancient DNA specimens or samples that have been highly degraded often fail to produce results with nuclear DNA typing systems. However, recovery of DNA information from environmentally damaged DNA is sometimes possible with mitochondrial DNA (mtDNA). While a nuclear DNA test is usually more valuable, a mtDNA result is better than no result at all. The probability of obtaining a DNA typing result from mtDNA is higher than that of polymorphic markers found in nuclear DNA particularly in cases where the amount of extracted DNA is very small, as in tissues such as bone, teeth, and hair. When remains are quite old or badly degraded, often bone, teeth, and hair are the only biological sources left from which to draw a sample.

The primary characteristic that permits its recovery from degraded samples is the fact that mtDNA is present in cells at a much higher copy number than the nuclear DNA from which STRs are amplified. In short, though nuclear DNA contains much more information, there are only two copies of it in each cell (one maternal and one paternal) while mtDNA has a bit of useful genetic information times hundreds of copies per cell. With a higher copy number, some mtDNA molecules are more likely to survive than nuclear DNA. Table 10.1 contains a comparison of some basic characteristics for nuclear DNA and mitochondrial DNA.

This chapter will review the characteristics of mitochondrial DNA, the steps involved in obtaining results in forensic casework, and issues important to interpreting mtDNA results.

CHARACTERISTICS OF MITOCHONDRIAL DNA

LOCATION AND STRUCTURE OF mtDNA

The vast majority of the human genome is located within the nucleus of each cell (see Figure 2.3, Table 10.1). However, there is a small, circular genome

Table 10.1

Comparison of human
nuclear DNA and
mitochondrial DNA
markers.

Characteristics	Nuclear DNA (nucDNA)	Mitochondrial DNA (mtDNA)
Size of genome	~3.2 billion bp	~16 569 bp
Copies per cell	2 (1 allele from each parent)	Can be >1000
Percent of total DNA content per cell	99.75%	0.25%
Structure	Linear; packaged in chromosomes	Circular
Inherited from	Father and Mother	Mother
Chromosomal pairing	Diploid	Haploid
Generational recombination	Yes	No
Replication repair	Yes	No
Unique	Unique to individual (except identical twins)	Not unique to individual (same as maternal relatives)
Mutation rate	Low	At least 5–10 times nucDNA
Reference sequence	Described in 2001 by the Human Genome Project	Described in 1981 by Anderson and co-workers

found within the mitochondria, the energy-producing cellular organelle residing in the cytoplasm. The number of mitochondrial DNA (mtDNA) molecules within a cell can vary tremendously. On average there are 4–5 copies of mtDNA molecules per mitochondrion with a measured range of 1–15 (Satoh and Kuroiwa 1991). Because each cell can contain hundreds of mitochondria (Robin and Wong 1988), mathematically there can be up to several thousand mtDNA molecules in each cell as in the case of ovum or egg cells. However, the average has been estimated at about 500 in most cells (Satoh and Kuroiwa 1991). It is this amplified number of mtDNA molecules in each cell that enables greater success (relative to nuclear DNA markers) with biological samples that may have been damaged with heat or humidity. Consider though that mtDNA makes up less than one percent (about 0.25%) of the total DNA content of a cell if we assume that there are 1000 copies of mtDNA (16 569 bp) in a cell and two copies of nuclear DNA (3.2 billion bp).

Mitochondrial DNA has approximately 16 569 base pairs and possesses 37 'genes' that code for products used in the oxidative phosphorylation process or cellular energy production. There is also a 1122 bp 'control' region that contains the origin of replication for one of the mtDNA strands but does not code for any gene products and is therefore referred to sometimes as the

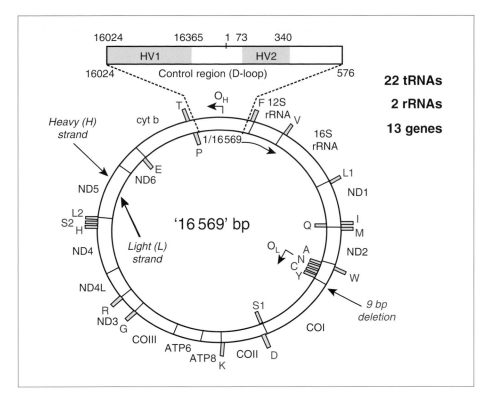

Figure 10.1

Schematic showing the circular mitochondrial DNA
genome (mtGenome). The
heavy (H) strand is represented by the outside line
and contains a higher
number of C-G residues
than the light (L) strand.
The 37 RNA and protein
coding gene regions are
abbreviated around the
mtGenome next to the
strand from which they
are synthesized (see Table
10.2). Most forensic
mtDNA analyses presently
examine only HV1 and
HV2 in the non-coding
control region or
displacement loop
shown at the top of the
figure. Due to insertions
and deletions that exist
around the mtGenome in
different individuals, it is
not always 16 569 bp.

'non-coding' region. The total number of nucleotides in a mitochondrial genome (mtGenome) can vary due to small mutations that are either insertions or deletions. For example, there is a dinucleotide repeat at positions 514–524, which in most individuals is ACACACACAC or $(AC)_5$ but has been observed to vary from $(AC)_3$ to $(AC)_7$ (Bodenteich *et al.* 1992, Szibor *et al.* 1997). The 37 transcribed 'genes' of mtDNA found in the 'coding region' include 13 proteins, two ribosomal RNAs (rRNA), and 22 transfer RNAs (tRNA) (Figure 10.1). The nucleotide positions for each coding and non-coding segment of the mtGenome are indicated in Table 10.2. Note that the genes are very tightly

Table 10.2

Mitochondrial DNA
information and genes.

Nucleotide Position	Strand Transcribed	Abbreviation	Description	Size (bp)	Non-coding
16024–16569, 1–576	D-loop		control region	1122	1122
16104–16569, 1–191	O_H		replication origin (H-strand)	658	
16158–16172			**D-loop termination signal**	15	
531–568			**H-strand transcription promoter**	38	

Continued

Nucleotide Position	Strand Transcribed	Abbreviation	Description	Size (bp)	Non-coding
577–647	H	F	tRNA phenylalanine	71	
648–1601	H	12S	12S rRNA	954	
1602–1670	H	V	tRNA valine	69	
1671–3229	H	16S	16S rRNA	1559	
3230–3304	H	L1	tRNA leucine 1	75	
3305–4263	H	ND1	NADH dehydrogenase 1	959	
4263–4331	H	I	tRNA isoleucine	69	
4329–4400	L	Q	tRNA glutamine	72	
4401	—		non-coding	1	1
4402–4469	H	M	tRNA methionine	68	
4470–5511	H	ND2	NADH dehydrogenase 2	1042	
5512–5579	H	W	tRNA tryptophan	68	
5580–5586	—		non-coding	7	7
5587–5655	L	A	tRNA alanine	69	
5656	—		non-coding	1	1
5657–5729	L	N	tRNA asparagine	73	
5730–5760		**O_L**	**L-strand origin**	**31**	**31**
5761–5826	L	C	tRNA cysteine	66	
5826–5891	L	Y	tRNA tyrosine	66	
5892–5900	—		non-coding	9	9
5901–7445	H	COI	Cytochrome c oxidase I	1545	
7445–7516	L	S1	tRNA serine 1	72	
7517	—		non-coding	1	1
7518–7585	H	D	tRNA aspartic acid	68	
7586–8294	H	COII	Cytochrome c oxidase II	709	
8295–8364	H	K	tRNA lysine	70	
8365–8572	H	ATP8	ATP synthase 8	208	
8527–9207	H	ATP6	ATP synthase 6	681	

Continued

Nucleotide Position	Strand Transcribed	Abbreviation	Description	Size (bp)	Non-coding
9207–9990	H	COIII	Cytochrome c oxidase III	784	
9991–10058	H	G	tRNA glycine	68	
10059–10404	H	ND3	NADH dehydrogenase 3	346	
10405–10469	H	R	tRNA arginine	65	
10470–10766	H	ND4L	NADH dehydrogenase 4L	297	
10760–12137	H	ND4	NADH dehydrogenase 4	1378	
12138–12206	H	H	tRNA histidine	69	
12207–12265	H	S2	tRNA serine 2	59	
12266–12336	H	L2	tRNA leucine 2	71	
12337–14148	H	ND5	NADH dehydrogenase 5	1812	
14149–14673	L	ND6	NADH dehydrogenase 6	525	
14674–14742	L	E	tRNA glutamic acid	69	
14743–14746		—	non-coding	4	4
14747–15887	H	Cyt b	Cytochrome b	1141	
15888–15953	H	T	tRNA threonine	66	
15954		—	non-coding	1	1
15955–16023	L	P	tRNA proline	69	

packed with only 55 nucleotides in the 15 447 bp of the coding region *not* being used to transcribe a protein, rRNA, or tRNA molecule. Thus, the genes within mtDNA are economically packaged with no introns and none or only a few non-coding nucleotides between the coding regions.

Table 10.2 (Continued)

An asymmetric distribution of nucleotides gives rise to 'light' and 'heavy' strands when mtDNA molecules are separated in alkaline CsCl gradients (Scheffler 1999). The 'heavy' or H-strand contains a greater number of guanine nucleotides, which have the largest molecular weight of the four possible nucleotides, than the 'light' or L-strand. Replication of mtDNA begins with the H-strand in the non-coding 'control region', also known as the displacement loop or D-loop (Figure 10.1). A total of 28 gene products are encoded from the H-strand while the L-strand transcribes eight transfer RNAs (tRNAs) and an enzyme called ND6 (Table 10.2).

Since the D-loop does not code for gene products, the constraints are less for nucleotide variability and polymorphisms between individuals are more abundant than in similar sized portions of the coding region. More simply, there can be differences in the D-loop region because the sequences do not code for any substances necessary for the cell's function. Most of the focus in forensic DNA studies to date has involved two hypervariable regions within the control region commonly referred to as HV1 and HV2, which will be described in more detail below.

REFERENCE SEQUENCE(S)

Human mtDNA was first sequenced in 1981 in the laboratory of Frederick Sanger in Cambridge, England (Anderson *et al.* 1981). For many years, the original 'Anderson' sequence (GenBank accession: M63933) was the reference sequence to which new sequences were compared. The Anderson sequence is also referred to as the Cambridge Reference Sequence (CRS). Typically laboratories report results in terms of variation compared to the L-strand of the CRS. Thus, the observation of a C nucleotide at position 16126, which contains a T in the Anderson sequence, would be reported as 16126C. If no other nucleotide variants are reported, then it is assumed that the remaining sequence contains the same sequence as the CRS.

In 1999, the original placenta material used by Anderson and co-workers to generate the CRS was re-sequenced (Andrews *et al.* 1999). The 1981 sequence was derived primarily from a single individual of European descent; however, it also contained some HeLa and bovine sequences to fill in gaps resulting from early rudimentary DNA sequencing procedures (Anderson *et al.* 1981). With improvements in DNA sequencing technology over the intervening two decades, it was felt that any original errors should be rectified to enable robust use of this reference sequence in the future.

The re-analysis effort confirmed all but 11 of the original nucleotides identified in the original published sequence (Table 10.3). One of these differences was the loss of a single cytosine residue at position 3106. An additional seven nucleotide positions were demonstrated to be accurate but represent rare polymorphisms. These sites were 263A, 311–315CCCCC, 750A, 1438A, 4769A, 8860A, and 15326A. Fortunately, no errors were observed in the widely used control region. Thus, the original Anderson sequence (Anderson *et al.* 1981) was found to be identical to the revised Cambridge reference sequence (Andrews *et al.* 1999) across the HV1 and HV2 regions that are widely used in forensic applications.

The revised Cambridge reference sequence (rCRS) is now the accepted standard for comparison (Andrews *et al.* 1999). However, the loss of a single C nucleotide at position 3106 means that the reference mtGenome is 16 568 bp rather than the traditionally accepted value of 16 569 bp. More critically the original nucleotide numbering would have to be updated for all previously

Nucleotide Position	Region of mtGenome	Original CRS	Revised CRS	Remarks
3106–3107	16S rRNA	CC	C	Error
3423	ND1	G	T	Error
4985	ND2	G	A	Error
9559	COIII	G	C	Error
11335	ND4	T	C	Error
13702	ND5	G	C	Error
14199	ND6	G	T	Error
14272	ND6	G	C	Error (bovine sequence inserted)
14365	ND6	G	C	Error (bovine sequence inserted)
14368	ND6	G	C	Error
14766	cyt b	T	C	Error (HeLa sequence inserted)

Table 10.3

Comparison of nucleotide differences observed between the original Cambridge Reference Sequence (Anderson et al. 1981) and the revised Cambridge Reference Sequence (Andrews et al. 1999) based on re-sequencing of the original placenta material. The true sequence at position 3106–3107 is only a single C making the entire mtGenome 16 568 bp rather than the originally reported 16 569 bp. However, to maintain the historical numbering, a deletion at position 3107 is used to serve as a placeholder (Andrews et al. 1999). Note that no differences exist between these sequences for the two hypervariable regions most commonly used in forensic applications that span positions 16 024–16 365 and 73–340.

identified sequence changes beyond nucleotide position 3106. Since this approach would have created an unacceptable amount of confusion and inability to easily correlate previous work, Andrews and co-workers (1999) recommended that the original numbering be retained in the rCRS with a deletion in the sequence at position 3107 to serve as a place holder. The 16 568 bp rCRS is available at the MITOMAP web site: http://www.mitomap.org/mitoseq.htm.

As a side note, however, it is probably worth noting that the official reference sequence used in the assembly of the human genome is not the rCRS. Rather the RefSeq mtGenome used by the National Center for Biotechnology Information in its official assembly of the human genome is contained in GenBank as accession NT_001807. This sequence, originally GenBank accession AF347015 sequenced by Ingman *et al.* (2000), is 16 571 bp and derived from an African (Yoruba) individual. Thus, difference reference sequences can and have been used for various purposes with mtDNA so it is important to note the one in use for a particular study.

MATERNAL INHERITANCE OF mtDNA

For forensic and human identification purposes, human mitochondrial DNA is considered to be inherited strictly from one's mother. At conception only the sperm's nucleus enters the egg and joins directly with the egg's nucleus.

The fertilizing sperm does not contribute other cellular components. When the zygote cell divides and a blastocyst develops, the cytoplasm and other cell parts save the nucleus are consistent with the mother's original egg cell. Thus, mitochondria with their mtDNA molecules are passed directly to all offspring independent of any male influence.

Eggs have been reported to have as many as 100 000 mtDNA molecules creating an extreme dilution for any paternal mtDNA molecules that may pass into the zygote (Chen *et al.* 1995). Furthermore, any sperm mitochondria that may enter a fertilized egg are selectively destroyed due to the presence of an ubiquitin tag added during spermatogenesis that appears to earmark sperm mitochondria for degradation by the newly formed embryo's cellular machinery (Sutovsky *et al.* 1999). Thus, barring mutation, a mother passes along her mtDNA type to her children, and therefore siblings and maternal relatives have an identical mtDNA sequence. Hence, an individual's mtDNA type is not unique to them.

An example family pedigree is shown in Figure 10.2 to demonstrate the inheritance pattern of mtDNA. In this example, unique mtDNA types exist solely for individuals 1, 5, 7, and 12. Note that individual 16 will possess the same mtDNA type as seven of the other represented individuals (e.g., 2, 3, 6, 8, 11, 13, and 15).

Figure 10.2

Illustration of maternal mitochondrial DNA inheritance for 18 individuals in a hypothetical pedigree. Squares represent males and circles females. Each unique mtDNA type is represented by a different letter.

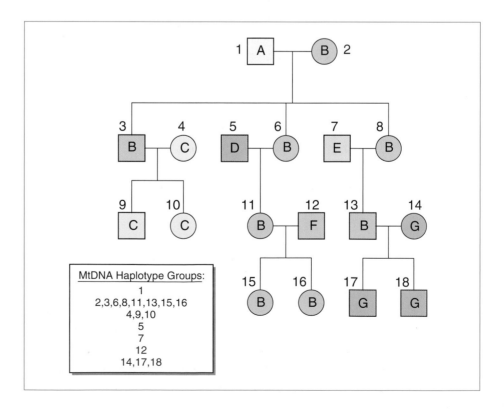

MtDNA Haplotype Groups:
1
2,3,6,8,11,13,15,16
4,9,10
5
7
12
14,17,18

This fact can be helpful in solving missing persons or mass disaster investigations but can reduce the significance of a match in forensic cases. Since even distantly related maternal relatives should possess the same mtDNA type, this extends the number of useful reference samples that may be used to confirm the identity of a missing person. Evidence from mtDNA has been helpful in linking families and solving historical puzzles such as identifying the Vietnam Unknown Soldier (D.N.A. Box 10.1) and the Romanov family (D.N.A. Box 10.2).

OTHER INTERESTING DIFFERENCES BETWEEN mtDNA AND NUCLEAR DNA

Mitochondrial DNA uses a different genetic code than nuclear DNA (Scheffler 1999). For example, the codon for mitochondrial-transcribed amino acid tryptophan is UGA while the universal (nuclear) genetic code for UGA is a stop codon. In the mtDNA genetic code, AUA codes for methionine instead of isoleucine and AGA and AGG both code for stops rather than arginine.

Fewer DNA repair mechanisms exist in mitochondria thereby leading to higher mutation rates compared to nuclear DNA. In addition, lack of proof-reading capabilities in the mtDNA polymerase increases mutations during replication. However, the 10-fold higher mutation rate (relative to nuclear DNA) helps introduce more variability in samples from identical maternal lineages that otherwise would not vary. This increased variability is a good thing for most applications in human identity testing although mutations can sometimes be a hindrance when trying to definitely establish familial relationships (e.g., when comparing remains to reference samples from distant maternal relatives).

The circular nature of mtDNA makes it less susceptible to exonucleases that break down DNA molecules needed to survive until forensic DNA testing can be completed. The presence of an increased number of mtDNA molecules per cell relative to the nuclear DNA chromosomes also enhances the mtDNA survival rate, as does the fact that they are encapsulated in a two-walled organelle.

VARIOUS APPLICATIONS FOR mtDNA TESTING

Mitochondrial DNA variation is extensively studied in several other disciplines besides forensic science. Medical scientists have linked a number of diseases to mutations in mtDNA (see Wallace *et al.* 1999). Evolutionary biologists examine human mtDNA sequence variation relative to other species in an effort to determine relationships. A good example of this application is the determination that Neanderthals are not the direct ancestors of modern humans based on control region sequences determined from ancient bones (Krings *et al.* 1997). Molecular anthropologists study differences in mtDNA sequences from various global population groups to examine questions of ancestry and migration of peoples throughout history (Relethford 2003). Hundreds of papers have been published in these fields over the past decade or two. Genetic genealogists are now using mtDNA and Y chromosome markers in an attempt to trace ancestry where paper trails run cold (Brown 2002).

In the past few years a number of interesting historical identifications have been performed with the aid of mtDNA testing. Remains from the Tomb of the Unknown Soldier associated with the Vietnam War have been identified as those of Michael Blassie (D.N.A. Box 10.1). Bones discovered in Russia in 1991 were demonstrated to be those of the Tsar Nicholas II (Gill *et al.* 1994, Ivanov *et al.* 1996) (D.N.A. Box 10.2). The claims of Anna Anderson Manahan as the Russian princess Anastasia were proven false (Stoneking *et al.* 1995). The remains of the outlaw Jesse James were linked to living relatives putting to rest a myth that he had somehow escaped death at the hands of Robert Ford (Stone *et al.* 2001).

D.N.A. Box 10.1

Identifying remains from the Tomb of the Unknown Soldier

On 30 June 1998, U.S. Secretary of Defense William Cohen announced to the world that DNA technology had been used to identify the Vietnam Unknown in the Tomb of the Unknown Soldier located in Arlington National Cemetery. The remains of First Lieutenant Michael J. Blassie, United States Air Force, were identified through the use of mitochondrial DNA. An exact match across 610 nucleotides of the polymorphic mtDNA control region was obtained between Jean Blassie, Michael's mother, and a sample extracted from the bone fragments removed from the Tomb of the Unknown Soldier. At the same time, eight other possible soldiers were excluded because family reference samples did not match.

Michael Blassie was an Air Force Academy graduate and the oldest of five children who grew up in St. Louis, Missouri. Lieutenant Blassie arrived in Vietnam in January 1972 and was flying his 132nd mission when his A-37B attack jet was shot down on 11 May 1972, outside An Loc, a hotly contested South Vietnamese village near the Cambodian border. Intense fighting in the area prevented the site from being searched and his remains were not recovered until almost five months later. By this time only four ribs, the right humerus and part of the pelvis remained along with some personal items, including Blassie's identification card. The remains were sent to the Army's Central Identification Laboratory in Hawaii where they remained for eight years designated as 'believed to be Michael Blassie.' In 1980, a military review board changed the designation on the remains to 'unknown' and the identification card found with the body had vanished.

The Tomb of the Unknown Soldier was first opened in 1921 to honor soldiers who had died in World War I. On the tomb are inscribed the words 'Here rests in honored glory an American soldier known but to God.' Within this hallowed ground lie four servicemen, the unknown soldiers of World War I, World War II, the Korean War and the Vietnam War. These unknown soldiers are guarded 24 hours a day at Arlington National Cemetery by a sentinel from the 3rd U.S. Infantry. The World War II and Korean War unknowns were selected from about 8500 and 800 unidentifiable remains, respectively, and were entombed on Memorial Day 1958. The Vietnam War casualty was authorized in 1973 for enshrinement, but it was not filled for 11 more years. To honor a Vietnam veteran on Memorial Day 1984 one of the few available unknown remains was selected for enshrinement and honored in a ceremony lead by President Ronald Reagan. There the remains of the Vietnam Unknown lay until 14 May 1998, when they were disinterred in a solemn ceremony and transported to the Armed Forces Institute of Pathology for investigation. So sacred is the tomb and the memory of the soldiers resting there, that it has only been opened four times: in 1921 for WW I, in 1958 for WW II and Korean, in 1984 for Vietnam, and in 1998 to remove the Vietnam remains for DNA testing.

Throughout the month of June 1998, mtDNA sequence information was recovered from the skeletal material (pelvis) and analyzed by scientists at the Armed Forces DNA Identification Laboratory (AFDIL) located in Rockville, Maryland. Maternal relatives from eight possible American casualties near An Loc were also evaluated as family reference samples. The mtDNA sequence content from positions 16024 to 16365 (HVI) and positions 73–340 (HVII) on the polymorphic control region were evaluated. Only a complete match was observed between Jean Blassie (Michael's mother) and the skeletal remains disinterred from the Tomb of the Unknown Soldier. Because of this positive identification, the Blassies were permitted to bury Lieutenant Blassie's remains at Jefferson Barracks National Cemetery located in St. Louis, Missouri. This ceremony was conducted on 11 July 1998, and brought closure to the Blassie family.

Source:
Holland, M.M. and Parsons, T.J. (1999) *Forensic Science Review*, 11, 21–50.

DIFFERENT METHODS FOR MEASURING mtDNA VARIATION

Over the past two decades, methods for measuring mtDNA variation have progressed in their ability to separate unrelated and closely related maternal lineages. The first studies with mtDNA in the 1980s involved low-resolution restriction fragment length polymorphism (RFLP) analysis using five or six restriction enzymes (see Richards and Macaulay 2001). Higher resolution restriction analysis involved polymerase chain reaction (PCR) amplification of typically nine overlapping fragments followed by digestion with 12 or 14 restriction enzymes. These restriction endonucleases included *Alu*I, *Ava*II, *Bam*HI, *Dde*I, *Hae*II, *Hae*III, *Hha*I, *Hinc*II, *Hinf*I, *Hpa*I, *Msp*I, *Mbo*I, *Rsa*I, and *Taq*I (Torroni *et al.* 1996).

Russian Czar (or Tsar) Nicholas II and his family were removed from power and murdered during the Bolshevik Revolution of 1918. They were shot by a firing squad, doused with sulfuric acid to render their bodies unrecognizable, and disposed of in a shallow pit under a road. Their remains were lost to history until July 1991 when nine skeletons were uncovered from a shallow grave near Ekaterinburg, Russia. A number of forensic tests were attempted involving computer aided reconstructions and odontological analysis, but as the facial areas of the skulls were destroyed, classical facial identification techniques were difficult at best and not conclusive.

The Chief Forensic Medical Examiner of the Russian Federation turned to the Forensic Science Service in the United Kingdom to carry out DNA-based analysis of the remains for purposes of identification. Five STR markers (VWA, TH01, F13A1, FES/FPS, and ACTBP2) were used to examine the nine skeletons. Approximately one gram of bone from each of the skeletons yielded about 50 pg of DNA, just enough for PCR amplification of several STR markers. The remains of the Romanov family members consisting of the Tsar, the Tsarina, and three children were distinguishable from those of three servants and the family doctor by their STR genotypes.

While the STR analysis served to establish family relationships between the remains through comparing matching alleles, a link still had to be made with a known descendant of the Romanov family to verify that the remains were indeed those of the Russian royal family. Mitochondrial DNA analysis was used to answer this question.

Mitochondrial DNA was extracted from the femur of each skeleton and sequenced. Blood samples were then obtained from maternally related descendants of the Romanov family and sequenced in the same manner. His Royal Highness Prince Philip, Duke of Edinburgh and husband of the present British Queen Elizabeth, is a grand nephew of unbroken maternal descent from Tsarina Alexandra. His blood sample thus provided the comparison to confirm the sibling status of the children and the linkage of the mother to the Tsarina's family. The sequences of all 740 tested nucleotides from the mtDNA control region matched between HRH Prince Philip and the putative Tsarina and the three children.

The mtDNA sequence from the putative Tsar was compared with two relatives of unbroken maternal descent from Tsar Nicholas II's grandmother, Louise of Hesse-Cassel. The two relatives had the same mtDNA sequence as the putative Tsar with the exception of a single nucleotide at position 16169. At this position, the putative Tsar's sample had a mixture of two nucleotides (T and C), a condition known as heteroplasmy, while the blood samples of relatives had only a T nucleotide.

To further confirm the putative Tsar's remains, the brother of Nicholas II, Grand Duke of Russia Georgij Romanov, was exhumed and tested by the Armed Forces DNA Identification Laboratory (Ivanov *et al.* 1996). Heteroplasmy was found again at the identical nucleotide site within the mtDNA sequence. Due to the extreme rarity of this heteroplasmy happening by chance between two unrelated individuals, the remains of Tsar Nicholas II and his family were

D.N.A. Box 10.2

(Continued)

declared authentic and laid to rest in Red Square with a funeral fit for a royal family. However, in spite of the matches made between multiple individuals that were confirmed in a second laboratory, controversy still arises from time-to-time regarding the Romanov remains (see Stone 2004).

Lineage of Romanov Family. The individuals represented by blue are maternal relatives of the Tsarina Alexandra while those shown in red are maternal relatives of Tsar Nicholas II. Living maternal relatives Prince Philip (for Tsarina) and Xenia Cheremeteff-Sfiri (for Tsar) served as family reference samples. The mtDNA mitotype for each reference sample is listed with the nucleotide changes relative to their position in the Anderson sequence. Tsar Nicholas II and his brother Georgij Romanov both exhibited a heteroplasmic T/C at mtDNA position 16169, which differed from the homoplasmic T found in Xenia Cheremeteff-Sfiri (Ivanov *et al.* 1996). Prince Philip's mitotype matched the remains of the Tsarina and her children while Xenia's mitotype matched the remains of the Tsar at all positions except the heteroplasmic position 16169 (Gill *et al.* 1994).

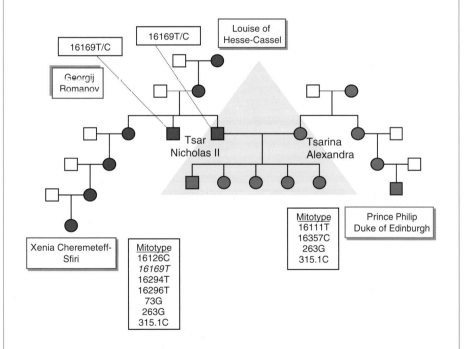

Sources:

Gill, P., Ivanov, P.L., Kimpton, C., Piercy, R., Benson, N., Tully, G., Evett, I., Hagelberg, E. and Sullivan, K. (1994) *Nature Genetics,* 6, 130–135.

Ivanov, P.L., Wadhams, M.J., Roby, R.K., Holland, M.M., Weedn, V.W. and Parsons, T.J. (1996) *Nature Genetics,* 12, 417–420.

Stone, R. (2004) *Science,* 303, 753.

Genetically different population types or haplotypes have been defined in the literature based on site losses or site gains with the various restriction enzymes. For example, haplogroup A, which is found in Asians and Native Americans, is defined by a site gain at position 663 with *Hae*III (listed as +663 *Hae*III). Haplogroup B was initially defined as a 9 bp deletion in the intergenic region between the COII and tRNA[LYS] genes (see Table 10.2). Individuals belonging to haplogroup A may also be defined by control region polymorphisms 16223T, 16290T, and 16319A while haplogroup B individuals differ from the Anderson reference sequence at 16189C and 16217C.

In the early 1990s, DNA sequence analysis from portions of the control region came into wide acceptance. Most population data outside of the forensic community continues to be collected for only hypervariable segment I (HVS-I) spanning approximately mtDNA nucleotide positions 16024 to 16365. As will be seen below, the forensic DNA typing community has standardized on specific portions of the control region for most of the data that currently exists.

December 2000 marked the beginning of the mtDNA population genomics era with the publication of 53 entire mtGenomes from a diverse set of individuals representing populations from around the world (Ingman *et al.* 2000). As of early 2004 about a thousand complete mtGenomes exists in public DNA databases (Ruiz-Pesini *et al.* 2004).

CONVENTIONAL FORENSIC MARKERS AND METHODS USED IN mtDNA TESTING

As mentioned earlier, the most extensive mtDNA variations between individuals in the human population are found within the control region, or displacement loop (D-loop). Two regions within the D-loop known as hypervariable region I (HV1, HVI, or HVS-I) and hypervariable region II (HV2, HVII, or HVS-II) are normally examined by PCR amplification followed by sequence analysis. Approximately 610 bp are commonly evaluated – 342 bp from HV1 and 268 bp from HV2 (Figure 10.3). The DNA sequence for each sample between nucleotide positions 16024 and 16365 in HV1 and 73 and 340 in HV2 is determined and then compared to the Anderson or the revised Cambridge Reference Sequence (as mentioned earlier, these reference sequences are equivalent for the control region). Differences are noted and reported with the nucleotide position and the altered base. Sometimes a third hypervariable region (HVIII) is examined that is 137 bp long and spans nucleotide positions 438–574. Additional polymorphic sites within HVIII can sometimes help resolve indistinguishable HVI/HVII samples (Lutz *et al.* 2000, Bini *et al.* 2003).

A number of different PCR and sequencing primers have been used to generate the DNA sequence data for HV1 and HV2. Various primer combinations will be discussed in the next section. The mtDNA control region has been estimated

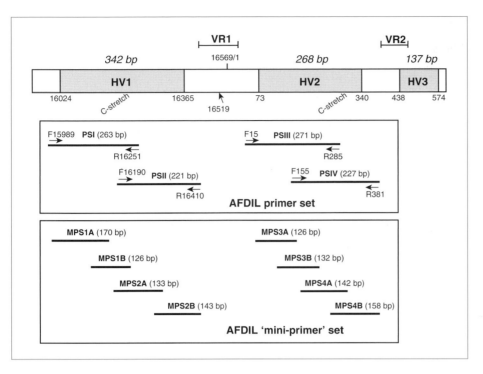

Figure 10.3

The three hypervariable (HV) regions of the mtDNA control region. HV1 spans nucleotide positions 16024–16365 (342 bp), HV2 spans positions 73–340 (268 bp), and HV3, which is rarely examined in forensic testing, spans positions 438–574 (137 bp). The general positions for variable regions VR1 and VR2 are noted although these are rarely used. PCR primer sets (PS) commonly used by the Armed Forces DNA Identification Laboratory (AFDIL) are illustrated. Primer nomenclature designates the 5′-nucleotide for each primer. PCR product sizes for each set of primers are noted in parentheses. The bottom section shows 'mini-primer' PCR product sizes that are used with highly degraded DNA samples to enable greater recovery of sequence information (see Gabriel et al. (2001a)).

to vary only about 1–2% (7–14 nucleotides out of the 610 bases examined is different) between unrelated individuals (Budowle *et al.* 1999). This variation is scattered throughout the HV1 and HV2 regions and is therefore best measured with DNA sequence analysis. However, there are 'hotspots' or hypervariable sites and regions where most of the variation is clustered (Stoneking 2000). Several methods for rapidly screening mtDNA variation have been developed that may be used for excluding samples that do not match. These methods often focus on measuring variation at the hypervariable hotspots and include using sequence-specific oligonucleotide probes (Stoneking *et al.* 1991), mini-sequencing (Tully *et al.* 1996), and denaturing gradient gel electrophoresis (Steighner *et al.* 1999) as well as a restriction digest assay for HV1 amplicons (Butler *et al.* 1998a) and a reverse dot blot or linear array assay approach (Comas *et al.* 1999, Gabriel *et al.* 2003).

MITOCHONDRIAL DNA SEQUENCING IN FORENSIC CASEWORK

In the following section, we describe the methodologies used for determining the sequence contained in mitochondrial DNA. Several nice overviews of forensic mtDNA analysis have been published and may be consulted for further information on this topic (Holland and Parsons 1999, Budowle *et al.* 2003, Isenberg 2004, Edson *et al.* 2004).

OVERVIEW OF METHODOLOGIES

The steps involved in performing mitochondrial DNA sequence comparisons are illustrated in Figure 10.4. Extraction of the mtDNA should be performed in a very clean laboratory environment because mtDNA is more sensitive to contamination than nuclear DNA since it is in a higher copy number per cell. Thus, it is preferable to analyze the reference samples after the evidence samples have been completely processed to avoid any potential contamination problems.

IMPORTANCE OF A CLEAN LABORATORY

The use of higher PCR cycle numbers (e.g., 36 or 42) and the already amplified copy number of mtDNA per cell necessitate great care to avoid contamination. The DNA templates under investigation are often damaged so they may not be as readily amplified as even low amounts of high quality DNA from laboratory

Figure 10.4

Process for evaluation of mtDNA samples. The evidence or question (Q) sample may come from a crime scene or a mass disaster. The reference or known (K) sample may be a maternal relative or the suspect in a criminal investigation. In a criminal investigation, the victim may also be tested and compared to the Q and K results.

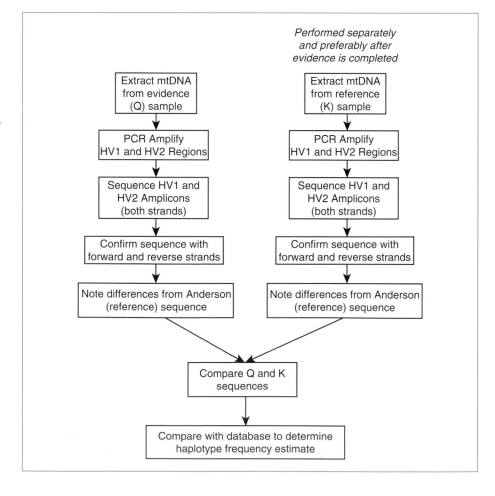

personnel or reference samples. Reference samples from the victim, the suspect, and maternal relatives are typically available as blood stains or buccal swabs and generally contain large amounts of high quality DNA.

Practices to reduce or minimize contamination often employed by forensic laboratories performing mtDNA testing include use of protective clothing such as disposable lab coats, frequent cleaning procedures with bleach and UV irradiation of hoods and lab bench surfaces, processing the question samples prior to the known samples, multiple glove changes during sample handling, using dedicated equipment for the mtDNA testing, and physically separating the pre- and post-amp spaces. During an analytical procedure only one item of evidence from a case is opened at a time (Isenberg and Moore 1999). Some laboratories even control movement of laboratory personnel between spaces. For example, a technician may not be permitted on the same day to return to a pre-amplification area after having entered a post-amp area. Vigilance on the part of all laboratory personnel is important to keep a forensic mtDNA laboratory clean. Reagent blanks and negative controls are also run to monitor levels of exogenous DNA in reagents, laboratory environment, or instruments.

SAMPLE EXTRACTION FOR mtDNA ANALYSIS

Mitochondrial DNA analysis typically involves materials where little DNA is present to begin with. Teeth, hair, and bones such as ribs and long bones (e.g., femur and humerus) are often materials used for mtDNA analysis in forensic cases. The mtDNA must be carefully extracted from these materials and often purified away from PCR inhibitors that can be co-extracted (Yoshii *et al.* 1992).

Because anthropological examination of a bone is often performed in addition to mtDNA testing, care must be taken to remove a section of the bone that will avoid destroying the physical features of the bone. Thus, an analyst might remove a small section from the middle of the bone without cutting all the way through the bone so that the overall length of the bone is not impacted. The same idea applies for teeth where odontological examinations are performed to aid an investigation. A tabulation of success rates for obtaining reportable mtDNA sequencing results across different skeletal materials found that ribs and femurs work best (Edson *et al.* 2004).

SPECIAL CONSIDERATIONS FOR HAIR EVIDENCE

Hair and fiber examiners can perform microscopic comparisons of hairs much more quickly than mtDNA can be analyzed and therefore can be used as an effective screening tool to reduce the amount of evidence processed through the steps of mtDNA sequencing. A correlation of microscopic and mitochondrial DNA hair comparisons found that the techniques can be complementary (Houck and Budowle 2002).

With hair evidence, the physical examination by a hair examiner must be performed prior to the mtDNA testing as the hair is destroyed during the extraction process. Typically for analysis of hair shafts, a tissue grinder is used to break down the keratin structure of the hair and release the mtDNA molecules (Wilson *et al.* 1995a). Usually 1–2 cm of hair shaft is ground up after carefully cleaning the outside of the hair (Jehaes *et al.* 1998).

Comparisons of head, pubic, and axillary hair shafts found the highest success rate with head hair shafts (Pfeiffer *et al.* 1999). The addition of bovine serum albumin or BSA (Giambernardi *et al.* 1998) helped reduce the PCR inhibitory effects of melanin previously noted by Yoshii *et al.* (1992) and Wilson *et al.* (1995a). A nested PCR amplification approach has successfully recovered mtDNA sequence information from as little as 33–330 femtograms of genomic DNA, which is equivalent to 10–100 copies of mtDNA (Allen *et al.* 1998).

ESTIMATING mtDNA QUANTITY

Many laboratories perform a nuclear DNA quantitation assay and then estimate the amount of mtDNA present assuming a fixed ratio between nuclear and mtDNA. For example, 50 or 500 pg of DNA template may be used in an mtDNA amplification based on a nuclear quantification result from Quantiblot (see Chapter 3). Newer approaches involving real-time PCR (see Chapter 4) have begun to appear in the literature (Meissner *et al.* 2000, Andreasson *et al.* 2002, von Wurmb-Schwark *et al.* 2002) that enable direct characterization of the number of mtDNA molecules in a cell.

PCR AMPLIFICATION

PCR amplification of mtDNA is usually done with 34–38 cycles. Protocols for highly degraded DNA specimens even call for 42 cycles (Gabriel *et al.* 2001a). Sometimes excess *Taq* is added to overcome PCR inhibitors such as melanin (Wilson *et al.* 1995a). It is important to keep in mind that sensitivity is maximized with mtDNA testing as it is usually only turned to as a last resort in efforts to obtain DNA results from a sample. The higher the sensitivity of any assay, the greater the chance for contamination and thus greater care is usually required with mtDNA work than with conventional STR typing.

SCIENCE OF DNA SEQUENCING

The Sanger method for DNA sequencing was first described almost 30 years ago (Sanger *et al.* 1977). This Nobel Prize winning technique is still employed in modern-day DNA sequencing. The process involves the polymerase incorporation of dideoxyribonucleotide triphosphates (ddNTPs) as chain terminators

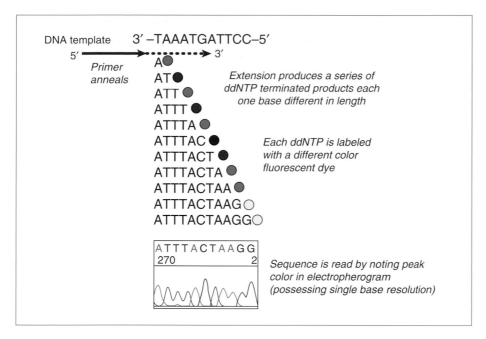

Figure 10.5

DNA sequencing process with fluorescent ddNTPs. A primer that has been designed to recognize a specific region of a DNA template anneals and is extended with a polymerase. Because a mixture of dNTPs and ddNTPs exist for each of the four possible nucleotides, some of the extension products are halted by incorporation of a ddNTP while other molecules continue to be extended. Each ddNTP is labeled with a different dye that enables each extension product to be distinguished by color. A size-based separation of the extension products permits the DNA sequence to be read provided that sufficient resolution is present to clearly see each base.

followed by a separation step capable of single nucleotide resolution. There is no hydroxyl group at the 3′-end of the DNA nucleotide (see Figure 2.1) with a ddNTP and therefore chain growth terminates when the polymerase incorporates a ddNTP into the synthesized strand. Extendable dNTPs and ddNTP terminators are both present in the reaction mix so that some portions of the DNA molecules are extended. At the end of sequencing reaction a series of molecules are present that differ by one base from one another.

Figure 10.5 illustrates the Sanger sequencing process. Each DNA strand is sequenced in separate reactions with a single primer. Often either the forward or reverse PCR primers are used for this purpose. Four different colored fluorescent dyes are attached to the four different ddNTPs. Thus, ddTTP (thymine) is labeled with a red dye, ddCTP (cytosine) is labeled with a blue dye, ddATP (adenine) is labeled with a green dye, and ddGTP (guanine) is labeled with a yellow dye although it is typically displayed in black for easier visualization. These are similar dyes as will be described in Chapter 13 for STR detection. Fluorescent dye labels have simplified DNA sequencing as have the widespread use of automated detection systems and capillary electrophoresis. The Human Genome Project was completed with these sequencing technologies.

Validation of various DNA sequencing chemistries has progressed over the past decade from a simple *Taq* polymerase, which often had high backgrounds and poor incorporation rates for many nucleotide combinations to the well-balanced Big Dye chemistries used today. Signal-to-noise ratios have improved with brighter dyes (Lee *et al.* 1997), which in turn now permits obtaining results from less material. As little as 1 ng of mtDNA PCR product can now be used for each DNA sequencing reaction (Stewart *et al.* 2003).

DNA sequencing of mtDNA is usually performed with the following steps: (1) PCR amplification of the entire control region or a portion of it with various primer sets as will be explained below; (2) removal of remaining dNTPs and primers from PCR through spin filtration using a Microcon 100 filter or enzymatic digestion with shrimp alkaline phosphatase and exonuclease I (Dugan *et al.* 2002); (3) determination of PCR product quantity (Wilson *et al.* 1995a, 1995b); (4) performance of DNA sequencing reaction to incorporate fluorescent ddNTPs as described above with each reaction containing a different primer to dictate which strand is sequenced; (5) removal of unincorporated fluorescent dye terminators from the completed sequencing reaction usually through spin column filtration; (6) dilution of purified sequencing reaction products in formamide and separation through electrophoresis in a capillary or gel system (see Chapters 12 and 14); and (7) sequence analysis of each reaction performed and interpretation of compiled sequence information as will be described below.

DNA sequencing may be reliably performed on a variety of platforms including the ABI 310, ABI 3100, and ABI 377 (Stewart *et al.* 2003). These various instrument platforms will be discussed in Chapter 14. The primary difference between STR analysis and mtDNA sequencing on these multi-color fluorescence detection instruments is that a separation medium capable of single base resolution is necessary for DNA sequence analysis while it is not always needed for STR typing. Thus, the separation medium POP™-6 is commonly used for DNA sequencing while POP™-4, a less viscous and lower resolution polymer, is used for STR typing.

PRIMERS USED FOR CONTROL REGION AMPLIFICATION AND SEQUENCING

PCR primers commonly used by the FBI Laboratory for mtDNA sequencing are shown as arrows on Figure 10.6 (Wilson *et al.* 1995b). Their primer nomenclature uses the strand corresponding to the primer (L for light and H for heavy) and the 3′ nucleotide position. Thus, primer A1 is designated as L15997 so it corresponds to the light strand of the Anderson reference sequence and ends at position 15997. Note that this nomenclature system does not indicate the 5′-end of the primer and therefore can make it more difficult to determine the overall PCR product size.

Another approach to mtDNA primer nomenclature is that used by the Armed Forces DNA Identification Laboratory. The primer positions for their primer sets (PS) I–IV are indicated in Figure 10.3. Strand designation in this case is by forward (F) and reverse (R) rather than light (L) and heavy (H). Also different is the fact that the 5′ nucleotide position is noted rather than the 3′ nucleotide as done by the FBI Laboratory. This approach permits an easier determination of the overall PCR product size defined by a primer pair. It is worth noting that two of the primers between the FBI and AFDIL sets are identical even though their names are different: FBI B1 (H16391) is the same primer as AFDIL R16410 used in PSII.

CHALLENGES WITH SEQUENCING BEYOND POLYMERIC C-STRETCHES

In Figure 10.6, a dotted box is found around a stretch of cytosine nucleotides in both the HV1 and HV2 regions. These regions are commonly referred to as 'C-stretches'. On the Anderson reference sequence that is shown in Figure 10.6, the HV1 C-stretch spans nucleotides 16184–16193 with a T at position 16189. In some samples position 16189 is a C giving rise to a stretch of 10 or more cytosines in a row (see Figure 10.7). The HV2 C-stretch region spans positions 303–315 on the reference sequence with a T at position 310 (Figure 10.6). This T can become a C in some samples leading to a homo-polymeric C-stretch.

Unfortunately, this homopolymeric stretch of cytosines creates problems for polymerases as they synthesize a complementary strand to the mtDNA template present in the reaction. Length heteroplasmy in HV1 between positions 16184 and 16193 can result in C-stretch lengths ranging from 8–14 cytosine residues (Bendall and Sykes 1995). Length heteroplasmy likely results from replication slippage after a T to C transition has occurred at position 16189. The mixture of length variants may already be present in the original DNA or generated in the sequencing reaction itself. Regardless of the source of the length variants, the impact of a 16189 T to C transition on sequencing results downstream of the C-stretch region can be seen in Figure 10.7.

A similar situation occurs with the HV2 C-stretch region when insertions of cytosines occur in the 303–310 area or a transition of T to C occurs at position 310 (Stewart *et al.* 2001). The presence of intra-individual variation in the number of cytosines observed when multiple hairs were tested from the same individual has led to the decision to not call an exclusion based solely on differences in the HV2 C-stretch region (Stewart *et al.* 2001). The issue of heteroplasmy and intra-individual variation will be discussed in more detail later in this chapter.

The ability to rapidly screen for the C-stretch prior to sequencing is advantageous and can be performed by noting the presence of extra heteroduplex peaks in quality control analyses of HV1 PCR products (Butler *et al.* 1998a).

Figure 10.6 (on following two consequent pages)

Annotation of the revised Cambridge Reference Sequence for mtDNA control region with primer positions and common sequence polymorphisms examined in screening assays (see Figure 10.10).

```
AATACCAACT  ATCTCCCTAA  TTGAAAACAA  AATACTCAAA  TGGGCCTGTC  CTTGTAGTAT
TTATGGTTGA  TAGAGGGATT  AACTTTTGTT  TTATGAGTTT  ACCCGGACAG  GAACATCATA
       15850        15860        15870        15880        15890        15900


AAACTAATAC  ACCAGTCTTG  TAAACCGGAG  ATGAAAACCT  TTTTCCAAGG  ACAAATCAGA
TTTGATTATG  TGGTCAGAAC  ATTTGGCCTC  TACTTTTGGA  AAAAGGTTCC  TGTTTAGTCT
       15910        15920        15930        15940        15950        15960
```

● Roche (F15975) →
FBI A1 (L15997) →

```
GAAAAAGTCT  TTAACTCCAC  CATTAGCACC  CAAAGCTAAG  ATTCTAATTT  AAACTATTCT
CTTTTTCAGA  AATTGAGGTG  GTAATCGTGG  GTTTCGATTC  TAAGATTAAA  TTTGATAAGA
       15970        15980        15990        16000        16010        16020
```

→ HV1

```
CTGTTCTTTC  ATGGGGAAGC  AGATTTGGGT  ACCACCCAAG  TATTGACTCA  CCCATCAACA
GACAAGAAAG  TACCCCTTCG  TCTAAACCCA  TGGTGGGTTC  ATAACTGAGT  GGGTAGTTGT
       16030        16040        16050        16060        16070        16080
```

Roche 16093   Roche IA

```
ACCGCTATGT  ATTTCGTACA  TTACTGCCAG  CCACCATGAA  TATTGTACGG  TACCATAAAT ●
TGGCGATACA  TAAAGCATGT  AATGACGGTC  GGTGGTACTT  ATAACATGCC  ATGGTATTTA
       16090        16100        16110        16120        16130        16140
```

FBI A2 (L16159) →   HV1 C-stretch

```
ACTTGACCAC  CTGTAGTACA  TAAAAACCCA  ATCCACATCA  AAACCCCCTC  CCCATGCTTA
TGAACTGGTG  GACATCATGT  ATTTTTGGGT  TAGGTGTAGT  TTTGGGGGAG  GGGTACGAAT
       16150        16160        16170 ←      16180        16190        16200
```
●

FBI A4 (L16209) →   FBI B4 (H16164)

```
CAAGCAAGTA  CAGCAATCAA  CCCTCAACTA  TCACACATCA  ACTGCAACTC  CAAAGCCACC
GTTCGTTCAT  GTCGTTAGTT  GGGAGTTGAT  AGTGTAGT    TGACGTTGAG  GTTTCGGTGG
       16210        16220        16230        16240        16250        16260
```
● ←
FBI B2 (H16236)   Roche IC

Roche IE

```
CCTCACCCAC  TAGGATACCA  ACAAACCTAC  CCACCCTTAA  CAGTACATAG  TACATAAAGC
GGAGTGGGTG  ATCCTATGGT  TGTTTGGATG  GGTGGGAATT  GTCATGTATC  ATGTATTTCG
       16270        16280        16290        16300        16310        16320
```

Roche ID   ← HV1

```
CATTTACCGT  ACATAGCACA  TTACAGTCAA  ATCCCTTCTC  GTCCCCATGG  ATGACCCCCC
GTAAATGGCA  TGTATCGTGT  AATGTCAGTT  TAGGGAAGAG  CAGGGGTACC  TACTGGGGGG
       16330        16340        16350        16360        16370        16380
```

```
TCAGATAGGG  GTCCCTTGAC  CACCATCCTC  CGTGAAATCA  ATATCCCGCA  CAAGAGTGCT
AGTCTATCCC  CAGGGAACTG  GTGGTAGGAG  GCACTTTAGT  TATAGGGCGT  GTTCTCACGA
       16390        16400        16410        16420        16430        16440
```
←   ●
FBI B1 (H16391)   Roche (R16418)

```
ACTCTCCTCG  CTCCGGGCCC  ATAACACTTG  GGGGTAGCTA  AAGTGAACTG  TATCCGACAT
TGAGAGGAGC  GAGGCCCGGG  TATTGTGAAC  CCCCATCGAT  TTCACTTGAC  ATAGGCTGTA
       16450        16460        16470        16480        16490        16500
```

16519

```
CTGGTTCCTA  CTTCAGGGTC  ATAAAGCCTA  AATAGCCCAC  ACGTTCCCCT  TAAATAAGAC
GACCAAGGAT  GAAGTCCCAG  TATTTCGGAT  TTATCGGGTG  TGCAAGGGGA  ATTTATTCTG
       16510        16520        16530        16540        16550        16560
```

```
ATCACGATG
TAGTGCTAC
       16569
```

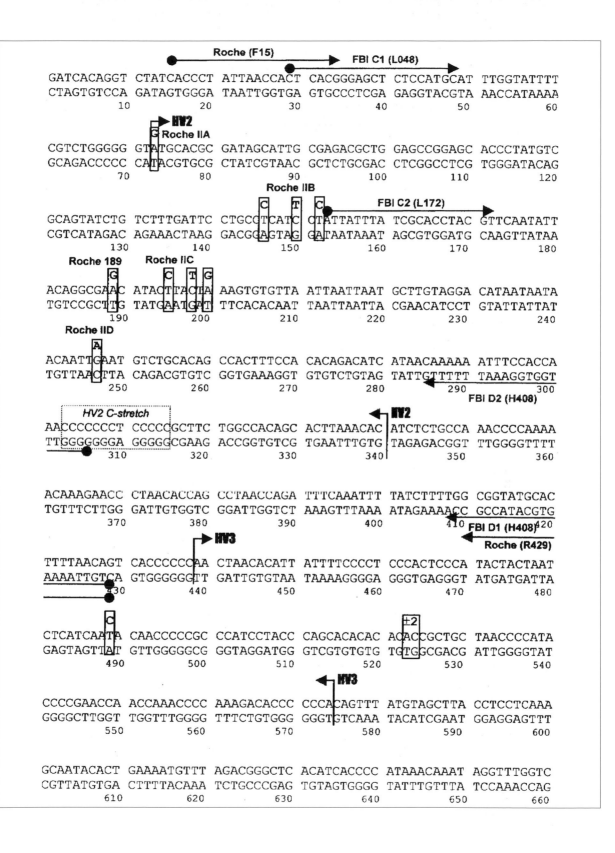

Figure 10.7

Comparison of a sample with (a) 16189T (no HV1 C-stretch) to (b) one with the C-stretch. Notice how the sequence quality quickly drops after the string of cytosine residues due to the presence of two or more length variants that creates a situation where the extension products are out of phase or register with one another. Different primer combinations are typically used on samples containing a C-stretch as illustrated in (c) to recover sequence information from both strands or to provide a double read of the same strand.

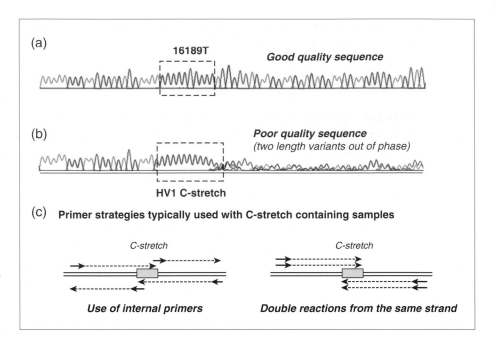

In the event that the C-stretch is present in a sample, different sequencing primers may be used to obtain reliable mtDNA sequence information downstream of the homopolymeric stretches (Rasmussen *et al.* 2002). For example, the FBI A4/B4 primer set (L16209 and H16164) shown in Figure 10.6 can be used on individuals possessing the HV1 C-stretch in order to recover sequence information from both sides of the homopolymeric stretch of cytosines (Wilson 1997). Alternatively the same strand may be examined twice in separate sequencing reactions to provide double coverage of all nucleotides (Figure 10.7c).

USE OF SMALL AMPLICONS TO IMPROVE AMPLIFICATION SUCCESS ON HIGHLY DEGRADED SAMPLES

As noted in Chapter 7 when encountering highly degraded DNA samples where the molecules have been fragmented to small sizes, the use of smaller sized PCR products improves recovery of information from the original DNA template. This is also the case with mitochondrial DNA and 'mini-primer sets' have been developed to amplify smaller portions of HV1 and HV2 (Gabriel *et al.* 2001a, Edson *et al.* 2004). The bottom portion of Figure 10.3 shows the relative position and PCR product sizes for eight mini-products ranging in size from 126 bp to 170 bp.

The use of these mini-amplicons that overlap one another is sometimes referred to as an 'ancient DNA' approach and is capable of recovering abundant DNA in a sample that might otherwise fail to produce results with a standard protocol (Gabriel *et al.* 2001a, Melton and Nelson 2001). This approach has been used to successfully recover information from Neanderthal bones that are thousands of years old (Krings *et al.* 1997, 1999).

USE OF POSITIVE AND NEGATIVE CONTROLS

As noted by Melton and Nelson (2001) the two primary goals in mtDNA testing are (1) to protect the integrity of the evidence by preventing contamination at any stage in the testing and (2) to collect the maximum amount of available mtDNA data inherent to any sample. Control samples that are processed in parallel with evidentiary samples through each step of the process serve to monitor performance and assess one's success with the two goals noted above.

Contamination assessment is performed with reagent blanks and negative controls. Reagent blanks monitor contamination from extraction to final sequence analysis while negative controls monitor contamination from amplification to final sequence analysis (SWGDAM 2003). All of the procedures performed on a sample are also performed on the reagent or extraction blank with the exception of adding DNA. Negative controls or amplification blanks are introduced at the PCR amplification step and use the same reagents as the sample with sterile water in place of the DNA template. If the reagent blank and/or the negative control associated with a particular amplification results in a sequence that is the same as that of the sample, all data for the sample must be rejected (Isenberg 2004). The analysis must then be repeated beginning with the re-amplification of the sample in question.

Reagent blank contamination is sometimes observed in spite of great efforts to keep the laboratory environment clean. Since mtDNA analysis is a very sensitive technique, the presence of low-level contamination is not uncommon (Isenberg 2004). For example, Mitotyping Technologies reported that reagent blanks resulted in amplification products in 29 of 1218 (2.4%) of PCR reactions performed in casework over a two year period of time (Melton and Nelson 2001). These contaminants did not match a staff member's type or the type of the recently handled sample a fact that suggests they are likely sporadic contaminants to a particular disposable tip or PCR tube. This type of contamination is not uncommon when working with low-copy number DNA as noted in Chapter 7.

If contamination is observed with either the reagent blank or the negative control, results from the unknown sample being run in parallel do not always have to be disregarded. Research with artificial sample mixtures has demonstrated that a threshold of background contamination can be set for still obtaining reliable sequence data. For example, the FBI Laboratory has established a 10:1 rule where any contamination seen in a reagent blank or negative control during post-PCR analysis must be less than one-tenth the amount of the sample being processed (Wilson *et al.* 1995a, 1995b). This sample-to-contamination ratio determination is possible due to the PCR product quantification analysis performed in their procedure. A more recent study demonstrated that the 10:1 rule is conservative and reliable (Stewart *et al.* 2003).

A positive control is a sample of known mtDNA sequence that serves to demonstrate that amplification and sequencing reaction components are working properly. This positive control is typically an extracted DNA sample that is processed through the steps of amplification, sequencing, and data analysis. For example, the FBI Laboratory uses the HL60 cell line as a positive control.

INTER-LABORATORY STUDIES

Inter-laboratory studies where laboratories perform testing on the same sample are valuable to demonstrate that a technique is reliable (see Chapter 16). As of early 2004, no manufacturers supply commercially available kits for the entire process of mtDNA sequencing, such as are available for STR typing. Thus, a number of different methods exist for mtDNA testing without a single universal protocol.

A number of inter-laboratory studies involving mtDNA sequencing have been conducted and have demonstrated that the same results can be successfully obtained in multiple laboratories using different protocols (Carracedo *et al.* 1998, Alonso *et al.* 2002, Prieto *et al.* 2003, Parsons *et al.* 2004, Tully *et al.* 2004).

STANDARD REFERENCE MATERIALS FOR mtDNA SEQUENCE ANALYSIS

Standard reference samples along with positive controls serve to demonstrate that mtDNA sequence analysis is being performed appropriately (see Szibor *et al.* 2003a). The U.S. National Institute of Standards and Technology (NIST) has developed two Standard Reference Materials (SRMs) to aid in confirming sequencing results with mtDNA (Levin *et al.* 1999, Levin *et al.* 2003). Information is available for the entire mtGenome on three samples (SRM 2392) and the cell line HL-60 (SRM 2392-I).

INTERPRETING AND REPORTING mtDNA RESULTS

DATA REVIEW AND EDITING

DNA sequencing is performed in both the forward and reverse directions so that the complementary strands can be compared to one another for quality control purposes. If it is not possible to get sequence from both strands, for example following a C-stretch, then the same strand can be sequenced twice in separate reactions. The goal is to have at least double coverage of every nucleotide being assessed either through sequencing the top and bottom strand or sequencing the same strand twice.

The sequencing process does not always lead to beautiful data that is unambiguous for each base. Some regions, such as the C-stretches, are challenging to decipher and may not even be included in the final interpretation (Stewart *et al.* 2001). Sequencing chemistries and instruments have improved in recent years leading to more even peaks, better sensitivity and less noise. However, experienced analysts must still manually review each nucleotide with the aid of computer software and then edit base calls when the base-calling algorithm has made an incorrect call. At present there is no publicly available software that can robustly evaluate mtDNA sequence data in a reliable and automated fashion without manual intervention.

The sequence editing process is aided by alignments from the multiple sequences generated over a region for the same sample. Computer programs such as Sequencer (GeneCodes, Ann Arbor, MI) align the forward and reverse sequencing reactions and allow the sequencing electropherograms for each reaction to be evaluated side-by-side. For casework samples that utilize smaller PCR products there is overlap between them (see Figure 10.3) that permits a further measure of quality assurance in the final compiled sequence. In addition, two forensic analysts must independently examine, interpret, and edit sequence matching results as a final quality assurance measure (Isenberg 2004).

REPORTING DIFFERENCES TO THE REVISED CAMBRIDGE REFERENCE SEQUENCE

For reporting purposes, sequences are listed in a minimum data format as differences relative to the rCRS. When differences are observed, the nucleotide position is cited followed by the base present at that site. For example in Figure 10.8a, differences are observed at positions 16093 and 16129 and are noted in Figure 10.8b in their minimum data format at 16093C and 16129A. In this format, all other nucleotides are assumed to be identical to the revised Cambridge Reference Sequence. Bases that cannot unambiguously be determined are usually coded N. At confirmed positions of ambiguity (e.g., sequence heteroplasmy), the International Union of Pure and Applied Chemistry (IUPAC) codes should be used, such as $A/G = R$ and $C/T = Y$ (SWGDAM 2003).

Insertions in a DNA sequence relative to the rCRS are described by noting the site immediately 5′ to the insertion as compared to the rCRS followed by a point and a '1' (for the first insertion), a '2' (if there is a second insertion), and so on, and then by the nucleotide that is inserted (Isenberg 2004). For example, 315.1C is a common observation where six Cs are observed following the T at position 310 in the rCRS. The rCRS contains only five Cs in positions 311–315 (Andrews *et al.* 1999). Therefore, the notation 315.1C describes the presence of five cytosines at positions 311–315 and an extra C as an insertion ('.1C') prior to position 316.

Deletions are noted by a dash ('-') or a 'D', 'd' or 'del' following the nucleotide position where the deletion was observed relative to the rCRS (e.g., 309D, 309-, or 309del). Some insertion and deletion combinations can lead to multiple possibilities for reporting a result in terms of differences from the reference sequence. Therefore, recommendations have been made for consistent treatment of length variants as will be described in the next section.

NOMENCLATURE ISSUES

Ambiguities with respect to mtDNA nomenclature can result in two different analysts calling the same sample differently. Likewise population databases could have multiple entries for the same mtDNA haplotype preventing an accurate estimate for the frequency of a particular type. Thus, standardization in designation of mtDNA sequences is important to have data that can easily be shared between laboratories.

Length variants present a challenge when alignments are made between a sample of interest and the Cambridge Reference Sequence. Treatments of insertions and deletions (gaps) can vary between laboratories causing some laboratories to code the same sequence differently. Mark Wilson and colleagues at the FBI Laboratory have made a number of recommendations to enable consistent treatment of length variants (Wilson *et al.* 2002a, 2002b). Three primary recommendations were made: (1) characterize profiles using the least number of differences from the reference sequence; (2) if there is more than one way to maintain the same number of differences with respect to the reference sequence, differences should be prioritized in the following manner: (a) insertions/deletions (indels), (b) transitions, and (c) transversions; (3) insertions and deletions should be placed 3′ with respect to the light strand. Insertions and deletions should be combined in situations where the same number of differences to the reference sequence is maintained. These recommendations are hierarchical meaning that recommendation (1) should take precedence over recommendation (2) and (3). A total of 41 specific examples are provided to demonstrate the need for consistent treatment of length variants in mtDNA sequence analysis and reporting (Wilson *et al.* 2002a, 2002b).

INTERPRETATION OF RESULTS

Following completion of mtDNA sequence analysis as described above, results from the edited and reviewed sequences for a question (Q) and a known (K) sample are compared as illustrated in Figure 10.8 for a portion of HV1. All 610 nucleotides (positions 16024–16365 and 73–340) are normally evaluated between samples being compared.

A comparison of the two sequences in question will either result in a perfect match or not. Samples are termed concordant if they match at every evaluated site.

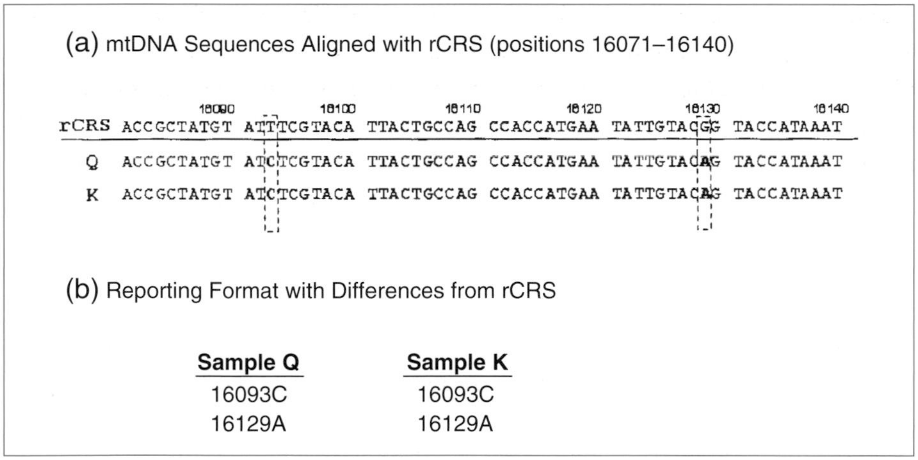

(a) mtDNA Sequences Aligned with rCRS (positions 16071–16140)

```
              16090        16100        16110        16120        16130        16140
rCRS ACCGCTATGT ATTTCGTACA TTACTGCCAG CCACCATGAA TATTGTACGG TACCATAAAT

  Q  ACCGCTATGT ATCTCGTACA TTACTGCCAG CCACCATGAA TATTGTACAG TACCATAAAT

  K  ACCGCTATGT ATCTCGTACA TTACTGCCAG CCACCATGAA TATTGTACAG TACCATAAAT
```

(b) Reporting Format with Differences from rCRS

Sample Q	Sample K
16093C	16093C
16129A	16129A

Figure 10.8

(a) Comparison of sequence alignments for hypothetical Q and K samples with (b) conversion to the revised Cambridge Reference Sequence (rCRS) differences for reporting purposes.

However, interpretation of results is not always so cut and dry. Laboratories must develop interpretation guidelines as noted below.

Results can generally be grouped into three categories: exclusion, inconclusive, or failure to exclude. The Scientific Working Group on DNA Analysis Methods (SWGDAM) Guidelines for Mitochondrial DNA (mtDNA) Nucleotide Sequence Interpretation lists the following commonly used recommendations (SWGDAM 2003):

- *Exclusion* – if there are two or more nucleotide differences between the questioned and known samples, the samples can be excluded as originating from the same person or maternal lineage.
- *Inconclusive* – if there is one nucleotide difference between the questioned and known samples, the result will be inconclusive.
- *Cannot Exclude (Failure to Exclude)* – if the sequences from questioned and known samples under comparison have a common base at each position or a common length variant in the HV2 C-stretch, the samples cannot be excluded as originating from the same person or maternal lineage.

A common base is defined as a shared base in the case of ambiguity (e.g., heteroplasmy) in the sequence (Isenberg 2004). For example, if one sequence possesses heteroplasmy at a site and another does not (see Figure 10.9), then they cannot be excluded from one another. A length variant alone especially in the HV2 homopolymeric C-stretch cannot be used to support an interpretation of exclusion (Stewart *et al.* 2001, SWGDAM 2003). Several examples are provided in Table 10.4 with their respective interpretations based on the SWGDAM guidelines.

The reason that a single base difference is classified in terms of an 'inconclusive result' is that mutations have been observed between mother and children (Parsons *et al.* 1997). For example, if a maternal relative is used for a reference

Table 10.4

Example mtDNA sequences and interpretations for known (K) and question (Q) sample pairs (adapted from Isenberg 2004).

Sequence Results	Observations	Interpretation
Q TATTGTACGG K TATTGTACGG	Sequences are fully concordant with common bases at every position	Cannot Exclude
Q TATTG<u>C</u>AC<u>A</u>G K TATTGTACGG	Sequences differ at two positions	Exclusion
Q TATT<u>N</u>TACGG K TATTGTACGG	A single unspecified base in one of the sequences; common base at every position	Cannot Exclude
Q TATT<u>N</u>TACGG K TATTGTAC<u>N</u>G	Ambiguous bases in both sequences at different positions; common base at every position	Cannot Exclude
Q TATTGTAC<u>A/G</u>G K TATTGTAC <u>G</u> G	Heteroplasmic mixture at a position in one sample that is not present in the other; common base at every position (G in both Q and K)	Cannot Exclude
Q TATTGTAC<u>A/G</u>G K TATTGTAC<u>A/G</u>G	Heteroplasmic mixture at the same site in both sequences; common base at every position	Cannot Exclude
Q TATTG<u>C</u>ACGG K TATTG<u>T</u>ACGG	Sequences identical at every position except one; no indication of heteroplasmy	Inconclusive

sample, the possibility of a single base difference may exist between two samples that are in fact maternally related. Often additional samples, such as more reference samples, are run if an inconclusive result is obtained in an attempt to clarify the interpretation. Hairs from an individual might be pooled in an attempt to detect heteroplasmy (Isenberg 2004).

REPORTING STATISTICS

When 'failure to exclude' is the interpretation for reference and evidence samples, then a statistical estimate of the significance of a match is needed. Mitochondrial DNA is inherited in its entirety from one's mother without recombination (discussed later in the chapter). Therefore individual nucleotide positions are inherited in a block and must be treated as a single locus haplotype the same as with Y chromosome information discussed in Chapter 9. The product rule applied to independently segregating STR loci found on separate chromosomes cannot be used with mtDNA polymorphisms.

The current practice of conveying the rarity of an mtDNA type among unrelated individuals involves counting the number of times a particular haplotype (sequence) is seen in a database (Wilson *et al.* 1993, Budowle *et al.* 1999). This approach is commonly referred to as the 'counting method' and depends entirely on the number of samples present in the database that is searched. Thus, the larger the number of unrelated individuals in the database, the better the statistics will be for a random match frequency estimate.

Population frequencies for most DNA types (around 60%) are not known presently because they occur only a single time in a database (Isenberg 2004). Based on available population information, confidence intervals can be used to estimate the upper and lower bounds of a frequency calculation (Holland and Parsons 1999, Tully *et al.* 2001) (see D.N.A. Box 10.3).

Other methods are sometimes used for calculating a statistical weight for an mtDNA match. For example, likelihood ratios have been proposed

---

D.N.A. Box 10.3

Calculation of mtDNA profile frequency estimates

In cases where an mtDNA profile is observed a particular number of times (X) in a database containing N profiles, its frequency (p) can be calculated as follows:

$$p = X/N$$

An upper bound confidence interval can be placed on the profile's frequency using:

$$p + 1.96\sqrt{\frac{(p)(1-p)}{N}}$$

In cases where the profile has not been observed in a database, the upper bound on the confidence interval is

$$1 - \alpha^{1/N}$$

where α is the confidence coefficient (0.05 for a 95% confidence interval) and N is the number of individuals in the database.

For example, the mtDNA type 16129A, 263G, 309d, 315.1C occurs twice in 1148 African-American profiles, twice in 1655 Caucasian profiles, and not at all in 686 Hispanic profiles when searched against the mtDNA Population Database (Monson *et al.* 2002). Using the equations above, calculations for the rarity of this profile in the respective sample sets are as follows:

For African-Americans: $p = 2/1148 + 1.96 [(2/1148)(1 - (2/1148))/1148]^{1/2}$
$= 0.0017 + 0.002 = 0.004 = 0.4\%$

For Caucasians: $p = 2/1655 + 1.96 [(2/1655)(1 - (2/1655))/1655]^{1/2} = 0.0012 + 0.0017$
$= 0.0029 = 0.29\%$

For Hispanics: $1 - (0.05)^{1/686} = 1 - 0.9956 = 0.0044 = 0.44\%$

These calculations demonstrate that the statistical weight can be similar whether or not a match is found to a few previously observed samples in a database.

Sources:

Evett, I.W. and Weir, B.S. (1998) *Interpreting DNA Evidence*. Sunderland, Massachusetts: Sinauer Associates, Inc., p. 142.

Tully, G., *et al.* (2001) *Forensic Science International*, 124, 83–91.

(Holland and Parsons 1999, Tully *et al.* 2001, Isenberg 2004). Regardless of the method used for calculating the rarity of an mtDNA profile, it is important to keep in mind that mtDNA can never have the power of discrimination that an autosomal STR marker can since its inheritance is uniparental.

LABORATORIES PERFORMING mtDNA TESTING IN THE UNITED STATES

The first efforts in mtDNA sequence analysis with a forensic applications focus were performed by the Forensic Science Service in England (Sullivan *et al.* 1991, Hopgood *et al.* 1992, Sullivan *et al.* 1992). Within the United States, the Armed Forces DNA Identification Laboratory and the FBI Laboratory have led the efforts in mtDNA analysis but in slightly different arenas.

ARMED FORCES DNA IDENTIFICATION LABORATORY (AFDIL)

The Armed Forces DNA Identification Laboratory is located in Rockville, Maryland and is charged with identifying the remains of military personnel (Holland *et al.* 1993). Bones recovered from Vietnam, Korea, and even World War II operations have been successfully analyzed with mtDNA (Holland *et al.* 1995, Holland and Parsons 1999). AFDIL also does some contract civilian work (see below) including mass disaster victim identification (see Chapter 24).

FBI LABORATORY

The FBI Laboratory focuses on the use of forensic evidence including mtDNA in criminal investigations. Two DNA units exist within the FBI Laboratory: DNA Unit I, which focuses exclusively on nuclear DNA, and DNA Unit II, which performs mtDNA analysis and aids missing persons investigations.

The FBI Laboratory first explored the feasibility of using mtDNA in human identity applications in the late 1980s (Budowle *et al.* 1990) and aggressively began researching analysis methods in 1992. The FBI Laboratory DNA Unit II has conducted mitochondrial DNA casework since June 1996. Their first case involving court testimony came in August 1996 with the State of Tennessee versus Paul William Ware, which involved mtDNA analysis of a single pubic hair found in the throat of young victim that matched the defendant (Marchi and Pasacreta 1997). Much of the mtDNA evidence processed by FBI involves shed hairs.

FBI REGIONAL LABORATORIES

The FBI announced in the fall of 2003 selection of four regional laboratories that would be funded by the FBI to perform mtDNA casework as an extension of their own operations. These labs are the Arizona Department of Public Safety (Phoenix, Arizona), the Connecticut State Police (Meriden, Connecticut), the Minnesota Bureau of Criminal Apprehension (St. Paul, Minnesota), and the New Jersey State Police (Trenton, New Jersey). These regional mtDNA laboratories should be fully operational by September 2005 and be able to analyze approximately 120 cases each on an annual basis. It is expected that the regional mtDNA laboratories will double the FBI's capacity to provide mtDNA analysis to the criminal justice system.

PRIVATE LABORATORIES CONDUCTING FORENSIC mtDNA CASEWORK

Several private laboratories in the United States have validated mtDNA procedures and offer mtDNA testing on a fee basis. These laboratories include Mitotyping Technologies, LLC (State College, Pennsylvania), ReliaGene Technologies, Inc. (New Orleans, Louisiana), Bode Technology Group (Springfield, Virginia), Orchid Cellmark (Dallas, Texas), the University of North Texas Health Sciences Center DNA Identity Lab (Ft. Worth, Texas), and Laboratory Corporation of America (Research Triangle Park, North Carolina). In addition, the Armed Forces DNA Identification Laboratory (Rockville, Maryland) has a consultative services branch called AFDIL[cs] that performs contract testing of civilian mtDNA cases. These laboratories typically charge around $2000 per sample for mtDNA testing in order to sequence the 610 nucleotides in HV1 and HV2.

It is interesting to consider some of the statistics noted by one of the private laboratories concerning their mtDNA analysis work. Mitotyping Technologies reported on processing 105 cases between February 1999 and February 2001 (Melton and Nelson 2001). These cases involved 199 questioned items of which 130 were hairs. A total of 137 known reference samples were also processed including 111 that were in the form of blood. Only 17 of their 199 questioned samples failed to yield any mtDNA amplification products. Length heteroplasmy was observed 15 times in the HV1 C-stretch region and 77 times in the HV2 C-stretch region with 17 samples having both HV1 and HV2 length heteroplasmy. Sequence site heteroplasmy was reported 19 times mostly at positions 16093 but also at nucleotide positions 16166, 16286, 72, 152, 189, 207, and 279. In 57 out of 105 cases (54.3%), the known reference sample could not be excluded as donor of a biological sample.

ISSUES IMPACTING INTERPRETATION

In this section, we will consider several issues that often arise when considering mtDNA evidence particularly in courts of law (see Walker 2003).

HETEROPLASMY

Heteroplasmy is the presence of more than one mtDNA type in an individual (Melton 2004). Two or more mtDNA populations may occur between cells in an individual, within a single cell, or within a single mitochondrion. It is now thought that all individuals are heteroplasmic at some level – many below the limits of detection in DNA sequence analysis (Comas *et al.* 1995, Bendall *et al.* 1996, Steighner *et al.* 1999, Tully *et al.* 2000). It is highly unlikely that millions of mtDNA molecules scattered throughout an individual's cells are completely identical given that regions of the mtGenome have been reported to evolve at 6–17 times the rate of single copy nuclear genes (see Brown *et al.* 1979, Wallace *et al.* 1987, Tully 1999). Consider that whereas only a single copy of each nuclear chromosome is present in an egg there are approximately 100 000 copies of the mtDNA genome present (Chen *et al.* 1995). Thus, for the transmission of a mtDNA mutation to become detectable it must spread to an appreciable frequency among a cell's mtDNA molecules.

Heteroplasmy may be observed in several ways: (1) individuals may have more than one mtDNA type in a single tissue; (2) individuals may exhibit one mtDNA type in one tissue and a different type in another tissue; and/or (3) individuals may be heteroplasmic in one tissue sample and homoplasmic in another tissue sample (Carracedo *et al.* 2000). Given that heteroplasmy happens, interpretation guidelines must take into account how to handle differences between known and questioned samples.

Both sequence and length heteroplasmy have been reported in the literature (Bendall and Sykes 1995, Bendall *et al.* 1996, Melton 2004). Length heteroplasmies often occur around the homopolymeric C-stretches in HV1 at positions 16184–16193 and HV2 at positions 303–310 (Stewart *et al.* 2001) (see Figure 10.6). Sequence heteroplasmy is typically detected by the presence of two nucleotides at a single site, which show up as overlapping peaks in a sequence electropherogram (Figure 10.9).

Heteroplasmy at two sites in the same individual, a condition known as 'triplasmy', has been reported (Tully *et al.* 2000), but occurs at lower frequencies than single site heteroplasmy. Since it is rare to find more than one heteroplasmic position in the 610 nucleotides sequenced for HV1 and HV2, a report of as many as six heteroplasmic sites in an individual mtDNA sequence (Grzybowski 2000) raised suspicions about the sequencing strategy used. The Grybowski study has been criticized as possibly containing contamination due to the excessive number of amplification cycles used (Budowle *et al.* 2002a, Brandstätter and Parson 2003). A re-analysis of the same samples used in the original Grybowski study with a direct rather than a nested PCR approach resulted in a reduction in the reported number of samples with heteroplasmic positions (Grzybowski *et al.* 2003).

One of the major challenges of heteroplasmic samples is that the ratio of bases may not stay the same across different tissues, such as blood and hair or

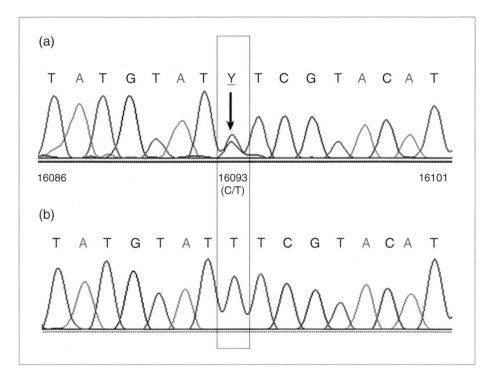

Figure 10.9

(a) Sequence heteroplasmy at position 16093 possessing both C and T nucleotides compared to (b) the same region (positions 16086–16101) on a different sample containing only a T at position 16093.

between multiple hairs (Sullivan *et al.* 1997, Wilson *et al.* 1997, Sekiguchi *et al.* 2003). Some mtDNA protocols now recommend sequencing multiple hairs from an individual in order to confirm heteroplasmy.

Hotspots for heteroplasmy include the following positions in HV1: 16093, 16129, 16153, 16189, 16192, 16293, 16309, and 16337 (Stoneking 2000, Tully *et al.* 2000, Brandstätter and Parson 2003) and 72, 152, 189, 207, and 279 in HV2 (Calloway *et al.* 2000, Melton and Nelson 2001). One study found that the frequency of heteroplasmy can differ across tissue types with muscle tissue being the highest and was statistically significant across different age groups suggesting that heteroplasmy increases with age (Calloway *et al.* 2000). Heteroplasmy has also been reported to remain stable over time in the same individuals and thus be inherited rather than age related (Lagerström-Fermér *et al.* 2001). While heteroplasmy can sometimes complicate the interpretation of mtDNA results, the presence of heteroplasmy at identical sites can improve the probability of a match, such as seen in the Romanov study (see D.N.A. Box 10.2).

SAMPLE MIXTURES

A major advantage of mtDNA in terms of sequencing is that it is haploid and therefore only a single type exists (barring detectable heteroplasmy) for analysis. However, mixed samples from more than one biological source are commonly encountered in forensic settings. Generally speaking attempts are not made to decipher samples containing a mixture of more than one individual due to the complexity of the sequencing signals that could arise. Peak height ratios for two different bases cannot be used for reliable quantification of the two components because incorporation rates are not always even. Thus, the ratio of an A:G mixed base might be 50:50 at a particular position but when the complementary strand is sequenced a 70:30 or 80:20 ratio for the T:C bases might be observed because the polymerase incorporates the fluorescently-labeled ddTTP and ddCTP with different efficiencies than the A and G dideoxynucleotides.

If three or more sites within the 610 bases evaluated across HV1 and HV2 are found to possess multiple nucleotides at a position (i.e., sequence heteroplasmy), then the sample can usually be considered a mixture – either by contamination or from the original source material. Presently mixture interpretation is not attempted in forensic laboratories performing routine casework.

Some researchers are making pursuing efforts to resolve mtDNA mixtures through cloning and sequencing the resulting HV1/HV2 regions from individual colonies (Bever *et al.* 2003, Walker *et al.* 2004). Theoretically, each individual colony produced during the process of cloning corresponds to the control region from a single individual or a single component of heteroplasmy. Interpretation of mixtures is being attempted with statistical analysis from multiple clones. A number of pitfalls exist with this approach including the possibility of over-estimating the number of contributors due to the occurrence of heteroplasmic mitochondria. The number of contributors will be under-estimated if individuals are closely related and members of the same mtDNA haplogroup (Walker *et al.* 2004). Denaturing HPLC has also been proposed as a possible approach to separating mtDNA amplicon mixtures (LaBerge *et al.* 2003), as has a mismatch primer-induced restriction site analysis method (Szibor *et al.* 2003b).

NUCLEAR PSEUDOGENES

Segments of the mtGenome are present in the human nuclear genome (Collura and Stewart 1995, Zischler *et al.* 1995, Wallace *et al.* 1997). These 'molecular fossils' or pseudogenes are rare events caused by migration and integration of a portion of the mtGenome into nuclear DNA. Zischler *et al.* (1995) reported that human chromosome 11 carries a portion of the mtDNA control region that reflects an ancient genetic transposition from the mitochondrion to the nuclear genome. This element differs from typical modern mtDNA sequences by approximately 7.5% and has not created problems with regular forensic casework (Morgan *et al.* 1998).

These nuclear fossils of the mtGenome can create the potential for complications in mtDNA human identity testing if they are amplified instead of the intended mtDNA target when a high number of PCR cycles are invoked to try and tease out mtDNA sequence information from a particularly difficult sample (Morgan *et al.* 1998). Under unique circumstances, nuclear pseudogenes could act to contaminate the true mtDNA sequence. Such was likely the case with the high degree of heteroplasmy reported on some hair samples that are amplified with a nested PCR approach involving a cumulative number of 60 cycles (Grzybowski 2000, Budowle *et al.* 2002a, Brandstätter and Parson 2003). However, with primer sets commonly used in forensic mtDNA testing and a direct PCR with fewer than 40 cycles, nuclear DNA sequences that are similar to mtDNA rarely cause a problem because their initial copy number is so much lower than that of mtDNA.

POSSIBILITIES OF RECOMBINATION OR PATERNAL LEAKAGE

Several years ago three papers were published suggesting the possibility of recombination in mtDNA or inheritance from the paternal rather than the maternal line (Hagelberg *et al.* 1999, Eyre-Walker *et al.* 1999, Awadalla *et al.* 1999). Paternal inheritance of mtDNA has been reported in mice (Gyllensten *et al.* 1991). The Hagelberg and Eyre-Walker papers created quite a stir in the mtDNA forensic and population genetic circles (Macaulay *et al.* 1999, Parsons and Irwin 2000, Kivisild and Villems 2000, Jorde and Bamshad 2000, Kumar *et al.* 2000). Hagelberg and co-workers later retracted their paper due to problems with the data (Hagelberg *et al.* 2000). Since there really appears to be no direct evidence to support either recombination within or between mtGenomes, this issue has been laid to rest for most scientists in the field (see Ingman *et al.* 2000, Elson *et al.* 2001, Wiuf 2001, Herrnstadt *et al.* 2002).

However, there has been a single report recently published of the transmission of a paternal human mtDNA type in skeletal muscle (Schwartz and Vissing 2002). This paternal haplotype was not found in any other tissues though. Several additional studies with individuals having a similar muscle disease failed to find any evidence of paternal transmission of mtDNA (Johns 2003, Filosto *et al.* 2003, Taylor *et al.* 2003). With tens of thousands of mtDNA samples demonstrating maternal inheritance over the past three decades (see Giles *et al.* 1980), it is safe to conclude that the central dogma of maternal inheritance for mtDNA is here to stay.

SIZE OF mtDNA POPULATION DATABASE AND THE QUALITY OF INFORMATION

There are now population databases with thousands of mtDNA profiles in them. The availability of population data for the HV1/HV2 regions that are sequenced in forensic mtDNA analysis will be discussed in more detail later in the chapter.

MOST COMMON TYPES

One of the biggest weaknesses of mtDNA analysis is that some haplotypes are rather common in various population groups. For example, in the FBI mtDNA Population Database of 1655 Caucasians there are 15 individuals that match at 263G, 315.1C and 153 additional profiles that have only a single difference. Thus, 168 out of 1655 (10.2%) of the Caucasian database would not be able to be excluded if a sample was observed with this common mtDNA type! As will be discussed later in the chapter, efforts are underway to gather additional sequence information from polymorphic sites around the entire mtGenome in order to better resolve these most common types (Parsons and Coble 2001, Coble *et al.* 2004).

SCREENING ASSAYS FOR mtDNA TYPING

Due to the effort both in terms of time and labor required to obtain full sequence information from mtDNA sequencing, screening approaches and rapid low-resolution typing assays can and have been used to eliminate the need for full analysis of samples that can be easily excluded from one another. Many times physical screening methods can put samples into context without having to indiscriminately perform mtDNA sequencing on all samples. For example, microscopic examinations of hair can help eliminate as many questioned hairs as possible leaving the mtDNA laboratory to concentrate their efforts on only key hairs (Houck and Budowle 2001). Likewise anthropological evaluations of bones or teeth can be important first screens prior to making the effort to analyze the mtDNA sequence (see Edson *et al.* 2004).

With the expense and effort required to obtain full mtDNA sequences across HV1 and HV2, the ability to rapidly screen out samples that do not match can be advantageous to overworked, understaffed, and poorly funded crime laboratories. Several assays have been developed and even validated for use in screening forensic casework (Table 10.5).

SSO PROBES AND LINEAR ARRAY TYPING ASSAYS

One of the most widely used screening assays for assessing mtDNA variation used to date are the sequence-specific oligonucleotide (SSO) probes originally designed by Mark Stoneking and colleagues in 1991. Rather than sequencing the entire HV1 and HV2 regions, the most polymorphic sites are examined through hybridization of PCR products to oligonucleotide probes designed to anneal to different variants. The original paper describes 23 probes across nine regions that permit evaluation of variation at 14 different nucleotide positions (Stoneking *et al.* 1991). The sites that are probed include 16126, 16129, 16217, 16223, 16304, 16311, 16362, 73, 146, 152, 195, 199, 247, and 309.1. A number

Table 10.5 (facing)

Methods for screening mtDNA variation (see Butler and Levin 1998 and Budowle et al. 2004).

Technique	Description	Reference
Sequence-specific oligonucleotide (SSO) dot blot assay	23 SSO probes testing 14 sites within nine regions from HV1 and HV2; 274 mtDNA types observed among 525 individuals from five ethnic groups	Stoneking et al. (1991); Melton et al. (2001)
Mini-sequencing	Single base primer extension with fluorescent ddNTPs and poly(T)-tailed primers to yield different electrophoretic mobilities; 10 substitution and two length polymorphisms measured in the control region; 65 haplotypes observed from 152 British Caucasian samples	Tully et al. (1996); Morley et al. (1999)
Single-strand conformational polymorphism (SSCP)	Differences in DNA secondary structure are detected on a native polyacrylamide gel; 25 mtDNA types observed among 45 Spanish individuals tested	Alonso et al. (1996)
Low-stringency single-specific-primer PCR (LSSP-PCR)	Following regular PCR, a single primer and a low annealing temperature are used to generate a 'signature' pattern; for 30 unrelated individuals, all signature patterns were different across the control region (1024 bp)	Barreto et al. (1996)
PCR-restriction fragment length polymorphism (PCR-RFLP)	A 199 bp region of HV1 is digested with RsaI; 19 unrelated mother-child pairs were examined with an 8% probability of a random match	Pushnova et al. (1994); Butler et al. (1998a)
Denaturing gradient gel electrophoresis (DGGE)	Two DNA samples are mixed and run on a denaturing gradient gel; heteroduplexes, which travel more slowly through the gel, may be separated from the homoduplexes, samples that differ at a single location have been resolved	Steighner et al. (1999); Tully et al. (2000)
Affymetrix high-density DNA chip hybridization array	135 000 probes complementary to the entire mtGenome are contained on a microchip for parallel processing through hybridization	Chee et al. (1996)
Pyrosequencing	Sequencing by synthesis over ~50 nucleotides per reaction through an enzyme cascade that produces visible light; a total of 4 HV1, 4 HV2, and 11 coding region reactions were run	Andreasson et al. (2002)
SNaPshot (minisequencing)	Allele-specific primer extension with 11 coding region SNPs combined into a single multiplex amplification and detection assay	Vallone et al. (2004)
Denaturing HPLC	HV1 and HV2 PCR products for a known and an unknown sample source are generated and then mixed together; samples that differ from one another by at least one nucleotide will form a heteroduplex on the HPLC	LaBerge et al. (2003)
Luminex 100 liquid bead array	30 SNPs within HV1 and HV2 are examined by allele-specific hybridization with SSO probes attached to different colored beads that are separated using flow cytometry	www.marligen.com
LINEAR ARRAYs	Reverse dot blot hybridization with lines instead of dots using 18 SNPs in the same general probe regions as Stoneking et al. (1991)	Gabriel et al. (2003)

of population studies have been conducted with these SSO probes including an examination of 2282 individuals from North America (Melton *et al.* 2001).

The original SSO probe assay required that the PCR products be attached through UV cross-linking to a nylon membrane and then each radioactively labeled probe was individually hybridized at different temperatures and finally exposed to autoradiographic film for several hours (Stoneking *et al.* 1991). Roche Molecular Systems (Alameda, CA) has converted the SSO probe assay into a more workable format involving colorimetric detection (e.g., Gabriel *et al.* 2001b). In a 'reverse dot blot' format, the SSO probes are attached to the nylon membrane in a linear array of spatially resolved lines of probes. Biotin-labeled PCR products are washed over nylon membrane strips containing immobilized SSO probes in the linear array and hybridized under uniform conditions. A streptavidin-horseradish peroxidase enzyme conjugate coupled with 3,3',5,5'-tetramethyl-benzidine creates a light blue colored precipitate using the same chemistry described for HLA-DQα reverse dot blot SSO probes (Saiki *et al.* 1989).

Figure 10.10 illustrates the probe layout for the LINEAR ARRAY Mitochondrial DNA HVI/HVII Region-Sequence Typing Kit now available from Roche Applied Sciences (Indianapolis, IN). The final linear array format examines 18 SNPs with 33 SSO probes present on 31 different lines. The Roche SSO probe sites are shown in Figure 10.6.

Two hypothetical results are illustrated in Figure 10.10 for non-matching K and Q samples. The K sample reported type of 1-1-1-1-1-1-1-1-1-1 is equivalent to the Cambridge Reference Sequence (see Figure 10.6). The Q sample possesses

Figure 10.10

(a) Results schematically displayed of a known (K) reference and a question (Q) sample that do not match one another using the Roche LINEAR ARRAY mtDNA HVI/ HVII Region-Sequence Typing Strips. (b) Types are reported as a string of numbers representing the LINEAR ARRAY probe results. Failure of the PCR product to bind to a probe region (e.g., HVIE in sample Q) is referred to as a 'blank', is reported as a zero in the string of numbers, and is due to polymorphisms in the sample near the probe site that disrupt hybridization. Weak signals such as indicated by the arrow for 189 in sample Q are also due to a closely spaced polymorphism that disrupts full hybridization of the PCR product to the sequence-specific probe present on the LINEAR ARRAY.

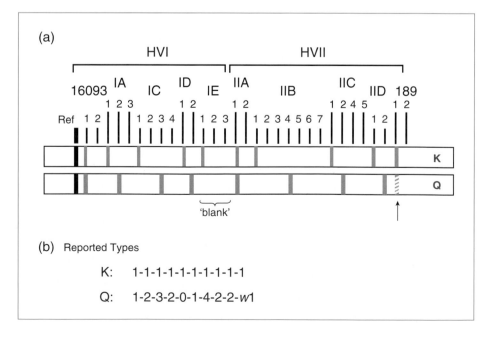

a different pattern and therefore can be excluded from the K sample. Notice that probe IE within HVI did not produce a signal from any of the three possible probes. This result is referred to as a 'blank' and occurs due to additional polymorphisms that are present in close proximity to the polymorphic sites designed for detection in the assay. These additional polymorphisms disrupt hybridization of the PCR product and therefore no signal is seen for any of the probes in HVIE. Likewise weak (w) signals such as the '*w*1' type are caused by mismatches between the PCR products and the SSO probes attached on the nylon strip.

Results from screening assays, such as the LINEAR ARRAY system described above, can be considered presumptive tests. They are useful in eliminating samples that can be excluded from one another. However, full HV1/HV2 sequencing would normally be performed to confirm any matches and see if differences outside of the SSO probe regions exist.

POPULATION DATABASES

Population databases play an important role in estimating the expected frequency of mtDNA haplotypes that are observed in casework when a suspect's mtDNA sequence matches that of an evidentiary sample. A great deal of effort has been expended to gather information from thousands of maternally unrelated individuals in various population groups around the world. Having high quality information in the database is also important in order to make a reliable estimate of the frequency for a random match.

AVAILABLE POPULATION DATABASES

MtDNA typing results on samples from unknown sources are most useful if they are evaluated in comparison to a known sample or a database. Databases of more than 1000 unrelated individuals now exist and have been compiled from multiple population groups (Handt *et al.* 1998, Budowle *et al.* 1999, Attimonelli *et al.* 2000, Wittig *et al.* 2000, Röhl *et al.* 2001, Monson *et al.* 2002). The size of the database is important because without recombination between mtDNA molecules, an mtDNA sequence is treated as a single locus (i.e., haplotype instead of genotype).

The largest compiled database described to date contains HV1 and HV2 sequences from 14 138 individuals (Röhl *et al.* 2001). This information was collated from 103 mtDNA publications prior to January 2000, 13 data sets published in 2000 and 2001, and two unpublished data sets. Authors of the original publications were contacted in an effort to confirm and correct sequence errors, eliminate duplications, and harmonize nomenclatures, but not every query was answered. Of the 116 publications, 90 required some kind of change to correct errors or adjust nomenclature illustrating the challenge of compiling accurate mtDNA sequence databases. The authors conclude that their annotated database probably still contains errors and that while it can be used for qualitative identification of relevant reference populations for a given mtDNA type, the determination of a 'legally defensible' frequency estimate of an mtDNA type within a population should be performed with higher-quality data yet to be produced (Röhl *et al.* 2001).

The FBI has compiled the mtDNA Population Database also known as CODIS[mt] (Monson *et al.* 2002) for the purpose of being able to determine a legally defensible frequency estimate. The CODIS[mt] database has a forensic and a published literature component to it (Miller and Budowle 2001) in order to separate data obtained from laboratories following validated forensic protocols and academic research laboratories where data quality is not reviewed as carefully prior to publication.

The forensic database contains 4839 mtDNA profiles from 14 different populations (Table 10.6). These samples have been sequenced and the electropherograms carefully reviewed across positions 16024–16365 for HV1 and positions 73–340 for HV2. Several publications have come out of analysis of mtDNA profiles contained within the various populations as summarized in Table 10.6.

Table 10.6

Summary of high quality forensic profiles present in the FBI Laboratory's mtDNA Population Database now called CODIS[mt] when it was released to the public in April 2002 (Monson et al. 2002).

Population Name	Number of Profiles	Data Analysis on Group
African-American	1148	Budowle *et al.* (1999)
Apache	180	Budowle *et al.* (2002b)
Caucasian	1655	Budowle *et al.* (1999), Allard *et al.* (2002)
China/Taiwan	356	Allard *et al.* (2004)
Egypt	48	
Guam	87	Allard *et al.* (2004)
Hispanic	686	Budowle *et al.* (1999)
India	19	
Japan	163	Budowle *et al.* (1999)
Korea	182	Allard *et al.* (2004)
Navajo	146	Budowle *et al.* (2002b)
Pakistan	8	
Sierra Leone	109	Budowle *et al.* (1999)
Thailand	52	Allard *et al.* (2004)
Total	**4839**	

An additional 6106 published profiles have been compiled from the literature with annotated population information (Miller *et al.* 1996, Miller and Budowle 2001). For classification of mtDNA profiles, a standard 14-character nucleotide sequence identifier was assigned to each profile where the first three characters represent the country of origin, the second three characters the group or ethnic affiliation, and the final six characters are sequential acquisition numbers (Miller and Budowle 2001, Monson *et al.* 2002).

Both of these databases were publicly released in April 2002 in a Microsoft Access format and can be downloaded from the FBI web site along with an analysis tool named MitoSearch (Monson *et al.* 2002). MitoSearch can examine the population data sets listed in Table 10.6 for specific mtDNA sequences, which are entered based on differences from the Cambridge Reference Sequence. The software returns the number of times that the specified profile appears in each population group. For example, the mtDNA type 16129A, 263G, 309del, 315.1C occurs twice in 1148 African-American profiles, twice in 1655 Caucasian profiles, and not at all in 686 Hispanic profiles.

The European forensic mtDNA sequencing community has also been actively engaged in developing new high quality population databases for forensic and human identity testing applications. A European DNA Profiling Group mitochondrial DNA population database project (EMPOP) has been launched to construct a high quality mtDNA database that eventually can be accessed online at http://www.empop.org.

ISSUES WITH SEQUENCE QUALITY

Concerns with mtDNA database sequence quality and the impact that it might have on accurately estimating frequency estimates for random matches have been raised by Peter Forster and Hans Bandelt in several recent publications (Röhl *et al.* 2001, Bandelt *et al.* 2001, Bandelt *et al.* 2002, Forster 2003, Dennis 2003). Using a statistical analysis tool called phylogenetics, the similarities and differences between multiple and closely related DNA sequences (i.e., from the same region) can be compared systematically (see Wilson and Allard 2004). Sequence alignments are created and compared to look for samples that are extremely different. Extreme or unusual differences may be an indication that the sample was contaminated or the sequence data was incorrectly recorded. For example, a laboratory may put HV1 data for a sample with another sample's HV2 sequence and thereby create an artificial recombinant or accidental composite sequence. Thus phylogenetic analyses can play a role in verifying sequence quality (Bandelt *et al.* 2001, Wilson and Allard 2004).

Errors that can creep into mitochondrial DNA population databases can be segregated into four different classes (Parson *et al.* 2004): (1) mistakes in the course of transcription of the results (i.e., clerical errors); (2) sample mix-up (e.g., putting data from HV1 on one sample together with data from HV2 on another sample); (3) contamination; and (4) use of different nomenclatures.

From a pilot collaborative study of 21 laboratories, 14 non-concordant haplotypes (16 individuals errors) were observed out of a total of 150 submitted samples/haplotypes representing the examination of approximately 150 000 nucleotides (Parson *et al.* 2004). Measures are being put into place for complete electronic transfer of data and base calling to avoid the primary problem of clerical errors when transferring information from raw sequence data to final report. In the future, mtDNA databases may require retention of raw data for population samples in order to more easily verify authenticity of results should an inquiry into the origin of sequence results be needed at a later date (Parson *et al.* 2004).

FUTURE DIRECTIONS IN mtDNA RESEARCH

WHOLE MITOCHONDRIAL GENOME SEQUENCING

The first description of a methodology for sequencing the entire mtGenome was by Deborah Nickerson's group at the University of Washington (Rieder *et al.* 1998). They used 24 pairs of primers to amplify PCR products ranging in size from 765 bp to 1162 bp. These primer pairs provide on average almost 200 bases of overlap between the various PCR products spanning the mtGenome. Ingman *et al.* (2000) used the Nickerson laboratory sequencing strategy to launch the era of mitochondrial population genomics when they sequenced 53 mtGenomes from diverse world population groups. Max Ingman maintains an mtGenome polymorphism database at http://www.genpat.uu.se/mtDB/.

In the past few years, a number of other methodologies have appeared in the literature for sequencing entire mtGenomes (Table 10.7). Regardless of the sequencing strategy used, the biggest challenge in conducting this work remains efforts to reduce and eliminate errors in sequence review (see Herrnstadt *et al.* 2003). Fortunately, the reference sequence (rCRS) was updated prior to the explosion of mtGenome information that began with Ingman *et al.* (2000).

The program MitoAnalyzer (Lee and Levin 2002) can be used to evaluate the location of an observed polymorphic nucleotide in the mtGenome. MitoAnalyzer is available online at http://www.cstl.nist.gov/biotech/strbase/mitoanalyzer.html.

RESOLVING MOST COMMON TYPES

One of the major challenges of mtDNA typing lies in the fact that many sequences fall into common groupings termed 'most common types'. For example, a review of the HV1/HV2 type distribution in 1655 Caucasians of U.S. and European descent (Monson *et al.* 2002) found that the most common mtDNA

Population	Number Sequenced	Reference (GenBank Accessions)	Approach Taken
Samples of diverse worldwide origin	53	Ingman et al. (2000) AF346963–AF347015	24 PCR reactions, 48 sequencing reactions
Samples of diverse worldwide origin	33	Maca-Meyer et al. (2001) AF381981–AF382013	32 PCR reactions, 64 sequencing reactions
African, Asian, European origin	560 (coding region only)	Herrnstadt et al. (2002) Sequences available at www.mitokor.com	68 PCR reactions, 136 sequencing reactions
East Asian lineages	48	Kong et al. (2003) AY255133–AY255180	15 PCR reactions, 47 sequencing reactions
Australian and New Guinean Aborigines and Polynesians	52	Ingman and Gyllensten (2003) AY289051–AY289102	24 PCR reactions, 48 sequencing reactions
Most common Caucasian types	241	Coble et al. (2004) AY495090–AY495330	12 PCR reactions, 95 sequencing reactions

Table 10.7

Summary of published mtGenome DNA sequencing efforts from December 2000 to February 2004 representing almost 1000 complete mtGenomes.

type, which matches the rCRS, occurred 7.1% of the time (Coble *et al.* 2004). Furthermore, it was observed that only 18 mtDNA types account for 20.8% of the total Caucasian data set (Coble *et al.* 2004). The presence of these most common types suggests that one of out every five times a mtDNA sequence analysis is performed on a Caucasian individual, the result would be expected to match numerous other individuals in a population database. While the same analysis revealed that approximately 50% of the 1655 individuals present in the European Caucasian population are 'unique in the database', having a sample that falls into one of these most common types can be present a disappointing statistic after all of the hard work taken to generate the full mtDNA HV1/HV2 sequence.

An extensive search for distinguishing single nucleotide polymorphisms in samples possessing the most common Caucasian types was recently undertaken (Parsons and Coble 2001). A total of 241 complete mtGenomes were sequenced from the 18 common European Caucasian HV1/HV2 types mentioned above (Coble 2004, Coble *et al.* 2004). The samples typed come from mtDNA haplogroups H, J, T, V, and K (see the next section for more discussion on haplogroups).

Examination of whole mtGenome sequence information expanded the 18 most common Caucasian HV1/HV2 types to 209 resolvable haplotypes (Coble *et al.* 2004). This almost 12-fold improvement in resolving power for these common HV1/HV2 types required about 27 times the amount of DNA sequencing – from 610 bases for just HV1/HV2 alone to ~16 569 for the entire mtGenome. Obviously, this approach is not a cost effective one. Furthermore, even with the expansion in sequence information, 32 of the 241 individuals matched one or more individuals across the entire mtGenome.

From their extensive sequencing information, Coble *et al.* (2004) selected a battery of SNP markers to aid in resolving the most common Caucasian mtDNA HV1/HV2 types without the costly and time-consuming venture of having to sequence the entire mtGenome. A total of 59 informative SNPs were placed into eight multiplex panels (Coble *et al.* 2004). The first panel provides maximum resolution of the most common Caucasian HV1/HV2 mtDNA type (i.e., that matching CRS) and examines the following nucleotides spanning the mtGenome: 477, 3010, 4580, 4793, 5004, 7028, 7202, 10211, 12858, 14470, and 16519. Vallone *et al.* (2004) combined these 11 SNP sites into a multiplex allele-specific primer extension or 'SNaPshot' assay (see Chapter 8) that can reliably type a sample that contains only a few hundred copies of mtDNA.

DEFINING mtDNA HAPLOGROUPS

Over the course of typing mtDNA samples from various populations, researchers have observed that individuals often cluster into haplogroups that can be defined by particular polymorphic nucleotides (see Wallace *et al.* 1999, Ruiz-Pesini *et al.* 2004). These haplogroups were originally defined in the late 1980s and 1990s by grouping samples possessing the same or similar patterns when subjected to a series of restriction enzymes that were used to separate various mtDNA types from diverse populations around the world (Table 10.8). Mitochondrial DNA haplogroups have now been correlated to HV1/HV2 polymorphisms as well as entire mtGenome variation. Haplogroups A, B, C, D, E, F, G, and M are typically associated with Asians while most Native Americans fall into haplogroups A, B, C, and D. Haplogroups L1, L2, and L3 are African, and haplogroups H, I, J, K, T, U, V, W, and X are typically associated with European populations (Wallace *et al.* 1999).

Along the same lines as the multiplex SNP detection assay described above for resolving samples containing the most common HV1/HV2 types, Brandstätter *et al.* (2003) described a multiplex SNP system for categorizing European Caucasian haplogroups. This approach involves the analysis of 16 coding region SNPs to aid assignment of individual samples into one of the nine major European Caucasian mtDNA haplogroups listed above. For example, the presence of a cytosine at position 7028 indicates that the sample can be grouped into haplogroup H as opposed to the other groups whose individuals possess a thymine at 7028.

Another SNP typing assay was recently reported to examine 17 coding region SNPs in a single multiplexed detection assay (Quintans *et al.* 2004). A SNaPshot reaction (see Chapter 8) is used to probe the following mtDNA nucleotide positions: 3010, 3915, 3992, 4216, 4336, 4529, 4580, 4769, 4793, 6776, 7028, 10398, 10400, 10873, 12308, 12705, and 14766. This assay was capable of

Haplogroup (Population)	Coding Region Polymorphisms	Control Region Polymorphisms (*not including 263G, 315.1C)
A (Asian)	663G	16233T, 16290T, 16319A, 235G
B (Asian)	9 bp deletion, 16159C	16217C, 16189C
C (Asian)	13263G	16233T, 16298C, 16327T
D (Asian)	2092T, 5178A, 8414T	16362C
H (Caucasian)	7028C, 14766C	73A and lack of CRS differences*
H1 (Caucasian)	3010A	73A and lack of CRS differences*
H2 (Caucasian)	1438A, 4769A	73A and lack of CRS differences*
H3 (Caucasian)	6776C	73A and lack of CRS differences*
H4 (Caucasian)	3992T	73A and lack of CRS differences*
H5 (Caucasian)	4336C	73A and lack of CRS differences*
H6 (Caucasian)	3915A	73A and lack of CRS differences*
H7 (Caucasian)	4793G	73A and lack of CRS differences*
I (Caucasian)	1719A, 8251A, 10238C	16223T, 199C, 204C, 250C
J (Caucasian)	4216C, 12612G, 13708A	16069T, 16126C, 295T
J1 (Caucasian)	3010A	462T
J2 (Caucasian)	7476T, 15257A	195C
K (Caucasian)	12372A, 14798C	16224C, 16311C
L1 (African)	2758A, 3594T, 10810C	16187T, 16189C, 16223T, 16278T, 16311C
L2 (African)	3594T	16223T, 16278T
L3 (African)	3594C	16223T
M (Asian)	10400T, 10873C	16223T, 16298C
T (Caucasian)	709A, 1888A, 4917G, 10463C, 13368A, 14905A, 15607G, 15928A, 8697A	16126C, 16294T
U5 (Caucasian)	3197C	16270T
V (Caucasian)	4580A, 15904T	16298C, 72C
W (Caucasian)	709A, 1243C, 8251A, 8697G, 8994A	16223T, 189G, 195C, 204C, 207A
X (Caucasian)	1719A, 6221C, 8251G, 14470C	16189C, 16223T, 16278T, 195C

Table 10.8

Major mitochondrial haplogroups and the specific polymorphisms in the coding region or control region that define them (see Finnila et al. 2001, Herrnstadt et al. 2002, Brandstatter et al. 2003, Kong et al. 2003, Allard et al. 2004, Quintans et al. 2004). Note that not all haplogroups, which have been defined in the literature, are listed here.

breaking 266 samples into 20 different mtDNA haplogroup designations and aided in resolving some of the most common type (i.e., 263G, 315.1C) haplogroup H samples from one another.

Forensic population databases have been analyzed in terms of haplogroup information to aid in quality control of samples contained within a population group (Allard *et al.* 2002, Budowle *et al.* 2003, Allard *et al.* 2004).

GENETIC GENEALOGY WITH MITOCHONDRIAL DNA

As mentioned at the end of the previous chapter, scientists have been using DNA for several decades to try to understand human migration patterns (Relethford 2001, Relethford 2003). Samples have been gathered from a number of individuals around the world often from isolated populations such as the Australian aborigines. The uniparental inheritance of mtDNA and Y chromosome markers (see Chapter 9) makes it easier to trace ancestral line-ages through multiple generations since the shuffling effects of recombination that promotes the diversity of autosomal DNA profiles is not present in haploid systems. The ability to successfully obtain mtDNA results from ancient bones is also useful, such as has been demonstrated with the recovery of HV1 and HV2 sequences from Neanderthal remains that are thousands of years old (Krings *et al.* 1997, 1999).

While the same DNA markers are being used in these types of studies as in forensic DNA typing, the sample groups are often analyzed differently since direct comparisons cannot usually be made. Rather the DNA information obtained is extrapolated over many generations between the various popula-tions tested. There is not a one-to-one unique match being made between a 'suspect' and 'evidence.' Instead scientists are often guessing at what genetic signatures existed in the past based on various assumptions – with a bit of 'story-telling' mixed in (see Goldstein and Chikhi 2002). However, large amounts of data are being collected in an attempt to better understand our heritage and travels as a human species (e.g., Helgason *et al.* 2003). Forensic DNA testing, disease diagnostics and anthropological and genealogical research efforts will all continue to be benefited by growth and developments in mitochondrial DNA analysis.

REFERENCES AND ADDITIONAL READING

Allard, M.W., Miller, K., Wilson, M., Monson, K. and Budowle, B. (2002) *Journal of Forensic Sciences*, 47, 1215–1223.

Allard, M.W., Wilson, M.R., Monson, K.L. and Budowle, B. (2004) *Legal Medicine*, 6, 11–24.

Allen, M., Engstrom, A.S., Meyers, S., Handt, O., Saldeen, T., von Haeseler, A., Paabo, S. and Gyllensten, U. (1998) *Journal of Forensic Sciences*, 43, 453–464.

Alonso, A., Martin, P., Albarran, C., Garcia, O. and Sancho, M. (1996) *Electrophoresis*, 17, 1299–1301.

Alonso, A., Salas, A., Albarran, C., Arroyo, E., Castro, A., Crespillo, M., di Lonardo, A.M., Lareu, M.V., Cubria, C.L., Soto, M.L., Lorente, J.A., Semper, M.M., Palacio, A., Paredes, M., Pereira, L., Lezaun, A.P., Brito, J.P., Sala, A., Vide, M.C., Whittle, M., Yunis, J.J. and Gomez, J. (2002) *Forensic Science International*, 125, 1–7.

Anderson, S., Bankier, A.T., Barrell, B.G., de Bruijn, M.H.L., Coulson, A.R., Drouin, J., Eperon, I.C., Nierlich, D.P., Roe, B.A., Sanger, F., Schreier, P.H., Smith, A.J.H., Staden, R. and Young, I.G. (1981) *Nature*, 290, 457–465.

Andreasson, H., Asp, A., Alderborn, A., Gyllensten, U. and Allen, M. (2002) *Biotechniques*, 32, 124–133.

Andreasson, H., Gyllensten, U. and Allen, M. (2002) *Biotechniques*, 33, 402–411.

Andrews, R.M., Kubacka, I., Chinnery, P.F., Lightowlers, R.N., Turnbull, D.M. and Howell, N. (1999) *Nature Genetics*, 23, 147.

Attimonelli, M., Altamura, N., Benne, R., Brennicke, A., Cooper, J.M., D'Elia, D., Montalvo, A., Pinto, B., De Robertis, M., Golik, P., Knoop, V., Lanave, C., Lazowska, J., Licciulli, F., Malladi, B.S., Memeo, F., Monnerot, M., Pasimeni, R., Pilbout, S., Schapira, A.H., Sloof, P. and Saccone, C. (2000) *Nucleic Acids Research*, 28, 148–152.

Awadalla, P., Eyre-Walker, A. and Smith, J. M. (1999) *Science*, 286, 2524–2525.

Bandelt, H.J., Lahermo, P., Richards, M. and Macaulay, V. (2001) *International Journal of Legal Medicine*, 115, 64–69.

Bandelt, H.J., Quintana-Murci, L., Salas, A. and Macaulay, V. (2002) *American Journal of Human Genetics*, 71, 1150–1160.

Barreto, G., Vago, A.R., Ginther, C., Simpson, A.J.G. and Pena, S.D.J. (1996) *American Journal of Human Genetics*, 58, 609–616.

Bendall, K. E. and Sykes, B. C. (1995) *American Journal of Human Genetics*, 57, 248–256.

Bendall, K. E., Macaulay, V. A., Baker, J. R. and Sykes, B. C. (1996) *American Journal of Human Genetics*, 59, 1276–1287.

Bever, R.A., Basalyga, F. and Thomas, J. (2003) Resolution of mixtures by cloning of the mitochondrial DNA control region. *Proceedings of the Fourteenth International Symposium on Human Identification*. Available online at : http://www.promega.com/geneticidproc/ussymp14proc/oralpresentations/Bever.pdf.

Bini, C., Ceccardi, S., Luiselli, D., Ferri, G., Pelotti, S., Colalongo, C., Falconi, M. and Pappalardo, G. (2003) *Forensic Science International*, 135, 48–52.

Bodenteich, A., Mitchell, L. G., Polymeropoulos, M. H. and Merril, C. R. (1992) *Human Molecular Genetics*, 1, 140.

Brandstätter, A. and Parson, W. (2003) *International Journal of Legal Medicine*, 117, 180–184.

Brandstätter, A., Parsons, T. J. and Parson, W. (2003) *International Journal of Legal Medicine*, 117, 291–298.

Brown, W. M., George, M., Jr. and Wilson, A. C. (1979) *Proceedings of the National Academy of Sciences U.S.A.*, 76, 1967–1971.

Brown, K. (2002) *Science*, 295, 1634–1635.

Budowle, B., Adams, D.E., Comey, C.C. and Merrill, C.R. (1990) Mitochondrial DNA – a possible genetic material suitable for forensic analysis. In Lee, H.C. and Gaensslen, R.E. (eds) *Advances in Forensic Sciences.* Chicago, Illinois: Year Book Medical Publishers, p. 76–97.

Budowle, B., Wilson, M.R., DiZinno, J.A., Stauffer, C., Fasano, M.A., Holland, M.M. and Monson, K.L. (1999) *Forensic Science International*, 103, 23–35.

Budowle, B., Allard, M.W. and Wilson, M.R. (2002a) *Forensic Science International*, 126, 30–33.

Budowle, B., Allard, M.W., Fisher, C.L., Isenberg, A.R., Monson, K.L., Stewart, J.E., Wilson, M.R. and Miller, K.W. (2002b) *International Journal of Legal Medicine*, 116, 212–215.

Budowle, B., Allard, M.W., Wilson, M.R. and Chakraborty, R. (2003) *Annual Reviews in Genomics and Human Genetics*, 4, 119–141.

Budowle, B., Planz, J.V., Campbell, R.S. and Eisenberg, A.J. (2004) *Forensic Science Review* 16, 21–36.

Butler, J.M., Wilson, M.R. and Reeder, D.J. (1998a) *Electrophoresis*, 19, 119–124.

Butler, J.M. and Levin, B.C. (1998b) *Trends in Biotechnology*, 16, 158–162.

Calloway, C.D., Reynolds, R.L., Herrin, G.L., Jr. and Anderson, W.W. (2000) *American Journal of Human Genetics*, 66, 1384–1397.

Carracedo, A., D'Aloja, E., Dupuy, B., Jangblad, A., Karjalainen, M., Lambert, C., Parson, W., Pfeiffer, H., Pfitzinger, H., Sabatier, M., Syndercombe-Court, D. and Vide, C. (1998) *Forensic Science International*, 97, 165–170.

Carracedo, A., Bar, W., Lincoln, P., Mayr, W., Morling, N., Olaisen, B., Schneider, P., Budowle, B., Brinkmann, B., Gill, P., Holland, M., Tully, G. and Wilson, M. (2000) *Forensic Science International*, 110, 79–85.

Chee, M., Yang, R., Hubbell, E., Berno, A., Huang, X.C., Stern, D., Winkler, J., Lockhart, D.J., Morris, M.S. and Fodor, S.P. (1996) *Science*, 274, 610–614.

Chen, X., Prosser, R., Simonetti, S., Sadlock, J., Jagiello, G. and Schon, E. A. (1995) *American Journal of Human Genetics*, 57, 239–247.

Coble, M.D. (2004) The identification of single nucleotide polymorphisms in the entire mitochondrial genome to increase the forensic discrimination of common HV1/HV2 types in the Caucasian population. PhD dissertation, Washington, DC: The George Washington University. Available online at: http://www.cstl.nist.gov/biotech/strbase/pub_pres/ Coble2004dis.pdf.

Coble, M.D., Just, R.S., O'Callaghan, J.E., Letmanyi, I.H., Peterson, C.T., Irwin, J.A. and Parsons, T.J. (2004) *International Journal of Legal Medicine,* 118, 137–146.

Collura, R.V. and Stewart, C.B. (1995) *Nature*, 378, 485–489.

Comas, D., Paabo, S. and Bertranpetit, J. (1995) *Genome Research*, 5, 89–90.

Comas, D., Reynolds, R. and Sajantila, A. (1999) *European Journal of Human Genetics*, 7, 459–468.

Dennis, C. (2003) *Nature*, 421, 773–774.

Dugan, K.A., Lawrence, H.S., Hares, D.R., Fisher, C.L. and Budowle, B. (2002) *Journal of Forensic Sciences*, 47, 811–818.

Edson, S.M., Ross, J.P., Coble, M.D., Parson, T.J. and Barritt, S.M. (2004) *Forensic Science Review*, 16, 63–90.

Elson, J. L., Andrews, R. M., Chinnery, P. F., Lightowlers, R. N., Turnbull, D. M. and Howell, N. (2001) *American Journal of Human Genetics*, 68, 145–153.

Eyre-Walker, A., Smith, N. H. and Smith, J. M. (1999) *Proceedings of the Royal Society of London B Biological Sciences*, 266, 477–483.

Filosto, M., Mancuso, M., Vives-Bauza, C., Vila, M.R., Shanske, S., Hirano, M., Andreu, A.L. and DiMauro, S. (2003) *Annals of Neurology*, 54, 524–526.

Finnila, S., Lehtonen, M.S. and Majamaa, K. (2001) *American Journal of Human Genetics*, 68, 1475–1484.

Forster, P. (2003) *Annals of Human Genetics*, 67, 2–4.

Gabriel, M.N., Huffine, E.F., Ryan, J.H., Holland, M.M. and Parsons, T.J. (2001a) *Journal of Forensic Sciences*, 46, 247–253.

Gabriel, M.N., Calloway, C.D., Reynolds, R.L., Andelinovic, S. and Primorac, D. (2001b) *Croatian Medical Journal*, 42, 328–335.

Gabriel, M.N., Calloway, C.D., Reynolds, R.L. and Primorac, D. (2003) *Croatian Medical Journal*, 44, 293–298.

Giambernardi, T.A., Rodeck, U. and Klebe, R.J. (1998) *Biotechniques*, 25, 564–566.

Giles, R.E., Blanc, H., Cann, H.M. and Wallace, D.C. (1980) *Proceedings of the National Academy of Sciences U.S.A.*, 77, 6715–6719.

Goldstein, D.B. and Chikhi, L. (2002) *Annual Reviews in Genomics and Human Genetics*, 3, 129–152. See quotation on p. 143.

Grzybowski, T. (2000) *Electrophoresis*, 21, 548–553.

Grzybowski, T., Malyarchuk, B.A., Czarny, J., Miscicka-Sliwka, D. and Kotzbach, R. (2003) *Electrophoresis*, 24, 1159–1165.

Gyllensten, U., Wharton, D., Josefsson, A. and Wilson, A.C. (1991) *Nature*, 352, 255–257.

Hagelberg, E., Goldman, N., Lio, P., Whelan, S., Schiefenhovel, W., Clegg, J.B. and Bowden, D.K. (1999) *Proceedings of the Royal Society of London B Biological Sciences*, 266, 485–492.

Hagelberg, E., Goldman, N., Lio, P., Whelan, S., Schiefenhovel, W., Clegg, J.B. and Bowden, D.K. (2000) *Proceedings of the Royal Society of London B Biological Sciences*, 267, 1595–1596.

Handt, O., Meyer, S. and von Haeseler, A. (1998) *Nucleic Acids Research*, 26, 126–129.

Helgason, A., Hrafnkelsson, B., Gulcher, J.R., Ward, R. and Stefansson, K. (2003) *American Journal of Human Genetics*, 72, 1370–1388.

Herrnstadt, C., Elson, J.L., Fahy, E., Preston, G., Turnbull, D.M., Anderson, C., Ghosh, S.S., Olefsky, J.M., Beal, M.F., Davis, R.E. and Howell, N. (2002) *American Journal of Human Genetics*, 70, 1152–1171.

Herrnstadt, C., Preston, G. and Howell, N. (2003) *American Journal of Human Genetics*, 72, 1585–1586.

Higuchi, R., von Beroldingen, C.H., Sensabaugh, G.F. and Erlich, H.A. (1988) *Nature*, 332, 543–546.

Holland, M.M., Fisher, D.L., Mitchell, L.G., Rodriquez, W.C., Canik, J.J., Merril, C.R. and Weedn, V.W. (1993) *Journal of Forensic Sciences*, 38, 542–553.

Holland, M.M., Fisher, D.L., Roby, R.K., Ruderman, J., Bryson, C. and Weedn, V.W. (1995) *Crime Laboratory Digest*, 22, 109–115.

Holland, M.M. and Parsons, T.J. (1999) *Forensic Science Review*, 11, 21–50.

Hopgood, R., Sullivan, K.M. and Gill, P. (1992) *Biotechniques*, 13, 82–92.

Houck, M.M. and Budowle, B. (2002) *Journal of Forensic Sciences*, 47, 964–967.

Ingman, M., Kaessmann, H., Paabo, S. and Gyllensten, U. (2000) *Nature*, 408, 708–713.

Ingman, M. and Gyllensten, U. (2001) *Journal of Heredity*, 92, 454–461.

Ingman, M. and Gyllensten, U. (2003) *Genome Research*, 13, 1600–1606.

Isenberg, A.R. and Moore, J.M. (1999) Mitochondrial DNA analysis at the FBI Laboratory. *Forensic Science Communications*, volume 1, number 2 [online]. Available at: http://www.fbi.gov/hq/lab/fsc/backissu/july1999/dnalist.htm.

Isenberg, A.R. (2004) Forensic mitochondrial DNA analysis. In Saferstein, R. (ed.): *Forensic Science Handbook. Volume II, Second Edition*. Upper Saddle River, New Jersey: Prentice-Hall, pp. 297–327.

Ivanov, P.L., Wadhams, M.J., Roby, R.K., Holland, M.M., Weedn, V.W. and Parsons, T.J. (1996) *Nature Genetics*, 12, 417–420.

Jehaes, E., Gilissen, A., Cassiman, J. J. and Decorte, R. (1998) *Forensic Science International*, 94, 65–71.

Johns, D.R. (2003) *Annals of Neurology*, 54, 422–424.

Jorde, L.B. and Bamshad, M. (2000) *Science*, 288, 1931.

Kivisild, T. and Villems, R. (2000) *Science*, 288, 1931.

Krings, M., Stone, A., Schmitz, R.W., Krainitzki, H., Stoneking, M. and Paabo, S. (1997) *Cell*, 90, 19–30.

Krings, M., Geisert, H., Schmitz, R.W., Krainitzki, H. and Paabo, S. (1999) *Proceedings of the National Academy of Sciences U.S.A.*, 96, 5581–5585.

Kumar, S., Hedrick, P., Dowling, T. and Stoneking, M. (2000) *Science*, 288, 1931.

Lagerstrom-Fermer, M., Olsson, C., Forsgren, L. and Syvanen, A.C. (2001) *American Journal of Human Genetics*, 68, 1299–1301.

Lee, L.G., Spurgeon, S.L., Heiner, C.R., Benson, S.C., Rosenblum, B.B., Menchen, S.M., Graham, R.J., Constantinescu, A., Upadhya, K.G. and Cassel, J.M. (1997) *Nucleic Acids Research*, 25, 2816–2822.

Lee, M.S. and Levin, B.C. (2002) *Mitochondrion*, 1, 321–326.

Levin, B.C., Cheng, H. and Reeder, D. J. (1999) *Genomics*, 55, 135–146.

Levin, B.C., Holland, K.A., Hancock, D.K., Coble, M., Parsons, T.J., Kienker, L.J., Williams, D.W., Jones, M.P. and Richie, K.L. (2003) *Mitochondrion*, 2, 387–400.

Linch, C.A., Whiting, D.A. and Holland, M.M. (2001) *Journal of Forensic Sciences*, 46, 844–853.

Lutz, S., Wittig, H., Weisser, H., Heizmann, J., Junge, A., Dimo-Simonin, N., Parson, W., Edelmann, J., Anslinger, K., Jung, S. and Augustin, C. (2000) *Forensic Science International*, 113, 97–101.

Macaulay, V., Richards, M. and Sykes, B. (1999) *Proceedings of the Royal Society of London B Biological Sciences*, 266, 2037–2039.

Marchi, E. and Pasacreta, R.J. (1997) *Journal of Capillary Electrophoresis*, 4, 145–156.

Meissner, C., Mohamed, S.A., Klueter, H., Hamann, K., von Wurmb, N. and Oehmichen, M. (2000) *Forensic Science International*, 113, 109–112.

Melton, T. and Nelson, K. (2001) *Croatian Medical Journal*, 42, 298–303.

Melton, T., Clifford, S., Kayser, M., Nasidze, I., Batzer, M. and Stoneking, M. (2001) *Journal of Forensic Sciences*, 46, 46–52.

Melton, T. (2004) *Forensic Science Review*, 16, 1–20.

Miller, K.W.P., Dawson, J.L. and Hagelberg, E. (1996) *International Journal of Legal Medicine*, 109, 107–113.

Miller, K.W. and Budowle, B. (2001) *Croatian Medical Journal*, 42, 315–327.

MITOMAP: A Human Mitochondrial Genome Database, Center for Molecular Medicine, Emory University, Atlanta, GA, USA, http://www.mitomap.org.

Monson, K.L., Miller, K.W.P., Wilson, M.R., DiZinno, J.A. and Budowle, B. (2002) The mtDNA population database: an integrated software and database resource, *Forensic Science Communications* [online]. Available at: http://www.fbi.gov/hq/lab/fsc/backissu/april2002/miller1.htm.

Morgan, M.A., Parsons, T.J. and Holland, M.M. (1998) *Proceedings from the Eighth International Symposium on Human Identification-1997*, p128. Madison, Wisconsin: Promega Corporation.

Morley, J.M., Bark, J.E., Evans, C.E., Perry, J.G., Hewitt, C.A. and Tully, G. (1999) *International Journal of Legal Medicine*, 112, 241–248.

Parson, W., Brandstatter, A., Alonso, A., Brandt, N., Brinkmann, B., Carracedo, A., Corach, D., Froment, O., Furac, I., Grzybowski, T., Hedberg, K., Keyser-Tracqui, C., Kupiec, T., Lutz-Bonengel, S., Mevag, B., Ploski, R., Schmitter, H., Schneider, P., Syndercombe-Court, Sorensen, E., Thew, H., Tully, G. and Scheithauer, R. (2004) *Forensic Science International*, 139, 215–226.

Parsons, T.J. and Irwin, J.A. (2000) *Science*, 288, 1931.

Parsons, T.J. and Coble, M.D. (2001) *Croatian Medical Journal*, 42, 304–309.

Parsons, T.J., Muniec, D.S., Sullivan, K., Woodyatt, N., Alliston-Greiner, R., Wilson, M.R., Berry, D.L., Holland, K.A., Weedn, V.W., Gill, P. and Holland, M.M. (1997) *Nature Genetics*, 15, 363–368.

Pfeiffer, H., Huhne, J., Ortmann, C., Waterkamp, K. and Brinkmann, B. (1999) *International Journal of Legal Medicine*, 112, 287–290.

Piercy, R., Sullivan, K.M., Benson, N. and Gill, P. (1993) *International Journal of Legal Medicine*, 106, 85–90.

Prieto, L., Montesino, M., Salas, A., Alonso, A., Albarran, C., Alvarez, S., Crespillo, M., di Lonardo, A.M., Doutremepuich, C., Fernandez-Fernandez, I., de la Vega, A.G., Gusmao, L., Lopez, C.M., Lopez Soto, M., Lorente, J.A., Malaghini, M., Martinez, C.A., Modesti, N.M., Palacio, A.M., Paredes, M., Pena, S.D., Perez-Lezaun, A., Pestano, J.J., Puente, J., Sala, A., Vide, M., Whittle, M.R., Yunis, J.J. and Gomez, J. (2003) *Forensic Science International*, 134, 46–53.

Pushnova, E.A., Akhmedova, S.N., Shevtsov, S.P. and Shwartz, E.I. (1994) *Human Mutation*, 3, 292–296.

Quintans, B., Alvarez-Iglesias, V., Salas, A., Phillips, C., Lareu, M.V. and Carracedo, A. (2004) *Forensic Science International,* 140, 251–257.

Rasmussen, E.M., Sorensen, E., Eriksen, B., Larsen, H.J., and Morling, N. (2002) *Forensic Science International,* 129, 209-213.

Relethford, J.H. (2001) *Genetics and the Search for Modern Human Origins*. New York: Wiley-Liss.

Relethford, J.H. (2003) *Reflections of Our Past: How Human History is Revealed in Our Genes*. Boulder, Colorado: Westview Press.

Reynolds, R. and Varlaro, J. (1996) *Journal of Forensic Sciences*, 41, 279–286.

Richards, M. and Macaulay, V. (2001) *American Journal of Human Genetics*, 68, 1315–1320.

Rieder, M.J., Taylor, S.L., Tobe, V.O. and Nickerson, D.A. (1998) *Nucleic Acids Research*, 26, 967–973.

Robin, E.D. and Wong, R. (1988) *Journal of Cell Physiology*, 136, 507–513.

Röhl, A., Brinkmann, B., Forster, L. and Forster, P. (2001) *International Journal of Legal Medicine*, 115, 29–39.

Ruiz-Pesini, E., Mishmar, D., Brandon, M., Procaccio, V. and Wallace, D.C. (2004) *Science*, 303, 223–226.

Saiki, R.K., Walsh, P.S., Levenson, C.H. and Erlich, H.A. (1989) *Proceedings of the National Academy of Sciences of the United States of America*, 86, 6230–6234.

Sanger, F., Nicklen, S. and Coulson, A.R. (1977) *Proceedings of the National Academy of Sciences of the United States of America*, 74, 5463–5467.

Satoh, M. and Kuroiwa, T. (1991) *Experimental Cell Research*, 196, 137–140.

Scheffler, I.E. (1999) *Mitochondria*. New York: Wiley-Liss.

Schwartz, M. and Vissing, J. (2002) *New England Journal of Medicine*, 347, 576–580.

Sekiguchi, K., Kasai, K. and Levin, B.C. (2003) *Mitochondrion*, 2, 401–414.

Steighner, R.J., Tully, L.A., Karjala, J.D., Coble, M.D. and Holland, M.M. (1999) *Journal of Forensic Sciences*, 44, 1186–1198.

Stewart, J.E., Fisher, C.L., Aagaard, P.J., Wilson, M.R., Isenberg, A.R., Polanskey, D., Pokorak, E., DiZinno, J.A. and Budowle, B. (2001) *Journal of Forensic Sciences*, 46, 862–870.

Stewart, J.E., Aagaard, P.J., Pokorak, E.G., Polanskey, D. and Budowle, B. (2003) *Journal of Forensic Sciences*, 48, 571–580.

Stone, A.C., Starrs, J.E. and Stoneking, M. (2001) *Journal of Forensic Sciences*, 46, 173–176.

Stoneking, M., Hedgecock, D., Higuchi, R.G., Vigilant, L. and Erlich, H.A. (1991) *American Journal of Human Genetics*, 48, 370–382.

Stoneking, M., Melton, T., Nott, J., Barritt, S., Roby, R., Holland, M., Weedn, V., Gill, P., Kimpton, C., Aliston-Greiner, R. and Sullivan, K. (1995) *Nature Genetics*, 9, 9–10.

Stoneking, M. (2000) *American Journal of Human Genetics*, 67, 1029–1032.

Sullivan, K.M., Hopgood, R., Lang, B. and Gill, P. (1991) *Electrophoresis*, 12, 17–21.

Sullivan, K.M., Hopgood, R. and Gill, P. (1992) *International Journal of Legal Medicine*, 105, 83–86.

Sullivan, K.M., Alliston-Greiner, R., Archampong, F.I.A., Piercy, R., Tully, G., Gill, P. and Lloyd-Davies, C. (1997) *Proceedings of the Seventh International Symposium on Human Identification-1996*, pp.126–130. Madison, Wisconsin: Promega Corporation.

Sutovsky, P., Moreno, R.D., Ramalho-Santos, J., Dominko, T., Simerly, C. and Schatten, G. (1999) *Nature*, 402, 371–372.

SWGDAM (2003) Guidelines for mitochondrial DNA (mtDNA) nucleotide sequence interpretation. Forensic Science Communications, 5 (2) [online]. Available at: http://www.fbi.gov/hq/lab/fsc/backissu/april2003/swgdammitodna.htm.

Szibor, R., Michael, M., Spitsyn, V.A., Plate, I., Ginter, E.K. and Krause, D. (1997) *Electrophoresis*, 18, 2857–2860.

Szibor, R., Edelmann, J., Hering, S., Plate, I., Wittig, H., Roewer, L., Wiegand, P., Cali, F., Romano, V. and Michael, M. (2003a) *Forensic Science International*, 138, 37–43.

Szibor, R., Michael, M., Plate, I., Wittig, H. and Krause, D. (2003b) *International Journal of Legal Medicine*, 117, 160–164.

Taylor, R.W., McDonnell, M.T., Blakely, E.L., Chinnery, P.F., Taylor, G.A., Howell, N., Zeviani, M., Briem, E., Carrara, F. and Turnbull, D.M. (2003) *Annals of Neurology*, 54, 521–524.

Torroni, A., Huoponen, K., Francalacci, P., Petrozzi, M., Morelli, L., Scozzari, R., Obinu, D., Savontaus, M.-L. and Wallace, D.C. (1996) *Genetics*, 144, 1835–1850.

Tully, G., Sullivan, K.M., Nixon, P., Stones, R.E. and Gill, P. (1996) *Genomics*, 34, 107–113.

Tully, G. Mitochondrial DNA: a small but valuable genome. (1999) *First International Conference on Forensic Human Identification.* Forensic Science Service.

Tully, G., Bar, W., Brinkmann, B., Carracedo, A., Gill, P., Morling, N., Parson, W. and Schneider, P. (2001) *Forensic Science International*, 124, 83–91.

Tully, G., Barritt, S.M., Bender, K., Brignon, E., Capelli, C., Dimo-Simonin, N., Eichmann, C., Ernst, C.M., Lambert, C., Lareu, M.V., Ludes, B., Mevag, B., Parson, W., Pfeiffer, H., Salas, A., Schneider, P.M. and Staalstrom, E. (2004) *Forensic Science International*, 140, 1–11.

Tully, L.A., Parsons, T.J., Steighner, R.J., Holland, M.M., Marino, M.A. and Prenger, V.L. (2000) *American Journal of Human Genetics*, 67, 432–443.

Vallone, P.M., Just, R.S., Coble, M.D., Butler, J.M., and Parsons, T.J. (2004) *International Journal of Legal Medicine*, 118, 147–157.

Walker, M.D. (2003) *Jurimetrics Journal*, 43, 427–440.

Walker, J.A., Garber, R.K., Hedges, D.J., Kilroy, G.E., Xing, J. and Batzer, M.A. (2004) *Analytical Biochemistry*, 325, 171–173.

Wallace, D.C., Ye, J. H., Neckelmann, S.N., Singh, G., Webster, K.A. and Greenberg, B.D. (1987) *Current Genetics*, 12, 81–90.

Wallace, D.C., Stugard, C., Murdock, D., Schurr, T. and Brown, M.D. (1997) *Proceedings of the National Academy of Sciences U.S.A*, 94, 14900–14905.

Wallace, D.C., Brown, M.D. and Lott, M.T. (1999) *Gene*, 238, 211–230.

Wilson, M.R. (1997) Update to: extraction, PCR amplification and sequencing of mitochondrial DNA from human hair shafts. In Gyllensten U. and Ellingboe J. (eds) *The PCR Technique: DNA Sequencing II*. Natick, Massachusetts: Eaton Publishing, pp. 322–328.

Wilson, M.R., Stoneking, M., Holland, M.M., DiZinno, J.A. and Budowle, B. (1993) *Crime Laboratory Digest*, 20, 68–77.

Wilson, M.R., Polanskey, D., Butler, J., DiZinno, J.A., Replogle, J. and Budowle, B. (1995a) *BioTechniques*, 18, 662–669.

Wilson, M.R., DiZinno, J.A., Polanskey, D., Replogle, J. and Budowle, B. (1995b) *International Journal of Legal Medicine*, 108, 68–74.

Wilson, M.R., Polanskey, D., Replogle, J., DiZinno, J.A. and Budowle, B. (1997) *Human Genetics*, 100, 167–171.

Wilson, M.R., Allard, M.W., Monson, K., Miller, K.W. and Budowle, B. (2002a) *Forensic Science International*, 129, 35–42.

Wilson, M.R., Allard, M.W., Monson, K., Miller, K.W.P. and Budowle, B. (2002b) Further discussion of the consistent treatment of length variants in the human mitochondrial DNA control region. *Forensic Science Communications* [online]. Available at: http://www.fbi.gov/hq/lab/fsc/backissu/oct2002/wilson.htm.

Wilson, M.R. and Allard, M.W. (2004) *Forensic Science Review*, 16, 37–62.

Wittig, H., Augustin, C., Baasner, A., Bulnheim, U., Dimo-Simonin, N., Edelmann, J., Hering, S., Jung, S., Lutz, S., Michael, M., Parson, W., Poetsch, M., Schneider, P.M., Weichhold, G. and Krause, D. (2000) *Forensic Science International*, 113, 113–118.

Wiuf, C. (2001) *Genetics*, 159, 749–756.

Wurmb-Schwark, N., Higuchi, R., Fenech, A.P., Elfstroem, C., Meissner, C., Oehmichen, M. and Cortopassi, G.A. (2002) *Forensic Science International*, 126, 34–39.

Yoshii, T., Tamura, K. and Ishiyama, I. (1992) *Nippon Hoigaku Zasshi*, 46, 313–316.

Zischler, H., Geisert, H., von Haeseler, A. and Paabo, S. (1995) *Nature*, 378, 489–492.

CHAPTER 11

NON-HUMAN DNA TESTING AND MICROBIAL FORENSICS

We urge as rapid development of new systems as is consistent with their validation before they are put into general use.

(NRCII, p. 59)

One's ideas must be as broad as Nature if they are to interpret Nature.

(Sherlock Holmes, *A Study in Scarlet*)

While the vast majority of forensic DNA typing performed for criminal investigations involves human DNA, it is not the only source of DNA that may be useful in demonstrating the guilt or innocence of an individual suspected of a crime (Sensabaugh and Kaye 1998). Domestic animals such as cats and dogs live in human habitats and deposit hair that may be used to place a suspect at the crime scene. Demonstration that a botanical specimen came from a particular plant can aid the linkage of a crime to a suspect or help demonstrate that the body of a deceased victim may have been moved from the murder site. DNA testing can now be used to link sources of marijuana. A large area of future application for forensic DNA typing involves identification of bio-terrorism materials such as anthrax. This chapter will briefly discuss each of these topics and the value of non-human DNA testing in forensic casework.

DOMESTIC ANIMAL DNA TESTING

The American Pet Products Manufacturers Association reported in April 2003 that over 64 million U.S. households own a pet (see http://www.appma.org). Their survey found 77.7 million cats and 65 million dogs in these households, which make up at least one-third of all U.S. residences. Since many of these domestic animals shed hair, these hairs could be picked up or left behind at the scene of a crime by a perpetrator. An assailant may unknowingly carry clinging cat hairs from a victim's cat away from the scene of a crime, or hair from the perpetrator's cat may be left at the scene.

The Veterinary Genetics Laboratory at the University of California-Davis (see http://www.vgl.ucdavis.edu/forensics) has been performing forensic animal

DNA analyses since 1996. They have found that there are three types of animal DNA evidence: (1) the animal as victim, (2) the animal as perpetrator, and (3) the animal as witness.

Animal abuse cases or the theft of an animal can sometimes be benefited by the power of DNA testing. The remains of a lost pet can be positively identified through genetic analysis. Typically genetic markers like short tandem repeats (STRs) and mitochondrial DNA (mtDNA) are examined in much the same way as with human DNA.

When animals are involved in an attack on a person, DNA typing may be used to identify the animal perpetrator (e.g., a Pit Bull). If the victim is deceased, then DNA evidence may be the only witness that an animal in custody committed the crime. Animal DNA testing can 'exonerate' innocent animals so that they are not needlessly destroyed.

Animal DNA has been used successfully to link suspects to crime scenes (see D.N.A. Box 11.1). A study on the transfer of animal hair during simulated criminal behavior found that hundreds of cat hairs or dog hairs could be transferred from the homes of victims to a burglar or an aggressor (D'Andrea *et al.* 1998). In fact, the number of hairs found was so high that the authors of this study felt that it is almost impossible to enter a house where a domestic animal lives without being 'contaminated' by cat and/or dog hairs even when the owner describes his or her animal as a poor source of hair (D'Andrea *et al.* 1998). Due to the fact that shed hairs often do not contain roots, nuclear DNA may not be present in sufficient quantities for STR typing. Mitochondrial DNA may be a more viable alternative for many of these types of shed hair transfers.

D.N.A. Box 11.1
Snowball's DNA

The identity of white cat hairs found on a bloodstained leather jacket left at a murder scene became a turning point in the case of Douglas Leo Beamish versus Her Majesty The Queen in the Providence of Prince Edward Island, Canada. The victim, Shirley Duguay, was discovered in a shallow grave in a wooded area eight months after she disappeared. Beamish, her former common law husband was charged with the crime. At the time he lived with his parents and a white cat named Snowball. Laboratory analysis of the bloodstains on the recovered jacket contained the victim's DNA profile. The white cat hairs matched Snowball at 10 STR loci. The defendant was convicted of murder based in part on this evidence.

Source:
Menotti-Raymond, M., *et al.* (1997) Pet cat hair implicates murder suspect. *Nature,* 386, 774.

CAT DNA

Cats have 18 pairs of autosomes and the sex chromosomes X and Y and genetic markers have been developed on each of the *Felis catus* chromosomes (Menotti-Raymond *et al.* 1999). A panel of STR markers dubbed the 'MeowPlex' has been developed that contains 11 STRs on nine different autosomes (Butler *et al.* 2002). A gender identification marker was also included in this assay through the addition of PCR primers that are specific for the SRY gene on the cat Y chromosome. The PCR products for this 12plex amplification fall in the size range of 100 bp to 400 bp and use three dye colors (Figure 11.1).

Feline STR allele frequencies from domestic cats have been published (Menotti-Raymond *et al.* 1997) for the purpose of demonstrating uniqueness of DNA profiles in forensic investigations, such as used in the Beamish case (D.N.A. Box 11.1). Population studies on over 1200 cats from 37 different breeds have been conducted by the Laboratory of Genomic Diversity at the National Cancer Institute-Frederick Cancer Research and Development Center in Frederick, Maryland. In an initial study of 223 cats from 28 different breeds, the MeowPlex exhibited an average composite locus heterozygosity of 0.73 across the breeds (Menotti-Raymond *et al.* 2003). The power of discrimination with this 11plex feline STR multiplex ranged from 5.5×10^{-7} to 3.3×10^{-13} across the various breeds.

A real-time quantitative polymerase chain reaction (PCR) assay (see Chapter 4) for estimating the DNA yield extracted from domestic cat specimens has been developed (Menotti-Raymond *et al.* 2003). This assay is capable of detecting down to 10 femtograms of feline genomic DNA and uses high-copy number short interspersed nuclear elements (SINEs) similar to the *Alu* repeats described in Chapter 8. Feline STRs and mtDNA testing is performed by Joy Halverson of QuestGen Forensics (http://www.animalforensics.com), which also does canine STR and mtDNA testing to aid forensic investigations.

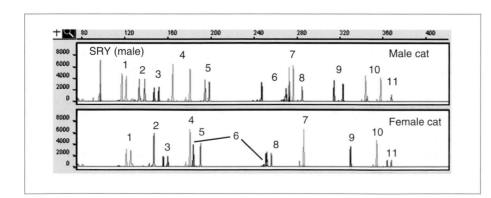

Figure 11.1

DNA profiles produced from male (top panel) and female (bottom panel) cat DNA using a multiplex STR typing assay dubbed the 'MeowPlex' (Butler et al. 2002). This test examines 11 autosomal STRs and a region of the SRY gene contained on the Y chromosome that can be used for sex determination.

DOG DNA

While cat DNA testing may be involved in situations where the animal hair acts as a silent witness to connecting a perpetrator to a crime scene, evidence from dogs is more frequently linked to situations where the animal is the perpetrator. Rottweilers, German Shepherds, Doberman Pinchers, and Pit Bulls can be trained as security animals and may attack, injure, or even kill people.

Canine mitochondrial DNA possesses two hypervariable regions (HV1 and HV2) similar to the human mtDNA described in Chapter 10. Savolainen *et al.* (1997) found 19 sequence variants across a 257 bp segment of the hypervariable region 1 of mtDNA control region in 102 domestic dogs of 52 different breeds. They concluded that on average 88 out of 100 tested animals could be excluded with this mtDNA sequence analysis. Another study that used a larger portion of the dog mtDNA control region found the overall exclusion capacity to be 0.93 in 105 dogs tested. By way of comparison in 100 British white Caucasians an exclusion capacity of 0.97 was observed (Piercy *et al.* 1993). Therefore, domesticated dog mtDNA is not as variable as human mtDNA yet it can still provide helpful clues in forensic cases (Savolainen and Lundeberg 1999, Schneider *et al.* 1999).

A number of STR markers have been mapped and characterized on the 38 pairs of autosomes and the X chromosome of *Canis familiaris*, the domestic dog (Neff *et al.* 1999). Recently, 15 canine STR loci have been characterized with sequenced alleles to define nomenclature for future work (Eichmann *et al.* 2004). A set of 10 dinucleotide repeat STRs has been used to aid investigations in illegal animal deaths (Padar *et al.* 2001) and a dog attack that resulted in the death of a seven-year-old boy (Padar *et al.* 2002). In addition, DNA profiling of human blood recovered from a dog's fur can associate or exonerate the animal from connection to an attack (Brauner *et al.* 2001).

SPECIES IDENTIFICATION

The remains of stolen animals or illegally procured meat (e.g., endangered species or poaching) can be identified through DNA testing (Giovambattista *et al.* 2001, Poetsch *et al.* 2001). The U.S. Fish and Wildlife Service have a forensic laboratory in Ashland, Oregon that does some species identification using DNA (see http://www.lab.fws.gov). Sequence analysis of the mtDNA cytochrome b gene is effective at identifying the species of origin for a biological sample (Bartlett and Davidson 1992, Parson *et al.* 2000, Hsieh *et al.* 2001, Branicki *et al.* 2003). Bataille *et al.* (1999) developed a multiplex amplification of a portion of the human mtDNA control and the cytochrome b gene to enable simultaneous human and species identification.

PLANT DNA

In the area of plant DNA testing, there are primarily two areas being investigated currently. The first is the linking of plant material to suspects or victims in order to make an association with a particular area where a crime was committed. The second is in linking marijuana to aid in forensic drug investigations. A review of some of the applications of forensic botany was published recently (Miller-Coyle *et al.* 2001).

LINKING PLANT MATERIALS TO SUSPECTS

Crimes often occur in localized areas containing a unique combination of botanical growth. If these plants, algae, or grass are sufficiently rare, then recovery of trace evidence from the clothing of a victim or the personal property of a suspect may be helpful in making an association that can link them to a crime scene (Szibor *et al.* 1998, Norris and Bock 2000, Horrocks and Walsh 2001).

Although it is not yet used routinely (Bock and Norris 1997), non-human DNA has helped link suspects to crime scenes and aided important investigations. In the first use of forensic botanical evidence, two small seedpods from an Arizona Palo Verde tree found in the back of pick-up truck were used to place an accused murderer at the crime scene (Yoon 1993). Genetic testing on the seeds showed that in a 'lineup' of 12 Palo Verde trees near the crime scene, DNA from the seeds matched only the tree under which the victim's body had been found. In *State versus Bogan*, the jury found the accused guilty based in large measure on the plant DNA evidence.

MARIJUANA DNA

The Connecticut State Forensic Science Laboratory has developed a sensitive DNA test for *Cannabis sativa* (marijuana) because it is an illegal substance associated with many crime scenes. In fact, marijuana is the most commonly identified drug tested by U.S. forensic laboratories in criminal investigations (see http://www.deadiversion.usdoj.gov/nflis). Marijuana DNA testing can link an individual to a sample, link growers, and help track distribution networks (Miller-Coyle *et al.* 2001). However, it is important to keep in mind that if the marijuana plants were propagated clonally rather than by seed, then they will have identical DNA profiles. Clonal propagation in marijuana is performed by taking cuttings from a 'mother' plant and rooting them directly in the soil to create large numbers of plants having identical DNA (Miller-Coyle *et al.* 2001).

Efficient extraction protocols have been developed that yield 125–500 ng of DNA per 100 mg of fresh plant tissue (Miller Coyle *et al.* 2003a). DNA testing of marijuana as with other plants has traditionally been performed with one of three methodologies: randomly amplified polymorphic DNA markers (RAPDs), amplified fragment length polymorphisms (AFLPs) or short tandem repeats (STRs). These techniques and their specific application to marijuana DNA typing have been reviewed (Miller-Coyle *et al.* 2003b).

Randomly amplified polymorphic DNA marker analysis utilizes short PCR primers consisting of random sequences usually in the size range of 8–15 nucleotides in length. Complex patterns of PCR products are generated as these random sequence primers anneal to various regions in an organism's genome. RAPD suffers from poor reproducibility between laboratories largely because of the requirement of consistent PCR amplification conditions including thermal cycler ramp speeds. The complex patterns of RAPD also prevent mixture interpretation and provide challenges in consistent scoring of electrophoretic images even in single source samples.

Patterns from amplified fragment length polymorphism markers can be generated with greater reproducibility compared to RAPDs. AFLPs are generated by first cutting a double-stranded DNA sample with one or more restriction enzymes (Vos *et al.* 1995, Ranamukhaarachchi *et al.* 2000). Specific 'adaptor' sequences are then ligated to the restriction cut sites. PCR primers that recognize these ligated adaptor sequences are used to amplify different sized DNA fragments that can then be separated using electrophoresis. The final result is a complex series of peaks usually in the 50–400 bp size range that can be scored with computer software and compared with other AFLP patterns from different marijuana plants. Even highly inbred individual plants can be distinguished by their AFLP patterns (Miller-Coyle *et al.* 2003b).

Several STR markers have been recently reported for *Cannabis sativa* (Hsieh *et al.* 2003, Gilmore *et al.* 2003, Alghanim and Almirall 2003). As with human STRs, marijuana STR markers are highly polymorphic, specific to unique sites in the genome, and capable of deciphering mixtures. A hexanucleotide repeat marker showed repeat units ranging from 3–40 in 108 tested marijuana samples, and primers amplifying this locus produced no cross-reactive amplicons from 20 other species of plants tested (Hsieh *et al.* 2003).

All of these molecular techniques for identifying marijuana plants need comparative databases to be effective tools for law enforcement purposes. In order to determine the possibility of a random match with marijuana seizure samples, it is important to have a database of seizure samples so their DNA profiles can be used for comparison (Miller-Coyle *et al.* 2003). More information regarding on-going research in the field of forensic botany and its application may be found at http://www.bodetech.com/research/botany_plant.html and http://www.plantdnatracker.com.

MICROBIAL FORENSICS

Unfortunately microbial forensics will likely become a larger part of DNA testing in the future with the threat of terrorism and the use of biological warfare agents. Microbial evidence can be from either real terrorist events or hoaxes. The efforts in this area will likely require forensic laboratories to build strong collaborations with academia, private sector and national laboratories. Important requirements of bio-threat detection assays are high sensitivity, high specificity in complex samples, fast measurement, compact design for portability and field use, and internal calibration and reference to ensure reliable results (Ivnitski *et al.* 2003).

In October 2001 a bio-terrorism attack impacted the United States as government offices and media outlets received anthrax-laden letters sent anonymously through the postal service. This attack resulted in 22 anthrax cases and five deaths. In addition, many people were afraid to open their mail for months afterwards. More than 125 000 samples were processed as part of this case in the two years following this attack and yet no one has been charged with the crime to date (Popovic and Glass 2003).

Several challenges arise when trying to gather evidence, identify the bio-crime organism(s), and trace the source of the organism(s). First responders to crime scenes where biological weapons have been dispersed have to be concerned about their own safety and the safety of others while maintaining chain of custody of any evidence collected from the crime scene, all the while trying to prevent contamination of the evidence and the environment. Databases need to be established for intrinsic background species and bio-threat strains. Reliable reference material is needed for comparison purposes. Proficiency and validation testing are necessary to estimate false-positive and false-negative rates (Kiem 2003).

The U.S. efforts in building a response to bio-terrorism have been announced in a policy paper (Budowle *et al.* 2003). The FBI has initiated a Scientific Working Group on Microbial Genetics and Forensics (SWGMGF) that will help develop guidelines related to the operation of microbial forensics (SWGMGF 2003). Currently there are an insufficient number of validated analytical tools to characterize and identify biological agents that might be used in a terrorist attack (Budowle 2003). Research efforts will continue to be made in this area.

Comparative genome sequencing promises to be a powerful tool for investigating infectious disease outbreaks as was performed with the whole-genome sequencing of *Bacillus anthracis* (anthrax) (Read *et al.* 2002a, 2002b). Phylogenetic analyses of viral strains of HIV have been admitted and used as evidence in court (Metzker *et al.* 2002). However, since bacteria and viruses reproduce asexually, clones are prevalent. A perfect match between evidence collected and a reference sample is much less definitive than with human identity testing where sexual reproduction shuffles genetic material each generation.

CHALLENGES WITH PRESENTING NON-HUMAN DNA IN COURT

Sensabaugh and Kaye (1998) consider several issues regarding whether a given application with non-human DNA is ready for court use. These issues include the novelty of the application, the validity of the underlying scientific theory, the validity of any statistical interpretations, and the relevant scientific community to consult in assessing the application. Many times new methods are applied for the first time in microbial forensics or animal or plant DNA testing that have not yet undergone the scrutiny of regular forensic DNA testing techniques. Reference DNA databases for comparison purposes and use in calculating the probability of a chance match take time to develop and may not be in place prior to an investigation. Finding appropriate experts to review the scientific soundness of a novel application can also be challenging. Nevertheless, the power and influence of forensic DNA testing will continue to grow as it is used in more and more diverse applications to solve crimes that were previously inaccessible.

REFERENCES AND ADDITIONAL READING

Alghanim, H.J. and Almirall, J.R. (2003) *Analytical and Bioanalytical Chemistry*, 376, 1225–1233.

Bartlett, S.E. and Davidson, W.S. (1992) *Biotechniques*, 12, 408–411.

Bataille, M., Crainic, K., Leterreux, M., Durigon, M. and de Mazancourt, P. (1999) *Forensic Science International*, 99, 165–170.

Beeching, N.J., Dance, D.A., Miller, A.R. and Spencer, R.C. (2002) *British Medical Journal*, 324, 336–339.

Bellis, C., Ashton, K.J., Freney, L., Blair, B. and Griffiths, L.R. (2003) *Forensic Science International*, 134, 99–108.

Bock, J.H. and Norris, D.O. (1997) *Journal of Forensic Sciences*, 42, 364–367.

Branicki, W., Kupiec, T. and Pawlowski, R. (2003) *Journal of Forensic Sciences*, 48, 83–87.

Brauner, P., Reshef, A. and Gorski, A. (2001) *Journal of Forensic Sciences*, 46, 1232–1234.

Budowle, B. (2003) Defining a new forensic discipline: microbial forensics. *Profiles in DNA*, 6 (1), 7–10.

Budowle, B., Schutzer, S.E., Einseln, A., Kelley, L.C., Walsh, A.C., Smith, J.A., Marrone, B.L., Robertson, J. and Campos, J. (2003) *Science*, 301, 1852–1853.

Butler, J.M., David, V.A., O'Brien, S.J. and Menotti-Raymond, M. (2002) *Profiles in DNA*, 5 (2), 7–10. Available online at: http://www.promega.com/profiles.

D'Andrea, F., Fridez, F. and Coquoz, R. (1998) *Journal of Forensic Sciences*, 43, 1257–1258.

Dimsoski, P. (2003) *Croatian Medical Journal*, 44, 332–335.

Eichmann, C., Berger, B. and Parson W. (2004) *International Journal of Legal Medicine*, 118, 249–266.

Fridez, F., Rochat, S. and Coquoz, R. (1999) *Science and Justice*, 39, 167–171.

Gilmore, S., Peakall, R. and Robertson, J. (2003) *Forensic Science International*, 131, 65–74.

Giovambattista, G., Ripoli, M.V., Liron, J.P., Villegas Castagnasso, E.E., Peral-Garcia, P. and Lojo, M.M. (2001) *Journal of Forensic Sciences*, 46, 1484–1486.

Horrocks, M. and Walsh, K.A. (2001) *Journal of Forensic Sciences*, 46, 947–949.

Hsieh, H.M., Chiang, H.L., Tsai, L.C., Lai, S.Y., Huang, N.E., Linacre, A. and Lee, J.C. (2001) *Forensic Science International*, 122, 7–18.

Hsieh, H.M., Hou, R.J., Tsai, L.C., Wei, C.S., Liu, S.W., Huang, L.H., Kuo, Y.C., Linacre, A. and Lee, J.C. (2003) *Forensic Science International*, 131, 53–58.

Ivnitski, D., O'Neil, D.J., Gattuso, A., Schllcht, R., Calidonna, M. and Fisher, R. (2003) *Biotechniques*, 35, 862–869.

Kiem, P. (2003) Microbial forensics: a scientific assessment. Washington, DC: American Academy of Microbiology. Available online at: http://www.asm.org/Academy/index.asp?bid=2093.

Menotti-Raymond, M., David, V.A. and O'Brien, S.J. (1997) *Nature*, 386, 774.

Menotti-Raymond, M., David, V.A., Stephens, J.C., Lyons, L.A. and O'Brien, S.J. (1997) *Journal of Forensic Sciences*, 42, 1039–1051.

Menotti-Raymond, M., David, V.A., Lyons, L.A., Schaffer, A.A., Tomlin, J.F., Hutton, M.K. and O'Brien, S.J. (1999) *Genomics*, 57, 9–23.

Menotti-Raymond, M., David, V., Wachter, L., Yuhki, N. and O'Brien, S.J. (2003) *Croatian Medical Journal*, 44, 327–331.

Metzker, M.L., Mindell, D.P., Liu, X.M., Ptak, R.G., Gibbs, R.A. and Hillis, D.M. (2002) *Proceedings of the National Academy of Sciences of the United States of America*, 99, 14292–14297.

Miller-Coyle, H., Ladd, C., Palmbach, T. and Lee, H.C. (2001) *Croatian Medical Journal*, 42, 340–345.

Miller-Coyle, H., Shutler, G., Abrams, S., Hanniman, J., Neylon, S., Ladd, C., Palmbach, T. and Lee, H.C. (2003a) *Journal of Forensic Sciences*, 48, 343–347.

Miller-Coyle, H., Palmbach, T., Juliano, N., Ladd, C. and Lee, H.C. (2003b) *Croatian Medical Journal*, 44, 315–321.

Neff, M.W., Broman, K.W., Mellersh, C.S., Ray, K., Acland, G.M., Aguirre, G.D., Ziegle, J.S., Ostrander, E.A. and Rine, J. (1999) *Genetics*, 151, 803–820.

Norris, D.O. and Bock, J.H. (2000) *Journal of Forensic Sciences*, 45, 184–187.

Padar, Z., Angyal, M., Egyed, B., Furedi, S., Woller, J., Zoldag, L. and Fekete, S. (2001) *International Journal of Legal Medicine*, 115, 79–81.

Padar, Z., Egyed, B., Kontadakis, K., Furedi, S., Woller, J., Zoldag, L. and Fekete, S. (2002) *International Journal of Legal Medicine*, 116, 286–288.

Parson, W., Pegoraro, K., Niederstatter, H., Foger, M. and Steinlechner, M. (2000) *International Journal of Legal Medicine*, 114, 23–28.

Piercy, R., Sullivan, K.M., Benson, N. and Gill, P. (1993) *International Journal of Legal Medicine*, 106, 85–90.

Poetsch, M., Seefeldt, S., Maschke, M. and Lignitz, E. (2001) *Forensic Science International*, 116, 1–8.

Popovic, T. and Glass, M. (2003) *Croatian Medical Journal*, 44, 336–341.

Ranamukhaarachchi, D.G., Kane, M.E., Guy, C.L. and Li, Q.B. (2000) *Biotechniques*, 29, 858–866.

Read, T.D., Salzberg, S.L., Pop, M., Shumway, M., Umayam, L., Jiang, L., Holtzapple, E., Busch, J.D., Smith, K.L., Schupp, J.M., Solomon, D., Keim, P. and Fraser, C.M. (2002) *Science*, 296, 2028–2033.

Read, T.D., Peterson, S.N., Tourasse, N., Baillie, L.W., Paulsen, I.T., Nelson, K.E., Tettelin, H., Fouts, D.E., Eisen, J.A., Gill, S.R., Holtzapple, E.K., Okstad, O.A., Helgason, E., Rilstone, J., Wu, M., Kolonay, J.F., Beanan, M.J., Dodson, R.J., Brinkac, L.M., Gwinn, M., DeBoy, R.T., Madpu, R., Daugherty, S.C., Durkin, A.S., Haft, D.H., Nelson, W.C., Peterson, J.D., Pop, M., Khouri, H.M., Radune, D., Benton, J.L., Mahamoud, Y., Jiang, L., Hance, I.R., Weidman, J.F., Berry, K.J., Plaut, R.D., Wolf, A.M., Watkins, K.L., Nierman, W.C., Hazen, A., Cline, R., Redmond, C., Thwaite, J.E., White, O., Salzberg, S.L., Thomason, B., Friedlander, A.M., Koehler, T.M., Hanna, P.C., Kolsto, A.B. and Fraser, C.M. (2003) *Nature*, 423, 81–86.

Savolainen, P., Rosen, B., Holmberg, A., Leitner, T., Uhlen, M. and Lundeberg, J. (1997) *Journal of Forensic Sciences*, 42, 593–600.

Savolainen, P. and Lundeberg, J. (1999) *Journal of Forensic Sciences*, 44, 77–81.

Savolainen, P., Arvestad, L. and Lundeberg, J. (2000) *Journal of Forensic Sciences*, 45, 990–999.

Schneider, P.M., Seo, Y. and Rittner, C. (1999) *International Journal of Legal Medicine*, 112, 315–316.

Scientific Working Group on Microbial Genetics and Forensics (SWGMGF) (2003) Quality assurance guidelines for laboratories performing microbial forensic work. Forensic Science Communications, 5 (4). Available online at: http://www.fbi.gov/hq/lab/fsc/backissu/oct2003/2003_10_guide01.htm.

Sensabaugh, G. and Kaye, D.H. (1998) *Jurimetrics Journal*, 38, 1–16.

Shutler, G.G., Gagnon, P., Verret, G., Kalyn, H., Korkosh, S., Johnston, E. and Halverson, J. (1999) *Journal of Forensic Sciences*, 44, 623–626.

Szibor, R., Schubert, C., Schoning, R., Krause, D. and Wendt, U. (1998) *Nature*, 395, 449–450.

Vos, P., Hogers, R., Bleeker, M., Reijans, M., van de, L.T., Hornes, M., Frijters, A., Pot, J., Peleman, J., Kuiper, M. and Zabeau, M. (1995) *Nucleic Acids Research*, 23, 4407–4414.

Wetton, J.H., Higgs, J.E., Spriggs, A.C., Roney, C.A., Tsang, C.S. and Foster, A.P. (2003) *Forensic Science International*, 133, 235–241.

Yoon, C.K. (1993) *Science*, 260, 894–895.

TECHNOLOGY

CHAPTER 12

DNA SEPARATION METHODS: SLAB GEL AND CAPILLARY ELECTROPHORESIS

It is a capital mistake to theorize before one has data. Insensibly one begins to twist facts to suit theories, instead of theories to suit facts.

(Sherlock Holmes, *A Scandal in Bohemia*)

INTRODUCTION

THE NEED FOR DNA SEPARATIONS

A polymerase chain reaction (PCR) reaction in which short tandem repeat (STR) alleles are amplified produces a mixture of DNA molecules that present a challenging separation problem. A multiplex PCR can produce 20 or more DNA fragments that must be resolved from one another. In addition, single base resolution is required to distinguish between closely spaced alleles (e.g., TH01 alleles 9.3 and 10). The typical separation size range where this single base resolution is needed is between 100 and 400 bp. Additionally, it is important that the separation method be reproducible and yield results that can be compared among laboratories.

In order to distinguish the various molecules from one another, a separation step is required to pull the different sized fragments apart. The separation is typically performed by a process known as electrophoresis and is either conducted in a slab-gel or capillary environment. This chapter will discuss the theory and background information on separation methods. Chapter 13 will cover how the bands in gel electrophoresis and the peaks in capillary electrophoresis are actually generated and detected. Chapter 14 will discuss specific techniques utilizing capillary electrophoresis and slab-gel electrophoresis that are widely used by forensic DNA typing laboratories.

ELECTROPHORESIS

PCR products from short tandem repeat DNA must be separated in a fashion that allows each allele to be distinguished from other alleles. Heterozygous alleles are resolved in this manner with a sized based separation method known as electrophoresis. The separation medium may be in the form of a slab gel or a capillary.

The word 'electrophoresis' comes from the Greek *electron* (charge) and the Latin *phore* (bearer). Thus, the process of electrophoresis refers to electrical charges carried by the molecules. In the case of DNA, the phosphate groups on the backbone of the DNA molecule have a negative charge. Nucleic acids are acids because the phosphate groups readily give up their H^+ ions making them negatively charged in most buffer systems. Under the influence of an electric field, DNA molecules will migrate away from the negative electrode, known as the cathode, and move towards the positive electrode, known as the anode. The higher the voltage, the greater the force felt by the DNA molecules and the faster the DNA moves.

The movement of ions in an electric field generates heat. This heat must be dissipated or it will be absorbed by the system. Excessive heat can cause a gel to generate bands that 'smile' or in very severe cases the gel can literally melt and fall apart. As will be described at the end of the chapter, performing electrophoresis in a capillary is an advantage because heat can be more easily dissipated from the capillary, which has a high surface area-to-volume ratio.

SLAB GELS

Figure 12.1

Schematic of a gel electrophoresis system. The horizontal gel is submerged in a tank full of electrophoresis buffer. DNA samples are loaded into wells across the top of the gel. These wells are created by a 'comb' placed in the gel while it is forming. When the voltage is applied across the two electrodes, the DNA molecules move towards the anode and separate by size. The number of lanes available on a gel is dependent on the number of teeth in the comb used to define the loading wells. At least one lane on each gel is taken up by a molecular weight size standard that is used to estimate the sizes of the sample bands in the other lanes.

Slab gels consist of a solid matrix with a series of pores and a buffer solution through which the DNA molecules pass during electrophoresis. Gel materials are mixed together and poured into a mold to define the structure of the slab gel. A sample 'comb' is placed into the gel such that the teeth of the comb are imbedded in the gel matrix. After the gel has solidified, the comb is removed leaving behind wells that are used for loading the DNA samples. The basic format for a gel electrophoresis system is shown in Figure 12.1.

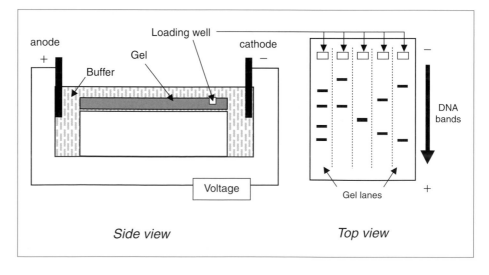

TYPES OF GELS

Two types of gels are commonly used in molecular biology and forensic DNA laboratories today to achieve DNA separations. Agarose gels have fairly large pore sizes and are used for separating larger DNA molecules while polyacrylamide gels are used to obtain high-resolution separations for smaller DNA molecules, usually below 500 or 1000 bp.

Forensic DNA typing methods use both types of gels. Restriction fragment length polymorphism (RFLP) methods use agarose gels to separate DNA fragments ranging in size from ~600 bp to ~23 000 bp. Low molecular weight DNA molecules are not well separated with agarose slab gels. On the other hand, PCR-amplified STR alleles, which range in size from ~100 bp to ~400 bp, are better served by polyacrylamide gels. In the case of some STR loci that contain microvariants, the high-resolution capability of polyacrylamide gels is essential for separating closely sized DNA molecules that may only differ by a single nucleotide.

AGAROSE GELS

Agarose is basically a form of seaweed and contains pores that are on the order of 2000 angstroms (Å) (200 nm) in diameter. Agarose gels are easily prepared by weighing out a desired amount of agarose powder and mixing it with the electrophoresis buffer. This mixture can be quickly brought to a boil by microwaving the solution whereupon the agarose powder goes into solution. After the solution cools down slightly, it is poured into a gel box to define the gel shape and thickness.

A comb is added to the liquid agarose before it cools to form wells with its teeth in the jelly-like substance that results after the gel 'sets.' Once the agarose has gelled the comb is removed leaving behind little wells that can hold 5–10 μL of sample or more depending on the size of the teeth and the depth at which they were placed in the agarose gel. The comb teeth define the number of samples that can be loaded onto the gel as well as where the lanes will be located. There are predominantly two types of combs: square-tooth and sharks-tooth.

Electrophoresis buffer is poured over the gel until it is fully submerged. Two buffers are commonly used with electrophoresis, Tris-acetate-EDTA (TAE) and Tris-borate-EDTA (TBE). Samples are mixed with a loading dye and carefully pipetted into each well of the submerged gel. This loading dye contains a mixture of bromophenol blue, a dark blue dye which helps to visually see the sample, and sucrose to increase the sample's viscosity and help it stay in the well prior to turning on the voltage and initiating electrophoresis.

The number of samples that can be run in parallel on the gel are defined by the number of teeth on the comb added to the gel before it sets (and hence the number of wells that will be created). Typically between eight and 24 samples are run at a time on an agarose gel. Molecular weight standards are run in some of the lanes in order to estimate the size of each DNA sample following electrophoresis.

After the samples are loaded, a cover is placed over the gel box containing the submerged gel and the electrodes on either end of the gel are plugged into a power source. The anode (positive electrode) is placed on the end of the gel furthest from the wells to draw the DNA molecules through the gel material. Typically 100–600 volts (V) are placed across agarose gels that are 10–40 cm in length, creating electric field strengths of approximately 1–10 V/cm.

As the DNA molecules are drawn through the gel, they are separated by size, the smaller ones moving more quickly and easily through the gel pores. It might help to think of the DNA molecules as marathon runners with different abilities. They all start together at the beginning and then separate during the 'race' through the gel. The smaller DNA molecules move more quickly than the larger ones through the obstacles along the gel 'race course' and thus are further along when the voltage is turned off and the 'race' completed. When the separation is completed, the gel is scanned or photographed to record the results for examination and comparison.

POLYACRYLAMIDE GELS

Polyacrylamide (PA) gels have much smaller pore sizes (~100–200 Å) than agarose gels (~1500–2000 Å). The average pore size of a gel is an important factor in determining the ability of a slab gel to resolve two similarly sized DNA fragments. The pores in PA gels are chemically created with a cross-linking process involving acrylamide and bisacrylamide (Figure 12.2).

Figure 12.2

Illustration of polyacrylamide gel polymerization. The pore size of a gel is controlled by its degree of cross-linking, which depends on the proportions of acrylamide and bisacrylamide in the gel. Polymerization is typically induced by free radicals resulting from the chemical decomposition of ammonium persulfate. TEMED, a free radical stabilizer, is also added to the gel mixture to stabilize the polymerization process. A polyacrylamide gel is poured between two glass plates that define its dimensions. A spacer is placed between the plates to define the thickness of the gel. A typical gel size is 17 cm × 43 cm × 0.4 cm. DNA molecules flow through the polyacrylamide gel matrix along the z-axis going into the page, much like flowing through a chain link fence with various sized holes.

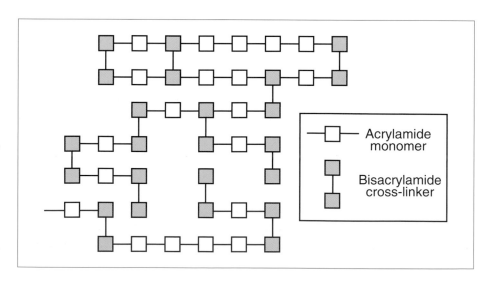

Polyacrylamide gels are chemically formed through polymerization of the monomeric acrylamide molecule in the presence of a variable quantity of the bisacrylamide cross-linker. The polymerization process is initiated by the generation of free radicals provided by ammonium persulfate and stabilized by the compound TEMED (N,N,N′,N′-tetramethylethylenediamine). This polymerization leads to the formation of long strands of acrylamide monomer with occasional cross-links provided by the bisacrylamide connector.

Polyacrylamide pore sizes can be decreased by increasing the overall concentration of acrylamide (both monomer plus cross-linker). The value of total acrylamide concentration in the gel solution is typically expressed as %T. The ratio of the monomer to the cross-linker may also be varied. The smallest pore sizes have been shown to occur when the cross-linker is 5% of the total acrylamide weight, or 5%C. A common gel solution used in STR allele separations is 5%T, 5%C. Another way this solution might be described is 5% acrylamide:bis (19:1).

One of the challenges in pouring or casting a slab gel is avoiding bubble formation. The polymerization process generates heat that can lead to bubbles forming in the gel. Gel mixtures are sometimes degassed under a vacuum for a short period of time prior to polymerization to remove any gases in the solution that might give rise to bubbles as the gel is solidifying. Sometimes bubbles occur in spite of great effort to avoid them. These gels may still be used as long as the bubbles are not in a lane where a sample will be run and/or do not interfere with the region of detection in fluorescence.

PA gels may be run in either a horizontal or a vertical format. The type of gel box defines the running format. Detection of DNA bands in polyacrylamide gels may be performed with fluorescent dyes or silver staining as will be described in Chapter 13.

NATIVE VERSUS DENATURING ELECTROPHORESIS CONDITIONS

Under normal conditions, the two complementary strands of DNA will remain together. Electrophoresis systems that perform the DNA separation while keeping the complementary strands together as double-stranded DNA are often referred to as 'native' or 'non-denaturing'. On the other hand, a separation system that possesses an environment capable of keeping the DNA strands apart as single-stranded DNA is usually referred to as a 'denaturing' system.

Generally better resolution between closely sized DNA molecules can be achieved with denaturing systems. This improved resolution is achieved because single-stranded DNA is more flexible than double-stranded DNA and therefore interacts with the sieving medium more effectively allowing closely sized molecules to be differentially separated. Additionally, natural conformation in DNA molecules, sometimes referred to as secondary structure, is eliminated in a denaturing environment.

To achieve a denaturing environment, chemicals, such as formamide and urea, may be used to keep the complementary strands of DNA apart from one another. The addition of six molar urea is a common technique for making a denaturing gel. Formamide and urea form hydrogen bonds with the DNA bases and prevent the bases from interacting with their complementary strand. The temperature of the separation or the pH of the solution may also be raised to aid in keeping the complementary strands of DNA apart.

A popular method for achieving denatured DNA strands (prior to electrophoresis) is to dilute the samples in 100% formamide. The samples are then heated to 95°C to denature the DNA strands, and then 'snap cooled' on ice by bringing them from the heated 95°C environment immediately to 0°C by placing them on ice.

PROBLEMS WITH GELS

The process of preparing a polyacrylamide gel involves a number of steps including cleaning and preparing the gel plates, combining the gel materials, pouring the gel, waiting for it to set-up, and finally removing the comb. These steps are time consuming and rather labor intensive and represent mundane tasks in the laboratory. In addition, the acrylamide gel materials are known neurotoxins and need to be handled with care.

Precast gels have also become popular due to the time and labor involved with preparing the gel plates, pouring the gel, and waiting for it to set. However, one still has to load the DNA samples very carefully into each well (to prevent contamination from adjacent wells). The development of capillary electrophoresis has excited many DNA scientists because the tedious processes of gel pouring and sample loading have been automated with this technique.

CAPILLARY ELECTROPHORESIS

Capillary electrophoresis (CE) is a relatively new addition to the electrophoresis family. The first CE separations of DNA were performed just over a decade ago in the late 1980s. Since the introduction of new CE instrumentation in the mid-1990s, the technique has gained rapidly in popularity and for good reason. While slab-gel electrophoresis has been a proven technique for over 30 years, there are a number of advantages to analyzing DNA in a capillary format.

ADVANTAGES OF CE OVER SLAB GELS

First and foremost, the injection, separation, and detection steps can be fully automated permitting multiple samples to be run unattended. In addition, only minute quantities of sample are consumed in the injection process and samples can be easily retested if needed. This is an important advantage for precious forensic specimens that often cannot be easily replaced.

Separation in capillaries may be conducted in minutes rather than hours due to higher voltages that are permitted with improved heat dissipation from capillaries. Another advantage is that quantitative information is readily available in an electronic format following the completion of a run. No extra steps such as scanning the gel or taking a picture of it are required. Lane tracking is not necessary since the sample is contained within the capillary, nor is there fear of cross-contamination from samples leaking over from adjacent wells with CE.

DISADVANTAGES OF CE

The one major disadvantage of CE instruments is throughput. Due to the fact that samples are analyzed sequentially one at a time, single capillary instruments are not easily capable of processing high numbers of samples or sample throughputs. As will be discussed in Chapters 14 and 17, however, capillary array systems have been developed to run multiple samples in parallel and those vastly improve the sample throughput.

CE instruments require a higher start-up cost (more than $50 000) than slab-gel electrophoresis systems and this fact prohibits some laboratories from using them. Nevertheless, CE instruments are quickly becoming the principal workhorses in a number of forensic DNA typing laboratories because of their automation and ease of use.

COMPONENTS OF CE

The primary elements of a CE instrument include a narrow capillary, two buffer vials, and two electrodes connected to a high-voltage power supply. CE systems also contain a laser excitation source, a fluorescence detector, an autosampler to hold the sample tubes, and a computer to control the sample injection and detection (Figure 12.3). CE capillaries are made of fused silica (glass) and typically have an internal diameter of 50–100 μm and a length of 25–75 cm.

The same buffers that are used in gel electrophoresis may also be used with CE. However, instead of a gel matrix through which the DNA molecules pass, a viscous polymer solution serves as the sieving medium. Larger DNA molecules are retarded more by the linear, flexible polymer chains than smaller DNA fragments, which leads to a size-based separation analogous to the DNA passing through the pores in the cross-linked polyacrylamide gels discussed above.

Prior to injecting each sample, a new gel is 'poured' by filling the capillary with a fresh aliquot of the polymer solution. The CE can be thought of as a long, skinny gel that is only wide enough for one sample at a time.

Figure 12.3

Schematic of capillary electrophoresis instruments used for DNA analysis. The capillary is a narrow glass tube approximately 50 cm long and 50 µm in diameter. It is filled with a viscous polymer solution that acts much like a gel in creating a sieving environment for DNA molecules. Samples are placed into a tray and injected onto the capillary by applying a voltage to each sample sequentially. A high voltage (e.g., 15 000 volts) is applied across the capillary after the injection in order to separate the DNA fragments in a matter of minutes. Fluorescent dye-labeled products are analyzed as they pass by the detection window and are excited by a laser beam. Computerized data acquisition enables rapid analysis and digital storage of separation results.

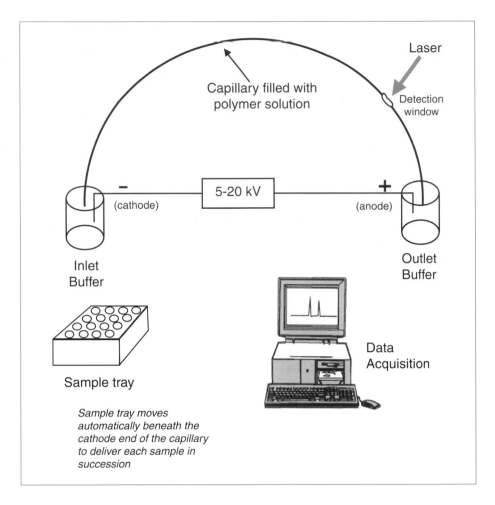

An important difference between CE and gels is that the electric fields are on the order of 10-to-100 times stronger with CE (i.e., 300 V/cm instead of 10 V/cm), which results in faster run times for CE.

Detection of the sample is performed automatically by the CE instrument through measuring the time span from sample injection to sample detection with a laser placed near the end of the capillary. Laser light is shined on to the capillary at a fixed position where a window has been burned in the coating of the capillary. DNA fragments are illuminated as they pass by this window in the capillary. As with gels, the smaller molecules will arrive at the detection point first followed by the larger molecules. Data from CE separations are plotted as a function of the relative fluorescence intensity observed from fluorescence emission of dyes passing the detector (see Chapter 13). The fluorescent emission signals from dyes attached to the DNA molecules can then be used to detect and quantify the DNA molecules passing the detector.

DNA SEPARATION MECHANISMS

Now that we have covered the two primary methods for DNA separations in use today, namely slab-gel electrophoresis and capillary electrophoresis, we will discuss briefly the theories behind DNA separations by electrophoresis.

With one phosphate group for every nucleotide unit, DNA molecules possess a constant charge-to-mass ratio. Thus, a piece of DNA that is 10 nucleotide units long will feel the same force pulling on it when an electric field is applied to it as a DNA oligomer that is 100 nucleotide units in length. In order to resolve DNA fragments that differ in size, a sieving mechanism is required. The separation of DNA is therefore accomplished with gels or polymer solutions that retard larger DNA molecules as they pass through the separation medium. The smaller molecules can slip through the gel pores faster and thus migrate ahead of longer DNA strands as electrophoresis proceeds (Figure 12.4).

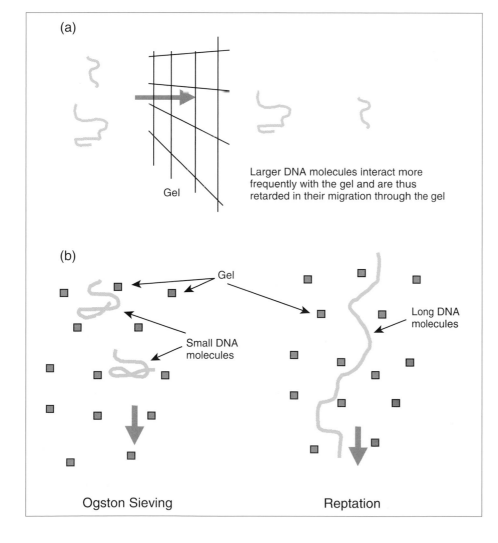

Figure 12.4

Illustration of DNA separation modes in gel electrophoresis. Separation according to size occurs as DNA molecules pass through the gel, which acts as a molecular sieve (a). Ogston sieving and reptation are the two primary mechanisms used to describe the movement of DNA fragments through a gel (b).

In the simplest sense, a gel may be considered as a molecular sieve with 'pores' that permit the DNA molecules to pass in a size-dependent manner because larger molecules are retarded more than smaller ones. Two primary mechanisms for DNA separations through gel pores have been described: the Ogston model and reptation. These two theories are complementary as they operate in different size regimes. The Ogston model describes the behavior of DNA molecules that are smaller than the gel pores while reptation describes the movement of larger DNA molecules (Figure 12.4).

OGSTON SIEVING

The Ogston model regards the DNA molecule as a spherical particle or coil like a small tangle of thread that is tumbling through the pores formed by the gel. Molecules move through the gel in proportion to their ability to find pores that are large enough to permit their passage. Smaller molecules migrate faster because they can pass through a greater number of pores. When DNA molecules are much larger than the mesh size of the gel-sieving medium, the Ogston model predicts that the mobility (movement) of the molecules will go to zero.

REPTATION

However, gel separations have been demonstrated with DNA fragments that are much larger than the predicted pore size of the gel. The reptation model for DNA separations views the DNA molecule as moving like a snake through the gel pores. DNA molecules become elongated like a straight length of thread and enter the gel matrix end on. Separation of sample components, such as two STR alleles, occurs as the DNA winds its way through the pores of the gel matrix.

ADDITIONAL COMMENTS ON ELECTROPHETIC SEPARATIONS

Electrophoresis is a relative rather than an absolute measurement technique. The position of a DNA band on a gel has no meaning without reference to a size standard containing material with known DNA fragment sizes. Thus, samples are run on a gel side-by-side with molecular weight markers. For example, a DNA restriction digest might be used with a half-dozen or more fragments ranging in size from 100–1000 bp. A visual comparison can then be made to estimate the fragment size of the unknown sample based on which band it comes closest to since the samples were subjected to identical electrophoretic conditions. Alternatively, in multi-color fluorescent systems, an internal sizing standard labeled with a different colored dye can be run with each sample to calibrate the migration times of the DNA fragments of interest with a sample of known size (see Chapters 13 and 15).

The separation media that the DNA passes through, as well as the overall shape of the molecule and the electric field applied to the sample, influences the molecular movement (i.e., speed of separation for each component). The exact technique that one uses to separate the DNA molecules in a particular sample is dependent on the resolution required. The resolution capability of a separation system is dependent on a number of factors including the type of separation medium used and the voltage applied.

ADDITIONAL READING

Butler, J.M. (1995) *Sizing and Quantitation of Polymerase Chain Reaction Products by Capillary Electrophoresis for Use in DNA Typing*. PhD Dissertation, University of Virginia.

Heller, C. (ed.) (1997) *Analysis of Nucleic Acids by Capillary Electrophoresis*. Braunschweig: Vieweg.

Martin, R. (1996) *Gel Electrophoresis: Nucleic Acids*. Oxford: Bios Scientific Publishers.

CHAPTER 13

DNA DETECTION METHODS: FLUORESCENT DYES AND SILVER-STAINING

It is an old maxim of mine that when you have excluded the impossible, then whatever remains, however improbable, must be the truth.

(Sherlock Holmes, *The Adventure of the Beryl Coronet*)

VARIOUS METHODS FOR DETECTING DNA MOLECULES

Over the years a number of methods have been used for detecting DNA molecules following electrophoretic separation. Early techniques involved radioactive labels and autoradiography. These methods were sensitive and effective but time consuming. In addition, the use of radioisotopes was expensive due to the need for photographic films and supplies and the extensive requirements surrounding the handling and disposal of radioactive materials.

Since the late 1980s, methods such as silver staining and fluorescence techniques have gained in popularity for detecting short tandem repeat (STR) alleles due to their low cost in the case of silver staining and their capability of automating the detection in the case of fluorescence. Table 13.1 reviews the various methods and instruments that have been used for detecting STR alleles. This chapter will focus primarily on fluorescence detection because it now dominates the forensic DNA community. Almost all commercially available STR typing kits involve the use of fluorescently labeled polymerase chain reaction (PCR) primers. However, we will briefly cover silver staining at the end of the chapter to provide what we hope will be a useful historical perspective for those who are using fluorescence detection.

FLUORESCENCE DETECTION

Fluorescence-based detection assays are widely used in forensic laboratories due to their capabilities for multi-color analysis as well as rapid and easy-to-use formats. Fluorescence measurements involve exciting a dye molecule and then detecting the light that is emitted from the excited dye. In the application to DNA typing with STR markers, the fluorescent dye is attached to a PCR primer that is incorporated into the amplified target region of DNA. Amplified STR

Technique/Instrumentation	Comments	Reference
Fluorescence/ABI 373 or 377	Four different color dyes are used to label PCR products; peaks are measured during electrophoresis as they pass a laser that is scanning across the gel	Edwards *et al.* (1991), Frazier *et al.* (1996)
Fluorescence/ABI 310	Four ABI dyes are used to label PCR products; capillary electrophoresis version of ABI 377; most popular method in use today among forensic labs	Buel *et al.* (1998), Lazaruk *et al.* (1998)
Fluorescence/ABI 3100 and ABI 3100 Avant	Five-dye colors available for detection with 16 capillaries or four capillaries in parallel	Sgueglia *et al.* (2003), Butler *et al.* (2004)
Fluorescence/FMBIO scanner	Gel is scanned following electrophoresis with a 532 nm laser; typically used with three different dyes and PowerPlex STR kits	Schumm *et al.* (1995), Lins *et al.* (1998)
Fluorescence/ALF Sequencer	Automated detection similar to the ABI 377 but with only single color capability	Decorte and Cassiman (1996)
Fluorescence/LICOR	Near-IR dyes are used to label PCR products for automated detection similar to the ABI 377	Roy *et al.* (1996)
Fluorescence/scanner	SYBR Green stain (intercalating dye) of gel following electrophoresis; gel is scanned with 488 nm laser	Morin and Smith (1995)
Fluorescence/Beckman CE	PCR products are labeled with an intercalating dye during CE separation for single color detection	Butler *et al.* (1994)
Fluorescence/capillary arrays	Laser scans across multiple capillaries to detect fluorescently labeled PCR products	Wang *et al.* (1995), Mansfield *et al.* (1998)
Silver staining	Following electrophoresis, gel is soaked in silver nitrate solution; silver is reduced with formaldehyde to stain DNA bands	Budowle *et al.* (1995), Micka *et al.* (1996)
Direct blotting electrophoresis	Following run, gel bands are blotted unto a nylon membrane, fixed with UV light, and detected with digoxygenin	Berschick *et al.* (1993)
Autoradiography	P32-labeled dCTP incorporated into PCR products	Hammond *et al.* (1994)

IR = infrared; UV = ultraviolet.

Table 13.1

Detection methods and instruments used for analysis of STR alleles. A wide variety of fluorescence detection instrument platforms are listed.

alleles are visualized as bands on a gel or represented by peaks on an electropherogram. In this section, we will first discuss some of the basics surrounding fluorescence and then follow with a review of the methods used today for labeling DNA molecules, specifically the PCR products produced from STR markers.

BASICS OF FLUORESCENCE

As mentioned above, fluorescence measurements involve exciting a dye molecule and then detecting the light that is emitted from the excited dye. A molecule that is capable of fluorescence is called a *fluorophore*. Fluorophores come in a variety of shapes, sizes, and abilities. The ones that are primarily used in DNA labeling are dyes that fluoresce in the visible region of the spectrum, which consists of light emitted in the range of approximately 400–600 nm.

The fluorescence process is shown in Figure 13.1. In the first step, a photon ($h\nu_{ex}$) from a laser source excites a fluorophore electron from its ground energy state (S_0) to an excited transition state (S'_1). This electron then undergoes conformational changes and interacts with its environment resulting in the relaxed singlet excitation state (S_1). During the final step of the process, a photon ($h\nu_{em}$) is emitted at a lower energy when the excited electron falls back to its ground state. Because energy and wavelength are inversely related to one another, the emission photon has a higher wavelength than the excitation photon.

The difference between the apex of the absorption and emission spectra is called the *Stokes shift*. This shift permits the use of optical filters to separate excitation light from emission light. Fluorophores have characteristic light absorption and emission patterns that are based upon their chemical structure and the environmental conditions. With careful selection and optical filters, fluorophores may be chosen with emission spectra that are resolvable from one another.

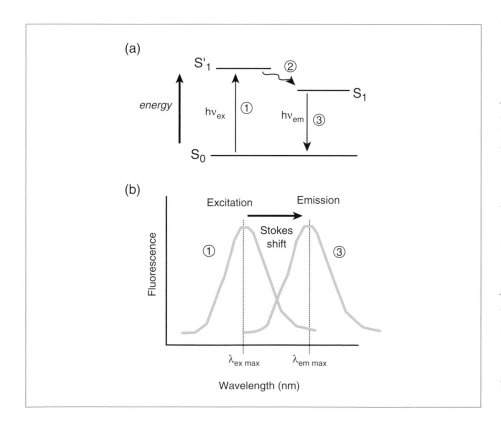

Figure 13.1

Illustration of the fluorescence process (a) and excitation/emission spectra (b). In the first step of the fluorescence process, a photon ($h\nu_{ex}$) from a laser source excites the fluorophore (dye molecule) from its ground energy state (S_0) to an excited transition state (S'_1). The fluorophore then undergoes conformational changes and interacts with its environment resulting in the relaxed singlet excitation state (S_1). During the final step of the process, a photon ($h\nu_{em}$) is emitted at a lower energy. Because energy and wavelength are inversely related to one another, the emission photon has a higher wavelength than the excitation photon.

As will be discussed later in the chapter, this capability permits the use of multiple fluorophores to measure several different DNA molecules simultaneously. The rate at which samples can be processed is much greater with multiple fluorophores than measurements involving a single fluorophore.

There are a number of factors that affect how well a fluorophore will emit light, or *fluoresce*. These factors include the following (Singer and Johnson 1997):

- *Molar extinction coefficient*: the ability of a dye to absorb light;
- *Quantum yield*: the efficiency with which the excited fluorophore converts absorbed light to emitted light;
- *Photo stability*: the ability of a dye to undergo repeated cycles of excitation and emission without being destroyed in the excited state, or experiencing 'photobleaching';
- *Dye environment*: factors that affect fluorescent yield include pH, temperature, solvent, and the presence of quenchers, such as hemoglobin.

The overall fluorescence efficiency of a dye molecule depends on a combination of these four factors. For example, fluorescein dyes have a lower molar extinction coefficient than rhodamine dyes yet the fluorescein dyes fluoresce well because they have higher quantum yields. Thus, fluorescein dyes do not absorb light as well but do a better job of converting the absorbed light into emitted light. This fact points out that the brightness of a fluorophore is proportional to the product of the molar extinction coefficient and the quantum yield.

SELECTING THE OPTIMAL FLUOROPHORE FOR AN APPLICATION

Optimal dye selection requires consideration of the spectral properties of fluorescent labels in relation to the characteristics of the instrument used for detection (Singer and Johnson 1997). The intensity of the light emitted by a fluorophore is directly dependent on the amount of light that the dye has absorbed. Thus, the excitation source is very important in the behavior of a fluorophore. Other important instrument parameters to be considered include optical filters used for signal discrimination and the sensitivity and spectral response of the detector.

Lasers are an effective excitation source because the light they emit is very intense and at primarily one wavelength. One of two different lasers is typically used to excite fluorescent dyes in the visible spectrum. The argon ion gas laser (Ar^+) produces light at 488 nm and 514.5 nm (see Chapter 14). This laser is by far the most popular for applications involving fluorescent DNA labeling because a number of dyes are available that closely match its excitation capabilities. The other laser that is used is the solid-state Nd:YAG laser that produces a beam of light at 532 nm (see Chapter 14).

A significant advantage of fluorescent labeling over other methods is the ability to record two or more fluorophores separately using optical filters and a fluorophore separation algorithm known as a *matrix*. With this multi-color capability, components of complex mixtures can be labeled individually and identified separately in the same sample. Fluorescent signals are differentiated by using filters that block out light from adjacent regions of the spectrum. Signal discrimination by software matrix deconvolution of the various dye colors will be discussed later in this chapter.

A fluorescence detector is a photosensitive device that measures the light intensity emitted from a fluorophore. Detection of low-intensity light may be accomplished with a photomultiplier tube (PMT) or a charge-coupled device (CCD). In both cases, the action of a photon striking the detector is converted to an electric signal. The strength of the resultant current is proportional to the intensity of the incident light. This light intensity is typically reported in arbitrary units, such as relative fluorescence units (RFUs).

METHODS FOR LABELING DNA

Fluorescent labeling of PCR products may be accomplished in one of three ways: (1) incorporating a fluorescent dye into the amplicon through a 5'-end labeled oligonucleotide primer; (2) incorporating fluorescently labeled deoxynucleotides (dNTPs) into the PCR product; and (3) using a fluorescent intercalating dye to bind to the DNA (Mansfield and Kronick 1993). These three methods are illustrated in Figure 13.2.

Each method of labeling DNA has advantages and disadvantages. Intercalating dyes may be used following PCR and are less expensive than the other two methods. However, they can only be used to analyze DNA fragments in a single color, which means that all of the molecules must be able to be separated in terms of size. On the other hand, dye labeled primers are popular because only a single strand of a PCR product is labeled, which simplifies data interpretation because the complementary DNA strand is not visible to the detector. Dye-labeled primers also enable multiple amplicons to be labeled simultaneously in an independent fashion.

The addition of a fluorescent dye to a DNA fragment impacts the DNA molecule's electrophoretic mobility. This is because the physical size and shape of the dye changes the overall size of the dye-DNA conjugate. The ionic charge, which is present on the dye, also alters the charge to size ratio of the nucleic acid conjugate. Fluorescent dyes that are covalently coupled to STR primers slightly alter the electrophoretic mobility of a STR allele PCR product moving through a gel or capillary. However, software corrections are used to mitigate this problem. In addition, genotyping of alleles is always performed relative to

Figure 13.2

Methods for fluorescently labeling DNA fragments. Double-stranded DNA molecules may be labeled with fluorescent intercalating dyes (a). The fluorescence of these dyes is enhanced upon insertion between the DNA bases. Alternatively a fluorescent dye may be attached to a nucleotide triphosphate and incorporated into the extended strands of a PCR product (b). The most common method of detecting STR alleles is the use of fluorescent dye labeled primers (c). These primers are incorporated into the PCR product to fluorescently label one of the strands.

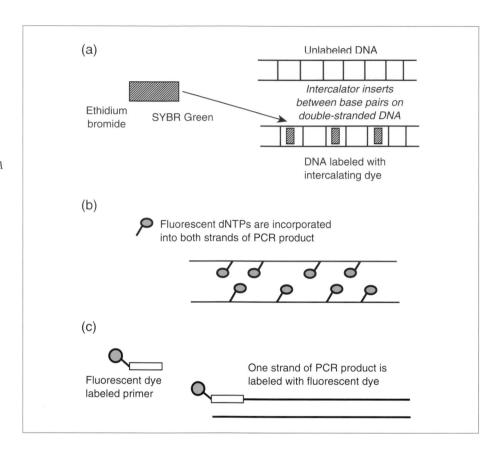

allelic ladders that are labeled with the *same* fluorescent dye so that differences in dye mobilities do not impact allele calls.

SUMMARY OF FLUORESCENCE DETECTION

A laser strikes a fluorophore (dye) that is attached to the end of a DNA fragment. The fluorophore absorbs laser energy and then emits light at a lower energy (higher wavelength). Filters are used to collect only emitted light at a particular wavelength or range of wavelengths. Photomultiplier tubes or charge-couple devices are used to collect and amplify the signal from the fluorophore and convert it to an electronic signal. These signals are measured in relative fluorescence units and make up the peaks seen in capillary electropherograms or bands on a gel image.

Advantages of fluorescence detection methods include higher sensitivity and a broader dynamic range than comparable colorimetric detection methods (e.g., silver-staining) and the capacity for simultaneous multi-parameter analysis of complex samples such as multiplex PCR products with different fluorescent labels.

FLUORESCENT DYES USED FOR STR ALLELE LABELING

Table 13.2 lists a number of fluorescent dyes that are commonly used to label PCR products for genotyping applications. The chemical names for the dyes are listed as well along with their excitation and emission wavelengths. AmpF/STR® kits from Applied Biosystems use PCR primers that are labeled with the NHS-ester dyes 5-FAM, JOE, or NED (Applied Biosystems 1998). GenePrint® PowerPlex® 1.1 and 2.1 kits from Promega Corporation use PCR primers labeled with fluorescein and tetramethyl rhodamine (TMR). Newer STR kits, such as Identifiler, include an expanded set of four dyes (6FAM™, VIC®, NED®, and PET®) for labeling PCR products in multiplex amplification reactions.

FAM and fluorescein fluoresces in the blue region of the visible spectrum, JOE in the green region, and NED and TMR in the yellow region. AmpF/STR® kits utilize a fourth dye named ROX that fluoresces in the red region to label an internal standard for DNA sizing purposes. PowerPlex® STR kits use the red

Dye	Chemical Name	Excitation Maximum (nm)	Emission Maximum (nm)
5-FAM	5-carboxy fluorescein	493	522
JOE	6-carboxy-2',7'-dimeoxy-4',5'-dichlorofluorescein	528	554
VIC	*Proprietary to Applied Biosystems*	538	554
NED	*Proprietary to Applied Biosystems*	546	575
PET	*Proprietary to Applied Biosystems*	558	595
LIZ	*Proprietary to Applied Biosystems*	638	655
ROX (CXR)	6-carboxy-X-rhodamine	587	607
Fluorescein (FL)	Fluorescein	490	520
TMR (TAMRA)	N,N,N',N'-tetramethyl-6-carboxyrhodamine	560	583
TET	4,7,2',7'-tetrachloro-6-carboxyfluorescein	522	538
HEX	4,7,2',4',5',7'-hexachloro-6-carboxyfluorescein	535	553
Rhodamine Red	Rhodamine Red™-X (Molecular Probes)	580	590
Texas Red	Texas Red®-X (Molecular Probes)	595	615
SYBR Green (intercalator)	*Proprietary to Molecular Probes*	497	520

Table 13.2

Characteristics of commonly used fluorescent dyes in STR kits and other genotyping applications.

dye carboxy-X-rhodamine (CXR) for labeling their internal size standard. ROX and CXR are essentially the same dyes (Singer and Johnson 1997).

The 5-FAM, JOE, and NED dyes are fluorescein derivatives that have spectrally resolvable fluorescent spectra. The structures of four commonly used fluorescent dyes from Applied Biosystems are shown in Figure 13.3. Because the structure of NED is proprietary to Applied Biosystems, TAMRA, or tetramethyl rhodamine, has been included in its place for a yellow dye. An additional dye named LIZ® is used to label the internal size standard run with 5-dye detection systems from Applied Biosystems. LIZ emits at ~650 nm and yet still has excellent sensitivity because it is an energy transfer dye (see D.N.A. Box 13.1). With a 5-dye detection system, the four dyes 6-FAM, VIC, NED, and PET are used for labeling PCR products and the fifth dye (LIZ) is used to label the internal size standard.

The fluorescent emission spectra of the four ABI dyes, 5-FAM, JOE, NED, and ROX, are shown in Figure 13.4. Each of the fluorescent dyes emits its maximum fluorescence at a different wavelength. This fact is used to design filters that

D.N.A. Box 13.1
Energy transfer dyes

The brightness of fluorescence dyes depends on a number of factors including how close a dye's absorbance maximum is to the excitation wavelength of the laser used to excite the dye. If two dyes – a donor and an acceptor – are in close physical proximity to one another and possess overlapping emission and absorbance spectra, then energy is transferred between the two dyes and can improve the amount of energy absorbed by the acceptor dye, which results in an increase in the fluorescence output of the acceptor dye.

Richard Mathies' group at the University of California-Berkeley has pioneered the development of a number of energy transfer (ET) dyes for the past decade to enable more sensitive detection of DNA with brighter dyes. This work influenced the development of the BigDye chemistries used by Applied Biosystems for DNA sequencing and the dye LIZ used in STR typing with 5-dye chemistry kits such as Identifiler™, SEfiler™, and Yfiler™. Note that LIZ has an emission wavelength of approximately 650 nm and yet is still sensitive (i.e., well-excited) with the 488 nm argon ion laser. This type of sensitivity with such a large spread in the excitation and emission wavelengths would not be possible without donor and acceptor energy transfer dyes.

Most ET dye-labeled primers carry a fluorescein derivative at the 5'-end as a common donor since it has an absorbance maximum that is near the 488 nm argon ion laser used in the ABI 310 and other common DNA detection platforms. Other fluorescein and rhodamine derivatives often serve as acceptor dyes and are attached to the primer within a few nucleotides of the donor dye. Efficiency of ET primers can generate 2- to 6-fold greater fluorescent signal than that of the corresponding primers or fragments labeled with single dyes.

Sources:

Ju, J. *et al.* (1995) Design and synthesis of fluorescence energy transfer dye-labeled primers and their application for DNA sequencing and analysis. *Analytical Biochemistry,* 231, 131–140.

Ju, J.Y., Glazer, A.N., Mathies, R.A. (1996) Energy transfer primers: a new fluorescence labeling paradigm for DNA sequencing and analysis. *Nature Medicine,* 2, 246–249.

Rosenblum, B.B. *et al.* (1997) New dye-labeled terminators for improved DNA sequencing patterns. *Nucleic Acids Research,* 25, 4500–4504.

capture the signal from each dye. The filters used to separate the various colors are shown as boxes centered on each of the four dye spectra. Note that there is considerable overlapping in color between several of the dyes in the filter regions. Blue and green have an especially high degree of overlap. This overlap influences the degree of 'pull-up' that can occur between two dye detection channels (see Chapter 15).

The dye sets for PowerPlex® kits that are detected on the Hitachi FMBIO II fluorescent scanner (see Chapter 14) are shown in Figure 13.5. Notice that the fluorescence emission spectra for these three dyes have less spectral overlap because they are further apart.

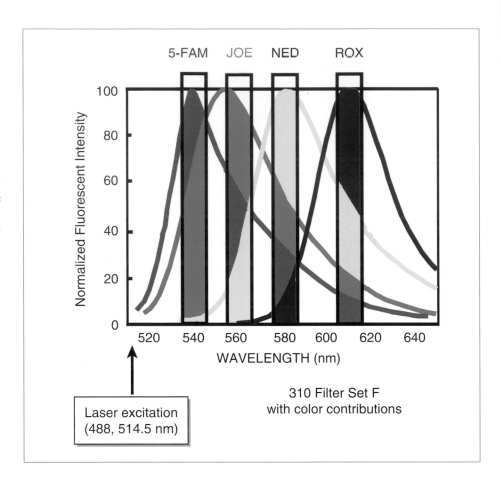

Laser excitation
(488, 514.5 nm)

Figure 13.4

Fluorescent emission spectra of ABI dyes used with AmpFlSTR kits. ABI 310 Filter Set F is represented by the four boxes centered on each of the four dye spectra. Each dye filter contains color contributions from adjacent overlapping dyes that must be removed by a matrix deconvolution. The dyes are excited by an argon ion laser, which emits light at 488 and 514.5 nm.

Each dye set used must be matched to the instrument optics involved in detection as well as the excitation source. For example the ABI Prism 310 uses an Argon ion (Ar^+) laser with excitation wavelengths at 488 nm and 514 nm while the FMBIO II uses a solid-state Nd:YAG laser with an excitation wavelength of 532 nm. PowerPlex® 1.1 fluorescein-labeled primers, designed for FMBIO detection, are present in higher concentration than the same primers in PowerPlex® 1.2, designed for ABI 310 detection, because of the different laser excitation wavelengths. Because the FMBIO laser is not as well suited for fluorescein dye excitation, more primer must be put in the PCR reaction to generate balanced product amounts compared to TMR-labeled amplicons.

When labeling DNA fragments with various dyes it is important to use appropriate concentrations of the dyes in order to obtain balanced signals between loci. For example, because NED has an excitation maximum that is further from the Ar^+ 488/514 nm lines more dye is required in order to obtain an equivalent signal to FAM.

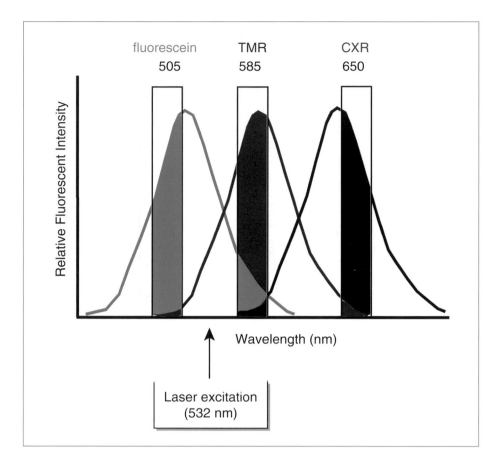

Figure 13.5

Schematic of three-color detection using the Hitachi FMBIO Scanner System. These dyes and spectral filters (shown with boxes) are used for detection of STR alleles amplified with the Promega PowerPlex systems.

ISSUES WITH FLUORESCENCE MEASUREMENTS

Multi-component analysis is performed with a mathematical matrix that subtracts out the contribution of other dyes in each measured fluorescent dye. 'Color deconvolution' is another phrase to describe what this matrix does. When raw data is collected from a fluorescence-based detection platform, spectral overlap of different dyes must be accounted for in order to gain the capability to see the results from each dye individually.

One method of performing multi-component analysis is to examine a standard set of DNA fragments labeled with each individual dye, known as matrix standard samples (Applied Biosystems 1998). Computer software then analyzes the data from each of the dyes and creates a matrix file to reflect the color overlap between the various fluorescent dyes. This matrix file table contains a table of numbers with four rows and four columns if there are four dyes that are being deconvoluted. Alternatively, a 5×5 matrix is generated with 5-dye chemistries on the ABI Prism 310 Genetic Analyzer.

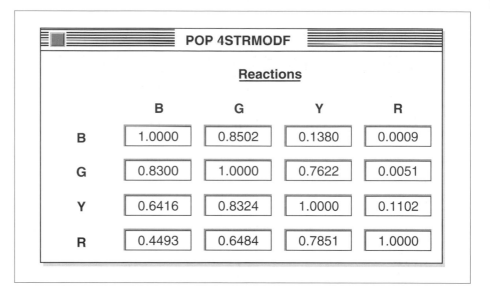

Figure 13.6

Example of a matrix file table from an ABI Prism 310 instrument. These values are used by the GeneScan® Analysis Software to separate the various dye colors from one another. The letters B, G, Y, and R represent the dye colors Blue, Green, Yellow, and Red, respectively. All matrix files should have values of 1.0 on the diagonal from top left to bottom right. The other values in the table should all be less than 1.0. These values represent the amount of spectral overlap observed for each dye in each virtual detection filter. Matrix file values differ between instruments and even run conditions on a single instrument. Thus, a unique matrix file must be made for an instrument and a particular set of run conditions.

An example matrix from an ABI Prism 310 instrument may be seen in Figure 13.6. The values in this table represent the amount of spectral overlap observed for each dye in each color: blue (B), green (G), yellow (Y), and red (R). In the case of AmpFℓSTR kits, the four fluorescent dyes are 5-FAM, JOE, NED, and ROX (Figure 13.4). Note that in the matrix file table (Figure 13.6), there are values of 1.0 on the diagonal from top left to bottom right and that all of the other values in the table are less than 1.0. These values represent the amount of spectral overlap observed for each dye. For example, the values in the B column are 1.000 (B), 0.8300 (G), 0.6416 (Y), and 0.4493 (R). Thus, in this example the most significant overlap is green into blue because 83% (0.8300) of the blue signal is made up of green on a normalized scale. Note that the emission spectra shown in Figure 13.4 also possess the most overlap between the blue and green dyes.

Matrix files differ between instruments and even different run conditions on the same instrument because fluorescence of a dye is affected by the dye's environment. If the environmental conditions, such as temperature, pH, etc., change even ever so slightly, then the fluorescence behavior of the dyes will be altered. A matrix file should therefore be generated frequently to ensure good dye color separation and certainly anytime instrument conditions are altered, such as running samples in a different buffer system. As long as the electrophoresis conditions are constant from run to run, then the emission spectra of the dyes will be reproducible and spectral overlap can be accurately deciphered.

If the matrix color deconvolution does not work properly than the baseline can be uneven or a phenomenon known as 'pull-up' can occur. *Pull-up* is the result of a color bleeding from one spectral channel into another, usually

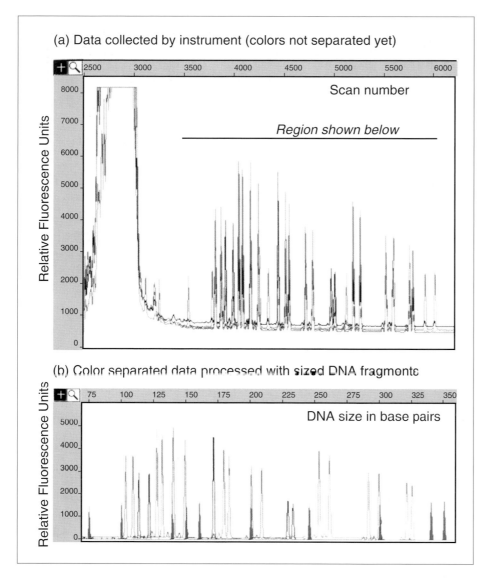

(a) Data collected by instrument (colors not separated yet)

Scan number

Region shown below

Relative Fluorescence Units

(b) Color separated data processed with sized DNA fragments

DNA size in base pairs

Relative Fluorescence Units

Figure 13.7
STR Data from ABI Prism 310 Genetic Analyzer. This sample was amplified with the AmpFlSTR SGM Plus kit. Raw data prior to color separation (a) compared with GeneScan 3.1 color separated allele peaks (b). The red-labeled peaks are from the internal sizing standard GS500-ROX.

because of off-scale peaks. The most common occurrence of pull-up involves small green peaks showing up under blue peaks that are off-scale. This occurs because of the significant overlap of the blue and green dyes seen in Figure 13.4. Samples can be diluted and analyzed again to reduce or eliminate the offending pull-up peak(s).

Raw data from a fluorescently labeled DNA sample is compared to the color separated processed data in Figure 13.7. DNA fragments labeled with the yellow dye NED are shown in black. Multi-component analysis is performed automatically with the GeneScan® Analysis Software using a mathematical matrix calculation.

FLUORESCENCE DETECTION PLATFORMS

As seen in Table 13.1, a number of fluorescence detection platforms exist and have been used for STR allele determination. The most popular detection platforms today for STR analysis in the United States are the ABI Prism 310 Genetic Analyzer, the ABI 3100 multi-capillary system, and the FMBIO II gel scanner. These three instruments will be reviewed more extensively in Chapter 14. The ABI 310 and FMBIO II instruments have fundamental differences in their approach to detecting fluorescently labeled PCR products. In addition, specific STR kits have been designed for using each approach.

With the ABI 310 approach, detection is performed during electrophoresis (Figure 13.8). Other instruments listed in Table 13.1 that use a similar detection format as the ABI 310 include the ABI 377, ABI 3100, ABI 3700, the ALF DNA sequencer, and the LICOR systems. Applied Biosystems has prepared the Profiler Plus™ and COfiler™ STR kits to work on any of the ABI detection platforms (i.e., ABI Prism 310, 377, 3100, and 3700). These two kits cover the 13 core CODIS STR loci with three markers in common between the sets for concordance purposes (Figure 13.9). The PowerPlex® 16 and Identifiler™ STR kits now enable amplification of all 13 CODIS loci plus two additional polymorphic STRs in a single reaction (see Figures 5.4 and 5.5).

Figure 13.8

Schematic illustration of the separation and detection of STR alleles with an ABI Prism 310 Genetic Analyzer.

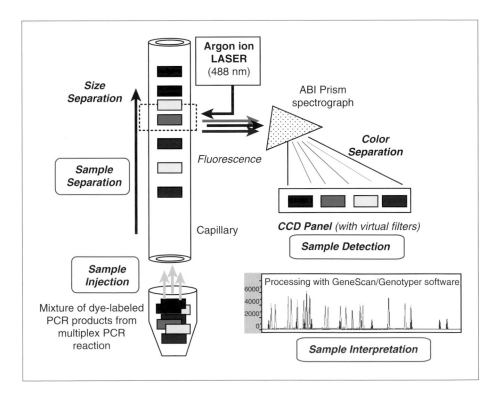

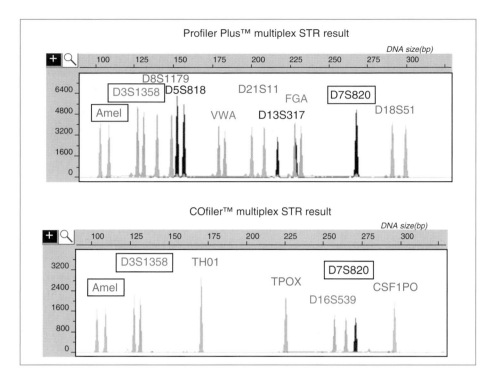

Figure 13.9

AmpFlSTR® Profiler Plus™ and COfiler™ STR Data Collected on an ABI 310 Capillary Electrophoresis System. The STR loci that are surrounded by a box are common to both multiplex mixes and are therefore useful as a quality assurance measure to demonstrate sample concordance.

With the FMBIO II or other gel scanning systems, detection is performed following electrophoresis (Figure 13.10). Thus, many gels can be run in separate gel rigs and detected via rapid scanning on a single FMBIO fluorescence imaging system. The Promega Corporation has created several multiplex STR kits to work with the FMBIO II detection platform: PowerPlex® 1.1, PowerPlex® 2.1, and PowerPlex® 16 BIO (Figure 13.11). These kits enable amplification of the 13 core CODIS STR loci with either the combination of PowerPlex® 1.1 and 2.1 or PowerPlex® 16 BIO, which amplifies all 13 CODIS loci plus Penta D and Penta E in a single multiplex reaction (Greenspoon *et al.* 2004).

These two separation and detection approaches have differing abilities to separate STR alleles of various size ranges. Every DNA fragment travels the same distance when the detector is at a fixed point relative to the injection of the sample, as with the ABI 310. On the other hand, when separation and detection are separate steps, as with the FMBIO gel scanner, DNA fragments of different sizes travel different distances through the gel. Smaller molecular weight PCR products (e.g., VWA) travel further through the gel and are thus better resolved from one another compared to the higher molecular weight species (e.g., FGA) that only move a short distance through the gel before the electrophoresis is stopped and the gel is scanned.

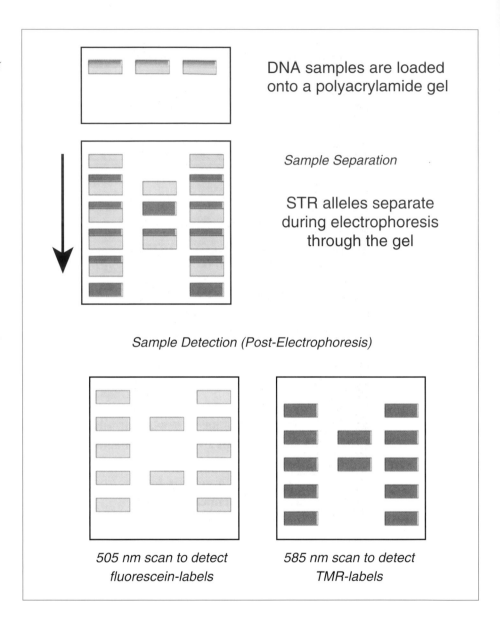

Figure 13.10
Schematic of gel separation and FMBIO II detection of STR alleles.

DNA samples are loaded onto a polyacrylamide gel

Sample Separation

STR alleles separate during electrophoresis through the gel

Sample Detection (Post-Electrophoresis)

505 nm scan to detect fluorescein-labels

585 nm scan to detect TMR-labels

SILVER STAINING

Silver staining of polyacrylamide gels has been useful for detecting small amounts of proteins and visualizing nucleic acids. Although not as commonly used today, silver staining procedures were used for the first commercially available STR kits from the Promega Corporation. Promega still supports silver-stain gel users although most of their customer base now uses fluorescent STR systems. Silver-stain detection methods are still quite effective for laboratories that want to perform DNA typing for a much smaller start-up cost. No expensive

Figure 13.11
PowerPlex® 16 BIO data
collected on a Hitachi
FMBIO III plus
Fluorescence Imaging
System. Color-separated
data for this same gel is
contained in Figure 14.9.
Figure courtesy of
Margaret Kline, NIST.

instruments are needed, simply a gel box for electrophoresis and some silver nitrate and other developing chemicals.

The procedure for silver staining is performed by transferring the gel between pans filled with various solutions that expose the DNA bands to a series of chemicals for staining purposes (Bassam *et al.* 1991). First, the gel is submerged in a pan of 0.2% silver nitrate solution. The silver binds to the DNA and is reduced with formaldehyde to form a deposit of metallic silver on the DNA molecules in the gel. A photograph is then taken of the gel to capture images of the silver-stained DNA strands and to maintain a permanent record of the gel. Alternatively, the gels themselves may be sealed and preserved.

ADVANTAGES AND DISADVANTAGES OF SILVER STAINING

Silver staining is less hazardous than radioactive detection methods although not as convenient as fluorescence methods. Most reagents for silver staining are harmless and thus require no special precautions for handling. The primary advantage of silver staining is that the technique is inexpensive. The developing chemicals are readily available at low cost. The PCR products do not need any special labels, such as fluorescent dyes. The staining may be completed within half an hour and with a minimal number of steps. Sensitivity is approximately 100 times higher than that obtained with ethidium bromide staining (Merril *et al.* 1998). However, a major disadvantage to data interpretation is that both DNA strands may be detected in a denaturing environment leading to two bands for each allele. In addition, only one 'color' exists, which makes PCR product size differences the only method for multiplexing STR markers.

REFERENCES AND ADDITIONAL READING

Applied Biosystems (1998) *AmpFlSTR Profiler Plus™ PCR Amplification Kit User's Manual.* Foster City, California: Applied Biosystems.

Bassam, B.J., Caetano-Anolles, G. and Gresshoff, P.M. (1991) *Analytical Biochemistry*, 196, 80–83.

Berschick, P., Henke, L. and Henke, J. (1993) *Proceedings of the Fourth International Symposium on Human Identification*, pp. 201–204. Madison, Wisconsin: Promega Corporation.

Budowle, B., Baechtel, F.S., Comey, C.T., Giusti, A.M. and Klevan, L. (1995) *Electrophoresis*, 16, 1559–1567.

Buel, E., Schwartz, M. and LaFountain, M.J. (1998) *Journal of Forensic Sciences*, 43, 164–170.

Butler, J.M., McCord, B.R., Jung, J.M. and Allen, R.O. (1994) *BioTechniques*, 17, 1062–1070.

Butler, J.M., Buel, E., Crivellente, F. and McCord, B.R. (2004) *Electrophoresis*, 25, 1397–1412.

Decorte, R. and Cassiman, J.-J. (1996) *Electrophoresis*, 17, 1542–1549.

Edwards, A., Civitello, A., Hammond, H.A. and Caskey, C.T. (1991) *American Journal of Human Genetics*, 49, 746–756.

Frazier, R.R.E., Millican, E.S., Watson, S.K., Oldroyd, N.J., Sparkes, R.L., Taylor, K.M., Panchal, S., Bark, L., Kimpton, C.P. and Gill, P.D. (1996) *Electrophoresis*, 17, 1550–1552.

Greenspoon, S.A., Ban, J.D., Pablo, L., Crouse, C.A., Kist, F.G., Tomsey, C.S., Glessner, A.L., Mihalacki, L.R., Long, T.M., Heidebrecht, B.J., Braunstein, C.A., Freeman, D.A., Soberalski, C., Nathan, B., Amin, A.S., Douglas, E.K. and Schumm, J.W. (2004) *Journal of Forensic Sciences*, 49, 71–80.

Hammond, H.A., Jin, L., Zhong, Y., Caskey, C.T. and Chakraborty, R. (1994) *American Journal of Human Genetics*, 55, 175–189.

Huang, N.E., Schumm, J.W. and Budowle, B. (1995) *Forensic Science International*, 71, 131–136.

Lazaruk, K., Walsh, P.S., Oaks, F., Gilbert, D., Rosenblum, B.B., Menchen, S., Scheibler, D., Wenz, H.M., Holt, C. and Wallin, J. (1998) *Electrophoresis*, 19, 86–93.

Lee, S.B., Buoncristiani, M., Schumm, J.W. and Wingeleth, D. (1995) *Proceedings of the Fifth International Symposium on Human Identification*, pp. 104–111. Madison, Wisconsin: Promega Corporation.

Lins, A.M., Micka, K.A., Sprecher, C.J., Taylor, J.A., Bacher, J.W., Rabbach, D., Bever, R.A., Creacy, S. and Schumm, J.W. (1998) *Journal of Forensic Sciences*, 43, 1178–1190.

Mansfield, E.S. and Kronick, M.N. (1993) *BioTechniques*, 15, 274–279.

Mansfield, E.S., Robertson, J.M., Vainer, M., Isenberg, A.R., Frazier, R.R., Ferguson, K., Chow, S., Harris, D.W., Barker, D.L., Gill, P.D., Budowle, B. and McCord, B.R. (1998) *Electrophoresis*, 19, 101–107.

Merril, C.R., Washart, K.M. and Allen, R.C. (1998) In Tietz, D. (ed) *Nucleic Acid Electrophoresis*. New York: Springer.

Micka, K.A., Sprecher, C.J., Lins, A.M., Comey, C.T., Koons, B.W., Crouse, C., Endean, D., Pirelli, K., Lee, S.B., Duda, N., Ma, M. and Schumm, J.W. (1996) *Journal of Forensic Sciences*, 41, 582–590.

Morin, P.A. and Smith, D.G. (1995) *BioTechniques*, 19, 223–227.

Roy, R., Steffens, D.L., Gartside, B., Jang, G.Y. and Brumbaugh, J.A. (1996) *Journal of Forensic Sciences*, 41, 418–424.

Schumm, J.W., Lins, A.M., Sprecher, C.J. and Micka, K.A. (1995) *Proceedings from the Sixth International Symposium on Human Identification*, pp. 10–19. Madison, Wisconsin: Promega Corporation.

Sgueglia, J. B., Geiger, S. and Davis, J. (2003) *Analytical and Bioanalytical Chemistry*, 376, 1247–1254.

Singer, V.L. and Johnson, I.D. (1997) *Proceedings of the Eighth International Symposium on Human Identification*, pp.70–77. Madison, Wisconsin: Promega Corporation.

Wang, Y., Ju, J., Carpenter, B.A., Atherton, J.M., Sensabaugh, G.F. and Mathies, R.A. (1995) *Analytical Chemistry*, 67, 1197–1203.

Worley, J.M., Ma, M., Lee, S.B., Lins, A.M., Schumm, J.W. and Mansfield, E.S. (1994) *Proceedings from the Fifth International Symposium on Human Identification*, pp. 109–117. Madison, Wisconsin: Promega Corporation.

CHAPTER 14

INSTRUMENTATION FOR STR TYPING: ABI 310, ABI 3100, FMBIO SYSTEMS

The world is full of obvious things, which nobody by any chance ever observes.

(Sherlock Holmes, *The Hound of the Baskervilles*)

This chapter will examine several electrophoresis instrumentation platforms commonly utilized for analysis of STR loci: the single capillary ABI Prism 310 Genetic Analyzer, the 16-capillary ABI Prism 3100, and the FMBIO Fluorescence Imaging System.

THE ABI PRISM 310 GENETIC ANALYZER

Since its introduction in 1995 by Applied Biosystems, the ABI Prism 310 Genetic Analyzer has been an increasingly popular method for short tandem repeat (STR) typing in forensic DNA laboratories. A vast majority of forensic DNA laboratories within the United States use the ABI 310 for performing STR genotyping (Steadman 2000). In addition, the FBI Laboratory in Washington D.C. performs its STR typing with the ABI 310.

The ABI 310 is a single capillary instrument with multiple color fluorescence detection that provides the capability of unattended operation. An operator simply loads a batch of samples in the 'autosampler,' places a capillary and a syringe full of polymer solution in the instrument, and starts the 'run.' The data and genotype information are serially processed at the rate of approximately one sample every 30 minutes of operation. A major advantage of the technique for forensic laboratories is that the DNA sample is not fully consumed and may be retested if need be.

Many forensic scientists are using the ABI 310 without background knowledge of capillary electrophoresis (CE). This chapter reviews the theory and practice of capillary electrophoresis with a particular focus on the capabilities of the ABI Prism 310 Genetic Analyzer for genotyping STR markers (Butler *et al.* 2004). A few troubleshooting tips for the ABI 310 are also included (see McCord 2003).

EARLY WORK WITH CE AND STR TYPING

The first CE separations of STR alleles were performed in late 1992 using non-denaturing conditions with the polymerase chain reaction (PCR) products in a double-stranded form (McCord *et al.* 1993a, 1993b). Fluorescent intercalating dyes were used to visualize the DNA and to promote the resolution of closely spaced alleles. Internal standards were used to bracket the alleles in order to perform accurate STR genotyping. An allelic ladder was first run with the internal standards to calibrate the DNA migration times followed by analysis of the samples with the same internal standards (Butler *et al.* 1994). This internal sizing standard method involving a single fluorescent wavelength detector had to be used because multiple color fluorescence CE instruments were not yet available. Since the commercialization of the ABI 310, internal standards labeled with a different color compared to the STR alleles can be used to perform the DNA size determinations and subsequent correlation to obtain the STR genotype.

Early on in the development of CE for DNA separations, one of the major concerns included sample preparation. PCR-amplified samples had to be dialyzed to remove salts that interfered with the injection of DNA fragments onto the CE column in order to observe the DNA with an ultraviolet (UV) detector. With the higher sensitivity of laser-induced fluorescence, sample preparation is no longer a major concern but does still play a role. Samples may be diluted in water or formamide and easily detected.

CAPILLARY ELECTROPHORESIS OF DNA

As discussed in Chapter 12, capillary electrophoresis involves the use of a narrow capillary filled with a polymer solution instead of a gel to perform the DNA size separation. The higher surface area-to-volume ratio in a capillary permits more efficient heat dissipation generated by the electrophoresis process and thus enables a higher separation voltage to be applied. Typical DNA separation times using CE are in the range of 5–30 minutes, compared to several hours for gel-based systems, because a higher voltage may be used. Most ABI 310 methods involve a separation voltage of 15 000 volts with a capillary length of 47 cm or 319 V/cm.

Polymer solutions have greatly aided DNA separations in capillaries. Prior to the injection of each new sample, a fresh portion of polymer solution is pumped into the capillary. This operation is analogous to pouring a new 'gel' automatically before each sample is loaded on the gel. The type and concentration of polymer solution used determines the resolution that may be obtained much in the same way that the percentage of cross-linking in polyacrylamide gels reflects the resolution capabilities of the electrophoretic system.

While CE is rapid on a per sample basis, it is a sequential technique where only one sample is analyzed at a time and is not as useful when trying to process large numbers of samples in parallel. Therefore, throughput is on the same time scale as, or even slower than, conventional gel electrophoresis methods.

As will be discussed in a future section, capillary array systems with 16 or 96 capillaries in parallel have been developed to aid in high-throughput operations.

DNA samples are loaded onto the capillary by applying a fixed voltage for a defined period of time or by applying pressure to the sample and forcing a plug of sample to enter the inlet end of the capillary. In the case of the ABI 310, only the voltage application or 'electrokinetic' injection mode is available for injecting DNA samples.

COMPONENTS OF THE ABI 310

The basic components of the ABI Prism 310 Genetic Analyzer are illustrated in Figure 14.1. A capillary is located between the pump block and the inlet electrode. The capillary is filled with polymer solution through the pump block. A heated plate is used to heat the capillary to a specified temperature. Samples are placed in an autosampler tray that moves up and down to insert the sample onto the capillary and electrode for the injection process.

Prior to running any samples, the ABI 310 CE system must be readied for analysis. This preparation generally involves four steps:

1. Putting a capillary in the instrument and aligning the detection window;
2. Loading the polymer solution into a syringe and priming the system to remove any bubbles;

Figure 14.1

Schematic of ABI Prism 310 Genetic Analyzer. The capillary stretches between the pump block and the injection electrode. The mechanical stepper motor pushes polymer solution in the syringe into the pump block where it enters and then fills the capillary. Samples placed in the autosampler tray are sequentially injected onto the capillary. Electrophoretic separation occurs after each end of the capillary is placed in the inlet and outlet buffer and a voltage is applied across the capillary. A laser (not shown) is used to detect fluorescently labeled DNA fragments as they pass by the capillary detection window.

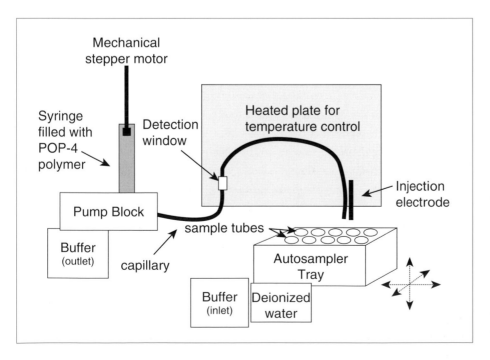

3. Filling the buffer reservoirs so that current can flow between the electrodes and separations occur across the capillary; and

4. Placing deionized water in tubes that will be used for keeping the end of the capillary wet and cleaning off buffer salts before and after the injection process.

Each of these components plays an important role in the CE process and will be described in greater detail below.

THE CAPILLARY

At the core of a CE system is the capillary, which is a narrow glass tube that has an inner diameter of 50, 75, or 100 μm. Capillaries typically used for STR separations on the ABI 310 have an inner diameter of 50 μm. The outside of the capillary is usually coated with a plastic polyimide jacket to allow users to handle the capillary without it breaking. However, this coating is opaque and thus inhibits optical detection of anything passing through the interior of the capillary. A small capillary window is therefore needed to observe the separation products and is usually generated by burning away several millimeters of the polyimide coating. Capillary with windows already created in them may be purchased or a user can buy a long roll of capillary and burn their own detection windows. Once the fused silica (glass) tube is exposed the window region must be handled carefully as it is easily breakable.

Two factors that are impacted by capillary length include peak resolution and separation time. Generally, the longer the capillary the better the resolution, but the greater the separation time. With CE, two capillary length measurements are important. The length-to-detector (L_{det}) is a measure of the distance from the capillary inlet where the DNA sample is injected to the detection window where the laser shines on the capillary and the fluorescent dyes are excited and detected. The total capillary length (L_{tot}) is the complete length of the capillary or in other words the distance from inlet to outlet buffer solutions when electrophoresis is occurring.

Varying the L_{tot} distance impacts the electric field strength that is applied to the CE system while shortening the L_{det} improves the separation speed. For optimal performance (i.e., highest resolution) with CE, it is best to have L_{det} as close to L_{tot} as possible. However, instrument space constraints keep L_{det} shorter than L_{tot} usually by 7 cm or more, primarily because the electrophoresis electrodes cannot be too close to the detection electronics. The detection window on the ABI 310 is fixed 11 cm from the outlet buffer. There is a minimum capillary length (L_{tot}) of approximately 41 cm (i.e., 30 cm L_{det}) with the ABI 310 Genetic Analyzer.

Applied Biosystems has made it easy for the user by supplying two lengths of capillary: 47 cm (36 cm to detector) and 61 cm (50 cm to detector). The shorter capillary is typically used for GeneScan applications, such as STR typing, where a faster separation speed is more important than resolution. The longer capillary is more effective for DNA sequencing where a longer size range with single-base resolution is more desirable than rapid run times.

Capillaries used with the ABI 310 may or may not possess a covalently attached internal wall coating. Many capillaries used for CE are 'coated' by chemical derivatization to prevent a process known as electro-osmotic flow (EOF). Above pH 6, silica is negatively charged and thus the inside wall of a fused silica capillary will be covered by negative charges. EOF results from positive charges from the solution inside the capillary that form along the negatively charged capillary wall in what is known as the 'double layer'. Upon the application of an electric field, the mobile cations in the double layer migrate toward the cathode (inlet) and pull solution molecules in the same direction.

EOF is in the opposite direction as electrophoretic migration of the DNA molecules and thus slows their progression through the capillary. EOF is highly dependent on environmental parameters, such as pH, temperature, voltage, and buffer viscosity. Thus, in order to obtain reproducible separations of DNA, EOF is removed by coating the inner wall of the capillary. There are two primary methods for coating the inner wall of a capillary: chemical derivatization of the charged silanol groups or dynamic coating with a viscous polymer solution.

The capillaries typically used in the ABI 310 CE system are 'uncoated' meaning that the interior surface of the capillary does not have any chemical modifications to cover the charged silanol groups. However, the POP-4 polymer solution used for DNA separations dynamically coats the inside capillary wall and prevents EOF.

POLYMER SOLUTION

Applied Biosystems currently sells two polymer formulations for use with the ABI Prism 310 Genetic Analyzer. POP-4™ and POP-6™, which stands for Performance Optimized Polymer, are 4% and 6% concentrations of linear, uncross-linked dimethyl polyacrylamide, respectively. A high concentration of urea is also present in the polymer solution to help create an environment in the capillary that will keep the DNA molecules denatured.

POP-4 is a commercially available preparation of a flowable polymer [poly (N,N-dimethylacrylamide); DMA]. POP-4 contains 4% DMA homopolymer, 8 M urea, 5% 2-pyrrolidinone, and 100 mM N-Tris-(hydroxymethyl) methyl-3-aminopropane-sulfonic acid (TAPS) adjusted to pH 8.0 with NaOH (Rosenblum et al. 1997).

POP-4 is most commonly used for STR typing while POP-6, which is more viscous and yields improved resolution at the expense of longer run times, is typically used for DNA sequencing applications. To prepare the CE system for use, approximately 600 or 700 μL are drawn up into a 1 mL syringe, which is placed in the ABI 310 instrument. Between 100 and 300 μL are often needed to prime the CE system and drive out any air bubbles. Several microliters are used between each DNA separation to refill the capillary for the next run. A full syringe can therefore last for 100 or more unattended injections prior to needing a refill.

ELECTROPHORESIS BUFFER

The electrophoresis buffer supplies the ions for conducting current across the capillary. If it is not properly replenished, the current can fluctuate affecting the DNA separation. Applied Biosystems supplies a 10X Genetic Analysis buffer with EDTA that is typically used in STR sample separations. The user simply dilutes the 10X buffer with nine times the volume of water to make a 1X solution. The 1X concentration of the Genetic Analysis buffer is 100 mM TAPs, 1 mM EDTA, pH 8.0 (Rosenblum *et al.* 1997). The electrophoresis buffer should be replaced after every set of about 100 sample injections.

OPERATION OF THE ABI 310 FOR GENOTYPING STR SAMPLES

The sample processing steps using the ABI 310 are illustrated in Figure 14.2. Samples are prepared and loaded into an autosampler tray. The user then enters the sample names and positions into a Sample Sheet where comments concerning each sample may be entered along with the fluorescent dyes (blue, green, yellow, and red) in the sample. An internal sizing standard is also included in each sample and the appropriate color (usually red or yellow) needs to be indicated at this point.

Following the completion of the sample sheet, an injection list is then created and the sample sheet information imported into it. On the injection list, an operation module is selected for each sample. The typical module used for STR typing is 'GS STR POP4 (1 mL) F' which includes a default injection of 15 kV for five seconds and a separation voltage of 15 kV for 24 minutes.

SAMPLE PREPARATION

Following PCR amplification of forensic DNA samples using a commercially available STR typing kit, the resulting fluorescently labeled STR alleles need to be separated, sized, and genotyped. Samples are prepared for the ABI 310 by diluting them in a denaturant solution that helps disrupt the hydrogen bonds between the complementary strands of the PCR products. An internal standard is also added to each sample for sizing purposes. The samples are then heat

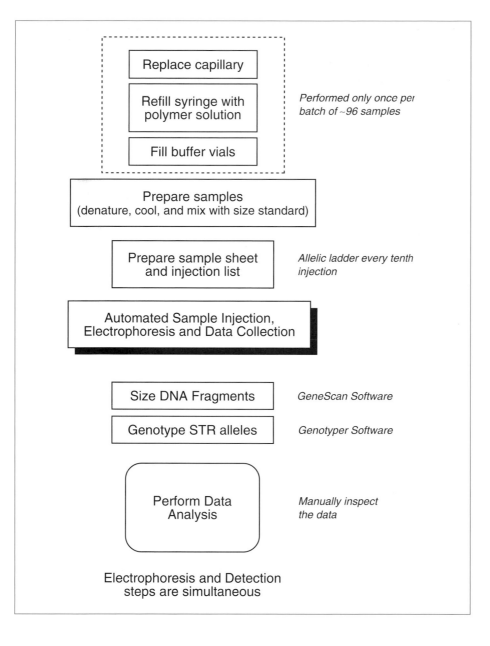

Figure 14.2
Sample processing steps using ABI 310 Genetic Analyzer.

denatured to separate the two strands of each PCR product and then loaded in the instrument for analysis.

A typical sample preparation method for the ABI 310 is as follows:

1. To a 0.5 mL tube, add 25 µL of deionized formamide.
2. Add 1 µL internal lane standard (GS500-ROX or ILS600).
3. Add 1 µL PCR product amplified with AmpF*l*STR kit or Promega PowerPlex 16.
4. Place grey septum on top of 0.5 mL tube.

5. Heat denature sample for 2–3 minutes at 95°C.

6. Snap-cool the sample on ice for 2–3 minutes.

7. Place the sample in the 48-position autosampler tray (96-position trays also exist).

8. Place the autosampler tray in the ABI 310 instrument.

9. Enter the sample names into the ABI 310 sample sheet.

To simplify the sample preparation process and remove step 2, an equivalent amount of labeled internal lane standard may be added to a batch of deionized formamide. For example, if 50 samples were being prepared at a time, then 1250 μL of deionized formamide and 50 μL of GS500-ROX could be combined and 26 μL aliquoted to each sample tube. Regardless of whether the formamide and ROX-labeled internal standard are added separately or together, each sample tube contains ~27 μL most of which is deionized formamide, and the PCR amplified sample has been diluted approximately 1:27 in the formamide. This dilution does two things. First, the high concentration of formamide helps keep the DNA stands denatured, especially after they are coaxed apart by heating to 95°C. Second, by diluting the PCR sample the salts are also diluted which aids in the sample injection process.

For the same reason, it is important that the formamide be deionized. A good method for deionizing the formamide is the addition of an Amberlite ion exchange resin to the formamide. It is also a good idea to prepare a large batch of formamide and then aliquot it into single use portions. Subjecting the formamide to freeze-thaw cycles can cause it to break down and form ionic byproducts that impact the injection process.

The salt content of a sample is very important in the CE electrokinetic injection process as will be discussed later. Contaminating salts can come from either the formamide or the sample if it is not diluted enough. It should also be noted that deionized water can be used in the place of deionized formamide with the only caveat that samples may not be as stable after several days in water compared to formamide. A description of a procedure involving water instead of formamide for ABI 310 sample preparation has been published (Biega and Duceman 1999).

SAMPLE INJECTION

On the ABI 310 Genetic Analyzer, DNA samples are loaded into the capillary with electrokinetic injection. Each sample is placed in an analysis tube and then a voltage is applied to the sample to help draw it into the capillary opening. Electrokinetic injections selectively introduce a sample's charged species into the capillary. More sample material can be introduced into the capillary by simply increasing the voltage or the time applied. In order for this type of injection to work properly, the capillary and the electrode must both extend deep enough into the sample tube to fully interact with the solution in order to establish current flow during the application of the injection voltage. The standard injection on the default STR typing module is 15 000 V for five seconds.

Signal intensity may be increased by lengthening the sample injection time (e.g., from 5 to 10 or 15 seconds), dialyzing the sample on a filter membrane to remove salt from the solution (McCord *et al.* 1993a), or suspending the sample in deionized water (Butler 1995). *The peaks in a sample's fluorescent signal should be kept between 150 and 6000 relative fluorescence units (RFUs) on the ABI 310 for optimal results.* If the peaks are off-scale (above ~7500 RFUs), then the sample can be simply re-injected for a lower amount of time, such as two seconds instead of the standard injection of five seconds, to bring the peaks back on scale.

Electrokinetic injections of DNA samples are highly dependent upon the levels of salt in the samples. These salt levels may be measured with a conductivity meter in terms of microsiemens (µS). The same sample with an identical amount of DNA that is diluted in formamide solutions with different conductivity results in vastly different sensitivity levels (Figure 14.3). As the salt level increases and the sample conductivity goes up, fewer DNA molecules are injected because they are competing with the salt ions to get onto the capillary. In fact, the amount of DNA injected is inversely proportional to the ionic strength of the sample (Butler *et al.* 2004). This differential sample injection due to salt content of the sample is from a process known as sample stacking.

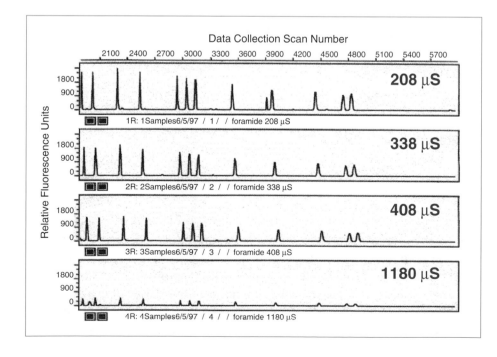

Figure 14.3

Results from ABI 310 using the same sample that has been diluted in different formamide solutions. Formamide solutions with higher conductivities (larger number of µS) result in less DNA being injected into the capillary. The sample is the GS350 ROX-labeled sizing standard. Figure courtesy of Bruce McCord, Ohio University.

Sample stacking is the process that results when samples are injected from a solution that has a lower ionic strength than the buffer inside the capillary. The buffer for example may be 100 mM in salts while the sample is less than 1 mM in ionic strength. When the electric field is applied during an electrokinetic injection, the resistance and field strength in the sample plug region increase because there are fewer ions to carry the current in the lower ionic strength sample. This causes the ions from the sample to migrate rapidly onto the capillary. As these sample ions enter a region where the polymer solution and buffer are at higher ionic strength, they stop moving as quickly and stack as a sharp band at the boundary between the sample plug and the electrophoresis buffer (Butler 1995).

One way to think of this sample stacking process is to imagine a flowing stream of water. When the banks are closer together, the water runs more rapidly than when the banks of the stream are further apart. By reducing the ionic strength of the DNA sample, the 'banks of the stream' are brought closer together and the sample rushes more quickly into the capillary. The amount of DNA that loads into the capillary during the period of time that voltage is applied during an electrokinetic injection is thus a function of the sample's ionic strength.

Hence, samples prepared and diluted in water or formamide with a lower conductivity will give the highest degree of stacking. Formamide conductivity has a dramatic influence on the amount of DNA injected onto the capillary and therefore the sensitivity of the STR typing assay. In addition, the quality of the formamide has been shown to impact the resolution of closely spaced alleles such as the TH01 9.3 and 10 alleles that are 1 bp apart (Buel *et al.* 1998).

TEMPERATURE CONTROL

Room temperature fluctuations cause problems with proper sizing of STR alleles when using the ABI 310. Electrophoretic separations of STR samples are performed at a temperature of 60°C in order to minimize the formation of DNA secondary structure or intrastrand hybrid structures that impact the DNA separation (Wenz *et al.* 1998). DNA sizing is less precise at a lower temperature because of the DNA secondary structure that forms when the strands are not kept fully denatured. Variation in the run temperature will result in relative migration differences between the internal sizing standard DNA fragments and the STR alleles being measured and thus a change in the calculated size of the STR alleles. Because samples are run in a sequential fashion on a capillary system, maintaining a high degree of precision is essential in order to compare the allele sizes in an allelic ladder to those in the samples being analyzed (Lazaruk *et al.* 1998).

The ABI 310 has a heated plate that is used to raise and maintain the temperature of the capillary. However, several centimeters of the capillary at both the inlet and outlet ends are exposed to the air and not directly in contact with the temperature-controlled plate. Maintaining the room temperature to a precision of less than $\pm 1°C$ will improve the precision of DNA separations on the ABI 310. Newer CE instruments, such as the 16-capillary ABI 3100 system, have improved temperature control that mitigates these thermal fluctuation problems.

CAPILLARY MAINTENANCE AND STORAGE

In order for capillaries to be effective, they must be properly maintained. Capillary maintenance includes storing the ends of the capillary in water or buffer. If the capillary inlet and outlet dry out, then urea or other salts from the buffer will form crystals. Because the capillary openings are so narrow even a very small crystal or particle from a solution can cause the capillary to clog. The best indicator of a clogged capillary is low current or no current when a voltage is placed across the capillary. The end of capillaries should be stored in deionized water when not in use for long periods of time.

BUFFER DEPLETION

Use of the same buffer over the course of a set of samples will result in a phenomenon known as buffer depletion. Ions move through the capillary due to the high voltage applied during electrophoresis. Positive ions will gather to the negatively charged electrode and negative ions will collect at the positively charged electrode during the course of electrophoresis. This ion movement results in an imbalance referred to as buffer depletion. To correct this imbalance, the buffer is replenished or replaced on a regular basis. The buffer needs to be changed every day or two if the instrument is being used to its full capacity.

If the buffer is not replenished frequently, the DNA fragments will not separate as well due to ion depletion effects. The current will drop when the buffer becomes too depleted. For example if the normal run current with a fresh buffer is $8\,\mu A$ at $15\,000\,V$ then it will drop to $4–5\,\mu A$ when the buffer becomes depleted. The current can thus serve as a useful diagnostic.

CAPILLARY FAILURE

Capillaries fail and need to be replaced after a number of sample injections. One of the primary reasons for capillary failure is that the dynamic coating on the inside wall of the capillary fails to work properly. Capillary failure is diagnosed by the presence of abnormally broad peaks that define a loss in resolution between closely spaced STR alleles. This loss in resolution is most likely the result of DNA and enzymes from the injected samples adhering to the capillary wall (Isenberg *et al.* 1998).

Capillaries can be removed from the ABI 310 and regenerated with consecutive washes of water, tetrahydrofuran, hydrochloric acid, and polymer solution (Madabhushi 1998). Margaret Kline at the National Institute of Standards and Technology (Gaithersburg, MD) has developed a capillary regeneration procedure that involves forcing several milliliters of deionized water and then a Tris-EDTA buffer through the capillary to remove any material that has bound to the inner wall. Of course the capillaries have to be removed from the instrument in order to perform this procedure. However, capillary lifetimes of over 500 injections have been repeatedly demonstrated when using these wash steps. Some labs though may find it more convenient to just replace the capillary at around 100 injections per the manufacturer's suggestion. Unfortunately, the ABI 310 does not permit an on-the-instrument wash that could be used to recondition a capillary and eliminate the need for frequent capillary replacement.

STEPS PERFORMED BY THE STANDARD MODULE

Instrument operation and data collection on the ABI 310 Genetic Analyzer is controlled by a series of steps and procedures that grouped together are referred to as a 'module'. The standard module used for STR typing is titled 'GS STR POP4 (1 mL) F'. The steps for this module are listed below with an explanation for the purpose of each procedure.

Prior to starting the regular cycle of filling the capillary with polymer solution and injecting and separating DNA samples, several steps are performed in the standard module. First, the temperature on the capillary heating plate is brought up to 60°C to thermally equilibrate the capillary. The laser is turned on to full power (~10 mW). The autosampler platform is moved around in order to verify that the instrument is working well. The following steps are then performed with each sample that is analyzed on the ABI 310 capillary system:

- *Capillary fill* – polymer solution is forced into the capillary by applying a force to the syringe; the syringe position moves down by 5–10 revolutions or steps per injection with POP-4 and a 47 cm capillary. If the syringe moves significantly more than 10 steps, then there is likely a leak in the pump block; if less, then the capillary may be plugged.
- *Pre-electrophoresis* – the separation voltage is raised to 10 000 V and run for five minutes. This step helps check for bubbles inside the capillary and helps to equilibrate the system for sample separation. If there are bubbles inside the capillary, the current will remain at zero when the voltage is raised because ions are not flowing through the capillary.
- *Water wash of capillary* – capillary is dipped several times in deionized water to remove buffer salts that would interfere with the injection process.

- *Sample injection* – the autosampler moves to position A1 (or the next sample in the sample set) and is moved up onto the capillary to perform the injection. A voltage is applied to the sample and a few nanoliters of sample are pulled onto the end of the capillary. The default injection is 15 kV for five seconds.
- *Water wash of capillary* – capillary is dipped several times in waste water to remove any contaminating solution adhering to the outside of the capillary.
- *Water dip* – capillary is dipped in clean water (position 2) several times.
- *Electrophoresis* – autosampler moves to inlet buffer vial (position 1) and separation voltage is applied across the capillary. The injected DNA molecules begin separating through the POP-4 polymer solution.
- *Detection* – data collection begins. Raw data files are collected with no spectral deconvolution of the different dye colors. The matrix is applied during Genescan analysis.

This entire process is accomplished in approximately 30 minutes per sample from one injection to the next assuming that the default time of 24 minutes for electrophoresis is used. The overall time for the capillary fill and pre-electrophoresis steps is about six minutes. DNA fragments up to approximately 400 bp in size should be through the capillary within 24 minutes of electrophoresis at 15 000 V on a 47 cm capillary (320 V/cm).

ALTERNATIVE SOLUTIONS FOR HIGHER THROUGHPUT CAPABILITIES

Each ABI 310 instrument is capable of routinely analyzing about 8000–10 000 sample injections per year. For laboratories desiring to process higher volumes of samples, multiple ABI 310 instruments or alternate analysis platforms may be used. Alternative electrophoresis instrumentation platforms from Applied Biosystems with the same multi-color detection technology include the ABI 377 slab gel system, the 16-capillary ABI 3100, the four-capillary ABI 3100-*Avant*, and the 96-capillary ABI 3700 and 3730.

ABI PRISM 377

The ABI PRISM 377 involves the use of a thin polyacrylamide gel to separate the DNA molecules. Originally the ABI 377 instrument was designed to run 36 samples in parallel although 64 lane and 96 lane upgrades are now available with the ABI 377XL. STR samples can be separated in runs of 2–3 hours duration. Thus, three runs could be performed per day per instrument, with a potential throughput of about 72 000 lanes of data per year. For many years, high-throughput laboratories, such as Myriad Genetics (Salt Lake City, UT) and the Forensic Science Service (Birmingham, England), have used dozens of ABI 377s to perform their high volume STR typing. However, many laboratories are moving to capillary array systems such as the ABI 3100 in order to avoid having to pour slab gels. In addition, Applied Biosystems will stop supporting the ABI 377 gel-based system in 2006.

CAPILLARY ARRAY ELECTROPHORESIS INSTRUMENTS

In Chapter 12, we discussed the advantages of capillary electrophoresis due to its capability for automated injection, separation, and detection of samples. However, one of the major disadvantages of single capillary instruments is that sample throughput is limited because samples are processed sequentially rather than in parallel as on a gel. Parallel CE separations may be performed though by placing a number of capillaries next to each other to form a capillary array electrophoresis (CAE) system. Each capillary in the array then would be analogous to a lane of a gel, although without the problems of lane tracking on the gel.

In 1999, two capillary 96-array electrophoresis systems became commercially available: the ABI 3700 from Applied Biosystems (Foster City, CA) and the MegaBACE from Molecular Dynamics/Amersham Pharmacia Biotech (Sunnyvale, CA). These instruments were developed to meet the large-scale sequencing needs of the Human Genome Project. Both the ABI 3700 and MegaBACE instruments have 96 capillaries in parallel and are capable of sequencing more than 500 nucleotides in each capillary every two or three hours. These CAE instruments are capable of analyzing more than 1000 samples every 24 hours and have been applied to high-throughput processing of forensic DNA database samples. The literature contains several published reports demonstrating effective STR typing with capillary array electrophoresis instruments (Wang *et al.* 1995, Mansfield *et al.* 1996, 1998, Gill *et al.* 2001, Sgueglia *et al.* 2003).

ABI PRISM 3100 GENETIC ANALYZER (16-CAPILLARY SYSTEM)

The multi-capillary ABI PRISM 3100 Genetic Analyzer became available in 2001 and offers a nice solution to higher throughputs with a very similar feel to the single capillary ABI 310 instrument. Both 96-well and 384-well plates of samples may be processed in the ABI 3100. With each run taking roughly 45–60 minutes, a 96-well plate can be analyzed in approximately 5–6 hours with six injections containing 16 samples each.

Several features of the ABI 3100 versus the ABI 310 are compared in Table 14.1. For example, plate records rather than sample sheets are used to designate sample positions with their names prior to collecting the STR data. Injections occur onto all 16 capillaries at once so if a user does not have 16 samples (two columns of eight samples in 96-well plate) then formamide should be put in the wells so as to not inject 'dry' which is not good for the capillaries. A photo of a capillary array as it appears inside the ABI 3100 instrument is shown in Figure 14.4.

	ABI 310	ABI 3100
Number of capillaries	1	16
Cost to replace capillary or array	$50	$625
Laser power	10 mW	25 mW
Temperature control	30–70°C with hot plate	30–70°C with enclosed oven
Computer	Macintosh; Windows NT or 2000	Windows NT or 2000
Data storage	Individual files in Run Folders	Oracle database
Sample information setup	Sample Sheet	Plate Record
Spectral calibration (multi-componenting matrix)	Can be applied to samples at any time	Must be applied prior to sample being run
Number of dyes available	4- or 5-dye chemistry	4- or 5-dye chemistry
Software to process STR data	GeneScan/Genotyper OR GeneMapper*ID*	GeneScan/Genotyper OR GeneMapper*ID*

Table 14.1

Comparison of several features on ABI Prism 310 and ABI Prism 3100 Genetic Analyzers.

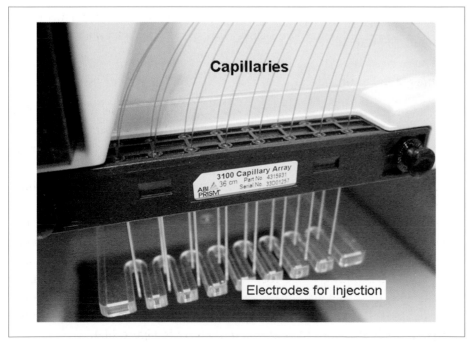

Figure 14.4

Photo of a 3100 capillary array illustrating the 16 glass capillaries (top) and the electrodes surrounding the capillaries (bottom) that enable electrokinetic injection of DNA samples from two columns of an 8 × 12 96-well microtiter plate.

Figure 14.5

Comparison of matrix samples used for spectral calibration in (a) ABI 310 and (b) ABI 3100 instruments.

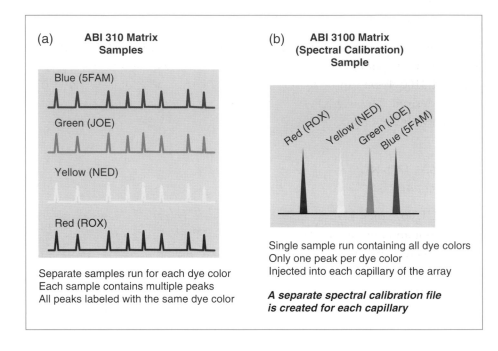

(a) **ABI 310 Matrix Samples**

Blue (5FAM)

Green (JOE)

Yellow (NED)

Red (ROX)

Separate samples run for each dye color
Each sample contains multiple peaks
All peaks labeled with the same dye color

(b) **ABI 3100 Matrix (Spectral Calibration) Sample**

Red (ROX)   Yellow (NED)   Green (JOE)   Blue (5FAM)

Single sample run containing all dye colors
Only one peak per dye color
Injected into each capillary of the array

A separate spectral calibration file is created for each capillary

An important difference between the ABI 310 and 3100 is that spatial and spectral calibrations are required prior to collecting data on the ABI 3100. The spatial calibration enables the CCD detector to know the location of each capillary while spectral calibration is essentially the same as a matrix file generated with ABI 310 runs (see Figure 13.6). A comparison of the format for the matrix standards used to enable dye color deconvolution is illustrated in Figure 14.5. With the ABI 310 separate samples are run representing each dye color that contain multiple peaks and the same dye color. The ABI 3100 on the other hand uses a single sample within each capillary that contains one peak per dye color (Figure 14.5). The spectral calibration (color deconvolution matrix) is applied during data collection on the ABI 3100 meaning that a sample must be rerun if a different matrix is desired for a particular sample.

Newer versions of ABI 3100 data collection software (versions 1.1, 2.0, and above) tweak the red and green color channels to balance signal with 5-dye chemistry kits such as Identifiler. Data files are also stored within an Oracle database on the ABI 3100 rather than just individual files such as on the ABI 310.

Precision studies conducted on the ABI 3100 (Sgueglia *et al.* 2003, Butler *et al.* 2004) and the ABI 3700 (Gill *et al.* 2001) have demonstrated that reliable results can be obtained with a multi-capillary CE system.

ABI PRISM 3100-AVANT GENETIC ANALYZER (FOUR-CAPILLARY SYSTEM)

For those who are not ready for the 'firepower' (or the price) of the ABI 3100, a four-capillary system known as the ABI 3100-*Avant* became available in 2002. The ABI 3100-*Avant* can be upgraded to a full 16-capillary ABI 3100 after purchase if desired. The new ABI 3100-*Avant* systems collect data that may then be analyzed with GeneMapper *ID* software (see Chapter 17).

ABI PRISM 3700 AND 3730 GENETIC ANALYZERS (96-CAPILLARY SYSTEMS)

The ABI 3700, which is a 96-capillary array system, offers even higher potential throughput and automation than the ABI 377. Sample injection of 96 samples is performed in parallel. In a 24-hour period of unattended operation, more than 750 samples can be injected. Thus, approximately 190 000 sample injections can be theoretically processed per year. The ABI 3730 is widely used in human genome centers for DNA sequencing and offers even higher throughput compared to the ABI 3700 even through both systems utilize arrays with 96-capillaries.

The ABI 3700 is used in several U.S. forensic laboratories including the Florida Department of Law Enforcement (Tallahassee, FL) and the New York State Police (Albany, NY) for performing STR typing of convicted offender samples going into DNA databanks.

HITACHI FMBIO II AND FMBIO III FLUORESCENCE IMAGING SYSTEMS

An alternative solution to capillary systems for processing STR samples is to run them on a polyacrylamide gel and then perform a post-electrophoresis fluorescent scan of the gel. The Promega Corporation's PowerPlex® 1.1, 2.1, and 16 BIO STR kits have been made compatible with the Hitachi FMBIO II and FMBIO III Fluorescence Imaging Systems. The STR amplicons are labeled with two different dyes, fluorescein and tetramethylrhodamine (usually referred to as TMR), in the case of PowerPlex® 1.1 and 2.1 or three different dyes for the PowerPlex® 16 BIO kit (fluorescein, JOE, and Rhodamine Red-X). An additional dye, carboxy-X-rhodamine or CXR, is attached to an internal lane standard for DNA fragment sizing purposes. The PowerPlex® 16 BIO kit uses an ILS600 size standard labeled with the dye Texas Red-X.

Unfortunately, in the fall of 2003 MiriaBio/Hitachi Genetic Systems decided to stop making the FMBIO instruments. Forensic laboratories that have an FMBIO II or FMBIO III instrument will be supported for several years but many of these laboratories are understandably looking to other instrument platforms for STR typing such as those described earlier in this chapter or technologies under development that are described in Chapter 17. Recognizing that this portion of the book may soon become obsolete for many working forensic laboratories that are currently using the FMBIO systems, we hope that this section will still provide a valuable historical perspective in the post-FMBIO world of the future.

The FMBIO II Fluorescence Imaging System from Hitachi Genetic Systems (Alameda, CA) consists of a scanning unit (Figure 14.6) that is controlled by a Macintosh or Windows computer and three software programs. The hardware for the system features a 20 mW solid-state Nd:YAG (neodinium yttrium aluminum garnet) laser that emits light at an excitation wavelength of 532 nm. The instrument can scan an area of 20 cm × 43 cm and has a reported linear dynamic range of four orders of magnitude. Newer instrument platforms include the FMBIO IIe, III, and III+ with multiple lasers to expand the possible excitation wavelengths to 635 nm and 488 nm and a larger scan area that is capable of examining two gels at once. Two photomultiplier tubes are used for simultaneous detection of emitted light from two or more different fluorescent dyes. Band pass interference filters are used to achieve multicolor imaging.

Up to four filters can be stored in the instrument at any one time and accessed through the data collection software. The FMBIO II takes approximately 15–20 minutes to scan two dyes at once. In a typical scan of PowerPlex samples, the gel image is produced after electrophoresis using a 505 nm band-pass filter to detect amplification products containing a fluorescein label and a 585 nm band-pass filter to detect amplicons labeled with TMR. A 650 nm band-pass filter is used to observe the CXR-labeled internal size standard DNA fragments. These images can then be overlaid into a three-color image or viewed separately by color for closer inspection of the data.

Figure 14.6

Schematic of FMBIO II Fluorescence Imaging System. Following electrophoretic separation of STR alleles, a gel is placed in the FMBIO and scanned to reveal the presence of fluorescently labeled PCR products. The FMBIO utilizes a solid-state, green laser (1) to excite fluorophores at 532 nm using a polygon scanning mirror (2). The resulting fluorescent light signals emitted from the excited fluorophores attached to DNA fragments are then collected by two optical fiber arrays (3). Specific fluorescent dye signals are isolated using separate interference filters (4) and are converted to electrical signals with two photomultiplier tubes (5). Figure used with permission from Hitachi Genetic Systems web page.

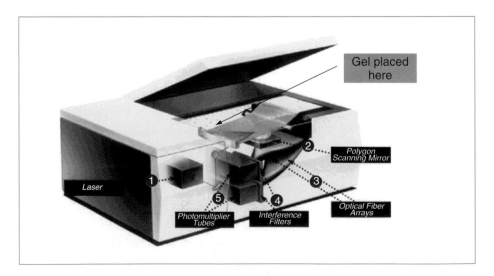

SOFTWARE PROGRAMS

The three software programs used in conjunction with the FMBIO scanner to perform STR genotyping include (1) Read Image, (2) FMBIO® Analysis Software, and (3) STaR Call™ Genotyping Software. *Read Image* controls the scan area, scan resolution, and photomultiplier sensitivity as the FMBIO II scanning unit generates the digital image of the gel. The user can indicate which fluorescent emission filters to use and add comments to the scanned image. The scanned image for each fluorescent wavelength of the experimental data is converted into a 16-bit digital TIFF file and stored for future data analysis.

The generated gel data images are next examined by the *FMBIO® Analysis Software*, which performs functions such as DNA fragment sizing and quantitation of peak height and area. DNA bands are sized through logarithmic comparison to size standards. Data can be displayed as either full gel images or electropherograms that are a virtual slice through one of the gel lanes. The gel images can also be examined one color at a time following application of the color separation matrix.

The analysis software includes a DNA band finding program. Because of fluorescence intensity variation between gels and even within samples on a gel, this step requires some user review and editing of the data. An analyst manually evaluates each called DNA band. Stutter bands can be highlighted and edited out of the processed data or removed based on user-defined criteria in the genotyping software described below.

Once the DNA bands have been sized, the STR alleles are genotyped using *STaR Call™ Genotyping Software*. Band sizes from STR alleles are compared to sized alleles from allelic ladders run in adjacent lanes and converted to the appropriate genotype. Band sizes calculated by FMBIO® Analysis Software are imported into STaR Call™ and compared to values for each STR locus in a multiplex set. STR 'lookup tables' are exported to a Microsoft® Excel worksheet for evaluation of genotypes and manual confirmation that the expected size ranges are obtained for each allele. A 'lookup table' typically includes the DNA band size, STR allele call, and band quantitation in the form of optical density (OD) units.

Based on comparison of DNA fragment sizes with allele ranges and allelic ladders, each band is assigned a locus name and repeat number. The program looks for bands with weaker intensities, assigns them as stutter products if they are one repeat unit less than a 'normal' allele, and appropriately excludes them from the final data output. All of these genotyping steps can be performed automatically by the software. However, the final genotype information is typically reviewed carefully in a manual fashion to insure that correct calls were made by the genotyping software.

SAMPLE PROCESSING ON THE FMBIO II

One of the downsides of performing sample processing with the FMBIO II is that the process involves the use of gel electrophoresis, which is more labor intensive than capillary electrophoresis. However, the FMBIO system can lead to higher sample throughputs per instrument. Unlike the ABI 310 capillary system that performs online detection during electrophoresis, a gel's image is captured *after* electrophoretic separation with the FMBIO system. In other words, gels are run separately from the detection portion of the analysis with the FMBIO approach (Figure 14.7). Therefore, with staggered start times, multiple gel electrophoresis systems can feed into a single FMBIO II/III Fluorescence Analysis System, leading to higher throughput for the cost of a single instrument. For example, in the late 1990s, the Bode Technology Group, a private contract DNA typing laboratory in Springfield, Virginia, ran on average 20 gels per day with each gel containing 25 samples plus allelic ladders and controls. Thus, in this case a single scanner processed 450 new samples every day.

Figure 14.7

Sample processing steps using the Hitachi FMBIO II or III. Multiple gels can be prepared simultaneously and then quickly scanned after electrophoresis to improve sample throughput.

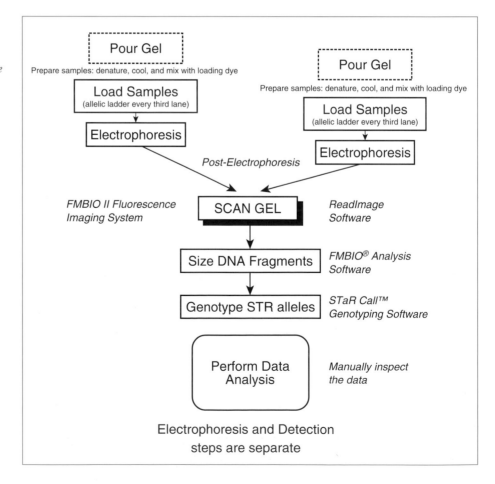

Of course this kind of throughput has been superseded with newer multi-capillary electrophoresis systems such as the ABI 3700 described earlier.

GEL ELECTROPHORESIS

Gels must be poured, samples loaded, and electrophoresis conducted prior to scanning the samples. Gel sizes are typically $17\,cm \times 43\,cm \times 0.4\,cm$ ('long gel') or $17\,cm \times 32\,cm \times 0.4\,cm$ ('short gel'). The denaturing polyacrylamide gel composition for PowerPlex 1.1 analysis is typically 4% polyacrylamide: bis (19:1), 7 M urea, and 0.5X Tris-Borate-EDTA buffer (Micka *et al.* 1999). Alternatively, a 5% Long Ranger™ (FMC BioProducts, Rockland, ME) gel may be used with 7 M urea and 1X TBE (Lins *et al.* 1998). Precast gels, such as the 4.5% R³™ Precast Gels, are also available from Hitachi Genetic Systems (Micka 1999). Gels can be re-used up to four times with similar performance if proper care is taken (Tereba *et al.* 1998).

The gels are usually run on a SA43 vertical electrophoresis apparatus. A pre-run at 60 W for 30–45 minutes is performed prior to loading the samples in order to warm the gel up to a plate temperature of 45°C to 50°C. The PCR-amplified STR alleles are then loaded onto the gel and separated for 60–90 minutes to resolve the DNA bands. A separation voltage of 60 V/cm is often used in order to resolve PowerPlex® 1.1 STR alleles (Schumm *et al.* 1997).

SAMPLE PREPARATION FOR GEL LOADING

Samples are prepared by mixing $2\,\mu L$ of the amplified sample with a $1\,\mu L$ aliquot of the CXR-labeled fluorescent internal lane standard and $3\,\mu L$ of Bromophenol blue loading solution (95% formamide, 0.05% bromophenol blue, 10 mM NaOH). This mixture is heated to 95°C for two minutes and then snap-cooled on ice to denature the DNA strands present in the sample. An aliquot of $2.5–3\,\mu L$ of this sample is then loaded onto the appropriate lane of the gel (Schumm *et al.* 1997, Promega Corporation 1999). Allelic ladders are prepared in a similar fashion and loaded onto the gel every five or six lanes. Note that while internal lane standard 400 works to size PowerPlex 1.1 STR loci, the extra high molecular weight DNA bands in the internal lane standard 600 are needed to properly size the large alleles in Penta E and FGA STR systems.

SOLUTION FOR 13 CODIS STR LOCI

The Promega solution to examining the 13 CODIS STR loci with the FMBIO instrument platform involves either the use of PowerPlex® 1.1 and 2.1 STR kits or a single megaplex amplification with PowerPlex® 16 BIO (see Chapter 5). DNA samples are amplified and genotyped with 8 STR markers in PowerPlex® 1.1

Figure 14.8

(left) PowerPlex 1.1 STR data collected from the 505 nm (fluorescein-labeled alleles) and 585 nm (TMR-labeled alleles) scans of a gel containing allelic ladders and PCR-amplified samples. (right) PowerPlex 2.1 STR data collected on the same two samples. The three marked loci – TH01, TPOX, and VWA – are common to both multiplex kits and are therefore useful as a quality assurance measure to demonstrate concordance from two separate PCR amplifications of the same DNA sample. Note that the allelic ladders for VWA and TH01 have different alleles for the PowerPlex 1.1 and 2.1 systems yet the samples shown here provide the same genotype and thus demonstrate full concordance of PCR amplifications from both kits. The 9.3 and 10 alleles for the PowerPlex 2.1 TH01 allelic ladder are resolved (see boxed area) indicating that single base resolution can be achieved in that portion of the gel electrophoretic separation. Original data courtesy of Hitachi Genetic Systems.

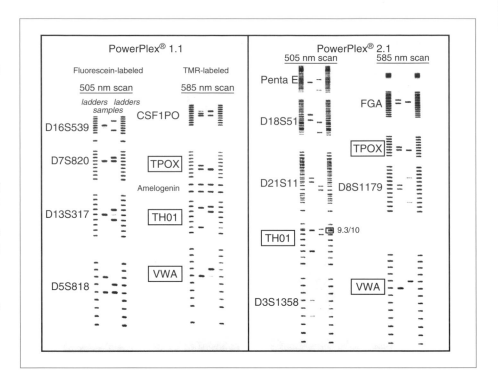

and 9 STR markers in 2.1 (Figure 14.8). There are three STR systems in common between these two kits, namely VWA, TH01, and TPOX, in order to help verify sample concordance and avoid sample shuffling. The STaR Call™ 3.0 genotyping software automatically compares the three overlapping loci between the PowerPlex® 1.1 and 2.1 systems and highlights any non-agreeing alleles as an internal quality control check.

Samples amplified with both PowerPlex® 1.1 and 2.1 are shown in Figure 14.8. The 505 nm scan detects the fluorescein-labeled PCR products while the 585 nm scan detects the tetramethyl rhodamine-labeled PCR products. The PowerPlex® 2.1 STR markers extend to a higher size range with the larger PCR products being almost 500 bp in size. Note that the position of the STR locus impacts its resolution on the gel. Alleles from the D3S1358 STR locus travel further through the gel matrix and are therefore better resolved from one another compared to FGA alleles that are near the top of the gel. Thus, it is more difficult to distinguish off-ladder alleles within the FGA locus compared to the D3S1358 locus because the alleles are not spread apart as well.

The data file size needed to capture the information from a fluorescence scan of a gel is quite large. Due to the different number of alleles and DNA size range, a typical gel scan of PowerPlex® 1.1 is approximately 36 Mb in size while a scan of a full PowerPlex® 2.1 gel takes up approximately 48 Mb for the storage

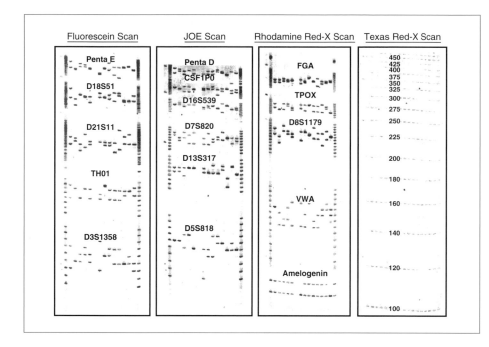

Figure 14.9

FMBIO III+ color-separated STR data collected from four scans of a gel containing PowerPlex® 16 BIO PCR-amplified samples. Each scan detects the PCR products labeled with one of three dyes: fluorescein, JOE, and Rhodamine Red-X. The ILS600 size standard is labeled with the fourth dye Texas Red-X and contains DNA fragments ranging from 100–600 bp in 20 or 25 bp increments (the upper bands are not captured on this gel). The STR loci amplified are indicated above the alleles observed. Allelic ladders are on the far left and right of each scan. The combined color image for the same gel may be seen in Figure 13.11. Figure courtesy of Margaret Kline, NIST.

of all information on that gel. A computer hard drive can fill up rather quickly with these image files necessitating the use of zip drives or magneto optical disk drives for data storage.

Figure 14.9 shows the spectrally separated gel images from a PowerPlex 16 BIO result on an FMBIO III+ instrument. The FMBIO III+ is more sensitive that the older FMBIO II instruments in part because it uses three lasers with excitation wavelengths of 488 nm, 532 nm, and 635 nm instead of just the 532 nm excitation present in the FMBIO II.

ISSUES WITH THE FMBIO II APPROACH

NUMBER OF SAMPLES PER GEL

The number of samples that may be run on each gel is flexible and depends on the comb used to form the sample loading wells. The Bode Technology Group uses a comb with 38 slots containing flat wells with intervening 'posts' that separate each loading well. Two slots are left empty on each end. Control lanes include a K 562 sample, a reagent blank, a negative control, and a blind control sample for quality control purposes. The remaining 30 lanes are composed of five sets of allelic ladders and 25 test samples. Internal lane standards are loaded into each lane to enable size determination and allele calling of PCR products amplified at each STR locus (shown in blue in Figure 13.11).

GEL RE-USE

Preparing, pouring, and polymerizing polyacrylamide gels can be tedious. Fortunately pre-cast gels are available including the Hitachi R³™ Precast Gel Electrophoresis System or Longer Ranger gels from Cambrex BioScience Rockland, Inc. (formerly FMC Bioproducts, Rockland, ME). In order to reduce the number of gels that have to be prepared to analyze a large set of samples, a method has been developed to re-use gels (Tereba *et al.* 1998).

After imaging a gel on the FMBIO II, it can be placed back in the gel box and electrophoresed in reverse for a short period of time to remove the DNA samples from the gel. Typically the gel reverse electrophoresis is performed for 15–30 minutes longer than the previous forward electrophoresis, in order to fully remove all of the DNA bands from the gel. The gel can then be reloaded with fresh samples and another set of data determined. The Promega scientists have published protocols for the re-use of gels up to four or more times (Tereba *et al.* 1998, Micka *et al.* 1999).

Amazingly the resolution of the gel does not degrade with this re-use. However, edge effects, such as 'frowning' of the outer lanes, become progressively worse with each run. With the use of internal lane standards though, the frowning effects do not impact the ability to accurately size the unknown DNA fragments in the gel. The re-use of gels for four or five times or even more over the period of several days translates to savings in terms of labor and reagent costs. The re-use of gels also minimizes the number of times that a laboratory has to clean the glass gel plates.

CLEANING GEL PLATES

A gel is poured between two glass plates. Any dust particles or other contaminants on the plates can interfere with the collection of fluorescence data by scattering the laser excitation or the fluorescence emission light. Cleaning glass plates is time consuming but essential to obtaining good fluorescent signal with low background. Special low fluorescence plates cost around $150 per pair compared to standard glass plates that are about $50 per pair.

When plates are cleaned, the gel has to be removed since it is bonded to one or both plates. The plates are often soaked to simplify the gel release. Cleaning usually takes about 5–10 minutes for a pair of plates once the gel has been removed.

DIFFERENT EXCITATION LASER

The FMBIO II uses a 532 nm laser wavelength to excite fluorescent dyes. The different laser excitation wavelength compared to the typical argon ion laser wavelengths of 488 nm and 514.5 nm means that the two dyes used to label STR alleles in the PowerPlex system are differentially excited causing the dye sensitivities to vary. This effect produces different relative intensities of the amplification products detected in the different instruments. Therefore, the primer concentrations were re-configured to make the amplification products from the PowerPlex systems compatible with the two instruments' different excitation/emission wavelengths (Lins *et al.* 1998). Hence, the primer concentrations in the PowerPlex™ 1.2 STR kit has been optimized for use on the ABI 310 while PowerPlex™ 1.1 and 2.1 kits have been designed for the FMBIO II.

COMPARISON OF FMBIO II AND ABI 310 INSTRUMENT PLATFORMS

There are advantages and disadvantages to any approach taken for DNA typing. The primary instrument platforms covered in this chapter, namely the FMBIO II Gel Imager and ABI Prism 310 Genetic Analyzer, are compared in Table 14.2. The steps surrounding the FMBIO are more labor intensive since the sample loading is not as automated as the ABI 310. However, the FMBIO is capable of about 10 times the throughput of the ABI 310 on a per instrument basis, and therefore on a similar throughput scale as the 16-capillary ABI 3100 instrument.

Parameter	FMBIO II/III Gel Scanner	ABI Prism 310 CE System
STR kit solution for 13 CODIS core loci	PowerPlex® 1.1 and 2.1 or PowerPlex® 16 BIO (Promega Corporation)	Profiler Plus™ and COfiler™ or Identifiler™ (Applied Biosystems) or PowerPlex® 16 (Promega Corporation)
Fluorescent dyes detected	Fluorescein, JOE, TMR, CXR, Rhodamine Red, Texas Red	5-FAM, JOE, VIC, NED, ROX, PET, LIZ
Laser wavelength used to excite fluorophores	532 nm (Nd:YAG solid-state laser); additional 635 nm and 488 nm with FMBIO III+	488 nm and 514.5 nm (Argon ion gas laser)
Instrument cost	~$80 000	~$60 000
Batch size (including allelic ladders)	25 per gel (depends on comb used)	48 or 96 per tray
Sample throughput	20 or more gels/scanner/day (~450 samples/day)	48 samples/24 hour period
Sample data collection time	15 minutes to scan gel	~30 minutes per sample
Computer type	Macintosh® or Windows NT/2000	Macintosh® or Windows NT/2000

Table 14.2

Comparison of FMBIO II and ABI 310 detection formats.

Table 14.2
(Continued)

Parameter	FMBIO II/III Gel Scanner	ABI Prism 310 CE System
Data collection software	ReadImage or FMBIO III Image Scanner	ABI 310 Data Collection
Peak sizing software	FMBIO® Analysis	GeneScan® (or GeneMapper*ID*)
Genotyping software	STaR Call™	Genotyper® (or GeneMapper*ID*)
Data file size	~30 Mb/gel (~1 Mb/sample)	~200 kb/sample (GeneScan) ~100 kb/sample (Genotyper)
Accessories needed	Gel electrophoresis apparatus, glass gel plates, acrylamide, loading dye, comb	Capillary, POP-4 polymer, tubes, septa tube caps
Lifetime of gel or capillary for re-use	2–5 runs	~100–300 runs
Primary advantage of approach sample	Capable of high volume sample processing because separation and detection are separate and many samples can be run in parallel on each gel	Automated sample processing with no gel pouring or loading required

REFERENCES AND ADDITIONAL READING

Biega, L.A. and Duceman, B.W. (1999) *Journal of Forensic Sciences*, 44, 1029–1031.

Buel, E., Schwartz, M. and LaFountain, M.J. (1998) *Journal of Forensic Sciences*, 43, 164–170.

Butler, J.M., McCord, B.R., Jung, J.M. and Allen, R.O. (1994) *BioTechniques*, 17, 1062–1070.

Butler, J.M. (1995) *Sizing and quantitation of polymerase chain reaction products by capillary electrophoresis for use in DNA typing*. PhD Dissertation, University of Virginia, Charlottesville.

Butler, J.M., Buel, E., Crivellente, F. and McCord, B.R. (2004) *Electrophoresis*, 25, 1397–1412.

Butler, J.M. (2004) Short tandem repeat analysis for human identity testing. *Current Protocols in Human Genetics*, John Wiley & Sons, Hoboken, NJ, Unit 14.8, (Supplement 41), pp.14.8.1–14.8.22.

Gill, P., Koumi, P. and Allen, H. (2001) *Electrophoresis*, 22, 2670–2678.

Isenberg, A.R., Allen, R.O., Keys, K.M., Smerick, J.B., Budowle, B. and McCord, B.R. (1998) *Electrophoresis*, 19, 94–100.

Lazaruk, K., Walsh, P.S., Oaks, F., Gilbert, D., Rosenblum, B.B., Menchen, S., Scheibler, D., Wenz, H.M., Holt, C. and Wallin, J. (1998) *Electrophoresis*, 19, 86–93.

Lins, A.M., Micka, K.A., Sprecher, C.J., Taylor, J.A., Bacher, J.W., Rabbach, D., Bever, R.A., Creacy, S. and Schumm, J.W. (1998) *Journal of Forensic Sciences*, 43, 1168–1180.

Mansfield, E.S., Vainer, M., Enad, S., Barker, D.L., Harris, D., Rappaport, E. and Fortina, P. (1996) *Genome Research*, 6, 893–903.

Mansfield, E.S., Robertson, J.M., Vainer, M., Isenberg, A.R., Frazier, R.R., Ferguson, K., Chow, S., Harris, D.W., Barker, D.L., Gill, P.D., Budowle, B. and McCord, B.R. (1998) *Electrophoresis*, 19, 101–107.

McCord, B.R., Jung, J.M. and Holleran, E.A. (1993a) *Journal of Liquid Chromatography*, 16, 1963–1981.

McCord, B.R., McClure, D.L. and Jung, J.M. (1993b) *Journal of Chromatography A*, 652, 75–82.

McCord, B.R. (2003) Troubleshooting capillary electrophoresis systems. *Profiles in DNA*, 6 (2); Available at: http://www.promega.com/profiles/602/ ProfilesInDNA_602_10.pdf.

Madabhushi, R.S. (1998) *Electrophoresis*, 19, 224–230.

Micka, K.A., Amiott, E.A., Hockenberry, T.L., Sprecher, C.J., Lins, A.M., Rabbach, D.R., Taylor, J.A., Bacher, J.W., Glidewell, D.E., Gibson, S.D., Crouse, C.A. and Schumm, J.W. (1999) *Journal of Forensic Sciences*, 44, 1243–1257.

Promega Corporation (1997) *GenePrint® Fluorescent STR Systems Technical Manual*. Madison, Wisconsin: Promega Corporation.

Promega Corporation (1999) *GenePrint® PowerPlex™ 2.1 System Technical Manual*. Madison, Wisconsin: Promega Corporation.

Rosenblum, B.B., Oaks, F., Menchen, S. and Johnson, B. (1997) *Nucleic Acids Research* 25, 3925–3929.

Schumm, J.W., Lins, A.M., Micka, K.A., Sprecher, C.J., Rabbach, D. and Bacher, J.W. (1996) *Proceedings from the First European Symposium on Human Identification*, pp. 90–104. Madison, Wisconsin: Promega Corporation.

Schumm, J.W., Sprecher, C.J., Lins, A.M., Micka, K.A., Rabbach, D.R., Taylor, J.A., Tereba, A. and Bacher, J.W. (1997) *Proceedings of the Eighth International Symposium on Human Identification*, pp. 78–84. Madison, Wisconsin: Promega Corporation.

Sgueglia, J. B., Geiger, S. and Davis, J. (2003) *Analytical and Bioanalytical Chemistry*, 376, 1247 1254.

Steadman, G.W. (2000) *Survey of DNA Crime Laboratories, 1998*. Bureau of Justice Statistics, Special Report February 2000, U.S. Department of Justice.

Tereba, A., Micka, K.A. and Schumm, J.W. (1998) *BioTechniques*, 25, 892–897.

Wang, Y., Ju, J., Carpenter, B.A., Atherton, J.M., Sensabaugh, G.F. and Mathies, R.A. (1995) *Analytical Chemistry*, 67, 1197–1203.

Wenz, H.M., Robertson, J.M., Menchen, S., Oaks, F., Demorest, D.M., Scheibler, D., Rosenblum, B.B., Wike, C., Gilbert, D.A. and Efcavitch, J.W. (1998) *Genome Research*, 8, 69–80.

CHAPTER 15

STR GENOTYPING ISSUES

The technology for DNA profiling and the methods for estimating frequencies and related statistics have progressed to the point where the reliability and validity of properly collected and analyzed DNA should not be in doubt.

(NRC II Report, p. 2)

In Chapters 12 and 13, we discussed how short tandem repeat (STR) amplification products labeled with fluorescent dyes are separated and detected. In Chapter 14, we examined several commonly used instrument approaches for collecting the STR data. However, the data collection process leaves the analyst with only a series of peaks in an electropherogram or bands on a gel. The peak information (DNA size and quantity) must be converted into a common language that will allow data to be compared between laboratories. This common language is the sample genotype. This chapter will review the process of taking multi-color fluorescent peak information and converting it into STR genotypes.

A locus genotype is the allele, in the case of a homozygote, or alleles, in the case of a heterozygote, present in a sample for a particular locus and is normally reported as the number of repeats present in the allele. *A sample genotype or STR profile is produced by the combination of all of the locus genotypes into a single series of numbers.* This profile is what is entered into a case report or a DNA database for comparison purposes to other samples. Chapter 21 will cover how statistical calculations, such as random match probabilities, are performed using a STR profile.

STR alleles from the same sample that are amplified with different primer sets or analyzed by different detection platforms will differ in size. However, by using locus-specific allelic ladders, such as those described in Table 5.4, allele peak sizes may be accurately converted into genotypes (Smith 1995). These genotypes then provide the universal language for comparing STR profiles.

THE GENOTYPING PROCESS

Sample data collected from the ABI 310 or other instruments described in Chapter 14 are usually represented in the form of peaks that correspond to the various STR alleles amplified from the DNA sample. These peaks are present at various locations in a sample's electropherogram and usually plotted as fluorescent signal intensity verses time passing the detector (in the case of the ABI 310 or 3100) or position on the gel (in the case of the FMBIO II gel imager). The steps for converting those fluorescent peaks into an allele call are shown in Figure 15.1. Computer programs, such as those listed on the right side of Figure 15.1, play an important role in this process (see Butler *et al.* 2004). Expert systems, such as True Allele and GeneMapper*ID*, will be discussed in Chapter 17.

The multiplex STR kits in use today take advantage of multiple fluorescent dyes that can be spectrally resolved (see Chapter 13). The various dye colors are separated and the peaks representing DNA fragments are identified and associated with the appropriate color. The DNA fragments are then sized by comparison to an internal sizing standard (Figure 15.2). Finally, the polymerase chain reaction (PCR) product sizes for the questioned sample are correlated to an allelic ladder that has been sized in a similar fashion with internal standards. The allelic ladder contains alleles of known repeat content and is used much like a measuring ruler to correlate the PCR product sizes to the number of

Figure 15.1

Genotyping process for STR allele determination. Software packages for DNA fragment analysis and STR genotyping perform much of the actual analysis, but extensive review of the data by trained analysts/examiners is often required.

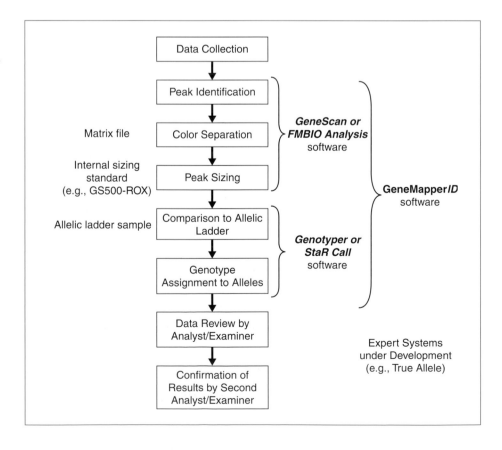

repeat units present for a particular STR locus. From this comparison of the unknown sample with the known allelic ladder, the genotype of the unknown sample is determined.

SIZING DNA FRAGMENTS

DNA fragments represented by peaks in capillary electropherograms or bands on a gel can be sized relative to an internal size standard that is mixed with the DNA samples. The internal size standard is typically labeled with a different colored dye so that it can be spectrally distinguished from the DNA fragments of unknown size. In the case of the ABI 310 and 4-dye chemistry AmpFℓSTR kits, the internal standard is usually the GS500-ROX (Figure 15.2). This size standard contains 16 DNA fragments, ranging in size from 35–500 base pairs (bp),

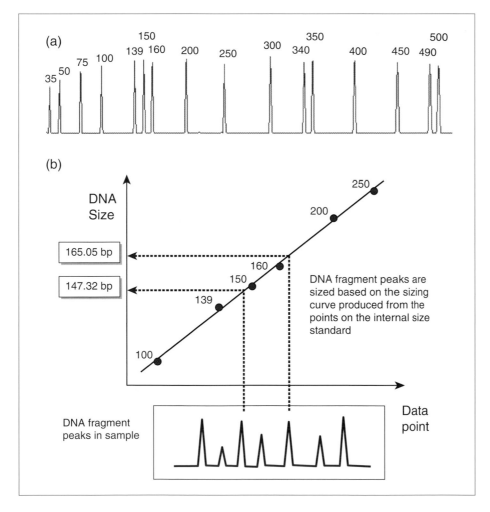

Figure 15.2

Peak sizing with DNA fragment analysis. An internal size standard, such as GS500-ROX (a), is analyzed along with the DNA sample and used to calibrate the peak data points to their DNA size (b). This standard is labeled with a different color fluorescent dye, in this case ROX (detected as red), so that it can be spectrally distinguished from the STR alleles which are labeled in other colors.

that have been labeled with the red fluorescent dye ROX. Alternatively, for 5-dye chemistries the DNA fragments are labeled with the orange dye LIZ to create the GS500-LIZ internal size standard.

For PowerPlex kit users, the internal lane standard ILS 600 is commonly employed (see Figure 5.5 bottom panel). This size standard contains 22 DNA fragments, ranging in size from 60–600 bp, that have been labeled with the red fluorescent dye CXR. The ILS 600 size marker is commonly used with the PowerPlex® 16 and PowerPlex® Y kits from Promega.

DNA FRAGMENT ANALYSIS AND GENOTYPING SOFTWARE

Fairly sophisticated software has been developed to take sample electrophoretic data rapidly through the genotyping process just described (Ziegle *et al.* 1992). For ABI 310 users (see Chapter 14), this is done in two steps by two different software programs. GeneScan® software is used to spectrally resolve the dye colors for each peak and to size the DNA fragments in each sample. The resulting electropherograms are then imported into the second software program, Genotyper®. This program determines each sample's genotype by comparing the sizes of alleles observed in a standard allelic ladder sample to those obtained at each locus tested in the DNA sample. As will be discussed in Chapter 17, Applied Biosystems also has a software package called GeneMapper *ID* that combines the functions of GeneScan and Genotyper with additional quality scores for the data.

Once GeneScan® processed electropherograms are imported into Genotyper®, a macro named 'Kazaam', that is specific for the STR loci and allelic ladders in each AmpFℓSTR® kit, is initiated in Genotyper® to actually perform the allele calling. At this point, the analyst usually examines the peaks that have been called and based on their experience may or may not edit the calls made by the software. An allele table may then be created from the edited allele calls. Finally, the alleles may be exported to a spreadsheet program, such as Microsoft Excel, for further data analysis or uploading into a DNA database.

FMBIO II/III users (see Chapter 14) perform automatic band calling, quantify peak heights and areas, and determine DNA band sizes through comparison to size standards with the FMBIO® Analysis Software. The analyst adjusts the color separation to obtain the best resolution between the dye labels. Gel bands in each color are then manually edited in order to make allele calls. Automated genotype calling from FMBIO STR data is performed with STaR Call™ Genotyping Software. This software takes the calculated STR allele base pair sizes and converts each peak into the appropriate allele call based on the fragment's calculated size compared to the calculated sizes of the alleles in the allelic ladder. Allele 'look-up' tables in StaR Call™, which are basically

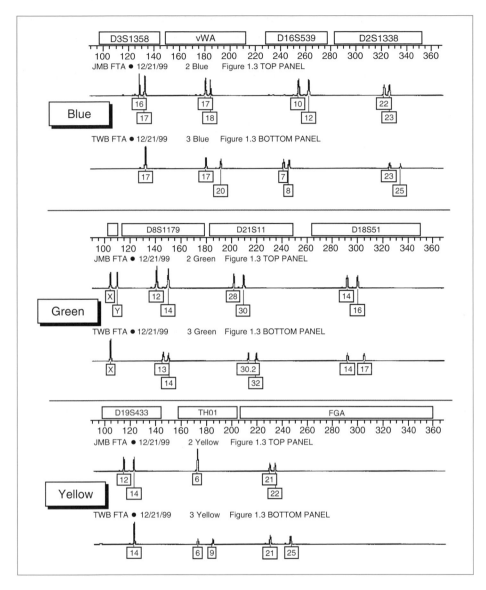

Figure 15.3

Genotype results on the two samples displayed in Figure 1.3 obtained with AmpFlSTR SGM Plus STR kit amplification and Genotyper 2.5 analysis.

Microsoft Excel spreadsheets, permit genotype information to be quickly reviewed and uploaded to a DNA database.

The genotypes for two samples shown earlier in the book (see Figure 1.3) are displayed in Figure 15.3. The output from Genotyper® is split into the three dye colors: blue, green, and yellow. STR genotypes for a batch of samples are typically examined in Genotyper® software first by locus, from smallest to largest within a dye color, and then by dye color (blue, green, and finally yellow).

MANUAL INTERVENTION IN STR GENOTYPE DETERMINATIONS

While STR allele calls may be made in an automated fashion with either Genotyper® or STaR Call™, the resulting genotype information needs to be manually examined by experienced analysts. In Chapter 17, we will discuss several expert systems (i.e., computer programs) that are under development to perform automated STR data review. Data analysis and review is essential for confirming STR results prior to making reports.

Software algorithms follow set parameters and criteria and hence can never be as effective at making difficult calls as a trained examiner. Strict guidelines for data interpretation should be in place to avoid problems with individual bias when the data is reviewed. However, there is always enough variation between data sets that not every situation can be covered by a pre-determined rule.

Laboratories typically have two independent reads of the data by different operators. The genotypes must agree with each other before results will be reported or passed on for uploading to a DNA database. Likewise, a match between two samples is only reported if the two DNA profiles display the same pattern.

FACTORS AFFECTING GENOTYPING RESULTS

There are a number of issues that are important to obtaining accurate genotype results. Some issues are biology related and some are technology related. For example, the amount of stutter or incomplete 3′ nucleotide addition present are biology issues related to the amount of DNA template used in the PCR amplification. On the other hand, 'pull-up' artifacts and threshold issues result from the fluorescent technology and software used for genotyping the samples.

Three parts of the genotyping process illustrated in Figure 15.1 are crucial to the success of genotyping samples. These include the matrix file, the internal size standard, and the allelic ladder sample.

The *matrix file* (termed a *spectral calibration* for ABI 3100 users) is critical for proper color separation in an electropherogram. If the observed peaks are not associated with the proper dye label, then the sample genotype cannot be correctly determined. Matrix files are established by running samples that contain each of the dyes individually. The results of the individual dye runs are combined to form a mathematical matrix that is used to subtract the contribution of other colors in the overlapping spectra (see Chapter 13). A matrix is most accurate under consistent environmental conditions. Thus, if the electrophoresis buffer is changed, a new matrix should be established in order to obtain the most accurate color deconvolution between the different dyes.

The *internal size standard* is necessary for properly sizing of DNA fragment peaks detected in an electropherogram. If any of the peaks in the size standard are below the peak detection threshold established in the data collection and analysis software, then the sizing algorithms will not work properly and STR alleles may be sized incorrectly. An analyst should check to make sure that the internal size standard peaks were all detected properly before proceeding to genotype the STR alleles in a sample.

The *allelic ladder* is the standard to which STR alleles are compared to obtain the sample genotype. The alleles in an allelic ladder need to be resolved from one another and above the peak detection threshold of the data collection and analysis software in order to correctly call STR alleles in unknown samples. The sizes obtained for each allele in the allelic ladder are used to make the final genotype determination in the unknown samples. Therefore, they must be determined correctly.

THE IMPORTANCE OF PRECISION IN ACCURATE GENOTYPING

STR genotyping is performed by comparison of the size of a sample's alleles to size of alleles in allelic ladders for the same loci being tested in the sample. A high degree of precision is needed between multiple runs in order to make an accurate comparison of data from two runs, where one run is the allelic ladder standard and the other run is the questioned sample. The precision for a measurement system is determined by analyzing replicate samples or allelic ladders under normal operating conditions.

Precision for the separation and detection platform must be less than +/− 0.5 bp to accurately distinguish between microvariant (partial repeat) alleles and complete repeat alleles that differ by a single nucleotide (Gill *et al.* 1996). In general the greater the molecular weight of the PCR products, the larger the measurement error. Thus, alleles from larger STR loci such as FGA and D18S51 will generally have a larger size measurement variation than smaller STR loci such as D3S1358 and TH01.

For ABI 310 users, there is a reliance on a high degree of precision for run-to-run comparisons since a number of samples are run in a sequential fashion through the capillary between each injection of the allelic ladder (Lazaruk *et al.* 1998). Even though the samples are analyzed in parallel on a slab gel, a high degree of precision between samples run in different lanes is also necessary since an allelic ladder is present only a few times on each gel. This same principle applies for multi-capillary array systems.

The precision on an ABI 310 instrument is typically better than 0.1 bp (Wallin *et al.* 1998, Applied Biosystems 1998). However, a temperature variation of as little as 2 or 3°C over the course of a number of runs can cause allele peaks to migrate slightly differently from the internal sizing standard and therefore size differently over time. To alleviate this problem, the allelic ladder may be run more frequently (e.g., every 10 injections instead of every 20 injections) and the samples can be typed to the allelic ladder sample that was injected nearest them.

SIZING ALGORITHM ISSUES

The sizing of DNA fragments with internal standards is performed as illustrated in Figure 15.2. The most common algorithm used for determining the DNA fragment size is known as the Local Southern method. This method uses the size of two peaks on either side of the unknown one being measured in order to make the calculations (Elder and Southern 1983). Using the example in Figure 15.2, the '165.05 bp' peak size is determined with Local Southern sizing by the position of the 150 and 160 bp peaks on the lower side and the position of the 200 and 250 bp peaks on the upper side.

The Local Southern method works very well for accurate sizing of DNA fragments over the 100–450 bp size range necessary for STR alleles. However, there are some caveats that should be kept in mind that depend upon the internal size standard used. Within the GS500-ROX and GS500-LIZ size standard, the 250 bp peak (and sometimes the 340 bp peak as well) does not size reproducibly especially when there are temperature fluctuations across or between runs. Therefore, the 250 bp peak is typically left out of analyses by not designating it as a standard peak (Moretti *et al.* 2001, Klein *et al.* 2003).

It is important to realize that unknown DNA fragment peaks cannot be accurately determined which are larger than the peaks present (or designated by the GeneScan software) in the internal sizing standard. Nor can peaks that fall near the edge of the region defined by the internal sizing standard. This is due to the fact that two peaks from the size standard are needed on either side of the unknown peak with Local Southern sizing. Therefore, with the GS500-ROX internal standard commonly used in conjunction with the AmpFℓSTR kits, any unknown peaks falling above 490 bp or below 50 bp will not be sized with the Local Southern method. Likewise, if the signal intensity for any of the calibration peaks in the internal sizing standard is too weak, then unknown peaks in that region will not be sized accurately. For this reason it is important to check that all peaks in the internal sizing standard are above the relative fluorescence threshold to be called as peaks and that these peaks are accurately designated by the software.

Two studies have found that a different sizing algorithm called the Global Southern method works well and maintains a better precision than Local Southern sizing in situations when temperature fluctuations can occur (Hartzell *et al.* 2003, Klein *et al.* 2003). Global Southern involves fitting all of the peaks in the size standard to form a best fit size calibration line rather than just using the two peaks above and below the peak of interest as is done with Local Southern. Regardless of which method is used it must be consistently applied to both the allelic ladders and samples being typed so that equivalent size comparisons may be made.

OFF-LADDER ALLELES

Occasionally a sample may contain an allele that does not fall within 0.5 bp of an allele from the corresponding locus-specific allelic ladder (Gill *et al.* 1996). These alleles are designed as 'off-ladder' alleles or microvariants (see Chapter 6). The off-ladder allele peak may be larger or smaller than the alleles spanning the allelic ladder range or it may fall in between the rungs on the allelic ladder.

If the allele is sized to be less than the ladder, it may be designated as smaller than the smallest allele in the ladder used for genotyping purposes. For example, a CSF1PO allele sized below allele 6, which is the smallest in the ladder, would be designated CSF1PO <6. Likewise an allele sized above allele 15, the largest in the ladder, would be designated CSF1PO >15. Because the alleles in allelic ladders differ between manufacturers, an allele designation of '>15' from an amplification using one STR kit could be equivalent to an allele designation '16' from another.

Alleles that are sized between the rungs on an allelic ladder are usually designated by the number of bases beyond the allele just smaller than it. For example, a TH01 allele sized three bases larger than allele 8 would be designated TH01 8.3. It could also be referred to less specifically as TH01 8.*x*.

Off-ladder alleles can be verified by re-running the amplified product, re-amplifying the sample, or by amplifying the sample with single-locus primers. Heterozygous samples with one 'normal' allele and one microvariant allele make it easy to confirm the microvariant. In this particular case, the normal allele with a full length repeat sequence will fall in an allele bin from the allelic ladder while the microvariant allele possessing a partial repeat sequence will fall between the allele bins created by the allelic ladder (see Figure 6.6). New microvariants are constantly being discovered as more samples are being analyzed around the world at various STR loci. As of April 2004, more than 220 variant 'off-ladder' alleles have been reported for all 13 of the CODIS core STR loci (see http://www.cstl.nist.gov/biotech/strbase/var_tab.htm).

PARTIAL STR PROFILES

If the genomic DNA in a sample is severely degraded or PCR inhibitors are present, only a partial STR profile may be obtained (see Chapter 7). Usually the larger STR loci in a multiplex reaction, such as D18S51 and FGA, will be the first to fail on a degraded DNA sample. When only a partial profile is obtained, the significance of a match will go down because there are not as many loci to compare (see Chapter 22). However, the use of miniplex STR systems with smaller PCR product sizes than those used in commercial STR kits can be used to recover information lost at the larger loci (Butler *et al.* 2003).

MIXTURE INTERPRETATION

Mixtures of DNA from two or more individuals are common in some forensic cases and must be dealt with in the interpretation of the DNA profiles. In evaluating the evidence, an analyst must decide whether the source of the DNA in the questioned sample is from a single individual or more than one person. This may be accomplished by examination of the number of alleles detected at each locus as well as peak height ratios and/or band intensities on a gel (see Chapters 7 and 22). Occasionally extra peaks occur in the data that should not be confused with true alleles.

EXTRA PEAKS OBSERVED IN THE DATA

Electropherograms may contain extra peaks besides the primary target alleles of interest. These peaks can arise from a number of sources related to the biology of STRs and the technology of detecting fluorescently labeled amplification products. It is important to recognize these peaks and not make a false exclusion because of the presence of supposedly spurious peaks in one of the samples.

A laboratory needs to establish criteria to identify a true allele because a DNA typing analyst must decide which peaks contribute to a donor(s) profile(s) and which are due to an artifact. The following material is intended as a helpful guide to some of the commonly seen artifacts and should not be considered a comprehensive list for troubleshooting purposes.

BIOLOGY RELATED ARTIFACT PEAKS

Stutter products are the most common source of additional peaks in an electropherogram of an STR sample. When STR loci are PCR-amplified a minor product peak four bases $(n-4)$ shorter than the corresponding main allele peak is commonly observed (see Chapter 6). Validation studies conducted in a laboratory help define maximum percent stutter for each locus. However, if the target allele peak is off-scale then the stutter product can appear larger than it really is in relationship to the corresponding allele peak (see Moretti *et al.* 2001). For data interpretation, an upper-limit stutter percentage interpretational threshold can be set for each locus as three standard deviations above the highest stutter percentage observed at that locus (Applied Biosystems 1998).

Incomplete 3′(A) nucleotide addition results with amplifications containing too much DNA template or thermal cycling conditions that affect the optimization of the PCR reaction. The *Taq* DNA polymerase used for amplifying STR loci will catalyze the addition of an extra nucleotide, usually an 'A', on the 3′-end of double-stranded PCR products (see Chapter 6). The commercially available multiplex STR kits have been optimized to favor complete adenylation. However, when incomplete 3′ nucleotide addition occurs 'split peaks' will result, sometimes referred to as +/−A, or N and N + 1 peaks, and the allele of interest will be represented by two peaks one base pair apart. Genotyping software may inadvertently call one of these peaks an 'off-ladder' (microvariant) allele.

Tri-allelic (three banded) patterns result from extra chromosomal fragments being present in a sample or the DNA sequence where the primers anneal being duplicated on one of the chromosomes. These rare anomalies are detected by an extra peak at a single locus, as opposed to multiple loci as would likely be seen in a mixture (see Chapter 7). The three peaks will commonly all be of equal intensity but do not have to be (Crouse *et al.* 1999).

Mixed sample results are observed if more than one individual contributed to the DNA profile. Mixtures are readily apparent when multiple loci are examined. An analyst looks for higher than expected stutter levels, more than two peaks at a locus of equivalent intensity, or a severe imbalance in heterozygote peak intensities of greater than 30% (see Chapter 7). It is usually difficult to detect the minor contributor below a level of 1:20 compared to the major donor in the DNA profile.

TECHNOLOGY RELATED ARTIFACT 'PEAKS'

Matrix (multi-component) failure, sometimes referred to as 'pull up' is a result of the inability of the detection instrument to properly resolve the dye colors used to label STR amplicons. This phenomenon is due to spectral overlap. A peak of another color is 'pulled-up' or 'bleeds through' (see Figure 15.4) as a result of exceeding the linear range of detection for the instrument (i.e., sample overloading). A matrix failure is observed as a peak-beneath-a-peak or as an elevation of the baselines for any color. Matrix standards may need to be rerun with the latest set of conditions and a new matrix generated by the software to correct this problem.

Dye blobs (artifacts) occur when fluorescent dyes come off of their respective primers and migrate independently through the capillary. These peaks are fairly broad and possess the spectrum of one of the dyes used for genotyping (see Figure 15.4). Dye artifacts can be removed following PCR using filtration columns (Butler *et al.* 2003).

Air bubbles, urea crystals, or voltage spikes can give rise to a false peak in the ABI 310. These peaks are usually sharp and appear equally intense throughout all four colors (see Figure 15.4). These peaks are not reproducible and should not appear in the same position if the sample is re-injected onto the capillary.

Sample contaminants. Materials which fluoresce in the visible region of the spectrum (~500–600 nm) may interfere with DNA typing when using fluorescent scanners or one of the ABI PRISM systems by appearing as identifiable peaks in the electropherogram. In some early studies conducted by the Forensic Science Service (FSS), a number of fluorescent compounds were examined to determine their apparent mobility when electrophoresed in a polyacrylamide gel (Urquhart *et al.* 1994).

Figure 15.4

*Hypothetical
electropherogram
displaying several
artifacts often observed
with STR typing.*

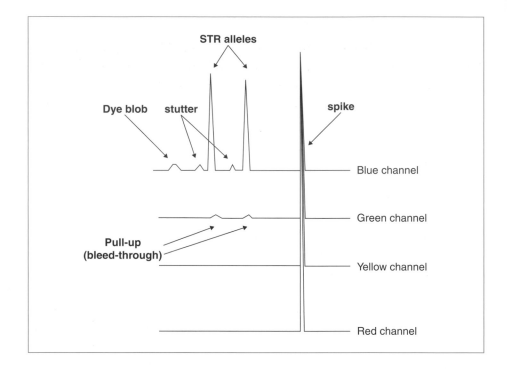

All of the compounds studied, which included antibiotics, vitamins, polycyclic aromatics, fluorescent brighteners, and various textile dyes, could be removed with an organic extraction (i.e., phenol/chloroform, as is commonly used to extract DNA from cells). Interestingly enough, the Chelex method of DNA extraction (see Chapter 3) failed to remove all of the contaminating fluorescent peaks (Urquhart *et al.* 1994). Fortunately, these interfering peaks were usually wide and possessed a broader fluorescent spectrum, which made it fairly easy to distinguish them from the fluorescent dye-labeled PCR products. The FSS researchers concluded that the use of appropriate substrate controls or negative controls in PCR should alleviate this potential problem.

Several possible forensic scenarios exist where sample contaminants may be possible (Urquhart *et al.* 1994):

1. Body fluid stains on dye materials from which the dye may leach during extraction;
2. Body fluid stains on plant material, from which chlorophyll may co-extract with DNA;
3. Blood or tissue samples from individuals with some pathological conditions, e.g., lead poisoning or some forms of porphyria, in which blood porphyrin levels are greatly elevated; and
4. Bone or tooth samples from individuals who were treated with tetracycline-group antibiotics in their youth as growing bones and teeth are known to incorporate and accumulate these antibiotics.

DEVELOPING AN INTERPRETATION STRATEGY

A forensic DNA laboratory should develop its own STR interpretation guidelines based upon their own validation studies and results reported in the literature
(SWGDAM 2000). Practical experience with instrumentation and results from
performing casework are also important factors in developing an interpretation
strategy.

■ Conduct necessary validation studies (see Chapter 16) and gain experience in your lab

■ Utilize analyst's experience

■ Use literature references as a resource in understanding if an 'off-ladder' allele has
been observed before. As a useful resource, we have included a list of all known alleles
for the 13 CODIS STR loci in Appendix I. A more up-to-date list can be found in the
STRBase STR fact sheets for each STR locus or the non-published variant allele page
(http://www.cstl.nist.gov/biotech/STRBase/var_tab.htm).

Validation studies will define observed stutter ratios for each locus, establish
minimum peak heights, and define heterozygous peak ratios within a locus.
When in doubt on a sample's correct result, the sample should be re-tested.
This may be as simple as re-injecting it on the ABI 310 or putting another
aliquot of the sample on the next gel. Even if sample re-testing involves
re-extracting and/or re-amplifying the 'problem' sample, it is worthwhile in
order to obtain an accurate result.

A MATCH OR NOT A MATCH: THAT IS THE QUESTION...

Generally, the process of comparing two or more samples is limited to one of
three possible outcomes that are submitted in a case report:

1. *Match* – Peaks between the compared STR profiles have the same genotypes and no
   unexplainable differences exist between the samples. Statistical evaluation of the
   significance of the match is usually cited in the match report (see Chapter 21).
   Alternatives for presentation of a match range from statements of identity, to compu-
   tations of the likelihood ratio for the hypothesis that the defendant is the source, to
   descriptions of random-match probabilities in various populations, to a simple qualitative report of a match with no statistics behind its significance (see NRC II, p. 192).
2. *Exclusion* (Non-match) – The genotype comparison shows profile differences that can
   only be explained by the two samples originating from different sources.
3. *Inconclusive* – The data does not support a conclusion as to whether the profiles
   match. This finding might be reported if two analysts remain in disagreement after
   review and discussion of the data and it is felt that insufficient information exists to
   support any conclusion.

If a match is observed between a suspect and crime-scene evidence, then three possibilities exist: (1) the suspect deposited the sample, (2) the suspect did not provide the sample but has the profile by chance, and (3) the suspect did not provide the sample and the matching result is a false positive due to laboratory error. The first explanation is the basis behind the use of DNA testing in the criminal justice system. The second possibility depends on population genetic principles that are covered in Chapters 19–21 from which the probability of a random match is determined. The third explanation of why a match might occur concerns the possibility of laboratory mistakes. Chapter 16 discusses laboratory validation and proficiency tests that are in place to prevent or reduce the possibility of error in performing DNA testing. Generally speaking a great deal of effort goes into insuring reliable forensic DNA testing although laboratory errors have been reported.

In forensic DNA typing, if any one STR locus fails to match when comparing the genotypes between two or more samples, then the profiles between the questioned and reference sample will be declared a non-match, regardless of how many other loci match.

Paternity testing is an exception to this because of the possibility of mutational events (see Chapters 6 and 23). When analyzing and reporting the results of parentage cases, an allowance for one or even two possible mutations is often made. In other words, if 13 loci are used and the questioned parentage is included for all but one locus, the data from the non-inclusive allele will be attributed to a possible mutation.

In the end, interpretation of results in forensic casework is a matter of professional judgment and expertise. Interpretation of results within the context of a case is the responsibility of the case analyst with supervisors or technical leaders conducting a follow-up verification of the analyst's interpretation of the data as part of the technical review process. When coming to a final conclusion regarding a match or exclusion between two or more DNA profiles, laboratory interpretation guidelines should be adhered to by both the case analyst and the supervisor. However, as experience using various analytical procedures grows, interpretation guidelines may evolve and improve. These guidelines should always be based on the use of proper controls and validated methods as described in the next chapter.

REFERENCES AND ADDITIONAL READING

Applied Biosystems (1998) *AmpFlSTR® Profiler Plus™ PCR Amplification Kit User's Manual*. Foster City, California: Applied Biosystems.

Butler, J.M., Shen, Y. and McCord, B.R. (2003) *Journal of Forensic Sciences*, 48, 1054–1064.

Butler, J.M., Buel, E., Crivellente, F. and McCord, B.R. (2004) *Electrophoresis*, 25, 1397–1412.

Crouse, C., Rogers, S., Amiott, E., Gibson, S. and Masibay, A. (1999) *Journal of Forensic Sciences*, 44, 87–94.

Elder, J.K. and Southern, E.M. (1983) *Analytical Biochemistry*, 128, 227–231.

Gill, P., Kimpton, C.P., Urquhart, A., Oldroyd, N.J., Millican, E.S., Watson, S.K. and Downes, T.J. (1995) *Electrophoresis*, 16, 1543–1552.

Gill, P., Urquhart, A., Millican, E.S., Oldroyd, N.J., Watson, S., Sparkes, R. and Kimpton, C.P. (1996) *International Journal of Legal Medicine*, 109, 14–22.

Hartzell, B., Graham, K. and McCord, B. (2003) *Forensic Science International*, 133, 228–234.

Klein, S.B., Wallin, J.M. and Buoncristiani, M.R. (2003) *Forensic Science Communications*, 5. Available at: http://www.fbi.gov/hq/lab/fsc/backissu/jan2003/klein.htm.

Lazaruk, K., Walsh, P.S., Oaks, F., Gilbert, D., Rosenblum, B.B., Menchen, S., Scheibler, D., Wenz, H.M., Holt, C. and Wallin, J. (1998) *Electrophoresis*, 19, 86–93.

Moretti, T.R., Baumstark, A.L., Defenbaugh, D.A., Keys, K.M., Smerick, J.B. and Budowle, B. (2001) *Journal of Forensic Sciences*, 46, 647–660.

National Research Council (1996) *NRC II: The Evaluation of Forensic DNA Evidence*. Washington, DC: National Academy Press.

Scientific Working Group on DNA Analysis Methods (SWGDAM) (2000) Short tandem repeat (STR) interpretation guidelines. *Forensic Science Communications* 2 (3); Available online at: http://www.fbi.gov/hq/lab/fsc/backissu/july2000/strig.htm.

Smith, R.N. (1995) *BioTechniques*, 18, 122–128.

Urquhart, A., Chiu, C.T., Clayton, T.M., Downes, T., Frazier, R.R.E., Jones, S., Kimpton, C.P., Lareu, M.V., Millican, E.S., Oldroyd, N.J., Thompson, C., Watson, S., Whitaker, J. P. and Gill, P. (1994) *Proceedings from the Fifth International Symposium on Human Identification*, pp. 73–83. Madison, WI: Promega Corporation.

Wallin, J.M., Buoncristiani, M.R., Lazaruk, K., Fildes, N., Holt, C. and Walsh, P.S. (1998) *Journal of Forensic Sciences*, 43, 854–870.

Ziegle, J.S., Su, Y., Corcoran, K.P., Nie, L., Mayrand, P.E., Hoff, L.B., McBride, L.J., Kronick, M.N. and Diehl, S.R. (1992) *Genomics*, 14, 1026–1031.

CHAPTER 16

LABORATORY VALIDATION

Laboratories should adhere to high quality standards (such as those defined by TWGDAM and the DNA Advisory Board) and make every effort to be accredited for DNA work (by such organizations as ASCLD-LAB).

(NRC II Report, p. 4. Recommendation 3.1)

INTRODUCTION

IMPORTANCE OF QUALITY CONTROL

Any scientific test which results in information that may lead to the loss of liberty for an individual accused of a crime needs to be performed with utmost care. DNA typing is no exception. It is a multi-step, technical process that needs to be performed by qualified and effectively trained personnel to ensure that accurate results are obtained and interpreted correctly. When the process is conducted properly, DNA testing is a capable investigative tool for the law enforcement community with results that stand up to legal scrutiny in court. When laboratories do not follow validated protocols, problems can arise (see D.N.A. Box 16.1).

Two topics are commonly referred to when discussing the importance of maintaining good laboratory practices to obtain accurate scientific results: quality assurance and quality control. *Quality assurance* (QA) refers to those planned or systematic actions necessary to provide adequate confidence that a product or service will satisfy given requirements for quality. *Quality control* (QC), on the other hand, usually refers to the day-to-day operational techniques and the activities used to fulfill requirements of quality.

Thus, an organization plans QA measures and performs QC activities in the laboratory. The forensic DNA community has long recognized the importance of quality control and since early in the development of forensic DNA technology, has established organizations to recommend and oversee quality assurance guidelines and quality control measures.

DEFINITIONS

As we begin our discussion of laboratory validation, it is important to define several words that will be frequently used throughout the chapter. These words

An alarming audit of the Houston Police Department (HPD) Crime Laboratory in December 2002 found that a number of problems abounded in this unaccredited laboratory. In a poorly funded and managed environment, laboratory personnel were not adequately trained, evidence was often consumed, and even the roof in the evidence storage area leaked from rain damage. The city of Houston shut down operations in the HPD laboratory and outsourced hundreds of cases for review to a private Houston-based laboratory named Identigene. While most of the retesting supported the original conclusions that the suspect in a case could be included in contributing the crime sample, unfortunately errors in data interpretation by the HPD laboratory led to the false conviction and incarceration of a young man accused of a 1998 rape.

In March 2003 Josiah Sutton's case made national headlines when it was revealed that DNA tests performed by Identigene found that he could not have committed the crime for which he had been incarcerated for more than four years based on DNA evidence originally analyzed by HPD. With this news also came the stigma that DNA testing was fallible.

It is important to point out that many of the problematic tests performed by the HPD laboratory involved DNA mixtures and the use of an earlier, low-resolution PCR-based test known as HLA-DQA1 rather than the current and more precise method of STR typing. Since only six alleles are possible with DQA1 typing, it is inherently poor at separating mixture components.

As of mid-2004, over $4.6 million has been allocated for retesting of samples from almost 400 cases originally handled by the HPD laboratory. Thus, failure to achieve laboratory accreditation, properly train personnel, maintain adequate facilities, and follow guidelines for data interpretation can cost significantly more than just a laboratory's reputation. In May 2004, the HPD crime laboratory finally applied for accreditation with ASCLD/LAB. Hopefully in the coming months this laboratory will join the ranks of the careful laboratories around the world conducting quality forensic DNA testing.

The FBI DNA Laboratory has also come under fire in recent years largely due to the deceitful actions of a forensic biologist named Jacqueline Blake. Ms. Blake apparently ran over 100 cases in the FBI's DNA Analysis Unit I without performing testing of her negative control samples – and then falsified documents to make it appear as though she had followed the standard operating procedure. The Department of Justice's Office of the Inspector General issued a report in May 2004 reviewing the protocol and practice vulnerabilities of the FBI DNA Laboratory so that this type of failure is not observed again.

It is important to keep in mind that these two cases represent the rare exception rather than the rule as the vast majority of forensic laboratories work hard to be accredited, maintain analyst training and proficiency, carefully validate methods, and follow standard operating procedures. The science itself is sound and reliable when performed correctly. These situations simply illustrate the need for consistent internal quality assurance and external oversight to ensure procedural accuracy within a laboratory.

Source:
http://www.chron.com/content/chronicle/special/03/crimelab/index.html (Houston Chronicle); 'The FBI DNA Laboratory: A Review of Protocol and Practice Vulnerabilities' (Dept of Justice Office of the Inspector General, May 2004), see http://www.usdoj.gov/oig/special/0405/final.pdf

include validation, proficiency testing, laboratory accreditation, and the terms robust, reliable, and reproducible.

Validation refers to the process of demonstrating that a laboratory procedure is robust, reliable, and reproducible in the hands of the personnel performing the test in that laboratory. A *robust method* is one in which successful results are obtained a high percentage of the time and few, if any, samples need to be repeated. A *reliable method* refers to one in which the obtained results are accurate and correctly reflect the sample being tested. A *reproducible method* means that the same or very similar results are obtained each time a sample is tested. All three types of methods are important for techniques performed in forensic laboratories.

A *proficiency test*, as it relates to the DNA typing field, is an evaluation of a laboratory's performance in conducting DNA analysis procedures. These tests are performed periodically, usually on a semi-annual basis, for each DNA analyst or examiner. In fact, the DNA Advisory Board Standard 13.1 requires that each DNA analyst undergo an external proficiency test at regular intervals not exceeding 180 days (see Appendix IV). Biological specimens with a previously determined DNA profile are submitted to the laboratory personnel being tested. The purpose of the test is to evaluate their ability to obtain a concordant result using the laboratory's approved *standard operating protocols* (SOPs).

The tests may be administered by someone else in the laboratory (*internal proficiency test*) or by an external organization (*external proficiency test*). If the test administered by an external organization is performed such that the laboratory personnel do not know that a test is being conducted, then it is termed a *blind external proficiency test*. A blind external proficiency test is generally considered the most effective at monitoring a laboratory's abilities but can be rather expensive and time-consuming to arrange and conduct (Peterson *et al.* 2003a, 2003b). Participation in a proficiency-testing program is an essential part of a successful laboratory's quality assurance effort. Forensic laboratories develop their own proficiency-testing program or establish one in cooperation with other laboratories (see Rand *et al.* 2002, 2004). The German DNA profiling group (GEDNAP) has established a successful blind proficiency-testing program (D.N.A. Box 16.2).

A *laboratory audit* evaluates the entire operation of a laboratory. It is a systematic examination that may be conducted by the laboratory management or by an independent organization according to pre-established guidelines. A laboratory must possess standard operating protocols and adhere to them. Likewise, instruments and other equipment vital to the successful completion of a forensic DNA case must be maintained properly and personnel must be appropriately trained to perform their jobs. Records of an audit are maintained and serve to describe the findings of the audit and a course of action that may be taken to resolve any existing problems.

The purpose of proficiency testing is to evaluate the performance of an analyst using a sample or set of samples that is unknown to the analyst but known to the test provider. Recommendation 3.2 of NRCII (see Appendix VI) states that: 'Laboratories should participate regularly in proficiency tests, and the results should be available for court proceedings.' Successful completion of this examination permits a degree of confidence to exist in how an analyst might perform on a real forensic case sample. Unfortunately, if analysts are aware that they are being tested, they might be more careful than they would when normally processing routine samples on a daily basis. Thus, the concept of blind proficiency has often been discussed in order to have a true test of the entire system because the analysts would not know that they were being tested. However, a number of challenges and costs are associated with blind proficiency tests.

Four models exist for blind proficiency testing (Peterson *et al.* 2003a): (1) Blind/Law Enforcement, where a law enforcement agency fully disguises the test as a routine case and participates in the deception of the target laboratory; (2) Blind/Conduit Lab, where another laboratory perhaps part of a multi-laboratory state system submits appropriate specimens for a case and is part of the deception of the target laboratory; (3) Blind Analyst, where only the DNA analysts are in the dark about the test while laboratory QA coordinators administer the test, and (4) Random Audit/Re-analysis, where a re-examination of a case is performed by another analyst or an auditor external to the laboratory to review and even re-analyze the samples. Cost estimates for these various forms of blind proficiency range from $1400–$10 000 per test with the blind analyst approach (number 3) being the least expensive and easiest to implement. Thus, running a program that tests say 150–200 DNA laboratories in the United States with two tests per year would become rather expensive.

Another challenge besides cost is the fact that deception of the other party is necessary in an effort to create a 'real' case situation. When laboratories and law enforcement agencies are trying to build trust, deception for the purpose of quality assurance may seem a bit extreme. In addition, protection of specimen donors is important and if a 'case' is entered into CODIS, can the profiles for the innocent donors be purged? Likewise, if only a few donors are used in the proficiency test, then DNA laboratories might figure out in a short period of time which profiles are part of the blind proficiency test. Based upon these and other considerations, implementation of a large national blind proficiency-testing program was not recommended in the U.S. (Peterson *et al.* 2003b).

In Europe, the German DNA profiling group (GEDNAP) has developed a blind trial concept, which is really a 'graded' inter-laboratory test. The primary requirement of this blind trial is that all participants receive exactly the same material to be tested enabling a direct comparison with the known standard as well as an inter-laboratory comparison to be carried out (Rand *et al.* 2002). Samples are prepared to be as close to real casework situations as possible.

The GEDNAP trials have four purposes: (1) standardization of methods and procedures; (2) standardization of nomenclature; (3) evaluation of the competence of a laboratory to obtain the correct result, and (4) elimination of errors in typing. GEDNAP trial 22 and 23 conducted in 2001 had 122 participating laboratories from 28 European countries while participation grew to 160 laboratories for trial 26 and 27 held in 2003 (Rand *et al.* 2004).

Laboratories are each assigned a code number that enables anonymity throughout the inter-laboratory process. Typically seven samples are provided with each GEDNAP trial. When results are returned to the organizing laboratory in Münster, each allele call is classified into one of four categories: (1) no errors; (2) mixture not detected; (3) error in typing but would not be reported; and (4) error in typing which would be reported. Finally, a certificate is issued by the organizing laboratory, which states that the laboratory in question has successfully completed the blind trial for the particular loci examined (Rand *et al.* 2002).

The types of errors observed in the GEDNAP trials show that human carelessness is the predominant source of error with transposition of samples and transcription errors (Rand *et al.* 2004). The error rate over the past few years has held relatively constant at 0.4–0.7%.

Sources:

Peterson, J. *et al.* (2003) The feasibility of external blind DNA proficiency testing. I. Background and findings. *Journal of Forensic Sciences*, 48, 21–31.

Peterson, J. *et al.* (2003) The feasibility of external blind DNA proficiency testing. II. Experience with actual blind tests. *Journal of Forensic Sciences*, 48, 32–40.

Rand, S. *et al.* (2002) The GEDNAP (German DNA profiling group) blind trial concept. *International Journal of Legal Medicine*, 116, 199–206.

Rand, S. *et al.* (2004) The GEDNAP blind trial concept part II. Trends and developments. *International Journal of Legal Medicine*, 118, 83–89.

Laboratory accreditation results from a successful completion of an inspection or audit by an accrediting body. A list of major accrediting organizations that are recognized by the forensic DNA community is contained in the next section. Accreditation requires that the laboratory demonstrates and maintains good lab practices including chain-of-custody and evidence handling procedures.

The accreditation process generally involves several steps such as a laboratory self-evaluation, filing application and supporting documents to initiate the accreditation process, on-site inspection by a team of trained auditors, an inspection report, and an annual accreditation review report. The inspection evaluates the facilities and equipment, the training of the technical staff, the written operating and technical procedures, and the casework reports and supporting documentation of the applicant laboratory.

According to the DNA Advisory Board Standards accepted as the national standards in the United States (see Appendix IV), all examiners who are actively engaged in DNA analysis need to undergo proficiency tests on at least a semi-annual basis (standard 13.1). Likewise, laboratory audits need to be conducted on an annual basis by the laboratory (standard 15.1) and on a bi-annual basis by an outside agency (standard 15.2).

ORGANIZATIONS INVOLVED IN ENSURING QUALITY AND UNIFORMITY OF DNA TESTING

A number of organizations exist around the world that work on a local, national, or international level to ensure that DNA testing is performed properly. The organizations are made up primarily of working scientists who want to coordinate their efforts to benefit the DNA typing community as a whole.

ORGANIZATIONS BASED IN THE UNITED STATES

The *Technical Working Group on DNA Analysis Methods* (TWGDAM) was established in November 1988 under FBI Laboratory sponsorship to aid forensic DNA scientists throughout North America. The first meeting consisted of 31 scientists representing 16 forensic laboratories in the United States and Canada and two research institutions. TWGDAM meetings were originally held twice a year at the FBI Academy in Quantico, Virginia, usually in January and July. More recently, public meetings have also been held in conjunction with scientific meetings such as the International Symposium on Human Identification, sponsored each fall by the Promega Corporation.

The original TWGDAM chairman was James Kearney of the FBI Laboratory who was followed by Bruce Budowle also of the FBI Laboratory. In 1998 the TWGDAM name was changed to the Scientific Working Group on DNA Analysis Methods or SWGDAM. In October 2000, the SWGDAM chairman became Richard Guerrieri of the FBI Laboratory's DNA Analysis Unit I, who will serve for a term of six years.

Over the years, several subcommittees have operated to bring recommendations before the SWGDAM group. These subcommittees have included the restriction fragment length polymorphism (RFLP), polymerase chain reaction (PCR), Combined DNA Index System (CODIS), mitochondrial DNA, short tandem repeat (STR) interpretation, training, validation, Y chromosome, expert systems, and quality assurance working groups. TWGDAM issued guidelines for quality assurance in DNA analysis in 1989, 1991, and 1995. Updated validation guidelines were approved by SWGDAM in July 2003.

Evolving technology and laboratory practices made it necessary to issue revisions in the quality assurance standards for DNA testing. These QA guidelines were originally intended to serve as a guide to laboratory managers in establishing their own QA program. However, the 1995 'Guidelines for a Quality

Assurance Program for DNA Analysis' served as the *de facto* standards for forensic DNA testing until October 1998, when the ensuing DNA Advisory Board standards went into effect (see Appendix IV).

The *DNA Advisory Board* (DAB) is a congressionally mandated organization that was created and funded by the United States Congress DNA Identification Act of 1994. The first meeting of the DAB was held on 12 May 1995, and chaired by Nobel laureate Dr. Joshua Lederberg. The DAB consists of 13 voting members that include scientists from state, local, and private forensic laboratories; molecular geneticists and population geneticists not affiliated with a forensic laboratory; a representative from the National Institute of Standards and Technology; the chair of TWGDAM; and a judge. The DAB was created for a five-year period to issue standards for the forensic DNA community. Following conclusion of the DAB's responsibilities in 2000, SWGDAM now operates as the group responsible for offering recommendations to the forensic community within the United States.

The *American Society of Crime Laboratory Directors* (ASCLD) and its Laboratory Accreditation Board (ASCLD/LAB) play an important role in the United States as well as internationally for laboratory accreditation programs. The ASCLD/LAB motto is 'quality assurance through inspection.' The Crime Laboratory Accreditation Program is a voluntary program in which any crime laboratory may participate to demonstrate that its management, operations, personnel, procedures and instruments meet stringent standards. The goal of accreditation is to improve the overall service of forensic laboratories to the criminal justice system. If a forensic laboratory is interested in becoming accredited, an ASCLD/LAB Accreditation Manual is available from the Executive Secretary for a fee. As of April 2004, a total of 265 crime laboratories are accredited by ASCLD/LAB although not all of them are doing DNA testing. For additional information on ASCLD, visit its web site: www.ascld-lab.org.

The *National Forensic Science Training Center* (NFSTC) located in Largo, Florida, has an accreditation program and offers to certify laboratories that comply with the SWGDAM/DAB guidelines. NFSTC accreditation is generally sought by contract service laboratories doing DNA database work since ASCLD/LAB accreditation is only available to forensic laboratories performing casework. For additional information on NFSTC, visit its web site: www.nfstc.org.

The *American Association of Blood Banks* (AABB) sets the national standards for laboratories performing DNA parentage testing. AABB provides accreditation for paternity testing laboratories. As of April 2004, there were 41 accredited paternity testing laboratories in the United States. For more information on AABB, visit their web site: www.aabb.org.

The *College of American Pathologists* (CAP) offers external proficiency testing to forensic and paternity testing labs as well as clinical laboratories. For further information on CAP, visit their web site: www.cap.org.

Cellmark Diagnostics, now *Orchid Cellmark*, a forensic DNA testing laboratory, provides a proficiency test to help ensure on-going laboratory quality. Their International Quality Assessment Scheme (IQAS) DNA Proficiency Test Program is designed for all laboratories conducting forensic DNA analysis. The proficiency tests consist of simulated forensic evidence case samples that are distributed four times a year. The Cellmark tests include questioned bloodstain and semen stain evidence along with known samples of blood. For more information on Orchid Cellmark, visit their web site: www.cellmark-labs.com.

Collaborative Testing Services, Inc. (CTS) is an ASCLD/LAB proficiency test provider offering six different tests in its forensic biology program. For more information on CTS, visit their web site: www.cts-interlab.com.

The *Human Identity Trade Association* (HITA) is a non-profit organization that represents the interests of DNA companies and suppliers within the human identity market. HITA generally meets in conjunction with the International Symposium on Human Identification each Fall. For additional information on HITA, visit the organization's web site: www.hita.org.

The *National Institute of Standards and Technology* (NIST) develops standard reference materials (SRMs) that may be used by forensic laboratories to calibrate and verify their analytical procedures. Under the DAB standards, a laboratory is required to check its DNA procedures annually or whenever substantial changes are made to the protocol(s) against an appropriate and available NIST standard reference material or a standard traceable to a NIST standard (DAB standard 9.5, see Appendix IV). The various SRMs available from NIST are described below in the section on DNA standards. For additional information regarding NIST, visit its web site: www.nist.gov.

Quality Forensics, Inc., is an ASCLD/LAB proficiency test provider offering sets of samples to assess DNA casework, DNA database, and mitochondrial DNA proficiency. For additional information regarding Quality Forensics, visit their web site: www.qualityforensics.com.

Serological Research Institute (SERI) is another ASCLD/LAB proficiency test provider with body fluid identification and mock case proficiencies offered to forensic laboratories. For more information on SERI, visit their web site: www.serological.com.

ORGANIZATIONS BASED IN EUROPE

The International Society for Forensic Genetics (ISFG) formerly known as the *International Society of Forensic Haemogenetics* (ISFH) is an international organization responsible for the promotion of scientific knowledge in the field of genetic markers analyzed with forensic purposes. The ISFG includes more than 800 members from 50 countries. Regular meetings are typically held bi-annually on an international level. Conference volumes were published under the title 'Advances in Forensic Haemogenetics' and are now titled 'Progress in Forensic Genetics'. Recommendations on forensic genetic analysis are also proposed as needed by an *ad hoc* DNA Commission. Several formal recommendations have come from this group regarding the use of STR markers (DNA Commission 1992, Bar *et al.* 1994, Bar *et al.* 1997), mitochondrial DNA (Carracedo *et al.* 2000), and Y chromosome markers (Gill *et al.* 2001). For additional information on ISFG, visit its web site: http://www.isfg.org.

The *European Network of Forensic Science Institutes* (ENFSI) was started in 1995 to set standards for exchange of data between European member states and to be an accrediting body through conducting laboratory audits. Within the ENFSI, there is a DNA working group that meets twice a year to discuss forensic DNA protocols and research in much the same fashion as SWGDAM does within the United States.

The European forensic DNA community has another organization similar to SWGDAM named *EDNAP* (*European DNA Profiling Group*). EDNAP is effectively a working group of the international Society of Forensic Genetics and consists of representatives from more than a dozen European nations. EDNAP has conducted a series of inter-laboratory studies on various STR markers to investigate the reproducibility of multiple laboratories in testing the same samples (Table 16.2). These studies have demonstrated that with the proper quality control measures excellent reproducibility can be seen between forensic laboratories. For additional information on EDNAP, visit its web site: http://www.isfg.org/ednap/ednap.htm.

Another European organization for standardizing forensic DNA methods is *STADNAP*, which is an acronym for Standardization of DNA Profiling Techniques in the European Union. The goals of STADNAP include defining criteria for the selection of forensic DNA typing systems used among the European countries, exchanging and comparing methods for unifying protocols used for DNA typing, and compiling reference allele frequency databases for European populations. For more information on STADNAP, visit its web site: www.stadnap.uni-mainz.de.

The *Interpol European Working Party on DNA Profiling* (IEWPDP) consists of DNA experts from Belgium, the Czech Republic, Germany, Hungary, Italy, Netherlands, Norway, Slovakia, Spain, and the United Kingdom. IEWPDP makes recommendations concerning the use of DNA evidence in criminal investigations with the goal of facilitating a wider use of this technique in Europe. For example, Interpol recommended that the European standard set of STR loci include FGA, TH01, VWA, and D21S11. For more information on Interpol's DNA efforts, visit their web site: http://www.interpol.int/Public/Forensic/dna/default.asp.

ORGANIZATIONS FOR LATIN AMERICA AND THE IBERIAN PENINSULA

The *Grupo Iberoamericano de Trabajo en Analisis de DNA* (GITAD) was organized in 1998 to serve the needs of forensic DNA laboratories and institutions in Latin America and the two countries of the Iberian Peninsula – Spain and Portugal. Much like SWGDAM and ENSFI, the primary objectives of GITAD include standardizing techniques, implementing a quality assurance/quality control system, and facilitating communication and training of laboratory personnel. For additional information regarding GITAD, visit their web site: www.gitad.org.

VALIDATION

Validation is a very important part of forensic DNA typing. Defense lawyers today rarely challenge the science behind DNA typing – rather they challenge the process by which the laboratory performs the DNA analysis. Thus, the scientific community must carefully document the validity of new techniques and technologies to ensure that procedures performed in the laboratory accurately reflect the examined samples. In addition, a laboratory must carefully document their technical procedures and policies for interpretation of data and follow them to guarantee that each sample is handled and processed appropriately.

STR VALIDATION

There are generally considered to be two stages to validation: developmental validation and internal validation. *Developmental validation* involves the testing of new STR loci or STR kits, new primer sets, and new technologies for detecting STR alleles. *Internal validation*, on the other hand, involves verifying that established procedures examined previously under the scrutiny of developmental validation (often by another laboratory) will work effectively in one's own laboratory. Developmental validation is typically performed by commercial STR kit manufacturers and large laboratories such as the FBI Laboratory while internal validation is the primary form of validation performed in smaller local and state forensic DNA laboratories.

DEVELOPMENTAL VALIDATION STUDIES

The standard studies that are conducted to become 'SWGDAM Validated' are listed below with the original TWGDAM validation guidelines (1995) numerical headings shown in parentheses. The purpose of each study is also enumerated. Although SWGDAM approved an updated version of these validation guidelines in July 2003 (SWGDAM 2004), the 1995 version is the basis for most validation studies to date with STR typing kits.

■ *Standard specimens* (4.1.5.1). DNA is isolated from different tissues and body fluids coming from the same individual and tested to make sure that the same type is observed. These studies are important because the blood from a suspect might be used to try and match semen found at a crime scene.

■ *Consistency* (4.1.5.2). The measurement technique is evaluated repeatedly to assess the reproducibility of the method within and between laboratories. The power of DNA testing is only fully realized when results can be compared between laboratories in different areas or database samples that were analyzed some time before a crime was committed. Thus, results must be comparable across both distance and time. The use of internal sizing standards and allelic ladders has greatly improved the consistency of STR typing.

■ *Population studies* (4.1.5.3). A set of anonymous samples that have been grouped by ethnicity is analyzed to determine allele frequencies for each major population group that exists in a forensic laboratory's vicinity. These allele frequencies are then used in reporting population statistics and calculating the probability of a random match (see Chapter 21).

■ *Reproducibility* (4.1.5.4). Dried blood and semen stains are typed and compared to DNA profiles obtained from liquid samples. Samples from the same source should match. Obviously, this fact is important since a crime scene stain should match the reference blood sample of a suspect if he or she is the perpetrator of the crime.

■ *Mixed specimen studies* (4.1.5.5). The ability of the DNA typing system to detect the various components of mixed specimens is investigated. Evidence samples in forensic cases often originate from more than one individual and thus it is essential that typing systems can detect mixtures. Several studies are typically conducted to define the limitations of the DNA typing system. Genomic DNA from two samples of known genotype is often mixed in various ratios ranging from 50:1 to 1:50. The limit of detection for the minor component is determined by examining the profiles of the mock mixtures (see Chapter 7). Studies are also performed to examine the peak height ratios of heterozygote alleles within a locus and to determine the range of stutter percentages for each allele of each locus (see Chapter 6). The results of these relative peak height measurements can then be used to establish guidelines for separating a minor component of a mixture from the stutter product of a single source sample.

■ *Environmental studies* (4.1.5.6). Samples of known genotype are environmentally stressed and examined to verify that the correct genotype is obtained. The environmental studies reflect the situations typical of a forensic case (i.e., exposure to sunlight, humidity, and temperature fluctuations).

■ *Matrix studies* (4.1.5.7). Samples of known genotype are examined after contact with a variety of substrates commonly encountered in forensic cases. For example, blood and semen may be deposited on leather, denim, glass, metal, wool, or cotton as well as mixed with dyes and soil. DNA profiles from samples exposed to these substrates are carefully examined for non-specific artifacts and amplification failure at any of the loci studied.

■ *Non-probative evidence* (4.1.5.8). DNA profiles are obtained from samples that are from forensic cases that have already been closed. These samples demonstrate that the DNA typing system being examined can handle real casework situations.

■ *Non-human studies* (4.1.5.9). The DNA typing system being evaluated is subjected to non-human DNA to see if other biological sources could interfere with the ability to obtain reliable results on samples recovered from crime scenes. Primates, such as gorillas and chimpanzees, are typically tested along with domestic animals, such as horses, cattle, dogs, and cats. Bacteria and yeast, which can be prevalent at some crime scenes, are also tested. Most STR loci used for human identity testing are primate-specific, that is, they amplify in gorillas and chimps but not dogs or cats. Bacteria, yeast, and most non-primates do not yield any detectable products with the STR kits currently available (Applied Biosystems 1998, Micka *et al.* 1999). The sex-typing marker amelogenin does amplify in a number of other species but with DNA fragments that are slightly smaller in size than the standard 106 and 112 bp for human X and Y alleles (Buel *et al.* 1995).

■ *Minimum sample* (4.1.5.10). The minimum quantity of genomic DNA needed to obtain a reliable result is typically determined by examining a dilution series of a sample with a known genotype. For example, 10 ng, 5 ng, 2 ng, 1 ng, 0.5 ng, 0.25 ng and 0.1 ng might be evaluated. Most protocols call for using at least 0.25–0.5 ng genomic DNA for PCR amplification to avoid allele dropout from stochastic effects during the PCR step or poor sensitivity during the detection phase of the analysis.

ADDITIONAL DEVELOPMENT AND INTERNAL VALIDATION STUDIES

■ *Precision studies.* The calculated base pair sizes for STR allele amplification products are measured. All measured alleles should fall within a ± 0.5 bp window around the measured size for the corresponding allele in the allelic ladder.

■ *Stutter studies.* The percentage of observed stutter at each STR locus is examined by calculating the ratio of the stutter peak area and/or peak height compared to the corresponding allele peak area and height. Stutter values are derived from homozygotes and heterozygotes with alleles separated by at least two repeat units. The upper levels of stutter observed for each locus are then used to develop interpretation guidelines. Because the levels of stutter for each of the 13 CODIS STR loci have been described and usually fall below 10% of the allele peak area and height, many labs just use a standard 15% cut-off for interpreting stutter products. If the stutter peak is below 15% of the allele peak, it is ignored as a biological artifact of the sample. However, if it is above 15% then a possible mixture could be present in the sample (see Chapter 7).

■ *Heterozygous peak height balance.* The peak heights of the smaller and the larger allele are compared typically by dividing the smaller sized allele peak height by the larger sized allele peak height. In other words, the height of the lower peak in relative fluorescence units (RFU) is divided by the height of the higher peak (in RFU). This peak height ratio is expressed as a percentage. The average heterozygote peak height ratio is usually greater than 90% meaning that a heterozygous individual generally possesses well-balanced peaks. Ratios below 70% are rare in normal, unmixed samples (although primer point mutations can cause one of the alleles to not amplify as well, i.e., a partial null allele – see discussion on null alleles in Chapter 6).

- *Annealing temperature studies.* These studies are conducted by running the amplification protocol with the annealing temperature either two degrees above or two degrees below the optimal temperature. Annealing temperature studies are important because thermal cyclers might not always be calibrated accurately and can drift over time if not maintained properly. Thus, an operator might think that the annealing temperature during each cycle is 59°C when in fact the thermal cycler is running hotter at 61°C. If any primers in the multiplex mix are not capable of withstanding slight temperature variation (i.e., they do not hybridize as well), then a locus could dropout or non-specific amplification products could arise.

- *Cycle number studies.* The optimal PCR conditions (i.e., denaturing, annealing, and extension temperatures and times) are examined with a reduced number of cycles as well as a higher number of cycles than the standard protocol calls for to evaluate the performance of the STR multiplex system. The sensitivity of detection of alleles for each locus is dependent of course on the quality of input DNA template. The cycle number studies permit a laboratory to determine the tolerance levels of a STR multiplex system with various amounts of DNA template. While a higher number of PCR cycles (e.g., 34 instead of 28) might be able to better amplify very low levels of genomic DNA, the likelihood of nonspecific amplification products arising increases with higher numbers of PCR cycles.

PUBLICATION AND PRESENTATION OF VALIDATION RESULTS

Results of developmental validation studies are shared as soon as possible with the scientific community either through presentations at scientific professional meetings or publication in peer-reviewed journals. Rapid dissemination of information about these studies is important to the legal system that forensic science serves because the courts rely on precedence when ruling if DNA evidence is admissible.

The four most commonly used scientific journals for publishing validation studies and population results are the *Journal of Forensic Sciences,* the *International Journal of Legal Medicine, Forensic Science International,* and *Legal Medicine* (Figure 16.1). Scientific meetings where DNA typing validation studies are presented include the International Symposium on Human Identification (sponsored each Fall by the Promega Corporation), the American Academy of Forensic Sciences (held each February), the Congress of the International Society of Forensic Genetics (held in late summer bi-annually), and the International Association of Forensic Sciences (held every

three years in August or September). A number of other regional meetings are also held each year.

Several thorough developmental validation papers have been published that discuss results obtained from conducting the studies listed above for the AmpFlSTR® STR kits (Wallin *et al.* 1998, Holt *et al.* 2002), the PowerPlex® 1.1/amelogenin STR multiplex system (Micka *et al.* 1999), and the PowerPlex® 16 kit (Krenke *et al.* 2002). In addition, validation studies have been published on the Forensic Science Service quadruplex (Lygo *et al.* 1994) and Second Generation Multiplex (Sparkes *et al.* 1996) as well as multiplex systems developed by the Royal Canadian Mounted Police (Fregeau *et al.* 1999). Table 16.1 lists a number of validation studies that have been conducted with STR typing kits. In all cases, multiplex STR profiling systems were found to yield reliable, reproducible, and robust results.

INTERNAL VALIDATION OF ESTABLISHED PROCEDURES

In order to meet DAB/SWGDAM Guidelines for quality assurance, forensic DNA laboratories conduct the tests described above as part of the process of becoming 'validated'. These studies demonstrate that DNA typing results can be consistently and accurately obtained by the laboratory personnel involved in the testing.

Validation studies are performed with each new DNA typing system that is developed and used. For example, a lab may be validated with the PowerPlex® 1.1 kit but it would need to perform additional validation studies when expanding its capabilities to amplifying the STR loci included in the PowerPlex® 16 kit.

STR Typing Kit	Reference
AmpFlSTR Blue	Wallin et al. (1998)
SGM Plus	Cotton et al. (2000)
Profiler Plus	Frank et al. (2001)
Profiler Plus	Fregeau et al. (2003)
Profiler Plus ID	Leibelt et al. (2003)
Profiler Plus, COfiler	LaFountain et al. (2001)
Profiler Plus, COfiler	Moretti et al. (2001a)
Profiler Plus, COfiler, Blue, Green I, PowerPlex 1.1, PowerPlex 1.2	Moretti et al. (2001b)
Profiler Plus, COfiler, Green I, Profiler	Holt et al. (2002)
Profiler Plus, COfiler	Buse et al. (2003)
Profiler Plus, COfiler, PowerPlex 16, PowerPlex 1.1, PowerPlex 2.1	Tomsey et al. (2001)
PowerPlex 1.1	Micka et al. (2001)
PowerPlex 1.1	Greenspoon et al. (2001)
PowerPlex 2.1	Levedakou et al. (2002)
PowerPlex 16	Krenke et al. (2002)
PowerPlex 16 BIO	Greenspoon et al. (2004)
Y-PLEX 6	Sinha et al. (2003a)
Y-PLEX 5	Sinha et al. (2003b)
genRES MPX-2	Junge et al. (2003)

Table 16.1

Validation studies conducted using commercial STR kits.

Typical studies for an internal validation include reproducibility, precision measurements for sizing alleles, and sensitivity (e.g., 50 ng down to 20 pg) studies along with mixture analysis and non-probative casework samples. Sizing precision studies are also conducted, where the calculated allele sizes in base pairs is plotted against the size deviation from the corresponding allele in the allelic ladder with which the genotype was determined (Wallin *et al.* 1998, Holt *et al.* 2002, Krenke *et al.* 2002). If a high degree of precision cannot be maintained due to laboratory conditions such as temperature fluctuations, then samples may not be able to be genotyped accurately.

Each forensic laboratory develops or adopts *standard operating protocols* (SOPs) that give a detailed listing of all the materials required to perform an assay as well as the exact steps required to successfully complete the experiment. In addition, SOPs list critical aspects of the assay that must be monitored carefully. SOPs are followed exactly when performing forensic DNA casework.

INTER-LABORATORY TESTS

Inter-laboratory tests are the means by which multiple laboratories compare results and demonstrate that the methods used in one's own laboratory are reproducible in another laboratory. These tests are essential to demonstrate consistency in results from multiple laboratories especially since DNA databases are now used where many laboratories contribute to the DNA profile information (see Chapter 18).

Since 1994, the European DNA Profiling Group (EDNAP) has conducted a series of inter-laboratory evaluations on various STR loci and methodologies used for analyzing them. A listing of eight published EDNAP reports that deal with STR markers may be found in Table 16.2. Each study involved the examination of 5–7 bloodstains that were distributed to multiple laboratories (usually a dozen or more) to test their ability to obtain consistent results. In all cases where simple STR loci were tested, consistent results were obtained. However, complex STR markers, such as ACTBP2 (SE33), often gave inconsistent results (Gill *et al.* 1994, 1998). Thus, STRs with complex repeat structures were not recommended for use in DNA databases at that time where results are submitted from multiple laboratories. However, the availability of commercial kits and allelic ladders that enable consistent amplification and typing of SE33 now mean that obtaining reproducible results is more feasible.

In the United States several inter-laboratory studies have been performed. The first large test with commercial kits was conducted by the National Institute of Standards and Technology (NIST) and involved 34 laboratories that evaluated the three STRs TH01, TPOX, and CSF1PO in a multiplex amplification format (Kline *et al.* 1997). This study concluded that as long as locus-specific allelic ladders were used, a variety of separation and detection methods could be used to obtain equivalent genotypes for the same samples.

As described in Chapter 5, the FBI Laboratory sponsored a STR Project from which the 13 core STR loci where chosen for inclusion in the Combined DNA Index System. A total of 22 DNA typing laboratories were involved in this project where a series of samples was systematically examined by the various laboratories. More recently, DNA quantitation and mixture studies were conducted by NIST to evaluate the differential extraction capabilities of 45 participating laboratories (Duewer *et al.* 2001) and quantification reproducibility with 74 laboratories (Kline *et al.* 2003a). These inter-laboratory studies all demonstrate that consistent results can be obtained between participating laboratories thus

Table 16.2

European DNA Profiling Group (EDNAP) collaborative studies regarding DNA typing with STR markers. The purpose of these studies was to explore whether uniformity of DNA profiling results could be achieved between European laboratories.

Study #	STR Loci Examined	Protocols Provided?	Primers/Ladders Provided?	Number of Laboratories Involved	Reference
1	TH01, ACTBP2	For PCR	Both	14	Gill *et al.* (1994)
RESULT: TH01 worked well in all labs, ACTBP2 exhibited variable sizing with different electrophoresis systems					
2	TH01, VWA, FES/FPS, F13A1	For PCR	Both	30	Kimpton *et al.* (1995)
RESULT: Fluorescent multiplex results were robust but problems existed with allele designations at FES/FPS and F13A1 when alternative detection methods were used					
3	TH01, VWA	No	No	16	Andersen *et al.* (1996)
RESULT: All allele designations matched those of the originating laboratory; achieved despite variation in amplification, electrophoresis, and detection systems used					
4	D21S11, FGA	No	Ladders only	16	Gill *et al.* (1997)
RESULT: Comparable results were obtained from all laboratories despite the fact that various primers and protocols were utilized; the key to standardization with more complex STR loci is to use a common allelic ladder					
5	ACTBP2, APOAI1, D11S554	Yes	Primers and ladder for ACTBP2	7	Gill *et al.* (1998)
RESULT: ACTBP2 showed good reproducibility between laboratories (<0.15 bp measured size difference); greater than expected variation with APOAI1 and D11S554 – they need locus-specific ladders					
6	D12S391, D1S1656	Yes	Both	7, 12	Gill *et al.* (1998)
RESULT: Excellent reproducibility between seven laboratories for D12S391 and 12 laboratories for D1S1656; demonstrated the need to use ±0.5 bp windows centered on the appropriate allelic ladder marker					
7	DYS385	Yes	Both	14	Schneider *et al.* (1999)
RESULT: Reproducible results may be obtained with a variety of separation and detection systems					
8	DYS19, DYS389I/II, DYS390, DYS393	Yes	Yes	18	Carracedo *et al.* (2001)
RESULT: Reproducible results may be obtained with a pentaplex format for single source males and mixed stains containing a male component.					

helping support the conclusion that forensic DNA typing methods are reliable and reproducible.

DNA STANDARD REFERENCE MATERIALS

Reference DNA samples are crucial to the validation of any DNA testing procedure (see Szibor *et al.* 2003). The National Institute of Standards and Technology, part of the U.S. Department of Commerce, is responsible for developing national and international standard reference materials (SRMs). These SRM sets are generally used to validate a laboratory's measurement capability, calibrate instrumentation, and troubleshoot protocols (Reeder 1999). Hundreds of SRMs are available from NIST but four in particular apply directly to the forensic DNA typing community. These are SRM 2390, SRM 2391b, SRM 2392-I, and SRM 2395.

The recently issued DNA Advisory Board standard 9.5 (1998) states: 'The laboratory shall check its DNA procedures annually or whenever substantial changes are made to the protocol(s) against an appropriate and available NIST standard reference material or standard traceable to a NIST standard' (see Appendix IV). A review of the various SRM materials that are now available to aid the forensic DNA typing community is given below.

RFLP TESTING STANDARD: SRM 2390

SRM 2390, titled DNA Profiling Standard (RFLP-based typing methods), was released in August 1992 and is intended for use in standardizing forensic and paternity testing quality assurance procedures for restriction fragment length polymorphism (RFLP) testing that uses *Hae*III restriction enzymes as well as instructional law enforcement and non-clinical research purposes.

It contains two well-characterized human DNA samples: a female cell line (K562) and a male source (TAW). Both samples are available in three forms: as a cell pellet (3×10^6 cells), an extracted genomic DNA (~200 ng/μL), and a *Hae*III restriction digest (pre-cut DNA; 25 ng/μL). A molecular weight marker for DNA sizing purposes and six quantitation standards (250 ng, 100 ng, 50 ng, 25 ng, 12.5 ng and 6 ng) are also included as is agarose for slab gel preparation.

Certified values for the DNA band sizes are available for five commonly used RFLP markers. These markers (and VNTR probes) are D2S44 (YNH24), D4S139 (PH30), D10S28 (TBQ7), D1S7 (MS1), and D17S79 (V1). The certified values represent the pooled results from analyses performed at NIST and 28 collaborating laboratories and come with calculated uncertainties (see Duewer *et al.* 2000).

PCR-BASED TESTING STANDARD: SRM 2391b

SRM 2391b, titled PCR-based DNA Profiling Standard, was re-issued in 2002. It is an update of the original SRM 2391 that became available in 1995 and includes certified values for new STR loci. SRM 2391b is intended for use in standardizing forensic and paternity testing quality assurance procedures involving polymerase chain reaction (PCR)-based genetic testing as well as instructional law enforcement and non-clinical research purposes.

It contains 12 components of well-characterized DNA in two forms: genomic DNA and DNA to be extracted from cells spotted on filter paper. There are 10 genomic DNA samples all at a concentration of $1 \, ng/\mu L$ ($20 \, \mu L$ volume). Cell lines 9947A and 9948 are included on a 6 mm Schleicher & Schuell 903 filter paper circle spotted with 5×10^4 cells. The cells permit a laboratory to test its ability to perform DNA extraction while the genomic DNA materials may be used to verify reliable PCR amplification and detection technologies.

Certified genotype values for the 12 SRM components are listed for the FBI's CODIS 13 STR loci (CSF1PO, D3S1358, D5S818, D7S820, D8S1179, D13S317, D16S539, D18S51, D21S11, FGA, TH01, TPOX, and VWA) as well as additional STR loci F13A01, F13B, FES/FPS, LPL, Penta D, Penta E, D2S1338, D19S433, and SE33. These STR markers are all available in commercial kits from either the Promega Corporation or Applied Biosystems. Certified values for the genetic loci HLA-DQA1, PolyMarker, D1S80, and amelogenin are also described for the 12 components.

MITOCHONDRIAL DNA TESTING STANDARD: SRM 2392-I

SRM 2392-I, titled Mitochondrial DNA Sequencing (Human) Standard, was released in June 2003. This SRM is intended to provide quality control when performing the polymerase chain reaction (PCR) and sequencing human mitochondrial DNA (mtDNA) for forensic investigations, medical diagnosis, or mutation detection as well as to serve as a control when PCR amplifying and sequencing any DNA sample.

SRM 2392-I contains extracted DNA from the human cell line HL-60 ($65 \, \mu L$ DNA at $1.4 \, ng/\mu L$) that has been sequenced across the entire mtDNA genome (Levin *et al.* 2003). A list of 58 unique primer sets that were designed to amplify any portion or the entire human mtDNA genome is also included.

HUMAN Y CHROMOSOME DNA PROFILING STANDARD: SRM 2395

SRM 2395 was released in July 2003 for use with verifying results involving Y chromosome STR testing (see Chapter 9). SRM 2395 includes five male DNA samples (50 μL at 2 ng/μL) selected to exhibit a diverse set of alleles across 31 commonly used Y chromosome STR and 42 single nucleotide polymorphism (SNP) markers. A female DNA sample is also included to serve as a negative control for male-specific DNA tests. In addition to the typing results from all commercially available Y-STR kits, the five male samples in SRM 2395 have been sequenced at 22 Y-STR loci to confirm allele calls (Kline *et al.* 2003b).

QUALITY CONTROL FOR COMMERCIAL SOURCES OF MATERIALS

In the early days of STR typing, forensic laboratories put together their own PCR mixes, primer sets, and allelic ladders. This meant that variation existed in the materials used for various laboratories and sometimes in the interpretation of a sample's genotype. Laboratories often had to spend a significant amount of time preparing the allelic ladders and verifying that each lot of primer mix worked appropriately. Today, most forensic DNA laboratories use commercially available STR kits that provide a uniform set of materials and protocols for the community. Thus, the primary responsibility of performing quality control on the STR amplification reagents has fallen on the commercial manufacturers.

Production of STR kits by commercial manufacturers requires extensive quality control. A fluorescent dye is attached to one primer for each locus amplified by the multiplex STR kit. Each primer must be purified and combined in the correct amount in order to produce a balanced amplification. Variation in this primer mix production can affect locus-to-locus balance in the multiplex amplification. In addition, allelic ladders must also be produced on a large scale and be well characterized since they serve as the standard for performing the DNA typing experiments with unknown samples. As they may be needed in some situations, *Certificates of Analysis* are available upon request from STR kit manufacturers for court purposes. The certificate confirms that the specific combination of components that comprise a given kit lot number perform together to meet the stated performance (Applied Biosystems 1998).

REFERENCES AND ADDITIONAL READING

Andersen, J.F., Martin, P., Carracedo, A., Dobosz, M., Eriksen, B., Johnsson, V., Kimpton, C.P., Kloosterman, A.D., Konialis, C., Kratzer, A., Phillips, P., Mevag, B., Pfitzinger, H., Rand, S., Rosen, B., Schmitter, H., Schneider, P.M. and Vide, M.C. (1996) *Forensic Science International*, 78, 83–93.

Applied Biosystems (1998) *AmpFlSTR® Profiler Plus™ PCR Amplification Kit User's Manual*. Foster City, California: Applied Biosystems.

Bar, W., Brinkmann, B., Lincoln, P., Mayr, W.R. and Rossi, U. (1994) *International Journal of Legal Medicine*, 107, 159–160.

Bar, W., Brinkmann, B., Budowle, B., Carracedo, A., Gill, P., Lincoln, P., Mayr, W. and Olaisen, B. (1997) *International Journal of Legal Medicine*, 110, 175–176.

Buel, E., Wang, G. and Schwartz, M. (1995) *Journal of Forensic Sciences*, 40, 641–644.

Buse, E.L., Putinier, J.C., Hong, M.M., Yap, A.E. and Hartmann, J.M. (2003) *Journal of Forensic Sciences*, 48, 348–357.

Carracedo, A., Bar, W., Lincoln, P., Mayr, W., Morling, N., Olaisen, B., Schneider, P., Budowle, B., Brinkmann, B., Gill, P., Holland, M., Tully, G. and Wilson, M. (2000) *Forensic Science International*, 110, 79–85.

Carracedo, A., Beckmann, A., Bengs, A., Brinkmann, B., Caglia, A., Capelli, C., Gill, P., Gusmao, L., Hagelberg, C., Hohoff, C., Hoste, B., Kihlgren, A., Kloosterman, A., Myhre, D.B., Morling, N., O'Donnell, G., Parson, W., Phillips, C., Pouwels, M., Scheithauer, R., Schmitter, H., Schneider, P.M., Schumm, J., Skitsa, I., Stradmann-Bellinghausen, B., Stuart, M., Syndercombe, C.D. and Vide, C. (2001) *Forensic Science International*, 119, 28–41.

Cotton, E.A., Allsop, R.F., Guest, J.L., Frazier, R.R., Koumi, P., Callow, I.P., Seager, A. and Sparkes, R.L. (2000) *Forensic Science International*, 112, 151–161.

Crouse, C. and Schumm, J.W. (1995) *Journal of Forensic Sciences*, 40, 952–956.

DNA Advisory Board, Federal Bureau of Investigation, U.S. Department of Justice (1998) *Quality Assurance Standards for Forensic DNA Testing Laboratories.*

DNA Recommendations (1992) *International Journal of Legal Medicine*, 105, 63–64.

Duewer, D.L., Richie, K.L. and Reeder, D.J. (2000) *Journal of Forensic Sciences*, 45, 1093–1105.

Duewer, D.L., Kline, M.C., Redman, J.W., Newall, P.J. and Reeder, D.J. (2001) *Journal of Forensic Sciences*, 46, 1199–1210.

Frank, W.E., Llewellyn, B.E., Fish, P.A., Riech, A.K., Marcacci, T.L., Gandor, D.W., Parker, D., Carter, R.R. and Thibault, S.M. (2001) *Journal of Forensic Sciences*, 46, 642–646.

Fregeau, C.J., Bowen, K.L. and Fourney, R.M. (1999) *Journal of Forensic Sciences*, 44, 133–166.

Fregeau, C.J., Bowen, K.L., Leclair, B., Trudel, I., Bishop, L. and Fourney, R.M. (2003) *Journal of Forensic Sciences*, 48, 1014–1034.

Gill, P., Kimpton, C., D'Aloja, E., Andersen, J.F., Bär, W., Brinkmann, B., Holgerssen, S., Johnsson, V., Kloosterman, A.D., Lareu, M.V., Nellemann, L., Pfitzinger, H., Phillips, C.P., Schmitter, H., Schneider, P.M. and Stenersen, M. (1994) *Forensic Science International*, 65, 51–59.

Gill, P., D'Aloja, E., Andersen, J., Dupuy, B., Jangblad, M., Johnsson, V., Kloosterman, A.D., Kratzer, A., Lareu, M.V., Meldegaard, M., Phillips, C., Pfitzinger, H., Rand, S., Sabatier, M., Scheithauer, R., Schmitter, H., Schneider, P. and Vide, M.C. (1997) *Forensic Science International*, 86, 25–33.

Gill, P., D'Aloja, E., Dupuy, B., Eriksen, B., Jangblad, M., Johnsson, V., Kloosterman, A.D., Kratzer, A., Lareu, M.V., Mevag, B., Morling, N., Phillips, C., Pfitzinger, H., Rand, S., Sabatier, M., Scheithauer, R., Schmitter, H., Schneider, P., Skitsa, I. and Vide, M.C. (1998) *Forensic Science International*, 98, 193–200.

Gill, P., Brenner, C., Brinkmann, B., Budowle, B., Carracedo, A., Jobling, M.A., de Knijff, P., Kayser, M., Krawczak, M., Mayr, W.R., Morling, N., Olaisen, B., Pascali, V., Prinz, M., Roewer, L., Schneider, P.M., Sajantila, A. and Tyler-Smith, C. (2001) *International Journal of Legal Medicine*, 114, 305–309.

Greenspoon, S.A., Lylte, P.J., Turek, S.A., Rolands, J.M., Scarpetta, M.A. and Carr, C.D. (2000) *Journal of Forensic Sciences*, 45, 677–683.

Greenspoon, S.A., Ban, J.D., Pablo, L., Crouse, C.A., Kist, F.G., Tomsey, C.S., Glessner, A.L., Mihalacki, L.R., Long, T.M., Heidebrecht, B.J., Braunstein, C.A., Freeman, D.A., Soberalski, C., Nathan, B., Amin, A.S., Douglas, E.K. and Schumm, J.W. (2004) *Journal of Forensic Sciences*, 49, 71–80.

Holt, C.L., Buoncristiani, M., Wallin, J.M., Nguyen, T., Lazaruk, K.D. and Walsh, P.S. (2002) *Journal of Forensic Sciences*, 47, 66–96.

Junge, A., Lederer, T., Braunschweiger, G. and Madea, B. (2003) *International Journal of Legal Medicine*, 117, 317–325.

Kimpton, C.P., Gill, P., D'Aloja, E., Andersen, J.F., Bar, W., Holgersson, S., Jacobsen, S., Johnsson, V., Kloosterman, A.D., Lareu, M.V., Nellemann, L., Pfitzinger, H., Phillips, C.P., Rand, S., Schmitter, H., Schneider, P.M., Sternersen, M. and Vide, M.C. (1995) *Forensic Science International*, 71, 137–152.

Kline, M.C., Duewer, D.L., Newall, P., Redman, J.W., Reeder, D.J. and Richard, M. (1997) *Journal of Forensic Sciences*, 42, 897–906.

Kline, M.C., Duewer, D.L., Redman, J.W. and Butler, J.M. (2003a) *Analytical Chemistry*, 75, 2463–2469.

Kline, M.C., Schoske, R., Vallone, P.M., Redman, J.W. and Butler, J.M. (2003b) NIST SRM 2395 and other Y chromosome work. *Proceedings of the Fourteenth International Symposium on Human Identification*. Promega Corporation: Madison, Wisconsin. Available at: http://www.promega.com/geneticidproc/ussymp14proc/posterpresentations/Kline1.pdf.

Krenke, B.E., Tereba, A., Anderson, S.J., Buel, E., Culhane, S., Finis, C.J., Tomsey, C.S., Zachetti, J.M., Masibay, A., Rabbach, D.R., Amiott, E.A. and Sprecher, C.J. (2002) *Journal of Forensic Sciences*, 47, 773–785.

LaFountain, M.J., Schwartz, M.B., Svete, P.A., Walkinshaw, M.A. and Buel, E. (2001) *Journal of Forensic Sciences*, 46, 1191–1198.

Leibelt, C., Budowle, B., Collins, P., Daoudi, Y., Moretti, T., Nunn, G., Reeder, D. and Roby, R. (2003) *Forensic Science International,* 133, 220–227.

Levedakou, E.N., Freeman, D.A., Budzynski, M.J., Early, B.E., Damaso, R.C., Pollard, A.M., Townley, A.J., Gombos, J.L., Lewis, J.L., Kist, F.G., Hockensmith, M.E., Terwilliger, M.L., Amiott, E., McElfresh, K.C., Schumm, J.W., Ulery, S.R., Konotop, F., Sessa, T.L., Sailus, J.S., Crouse, C.A., Tomsey, C.S., Ban, J.D. and Nelson, M.S. (2002) *Journal of Forensic Sciences*, 47, 757–772.

Levin, B.C., Holland, K.A., Hancock, D.K., Coble, M., Parsons, T.J., Kienker, L.J., Williams, D.W., Jones, M.P. and Richie, K.L. (2003) *Mitochondrion*, 2, 387–400.

Lygo, J.E., Johnson, P.E., Holdaway, D.J., Woodroffe, S., Whitaker, J.P., Clayton, T.M., Kimpton, C.P. and Gill, P. (1994) *International Journal of Legal Medicine*, 107, 77–89.

Micka, K.A., Amiott, E.A., Hockenberry, T.L., Sprecher, C.J., Lins, A.M., Rabbach, D.R., Taylor, J.A., Bacher, J.W., Glidewell, D.E., Gibson, S.D., Crouse, C.A. and Schumm, J.W. (1999) *Journal of Forensic Sciences*, 44, 1243–1257.

Moretti, T.R., Baumstark, A.L., Defenbaugh, D.A., Keys, K.M., Smerick, J.B. and Budowle, B. (2001a) *Journal of Forensic Sciences*, 46, 647–660.

Moretti, T.R., Baumstark, A.L., Defenbaugh, D.A., Keys, K.M., Brown, A.L. and Budowle, B. (2001b) *Journal of Forensic Sciences*, 46, 661–676.

National Research Council (1996) *NRC II: The Evaluation of Forensic DNA Evidence*. Washington, DC: National Academy Press.

Peterson, J.L., Lin, G., Ho, M., Chen, Y. and Gaensslen, R.E. (2003a) *Journal of Forensic Sciences*, 48, 21–31.

Peterson, J.L., Lin, G., Ho, M., Chen, Y. and Gaensslen, R.E. (2003b) *Journal of Forensic Sciences*, 48, 32–40.

Rand, S., Schurenkamp, M. and Brinkmann, B. (2002) *International Journal of Legal Medicine*, 116, 199–206.

Rand, S., Schurenkamp, M., Hohoff, C. and Brinkmann, B. (2004) *International Journal of Legal Medicine*, 118, 83–89.

Reeder, D.J. (1999) *Archives of Pathology and Laboratory Medicine*, 123, 1063–1065.

Schneider, P.M., D'Aloja, E., Dupuy, B.M., Eriksen, B., Jangblad, A., Kloosterman, A.D., Kratzer, A., Lareu, M.V., Pfitzinger, H., Rand, S., Scheithauer, R., Schmitter, H., Skitsa, I., Syndercombe-Court, D. and Vide, M.C. (1999) *Forensic Science International*, 102, 159–165.

Sinha, S.K., Budowle, B., Arcot, S.S., Richey, S.L., Chakrabor, R., Jones, M.D., Wojtkiewicz, P.W., Schoenbauer, D.A., Gross, A.M., Sinha, S.K. and Shewale, J.G. (2003a) *Journal of Forensic Sciences*, 48, 93–103.

Sinha, S.K., Nasir, H., Gross, A.M., Budowle, B. and Shewale, J.G. (2003b) *Journal of Forensic Sciences*, 48, 985–1000.

Sparkes, R., Kimpton, C., Watson, S., Oldroyd, N., Clayton, T., Barnett, L., Arnold, J., Thompson, C., Hale, R., Chapman, J., Urquhart, A. and Gill, P. (1996) *International Journal of Legal Medicine*, 109, 186–194.

SWGDAM (2004) *Forensic Science Communications*, July 2004, 6 (3). Available at: http://www.fbi.gov/hq/lab/fsc/current/backissu.htm.

Szibor, R., Edelmann, J., Hering, S., Plate, I., Wittig, H., Roewer, L., Wiegand, P., Cali, F., Romano, V. and Michael, M. (2003) *Forensic Science International*, 138, 37–43.

Tomsey, C.S., Kurtz, M., Kist, F., Hockensmith, M. and Call, P. (2001) *Croatian Medical Journal*, 42, 239–243.

TWGDAM (1989) *Crime Laboratory Digest*, 16, 40–59.

TWGDAM (1991) *Crime Laboratory Digest*, 18, 44–75.

TWGDAM (1995) *Crime Laboratory Digest*, 22, 21–43.

Wallin, J.M., Buoncristiani, M.R., Lazaruk, K.D., Fildes, N., Holt, C.L. and Walsh, P.S. (1998) *Journal of Forensic Sciences*, 43, 854–870.

CHAPTER 17

NEW TECHNOLOGIES, AUTOMATION, AND EXPERT SYSTEMS

There is no gene for the human spirit.

(From the movie GATTACA, 1997)

Technical advances in this field are very rapid. We can expect in the near future methods that are more reliable, less expensive, and less time-consuming than those in use today.

(NRC II, p. 59)

We live in an age of rapid discovery in biotechnology. New technologies that were only imagined a few years ago are now reality. Furthermore, DNA sequence information is becoming available at an unprecedented rate. This information is leading us to a better understanding of human genetic diversity. While it is impossible to predict where the forensic DNA community will be five or ten years from now, there are some new technologies that deserve recognition and that will be reviewed briefly in this chapter. We will also discuss the advantages of automation in large-scale testing operations such as are done for construction of DNA databases.

NEW DNA SEPARATION/GENOTYPING TECHNOLOGIES

Most DNA testing in the United States is currently performed in small public forensic laboratories, each consisting of less than a dozen scientists devoted to DNA analysis. However, with the development of several new technologies that will be discussed below it is conceivable that in the future large-scale operations might become more prevalent for DNA database work. In addition, some DNA tests may be performed at the crime scene with hand-held or portable devices. Thus, this section has been broken up into technological developments that will aid crime scene DNA testing and large-scale testing for DNA database development. Some methods are extensions of current fluorescence-based technology and other techniques involve completely novel technology.

PORTABLE DEVICES FOR POSSIBLE CRIME SCENE INVESTIGATIONS

Microfabrication techniques revolutionized the integrated circuit industry 20 years ago and have brought the world ever faster and more powerful computers. These same microfabrication methods are now being applied to develop miniature, microchip-based laboratories, or so-called 'labs-on-a-chip' (Paegel *et al.* 2003). Miniaturizing the sample preparation and analysis steps in forensic DNA typing could lead to devices that permit investigation of biological evidence at a crime scene or more rapid and less expensive DNA analysis in a more conventional laboratory setting. We will focus here on three areas of on-going research for miniature DNA analysis instruments: microchip capillary electrophoresis (CE) devices, miniature thermal cyclers, and hybridization arrays.

Microchip CE Devices

The primary advantage of analyzing DNA in a miniature CE device is that shorter channels, or capillaries, lead to faster DNA separations. Instead of using a 30 cm long glass capillary tube to perform the DNA separation, microchip CE devices are typically glass microscope slides with narrow channels etched into them that are 10–50 µm deep by 50 µm wide and several centimeters long. A glass coverplate is bonded on top of the etched channels in order to create a sealed separation channel (Woolley and Mathies 1994). Alternatively, injection molded plastic may be used (McCormick *et al.* 1997).

Separation speeds that are 10–100 times faster than conventional electrophoresis may be obtained with this approach. Using a 2 cm separation distance (compared to 36 cm for an ABI 310 capillary), tetranucleotide short tandem repeat (STR) alleles were separated in as little as 30 seconds (Schmalzing *et al.* 1997). Microchip CE systems are being developed with multi-color detection formats. These systems should therefore be compatible with commercially available STR kits. Figure 17.1 shows a simultaneous two-color analysis of polymerase chain reaction (PCR) products from the eight loci in the PowerPlex™ 1.1 STR kit (Schmalzing *et al.* 1999). This separation was performed in less than 2½ minutes.

In order to obtain ultrafast DNA separations, the injection plug must be narrow and short compared to the separation length. DNA separations in less than 30 seconds have also been demonstrated with short capillaries using a fast ramp power supply for rapid injections (Muller *et al.* 1998). Research is ongoing to improve separation speeds and ease of use with the hope that in the near future microchip CE devices will be used routinely for rapid DNA analyses (see Mitnik *et al.* 2002, Goedecke *et al.* 2004).

A capillary array microplate device has been constructed with 96 separation channels in order to scale up the number of samples processed at a single time. These devices are capable of separating 96 different samples in less than two minutes (Shi *et al.* 1999). Although the DNA separation parameters need

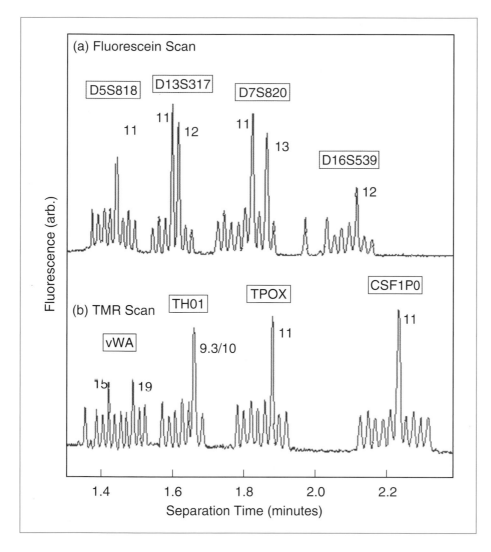

Figure 17.1

Rapid microchip CE separation of the eight STR loci from PowerPlex 1.1 (Schmalzing 1999). The electropherograms from scanning each color of a simultaneous two-color analysis are divided. The PCR-amplified sample is mixed with the allelic ladders prior to injection to provide a frame of reference for genotyping the sample. The allele calls for each locus are listed next to the corresponding peak. Figure courtesy of Dr. Daniel Ehrlich, Whitehead Institute.

improvement before this particular 96 channel microplate can resolve closely spaced STR alleles, the device demonstrates that rapid DNA separations are feasible in a highly parallelized format. Separations of STR alleles that have been demonstrated to date on multi-channel microchip devices (see Medintz *et al.* 2001, Mitnik *et al.* 2002, Goedecke *et al.* 2004) are performed at speeds that are really no different than what can already be accomplished with the conventional capillary array electrophoresis instruments described in Chapter 14.

Miniature Thermal Cyclers

Sample preparation devices are also shrinking in size (see Paegel *et al.* 2003). In particular, miniature thermal cyclers are being developed for performing PCR. These devices are being microfabricated with silicon reaction chambers (Northrup *et al.* 1998). A major advantage of miniaturizing the PCR thermal cycling process is the potential for lower reagent consumption and thus reduction in the cost of an analysis. In addition, more rapid thermal cycling times are possible because the PCR reaction mixture can be heated and cooled quickly. Since the reaction volume is smaller, it takes less time to thermally equilibrate the PCR reaction.

The miniature analytical thermal cycler instrument (MATCI) developed at Lawrence Livermore National Laboratory weighs only 35 pounds, fits in a medium-sized briefcase, and is powered by 13 rechargeable batteries. The MATCI system has been used to successfully amplify DQA1 alleles and a STR triplex consisting of the D3S1358, VWA, and FGA loci (Belgrader *et al.* 1998). A 25 μL PCR reaction was performed with 30 cycles in about 42 minutes using MATCI compared to 2.5 hours on a PE9600 thermal cycler. The MATCI has also been used in conjunction with time-of-flight mass spectrometry to complete a STR genotyping assay in less than 50 minutes (Ross *et al.* 1998).

Ultimately, the combination of sample preparation in a miniature thermal cycling device coupled to rapid DNA analysis on a microchip CE device may be the future (Lagally *et al.* 2001). There are certainly a number of applications for which developed capabilities of highly rapid human identification would be of value, such as biometrics (D.N.A. Box 17.1). In the first demonstration of coupling a microfabricated PCR reactor with a micro-capillary electrophoresis chip, a rapid PCR-CE analysis was performed in less than 20 minutes (Woolley *et al.* 1996). In this particular case, a PCR amplification of a single amplicon involving 30 cycles was performed in 15 minutes and was immediately followed by a high-speed CE chip separation in 83 seconds.

A similar type of online, automated DNA amplification and separation has also been demonstrated with a larger scale CE system (Swerdlow *et al.* 1997). In this case, the total time from extracted DNA to result was 20 minutes – eight minutes for thermal cycling, four minutes for purification, and eight minutes for electrophoresis. It may well be that in the not too distant future DNA results, from biological sample to STR profile, may be routinely obtained in under an hour. This type of rapid analysis would then enable DNA testing at or near crime scenes in mobile laboratories.

One major obstacle to seeing this type of rapid analysis achieved is that current STR multiplexes are not compatible with rapid thermal cycling. Rather, well-balanced amplification yields with the primer concentrations present in commercial STR kits are only achieved when ramp rates of 1°C per second are used rather than 5–10°C per second involved in rapid cycling.

STR Determination by Hybridization Arrays
Nanogen Inc. (San Diego, CA) has developed another microchip-based assay that initially appeared promising for rapid STR allele determination. *However, this approach is no longer being actively pursued by forensic DNA laboratories for STR typing.*

D.N.A. Box 17.1

DNA and biometrics

The field of biometrics involves developing automated methods of identifying a person or verifying the identity of a person based on a physiological or behavioral characteristic. Current modalities of biometrics include fingerprints, iris scans, hand and finger geometry, face recognition, voice recognition, and signature verification. So called 'smart cards' are being developed to provide increased security for a variety of applications. Biometric information is also being included on passports to help prevent falsification of passport identity.

Biometric authentication requires comparing a reference sample that is collected during the 'enrollment' phase against a newly captured biometric sample collected as part of identification. Access to a particular location or computer file is based on whether a match can be verified between the identification biometric signal and anyone of the authorized signatures (e.g., fingerprints, voice pattern, etc.) collected during enrollment.

The 1997 movie GATTACA is a futuristic vision of a world where rapid genetic testing is used to prevent access to secure locations by genetically 'imperfect' individuals. Currently this concept of using DNA for a biometric is still futuristic, primarily because of the cost and length of time required to perform DNA testing. However, as DNA analysis because less expensive and more rapid, it may become a valuable biometric to be used beyond the national DNA databases discussed in Chapter 18.

Source:
http://www.itl.nist.gov/div893/biometrics/Biometricsfromthemovies.pdf,
   http://www.biometrics.org.

The Nanogen assay involves the use of a silicon microchip composed of an arrayed set of electrodes that act as independent test sites (Figure 17.2). Electric potentials can be directed to each test site, which contains a unique DNA probe for hybridization (Sosnowski *et al.* 1997a).

A DNA hybridization assay is conducted by washing a DNA sample over the chip and seeing where it binds on the array. PCR-amplified samples will bind or hybridize to their complementary probe sequence. An 'electronic stringency' can then be applied to each probe site by simply adjusting the electric field strength. Samples that are not a perfect match for the probe will be denatured and driven away from the probe.

Since each test site is separate from the others on the spatial array of probes, a signal obtained from a particular position on the probe array indicates which sequence has bound to the probe. Fluorescent probes may be added to the DNA molecules for detection purposes. Software then reads the array position versus fluorescent signal and interprets the data to determine the sequence present in the DNA samples being measured.

In order to measure which STR alleles are present at a particular locus, a chip is prepared that contains known alleles for that locus. Each probe position on

Figure 17.2

Schematic of nanogen STR hybridization chip assay. The assay illustrated in (a) involves a capture probe oligonucleotide that is attached at a unique location on the chip (b). The capture probe hybridizes to the appropriate STR allele by binding to the repeat region and 30–40 bases of the flanking region. A reporter probe containing 1–3 repeat units, some flanking sequence and a fluorescent dye hybridizes to the STR allele and generates a fluorescent signal at the probe site that can be interpreted to yield the sample's genotype (c). Multiple STR loci may be probed on the same chip with one probe site existing for each allele.

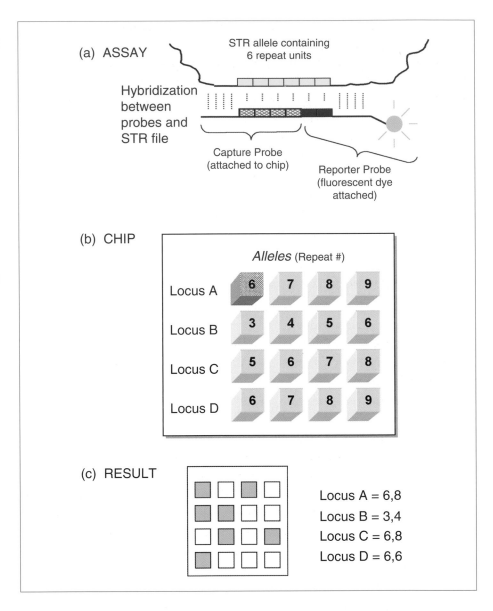

the chip has a different allele attached (Figure 17.2b). Thus, in order to have the capability of measuring eight alleles at the TPOX locus (e.g., 6–13 repeats), eight different probe sites are required.

The STR hybridization assay involves two probes that bind to the STR repeat and flanking regions. The 'capture probe' is attached to the chip at the test site and captures the PCR-amplified STR allele when the sample is added to the chip. The 'reporter probe' contains the fluorescent dye and thus enables the detection of the STR allele bound to the particular site defined by the capture probe (Figure 17.2a). The unique sequences on either side of the repeat make it possible to have discrimination between alleles of different STR loci with the same repeat sequence.

A sample's genotype is assessed by observing the positions that give a fluorescent signal once the PCR product has hybridized to its corresponding capture probe(s). The read-out provides a genotype that corresponds to the number of repeats present in the sample even though no size-based separation has been performed (Figure 17.2c). Thus, samples measured with this hybridization assay can be compared to results obtained from a conventional DNA separation of the STR alleles. Nanogen has conducted several successful sample correlation tests with the Bode Technology Group, a private forensic laboratory located in Springfield, Virginia (Sosnowski *et al.* 1997b).

INSTRUMENTS FOR LARGE-SCALE DNA DATABASE TESTING

As the demand for DNA typing results increases particularly with the national DNA databases being developed around the world (see Chapter 18), instrumentation capable of high volume sample processing will become more prevalent. Most laboratories are going the route of multi-capillary array electrophoresis systems (see Chapter 14), such as the ABI 3100 and ABI 3700, to meet high-throughput DNA testing needs. In this section, we will review another technology capable of high-throughput testing, *which is currently not in use for STR typing*.

MALDI-TOF Mass Spectrometry

Time-of-flight mass spectrometry is a technique capable of large-scale sample processing and has been demonstrated to work with STR analysis (Butler and Becker 2001). Mass spectrometry is a versatile analytical technique that involves the detection of ions and the measurement of their mass-to-charge ratio. Due to the fact that these ions are separated in a vacuum environment, the analysis times can be extremely rapid, on the order of seconds. Combined with robotic sample preparation, time-of-flight mass spectrometry offers the potential for processing thousands of DNA samples on a daily basis.

In order to get the DNA molecules into the gas phase for analysis in the mass spectrometer, a technique known as matrix-assisted laser desorption-ionization (MALDI) is used. When MALDI is coupled with time-of-flight mass spectrometry, this measurement technique is commonly referred to as MALDI-TOF-MS. Figure 17.3 shows a schematic of the MALDI-TOF-MS process.

The analysis of DNA using MALDI-TOF-MS proceeds as follows. A liquid DNA sample is combined with an excess of a matrix compound, such as 3-hydroxypicolinic acid (Butler *et al.* 1998). These samples are spotted onto a metal or silicon plate. As the sample air dries, the DNA and matrix co-crystallize. The sample plate is then introduced into the vacuum environment of the mass spectrometer for analysis. A rapid laser pulse initiates the ionization process.

Figure 17.3

Schematic of MALDI time-of-flight mass spectrometry. DNA samples are mixed with a matrix compound and dried as individual spots on a sample plate. The sample plate is then introduced to the vacuum environment of the mass spectrometer. A small portion of the sample is ionized by an ultraviolet (UV) laser pulse and the generated ions are accelerated through the ion optics. The ions separate by size as they pass down the flight tube. The mass-to-charge ratio (m/z) of each ion is detected as it impacts the detector.

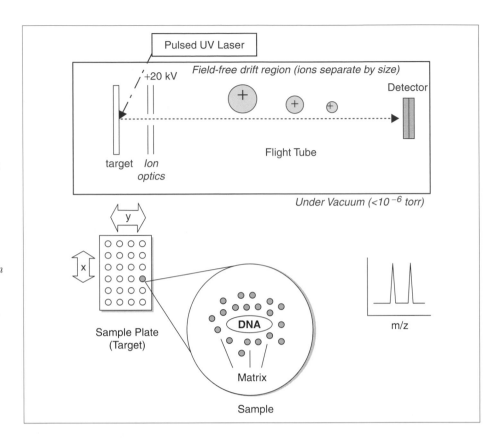

The matrix molecules that surround the DNA protect it from fragmentation during the ionization process.

Each pulse of the laser initiates ionization of the sample and the subsequent separation of ions in the flight tube (Figure 17.3). The DNA ions travel to the detector in a matter of several hundred microseconds as they separate based on their mass. However, it takes several seconds to analyze each sample because multiple laser pulses are taken and averaged to form the final mass spectrum. Samples are analyzed sequentially by moving the sample plate underneath a fixed laser beam. Sample plates are now commercially available that can hold 384 (or more) samples at a time. Each sample plate can be analyzed in less than one hour depending on the number of laser shots collected for each sample and the pulse rate of the laser.

Time-of-flight mass spectrometry has the potential to bring DNA sample processing to a new level in terms of high-throughput analysis. However, there are several challenges for analysis of PCR products, such as STRs, using MALDI-TOF-MS. The most significant problem is that resolution and sensitivity in the mass spectrometer are diminished when either the DNA size or the salt content of the sample is too large.

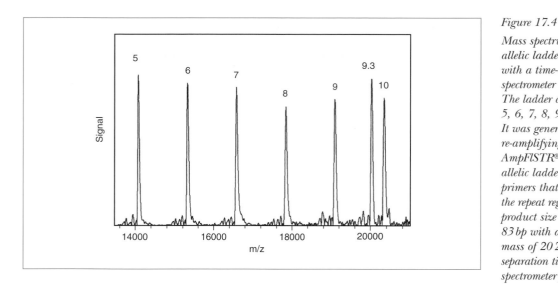

Figure 17.4

Mass spectrum of a TH01 allelic ladder obtained with a time-of-flight mass spectrometer (Butler 1998). The ladder contains alleles 5, 6, 7, 8, 9, 9.3 and 10. It was generated by re-amplifying an AmpFlSTR® Green I allelic ladder mix using primers that bind close to the repeat region. The PCR product size of allele 10 is 83 bp with a measured mass of 20 280 Da. The separation time in the mass spectrometer for allele 10 is only 204 microseconds! The allele 9.3 and allele 10 peaks, which are a single nucleotide apart, differ by 1.5 microseconds on a separation time scale and can be fully resolved with this method.

However, STR markers have been successfully analyzed via MALDI-TOF mass spectrometry by redesigning the PCR primers to be closer to the repeat region, and thereby reducing the size of the amplified alleles (Ross and Belgrader 1997, Ross *et al.* 1998, Butler *et al.* 1998). The mass spectrum of an allelic ladder for the STR locus TH01, shown in Figure 17.4, demonstrates that STR alleles may be effectively detected with MALDI-TOF-MS. These alleles are 105 bp smaller than corresponding alleles amplified with AmpFlSTR® kit primers for the TH01 STR locus.

Another benefit to MALDI-TOF-MS besides sample analysis speed is accuracy. In fact, the high degree of accuracy for sizing STR alleles using this technique permits reliable typing without the use of an allelic ladder (Butler *et al.* 1998). Allelic ladders as well as internal sizing standards are necessary in electrophoretic separation systems to adjust for minor variations in peak migration times due to fluctuations in temperature and voltage (see Chapter 12).

With mass spectrometry, the actual mass of the DNA molecule is being measured, making it a more accurate technique than a relative size measurement as in electrophoresis. In fact, STR allele measurements taken almost a year apart on different instruments produced virtually identical masses (Butler and Becker 2001). Furthermore, a comparison study of MALDI-TOF-MS results with over 1000 STR alleles measured by conventional fluorescent methods using an ABI 310 demonstrated an excellent correlation between the two methods (Butler and Becker 2001).

Unfortunately, the expense of the MALDI-TOF-MS system, which is on the order of several hundred thousand dollars, and the previous wide-scale acceptance of fluorescent methodologies will likely keep mass spectrometry from becoming a major player in forensic DNA analysis of STR markers. However, it is an effective means for analysis of single nucleotide polymorphisms (SNPs) and may have a role to play in forensic DNA analysis as SNPs become more widely accepted (see Chapter 8).

LABORATORY AUTOMATION

Laboratory automation is an important topic, especially since the demand for forensic DNA testing is increasing. Laboratories will take on more cases and have much larger amounts of samples to type because of DNA database laws. While the type of laboratory automation that is currently used by DNA typing laboratories varies widely from little to none, in the future automation will likely play an increasing role in primarily two areas: liquid handling and data analysis.

LIQUID HANDLING ROBOTS

There are a number of liquid handling tasks performed in DNA typing laboratories during the DNA extraction, PCR setup, and PCR amplification analysis steps. These liquid handling tasks are typically performed with manual pipettors by a DNA technician or analyst. Small volumes of liquids are repeatedly moved from one tube to another. These repetitive tasks can lead to mistakes as laboratory personnel get tired or become careless.

By introducing automated liquid handling with robotics, the level of human error can be greatly reduced. Computers and robotics do the same task the same way time after time without getting tired. The challenge though lies in setting up the automation and maintaining it (Hale 1999). The most likely place where liquid handling automation will be used in the future is with the high volume sample processing of convicted offender samples for computer DNA databases (see Chapter 18).

There are a number of popular liquid handling robotic systems that are commercially available. Beckman Coulter, Hamilton Company, MWG Biotech, QIAGEN, and Tecan market popular liquid handling robots. Appendix III contains contact information with these manufacturers. Each robotic system has different capabilities and should be carefully assessed in order to meet the needs and goals of one's own laboratory environment (see Crouse and Conover 2003).

The Beckman Biomek 2000 robot has been used in conjunction with DNA IQ and AluQuant chemistries (see Chapter 3) from the Promega Corporation to enable automated isolation and quantification of DNA samples (Tereba *et al.* 2003). Sexual assault (mixed stain) samples, cigarette butts, blood stains, buccal swabs, and various tissue samples were successfully extracted with the Biomek 2000 and the DNA IQ system without any evidence of contamination throughout extensive validation studies (Greenspoon *et al.* 2004). Hayn *et al.* (2004) demonstrated that the *Alu*Quant assay worked well for DNA quantification on the Biomek 2000. Robotic liquid handling for steps of DNA extraction, quantification, PCR amplification setup, and preparation of sample plates for STR typing will likely become more prevalent in forensic laboratories particularly as the need for higher-volume work increases.

SAMPLE TRACKING PROGRAMS

Managing large amounts of data becomes a problem for many laboratories as they scale up their efforts. Computer databases are often developed to aid in tracking samples and results obtained. Sample tubes can be bar-coded and tracked through the analysis process. An example of efforts in this area is the Overlord System developed at the Forensic Science Service (FSS; Hopwood *et al.* 1997). The FSS Overlord program is a laboratory information management system (LIMS) and aids sample tracking as well as overall control of the different robotic stations. LIMS systems are rather expensive and are typically used only by laboratories with very high sample volumes.

Commercial LIMS systems, such as the Crime Fighter B.E.A.S.T. (computerized Bar-coded Evidence Analysis, Statistics, and Tracking LIMS) from Porter Lee Corporation (Schaumburg, IL), are being used in a growing number of forensic laboratories to provide electronic case files and automated sample tracking capabilities. A LIMS manufacturer typically sets up their software and customizes it to accommodate protocols and processes within each customer laboratory.

The Armed Forces DNA Identification Laboratory (AFDIL, Rockville, MD) has worked in conjunction with Future Technologies Inc. (Fairfax, VA) to develop LISA, which stands for Laboratory Information Systems Applications. LISA contains a number of sub-systems that permit case accessioning and the ability to electronically track the life cycle of each evidence and reference sample. There are additional modules such as MFIMS (Mass Fatality Incident Management System) and ASAP (AFDIL Statistical Application Program) that manage victim and family reference data as well as easing the tedious process of reporting results.

STaCS™ (Sample Tracking and Control System) is a system co-developed by forensic DNA scientists at the Royal Canadian Mounted Police (RCMP) and Anjura Technology Corporation (Ottawa, Ontario) that integrates robotic sample processing with custom LIMS software. STaCS monitors instrument performance and can provide a variety of operational information reports to help make the process of DNA typing more efficient. This system has been set-up in several DNA databasing laboratories including the Florida Department of Law Enforcement (Tallahassee, FL) and the FBI Laboratory (Quantico, VA).

Fully integrated systems with robotic liquid handling are especially useful for DNA databasing of convicted offender samples, which are usually more uniform in nature (i.e., are all bloodstains or buccal swabs), single-source samples, and relatively concentrated in amount. In one laboratory, over 17 000 DNA samples were processed in a 20 month period using robotics and LIMS with an overall typing success-rate of 99.99% (Parson and Steinlechner 2001). While automation is being developed and implemented to robotically process and track samples through the steps of DNA extraction, quantitation, PCR amplification, and sample setup prior to electrophoretic separation, separate computer programs commonly referred to as 'expert systems' are being constructed to enable automatic interpretation of STR alleles from the resulting electropherograms.

EXPERT SYSTEMS FOR STR DATA INTERPRETATION

One of the highest labor efforts in the process of typing STRs is the data interpretation stage. For many high-throughput laboratories, data assessment and interpretation of STRs represents approximately 50% or more of the resource requirement to deliver final results for samples. In many cases, more time is actually spent evaluating the STR profiles than preparing and collecting the data on the sample. In order to reduce this resource requirement, software has been designed and implemented to replace the traditional manual assessment. Two of the first applications used operationally include STRess® developed by the Forensic Science Service in England and TrueAllele® developed by Mark Perlin of Cybergenetics (Pittsburg, PA).

Expert systems have conventionally been considered and designed to translate the electropherogram signal into a genotype compatible with a database. As these expert systems are developed and implemented, bottlenecks will shift to other areas in the DNA typing process and thus permit development of expert systems that can solve ever more complex and diverse problems. As will be discussed below, the Forensic Science Service has developed computer systems that perform a variety of roles within the convicted offender and crime stain analyses processes.

TRUEALLELE

TrueAllele® is a commercially available allele-calling program from Cybergenetics that uses quantitation and deconvolution algorithms to improve STR allele calls based upon quality measures (Palsson *et al.* 1999, Perlin *et al.* 2001, Perlin 2003). TrueAllele® is written in Matlab and runs with Macintosh, Windows, or UNIX-based systems. This program has an advantage over the Genotyper® or STaR Call™ software packages (see Chapter 15) in that TrueAllele® provides a quality measure for every allele call. The quality value assigned by TrueAllele® ranges between 0.0 and 1.0 and reflects a peak's height, shape, and stutter pattern (Palsson *et al.* 1999). The selection criteria used by the program are empirically derived through review of many STR profiles. TrueAllele® is being primarily used for microsatellite disease linkage studies where hundreds of thousands of genotypes are gathered to decipher chromosomal locations of disease genes. The first two forensic DNA laboratories to utilize TrueAllele® as an expert system for STR analysis are the Forensic Science Service and the New York State Police in Albany, New York. In a comparison of alleles calls from 2048 STR profiles between manual review with Genotyper® and automated review with TrueAllele®, only one significant difference was observed when the analyst using Genotyper® interpreted a spike as a DNA peak at D8S1179 but TrueAllele® correctly designated it as a spike (Kadash *et al.* 2004).

STR EXPERT SYSTEM SUITE (STRess) AND STRess2

In 1998, the Forensic Science Service (FSS) developed a data interpretation program called STRess (STR Expert System Suite) to aid their STR profile processing. Interpretation guidelines drawn from approximately 100 000 samples processed by the FSS and used by experienced operators were incorporated into the programming of STRess (Gill *et al.* 1996, Dunbar *et al.* 1998). FSS genotyping guidelines require that all samples are genotyped by two independent operators to ensure accuracy of DNA typing results followed by a third operator to review allele calls and confirm that they are concordant. The aim of STRess was to reduce the amount of manual effort needed to evaluate the STR data by replacing one of the genotype analysts. The FSS has estimated that incorporating the STRess program into routine analysis has resulted in a 10–20% time-savings at the interpretation stage with improved standardization and quality of interpretation (Martin Bill, personal communication).

The success of the original STRess program has spawned the development of a number of systems that automate interpretation and interact as a suite. The suite is currently being developed into a global package branded 'STRess2' that will organize the interaction between systems and be configurable for any multiplex (Figure 17.5). A breakdown of the software modules in STRess2 currently proposed by the FSS is given below (information kindly provided by Martin Bill, FSS):

- The *STRess2 core interpretation engine* is responsible for batch validation, ladder assessment and sizing, allele designation and interpretation. STRess2 has been designed and tested with well over five years of expert system operational experience. The application is designed as an 'open book' where the DNA unit can customize all the thresholds within the software to determine how 'confident' the software will behave. The FSS began making this software commercially available in late 2004. For details on how to obtain a copy of STRess2 or any of the other applications discussed, please contact Dr Chris N. Maguire (chris.maguire@fss.pnn.police.uk).

Figure 17.5

Modules in STRess2 showing data flow. Figure courtesy of Martin Bill, Forensic Science Service.

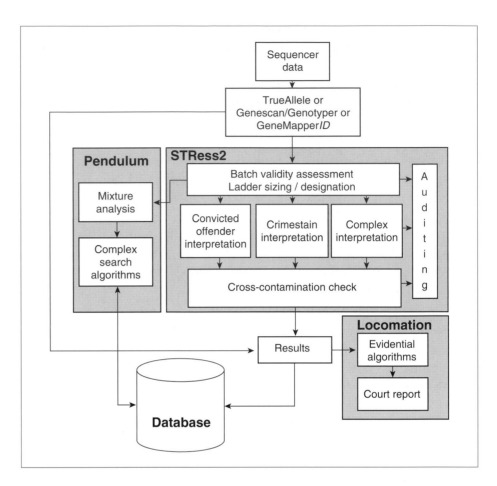

- The *cross-contamination module* offers a rapid screen for potential cross-contamination between samples. The program can search across both major and minor profiles performing many thousands of comparisons in seconds. A secondary search is performed to analyze samples within and between cases to look for 'intelligence links'. These links identified before loading to the database can increase the overall load rate. This application has subsequently been integrated with STRess2.

- A *contamination database* known as 'SPACE' allows the storage and comparison of staff and supplier profiles. Crime stains and convicted offender samples can be screened against this secondary database to help ensure the profile loaded is not a result of operator contamination.

- A *batch-processing module* enables reprocessing decisions, auditing and electronic reports along with interaction with the LIMS. This is a critical area of expert system development. As laboratories automate the allele interpretation steps, the bottleneck invariably moves to the auditing and case file handling stages thereby generating a requirement for a new automated solution. Functionality from this application has subsequently been integrated into STRess2.

■ A *mixture module* known as 'Pendulum' performs mixture analysis for databasing or court analysis. The program can deal with all two-person mixtures and is capable of deriving both contributors. The program can decipher mixtures when one of the profiles is from a known individual (e.g., victim in a rape kit test) or when both contributors are unknown (e.g., databasing). This program has delivered significant gains in quality and consistency. The program requires minimal operator intervention as it has an automated link to the expert system suite. STRess2 searches the batch for potential mixtures, any potential mixtures are sent to Pendulum for analysis, and the resultant profile is sent back to STRess2 to be incorporated with the rest of the batch results. All calculations performed by Pendulum are audited and presented graphically to the operator. The FSS offers training courses on this interpretation approach. The program also offers a complex search feed for profiles that cannot be interpreted using conventional theory. This information is sent to the 'search algorithms suite' for comparison against the database.

■ The *search algorithm module* enables high-end search algorithms to provide intelligence for samples not interpretable using conventional approaches. These applications offer ways of generating useful intelligence and increasing the overall 'value' of DNA forensic services. This will be a significant area of growth over the coming years.

Introduction of the suite of expert systems described above has resulted in a significant increase in efficiency and quality at the FSS with a large reduction in unit cost. Based largely on issues discovered during the development of STRess2, Martin Bill of the Forensic Science Service offers the following seven recommendations for expert system development:

1. *Integration.* Ensure that the information technology (IT) infrastructure, support and storage issues are considered when designing and developing expert systems rather than concentrating solely on the interpretation aspects. Solutions that are selected without considering these IT issues may result in most of the financial benefits of the expert systems being lost in future IT expenditure.

2. *External influences.* Consider potential changes to the supply chain and ensure the system will still be able to perform as required. On occasion external influences may require a change to interpretation. Any expert system solution must be flexible enough to work around such changes without causing significant problems.

3. *Process design.* Implement a new process that encompasses the expert system; do not simply implement an expert system using existing protocols. Process re-engineering is invariably required to maximize benefits when implementing the expert system. If this is not considered, the benefits of the software may disappear and it can be difficult to undo the damage.

4. *Benefits measurement.* There is a trite statement 'What gets measured gets done'. Decide how the benefits of an expert system can be measured and ensure the measurement process takes place. Make sure the correct units of measurement are used. Remember you are trying to measure the actual benefits realized not just the potential of the software. Unit costs are invariably better indicators than timing exercises.

5. *People and culture.* Expect cultural issues until the scientists gain confidence in the expert system and include this aspect of the implementation in the project plan. Do not be surprised or disappointed if people need time to become accustomed to the idea of automated interpretation.

6. *Success rate.* This factor is often overlooked yet it is one of the most critical areas to consider. As a DNA unit moves from manual to automated interpretation, the success rate of the process will change. Expert systems will probably never have total concordance with manual interpretation because the computer is following a rigid set of rules. The change in success rate should be closely monitored during the initial phase of deployment.

7. *Target setting.* Set realistic targets for the project. Analysts do not spend all of their time analyzing data and therefore it is impossible to realize 100% analyst reduction irrespective of how good the expert system is. This aspect closely links with the process re-engineering. The same problem exists with projected error rates. Many laboratories refuse to acknowledge that an error rate exists. We should recognize that there are many opportunities for error to occur, some within and some outside the control of the DNA unit. It is better to openly acknowledge that error can occur, as it is easier to look for solutions. No expert system will ever be designed that has an error rate of zero and therefore setting a target of zero is self-defeating. The real benefit of expert systems is that they behave predictably. It is this predictability and standardization that improves quality. As a starting point the objective should be to improve on the manual error rate (therefore making forward progress). When using expert systems, error rate and success rates are closely linked, one effectively determines the other. This level of control is extremely useful when attempting to optimize the output from a DNA unit.

GENEMAPPER ID

Applied Biosystems (Foster City, CA) released a computer program in November 2003 named GeneMapper*ID v.3.1* that combines the functions of GeneScan and Genotyper. Thus, GeneMapper*ID* designates peaks in electropherograms and calls alleles through size comparisons to allelic ladders. GeneMapper*ID* still requires manual review of the data but process quality values (PQVs) are generated to help provide confidence in allele calls and to aid troubleshooting efforts (Applied Biosystems 2003).

COMPARECALLS^{SM}: AN AUTOMATED ALLELE CONCORDANCE ANALYSIS SYSTEM

Myriad Genetics (Salt Lake City, Utah) has developed an automated allele concordance analysis system known as CompareCalls^{SM} that can aid in rapid review of single-source STR profiles (Ryan *et al.* 2004). DNA Advisory Board Standard 17.1.1 (see Appendix IV) states that public laboratories perform a '100% technical review' of data generated by contract or vendor laboratories prior to uploading these STR profiles to the National DNA Index System (NDIS) of CODIS (see Chapter 18). Two independent analysis pathways – typically two different human analyst reviewers – are generally required to demonstrate that STR allele calls are the same between the two analysis pathways with the expectation that concordant information will be reliable (see Chapter 15).

With the CompareCalls^{SM} approach, a human reviewer using Genotyper® and a human reviewer using a different analysis platform (in this case Myriad's SureLockID^{SM} allele calling software) generate data that is then compared for concordance. Thus, the CompareCalls^{SM} software is not making the allele calls but rather checking them to ensure quality results between two independent reads of the STR data. Validation studies with 290 676 STR markers found Myriad's CompareCalls^{SM} software to be at least as accurate as 100% human technical review of STR profiles (Ryan *et al.* 2004).

UNIQUE CHALLENGES WITH FORENSIC DNA AND NEW TECHNOLOGIES

DNA separations for the purposes of STR genotyping have been primarily conducted to date with electrophoresis, either in the form of slab gels or capillary instruments (see Chapters 12 and 14). However, a number of new methods are under development in research laboratories around the world. These methods involve techniques such as miniature electrophoresis separation systems, hybridization techniques, and mass spectrometry, all of which have been discussed briefly here. New genetic markers are also being investigated such as single nucleotide polymorphisms (see Chapter 8). We can expect that these new technologies will make DNA typing faster, cheaper, and easier to perform.

In the not too distant future, portable systems may be in use that would permit a rapid DNA test right at the crime scene. In addition, large laboratory centers are being established in many parts of the world that have the capability to perform thousands or even tens of thousands of DNA tests per day.

However, the adoption of new technology by the forensic DNA community takes time for several reasons. First and foremost, methods need to be carefully validated to ensure that results with a new technology are accurate and reproducible (see Chapter 16). Second, methods should yield comparable results to current technologies so that genotype information can be compared over time. The development of large DNA databases make it necessary to have a constant currency so that convicted offender samples have been analyzed with the same DNA markers as crime scene samples (see Chapter 18). A new set of markers or a new form of sample analysis, unless it gives an equivalent result to current technology, must have clear advantages and be very inexpensive to overcome legacy data in large DNA databases (Gill 2002, Gill *et al.* 2004). Now that millions of DNA profiles are present in national DNA databases (see Chapter 18) it is highly unlikely that the field will abandon the current STR loci in the next 5–10 years (National Commission on the Future of DNA Evidence 2000).

With the continued progress in biotechnology around the world will come better and better methods for DNA typing methods used in forensic DNA laboratories. We can expect that future DNA testing technologies will include the following desirable characteristics:

- Improved capabilities for multiplex PCR, i.e., the ability to amplify more regions of the DNA simultaneously in order to improve further the number of markers examined and therefore the discrimination power of the test;
- More rapid separation/detection technology;
- More automated sample processing and data analysis/interpretation;
- Less expensive sample analysis;
- Accurate, robust methods.

REFERENCES AND ADDITIONAL READING

Applied Biosystems (2003) *GeneMapper™ ID Software Version 3.1 Human Identification Analysis User Guide*. Foster City, California.

Belgrader, P., Smith, J.K., Weedn, V.W. and Northrup, M.A. (1998) *Journal of Forensic Sciences*, 43, 315–319.

Butler, J.M., Li, J., Shaler, T.A., Monforte, J.A. and Becker, C.H. (1998) *International Journal of Legal Medicine*, 112, 45–49.

Butler, J.M. and Becker, C.H. (2001) *Improved analysis of DNA short tandem repeats with time-of-flight mass spectrometry*. Washington, DC: National Institute of Justice. Available online at: http://www.ojp.usdoj.gov/nij/pubs-sum/188292.htm.

Crouse, C.A. and Conover, J. (2003) *Proceedings of the Fourteenth International Symposium on Human Identification*. Madison, Wisconsin: Promega Corporation. See http://www.promega.com/geneticidproc/ussymp14proc/oralpresentations/Crouse.pdf.

Dunbar, H.N., Sparkes, R.L., Hopwood, A.J., Pinchin, R. and Watson, S.K. (1998) *Proceedings of the Second European Symposium on Human Identification*, pp. 55–58. Madison, Wisconsin: Promega Corporation.

Gill, P., Urquhart, A., Millican, E., Oldroyd, N., Watson, S., Sparkes, R. and Kimpton, C. (1996) *International Journal of Legal Medicine*, 109, 14–22.

Gill, P. (2002) *BioTechniques*, 32, 366–372.

Gill, P., Werrett, D.J., Budowle, B. and Guerreri, R. (2004) *Science and Justice*, 44, 51–53.

Goedecke, N., McKenna, B., El-Difrawy, S., Carey, L., Matsudaira, P. and Ehrlich, D. (2004) *Electrophoresis*, 25, 1678–1686.

Greenspoon, S.A., Ban, J.D., Sykes, K., Ballard, E.J., Edler, S.S., Baisden, M. and Covington, B.L. (2004) *Journal of Forensic Sciences*, 49, 29–39.

Hale, A.N. (1999) Building realistic automated production lines for genetic analysis. In Craig, A.G. and Hoheisel, J.D.(eds) *Automation: Genomic & Functional Analyses, Methods in Microbiology*, Volume 28, Chapter 5, pp. 93–129. San Diego: Academic Press.

Hayn, S., Wallace, M.M., Prinz, M. and Shaler, R.C. (2004) *Journal of Forensic Sciences*, 49, 07–91.

Hopwood, A., Brookes, J., Shariff, A., Cage, P., Tatum, E., Mirza, R., Crook, M., Brews, K. and Sullivan, K. (1997) *Proceedings of the Eighth International Symposium on Human Identification*, pp. 20–24. Madison, Wisconsin: Promega Corporation.

Kadash, K., Kozlowski, B.E., Biega, L.A. and Duceman, B.W. (2004) *Journal of Forensic Sciences*, 49 (4), 660–667.

Lagally, E.T., Emrich, C.A. and Mathies, R.A. (2001) *Lab on a Chip*, 1, 102–107.

McCormick, R., Nelson, R.J., Alonso-Amigo, M.G., Benvegnu, D.J. and Hooper, H.H. (1997) *Analytical Chemistry*, 69, 2626–2630.

Medintz, I.L., Berti, L., Emrich, C.A., Tom, J., Scherer, J.R. and Mathies, R.A. (2001) *Clinical Chemistry*, 47, 1614–1621.

Mitnik, L., Carey, L., Burger, R., Desmarais, S., Koutny, L., Wernet, O., Matsudaira, P. and Ehrlich, D. (2002) *Electrophoresis*, 23, 719–726.

Muller, O., Minarik, M. and Foret, F. (1998) *Electrophoresis*, 19, 1436–1444.

National Commission on the Future of DNA Evidence (2000) *The Future of Forensic DNA Testing: Predictions of the Research and Development Working Group*. Washington, D.C.: National Institute of Justice. Available at: http://www.ojp.usdoj.gov/nij/pubs sum/183697.htm.

Northrup, M.A., Benett, B., Hadley, D., Landre, P., Lehew, S., Richards, J. and Stratton, P. (1998) *Analytical Chemistry*, 70, 918–922.

Paegel, B.M., Blazej, R.G. and Mathies, R.A. (2003) *Current Opinions in Biotechnology*, 14, 42–50.

Palsson, B., Palsson, F., Perlin, M., Gudbjartsson, H., Stefansson, K. and Gulcher, J. (1999) *Genome Research*, 9, 1002–1012.

Parson, W. and Steinlechner, M. (2001) *Forensic Science International*, 122, 1–6.

Perlin, M.W., Coffman, D., Crouse, C.A., Konotop, F. and Ban, J.D. (2001) Automated STR data analysis: validation studies. *Proceedings of the Twelfth International Symposium on Human Identification*. Madison, Wisconsin: Promega Corporation. Available at: http://www.promega.com/geneticidproc/ussymp12proc/contents/perlin.pdf.

Perlin, M.W. (2003) Simple reporting of complex DNA evidence: automated computer interpretation. *Proceedings of the Fourteenth International Symposium on Human Identification*. Madison, Wisconsin: Promega Corporation. Available at: http://www.promega.com/geneticidproc/ussymp14proc/oralpresentations/perlin.pdf.

Ross, P.L. and Belgrader, P. (1997) *Analytical Chemistry*, 69, 3966–3972.

Ross, P.L., Davis, P.A. and Belgrader, P. (1998) *Analytical Chemistry*, 70, 2067–2073.

Ryan, J.H., Barrus, J.K., Budowle, B., Shannon, C.M., Thompson, V.W. and Ward, B.E. (2004) *Journal of Forensic Sciences*, 49 (3), 492–499.

Schmalzing, D., Koutny, L., Adourian, A., Belgrader, P., Matsudaira, P. and Ehrlich, D. (1997) *Proceedings of the National Academy of Sciences USA*, 94, 10273–10278.

Schmalzing, D., Koutny, L., Chisholm, D., Adourian, A., Matsudaira, P. and Ehrlich, D. (1999) *Analytical Biochemistry*, 270, 148–152.

Shi, Y., Simpson, P.C., Scherer, J.R., Wexler, D., Skibola, C., Smith, M.T. and Mathies, R.A. (1999) *Analytical Chemistry*, 71, 5354–5361.

Sosnowski, R.G., Tu, E., Butler, W.F., O'Connell, J.P. and Heller, M.J. (1997a) *Proceedings of the National Academy of Sciences USA*, 94, 1119–1123.

Sosnowski, R.G., Canter, D., Duhon, M., Feng, L., Muralihar, M., Radtkey, R., O'Connell, J., Heller, M. and Nerenberg, M. (1997b) *Proceedings of the Eighth International Symposium on Human Identification*, pp. 119–125. Madison, Wisconsin: Promega Corporation.

Swerdlow, H., Jones, B.J. and Wittwer, C.T. (1997) *Analytical Chemistry*, 69, 848–855.

Tereba, A., Mandrekar, P.V., Flanagan, L., Olson, R., Mandrekar, M. and McLaren, R. (2003) *Proceedings of the Fourteenth International Symposium on Human Identification.* Madison, Wisconsin: Promega Corporation. See http://www.promega.com/geneticidproc/ussymp14proc/oralpresentations/Tereba.pdf.

Woolley, A.T. and Mathies, R.A. (1994) *Proceedings of the National Academy of Sciences USA*, 91, 11348–11352.

Woolley, A.T., Hadley, D., Landre, P., deMello, A.J., Mathies, R.A. and Northrup, M.A. (1996) *Analytical Chemistry*, 68, 4081–4086.

CHAPTER 18

COMBINED DNA INDEX SYSTEM (CODIS) AND THE USE OF DNA DATABASES

I think that everyone should give a DNA sample… Frankly, the remote possibility that Big Brother will one day be perusing my genetic fingerprint for some nefarious end worries me less than the thought that tomorrow a dangerous criminal may go free – perhaps only to do further evil – or an innocent individual may languish in prison for want of a simple DNA test.

(James Watson, *DNA: The Secret of Life*, p. 290)

On 13 October 1998, the Federal Bureau of Investigation (FBI) officially launched its nation-wide DNA database. By the end of 2003, this database, named the COmbined DNA Index System or CODIS, contained over 1.5 million short tandem repeat (STR) profiles and linked all 50 states in the United States with the capability to search criminal DNA profiles in a similar fashion as the FBI fingerprint database. Since the first national DNA database was established in the United Kingdom in 1995, DNA databases around the world have revolutionized the ability to use DNA profile information to link crime scene evidence to perpetrators.

These databases are effective because a majority of crimes are committed by repeat offenders. In fact, more than 60% of those individuals put in prison for violent offenses and subsequently released were re-arrested for a similar offense in less than three years (McEwen and Reilly 1994, Langan and Levin 2002, Langan *et al.* 2003). This chapter will discuss the DNA databases being used in the United States and throughout the world to stop violent criminals, such as the introductory case reviewed in Chapter 1.

VALUE OF DNA DATABASES

Information sharing has always been crucial to successful law enforcement. Good information can solve crimes and ultimately save lives. DNA databases are just beginning to serve as valuable tools in aiding law enforcement investigations. Their effectiveness will grow as the size of the database gets larger. These databases can be used to locate suspects in violent crime cases that would otherwise never have been solved. Consider the sexual assault case described in Chapter 1. Without the Virginia DNA database, the rapist would probably have avoided detection.

A second important role that DNA databases, or databanks, can serve is to make associations between groups of unsolved cases. Criminals do not honor the same geographical boundaries that law enforcement personnel do. Crimes committed in Florida can be linked to ones committed in Virginia through an effective national DNA database.

But, DNA profile information must be in the database for it to be of value. Today tremendous sample backlogs exist in the United States – meaning that samples have been collected but are waiting analysis and entry into CODIS. Hundreds of thousands of samples await short tandem repeat (STR) typing. Efforts are being made to correct this sample backlog problem. In March 2003, U.S. Attorney General John Ashcroft announced an initiative to invest $1 billion over the next five years into forensic DNA programs to reduce the backlogs of casework and convicted offender samples. Hopefully in a few years, crime scene samples can be quickly analyzed and uploaded for a rapid and effective search against a comprehensive national DNA database due to this increased funding. The establishment of an effective DNA database requires time and full cooperation between forensic DNA laboratories, the law enforcement community, and government policy makers. The investment though is worth the effort to society and especially to victims of crime (D.N.A. Box 18.1).

D.N.A. Box 18.1

The business case for using forensic DNA technology

National DNA databases, such as the Combined DNA Index System (CODIS), have opened an entirely new avenue of identifying repeat offenders and assisting in 'no suspect' sexual assault investigations. With limited budgets and difficult decisions being made by lawmakers on how best to prioritize funds to aid society, a business analysis of the expected return on an investment in forensic DNA technology was presented at the 2004 Annual Meeting of the American Academy of Forensic Sciences in Dallas, Texas. The numbers below come from this analysis by Ray Wickenheiser, Director of the Acadiana Criminalistics Laboratory (New Iberia, LA).

Within the United States, there are 366 460 sexual assaults reported each year (1992–2000 average). Since only 1/3 to 1/20 of sexual assaults are reported to the police, this number is fairly conservative. Approximately 34% of sexual assaults are committed by a stranger and would thus be termed 'no suspect'. These cases are normally unsolved without the power of DNA testing.

Studies have shown that 2/3 of offenders are repeat offenders. The average serial rapist commits eight sexual assaults prior to apprehension. Thus, seven of these offenses would be preventable if crime scene DNA testing was done on every case and the rapist's profile was in the DNA database to make the hit to the first sexual assault.

The cost of these crimes per offense committed is approximately $111 238 (adjusted from 1995 study to 2003 dollars). This figure includes the physical injury, hospitalization, lost time at work, counseling, and 'pain and suffering' incurred by the victim. The cost of investigating the crime and prosecuting and incarcerating the offender is not included in this number so it is probably pretty conservative.

There is approximately a 47.58% success rate of finding sperm and recovering a foreign DNA profile from sexual assault victims.

The Forensic Science Service in England has demonstrated that when a DNA database is sufficiently populated with criminal DNA profiles, a 42% hit rate can be obtained where a hit is made from a 'no suspect' case on a known offender present in the database.

Working through these numbers gives the following cost to crime:

366 460 × 34% = 124 596 reported 'no suspect' sexual assaults

124 596 × 2/3 = 83 056 of 'no suspect' sexual assaults are committed by repeat offenders

83 056 × 7 = 581 392 future sexual assaults that are preventable

581 392 × 47.58% = 276 626 unnecessary victims of preventable sexual assaults

276 626 × 42% = 116 183 estimated sexual assaults could be solved with DNA database hits

116 183 × $111 238 = **$12.9 billion saved** in terms of costs from prevented crimes

The cost to perform sexual assault testing in every case is approximately $366 million assuming a cost of $1000 per case and working all 366 460 sexual assaults. Thus, the return on investment is over 3500%. For every dollar invested in forensic DNA testing, this analysis shows over $35 would be saved in terms of expense to victims and society.

Source:

Ray Wickenheiser presentation at February 2004 American Academy of Forensic Sciences meeting (Dallas, TX); Wickenheiser, R.A. (2004) 'The Business Case for Using Forensic DNA Technology to Solve and Prevent Crime'. *Journal of Biolaw and Business*, 7 (3), 34–50.

ESTABLISHING A NATIONAL DNA DATABASE

Implementing a national DNA database is no small task. A number of features must be in place before the database can be established and actually be effective. These are listed below:

■ A commitment on the part of each state (and local) government to provide samples for the DNA database;

■ A common set of DNA markers or standard core set so that results can be compared between all samples entered into the database;

■ Standard software and computer formats so that data can be transferred between laboratories;

■ Quality standards so that everyone can rely on results from each laboratory.

The technology of forensic DNA databases basically involves three parts: (1) the collection of known specimens, (2) analyzing those specimens and placing their DNA profiles in a computer database, and (3) subsequent comparison of unknown profiles obtained from crime scene evidence with the known profiles in the computer database.

All 50 states within the U.S. have now enacted legislation to establish a DNA databank containing profiles from individuals convicted of specific crimes. The laws vary widely across the states concerning the scope of crimes requiring sample collection for DNA databank entry. However, more and more states are moving towards collecting samples from all felons. In January 2003, the state of Virginia began collecting and analyzing DNA samples from all those *arrested* of certain violent crimes (Ferrara and Li 2004). The trend towards broader coverage of criminal DNA databases will likely continue as these resources demonstrate their value to the criminal justice system.

Law enforcement agencies search these databanks for matches with DNA profiles from biological evidence of unsolved crimes. Using these databanks, law enforcement agencies have been successful in identifying suspects in cases that would likely be unsolvable by any other means (e.g., the Montaret Davis case described in Chapter 1).

A number of other countries around the world have also launched national DNA databases (Martin *et al.* 2001, Schneider and Martin 2001, Martin 2004). The earliest national DNA databank, and so far the most effective, was created in the United Kingdom in 1995. In the first five years, more than 500 000 DNA profiles were entered into the database and more than 50 000 criminal investigations were aided (Werrett and Sparkes 1998). As of 2004, the UK National DNA Database (NDNAD) contains more than two million profiles (see http://www.forensic.gov.uk, Asplen 2004).

COMBINED DNA INDEX SYSTEM (CODIS)

The FBI started CODIS as a pilot project in 1990 that served just 14 state and local laboratories. It took several years to gather enough DNA profiles from convicted offenders to reach the critical mass necessary to obtain matches for crime scene evidence. During the 1990s, the number of samples in CODIS grew to several hundred thousand. In addition, the number of laboratories submitting data increased. As of April 2004, CODIS software was installed in 175 public laboratories around the United States giving them the capability to submit DNA profiles to a national DNA database. In addition, 31 laboratories in 18 foreign countries also use the CODIS software.

Within CODIS there are two primary sample indexes: convicted offender samples and forensic casework samples. There is also a population file with DNA types and allele frequency data from anonymous persons intended to represent major population groups found in the United States (Budowle and Moretti 1998, Budowle *et al.* 2001a). These databases are used to estimate statistical frequencies of DNA profiles using the program PopStats (see Chapter 20). A missing persons index is also included in the U.S. national DNA database.

DNA profile information inputted into CODIS over the years has included restriction fragment length polymorphism (RFLP) loci and polymerase chain reaction (PCR)-based markers HLA-DQA1, PolyMarker, and D1S80. More recently, the 13 CODIS core STR loci are required for data entry into the national level of the U.S. DNA database (Budowle *et al.* 1998). These 13 STR markers were reviewed in Chapter 5 and provide a random match probability of approximately 1 in 100 trillion.

CODIS is not a criminal history information database but rather a system of pointers that provides only the information necessary for making matches. Only a unique identifier and the DNA profiles for a sample, such as the 13 STR loci shown in Table 18.1, are stored in CODIS. No personal information, criminal history information, or case-related information is contained within CODIS.

When CODIS identifies a potential match, the laboratories responsible for the matching profiles are notified and they contact each other to validate or refute the match (Niezgoda and Brown 1995, Niezgoda 1997). After the match has been confirmed by qualified DNA analysts, which often involves retesting of the matching convicted offender DNA sample, laboratories may exchange additional information, such as names and phone numbers of criminal investigators and case details. If a match is obtained with the Convicted Offender Index, the identity and location of the convicted offender is determined and an arrest warrant procured.

The primary metric for CODIS is the 'hit', which is defined as a match between two or more DNA profiles that provides the police with an investigative lead (or an 'investigation aided' lead) that would not otherwise have developed. Through April 2004, there have been more than 16 000 investigations aided using CODIS allowing thousands of crimes to be linked and solved around the United States. Because the number of hits is largely related to the size of the database, as CODIS continues to grow, so will its value. For example, England's national DNA database maintains a 40% chance of obtaining

Table 18.1

Example of the STR profile information stored in the CODIS DNA database for a single sample. Note that there is no personal information that can be used to link an individual to his or her DNA profile. The two alleles for each STR marker are placed in separate columns labeled value 1 and value 2. For markers with homozygous results, both value 1 and value 2 are the same, see for example CSF1PO. The information in the 'sample info' field can be related to a known individual only by the originating forensic DNA laboratory.

Sample Info	Sample #	Category	Tissue Type	Tissue Form	Population
F130	1	Convicted Offender	Blood	Stain	Caucasian
	Marker	**Value 1**	**Value 2**	**Date**	**Time**
	AMEL	X	Y	15-FEB-2000	17:38:30
	CSF1PO	10	10	15-FEB-2000	17:38:30
	D13S317	11	14	15-FEB-2000	17:38:30
	D16S539	9	11	15-FEB-2000	17:38:30
	D18S51	14	16	15-FEB-2000	17:38:30
	D21S11	28	30	15-FEB-2000	17:38:30
	D3S1358	16	17	15-FEB-2000	17:38:30
	D5S818	12	13	15-FEB-2000	17:38:30
	D7S820	9	9	15-FEB-2000	17:38:30
	D8S1179	12	14	15-FEB-2000	17:38:30
	FGA	21	22	15-FEB-2000	17:38:30
	TH01	6	6	15-FEB-2000	17:38:30
	TPOX	8	8	15-FEB-2000	17:38:30
	VWA	17	18	15-FEB-2000	17:38:30

a match between a crime scene profile and a 'criminal justice' (arrestee or suspect) profile loaded into the database (Asplen 2004).

LEVELS OF CODIS

The COmbined DNA Index System is composed of local, state, and national levels (Figure 18.1). All three levels contain the convicted offender and casework indexes and the population data file. The software is configurable to support any RFLP or PCR DNA markers although after 2000 only STR data is being added. At the local level or Local DNA Index System (LDIS), investigators can input their DNA profiles and search for matches with local cases. All forensic DNA records originate at the local level and are 'uploaded' or transmitted to the state and national levels.

Each participating state has a single laboratory that functions as the State DNA Index System (SDIS) to manage information at the state level. SDIS enables exchange and comparison of DNA profiles within a state and is usually operated by the agency responsible for maintaining a state's convicted offender DNA database program.

The National DNA Index System (NDIS) manages nationwide information in a single repository maintained by the FBI Laboratory. Participating states submit their DNA profiles in order to have searches performed on a national level. The role of NDIS is to search casework and offender indices, manage candidate matches, and return results of matches to the local LDIS level.

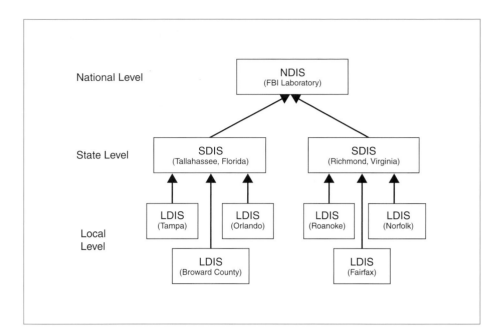

Figure 18.1

Schematic of the three tiers in the Combined DNA Index System (CODIS). DNA profile information begins at the local level, or Local DNA Index System (LDIS), and then can be uploaded to the state level, or State DNA Index System (SDIS), and finally to the national level, or National DNA Index System (NDIS). Each local or state laboratory maintains its portion of CODIS while the FBI Laboratory maintains the national portion (NDIS).

When NDIS was first activated in October 1998, there were 119 000 offender profiles and 5000 forensic casework profiles from nine states. By December 1999, a little over a year later, 21 states and the FBI had inputted 211 673 offender profiles and 11 112 forensic profiles. While many of the original DNA profiles were from RFLP markers, forensic DNA laboratories in the United States have now converted completely to the 13 core STR loci. Presumably all samples for the foreseeable future will be typed with these STRs (Gill *et al.* 2004). In April 2004, the total number of offender STR profiles stood at more than 1.6 million with around 80 000 forensic profiles present in NDIS.

In order for a state to have its DNA profiles included in the national DNA index system, a memorandum of understanding must be signed whereby the state DNA laboratories agree to adhere to the FBI issued quality assurance standards that are listed in Appendix IV. A complete DNA profile of four RFLP markers or the 13 core STR loci is required for submission of convicted offender information to the National DNA Index System. There must also be a minimum number of loci included in the casework DNA profile: at least three CODIS RFLP loci or 10 of the 13 CODIS STR loci before uploading the information to NDIS. The lower number of loci needed for casework DNA profiles comes from recognition that degraded DNA samples obtained from forensic cases may not yield results at every marker even though analysis is attempted with those DNA markers (see Chapter 7).

CJIS WAN

Public crime laboratories in the United States are connected via the FBI's Criminal Justice Information Services Wide Area Network (CJIS WAN) through T1 lines capable of transmitting 1.5 megabytes of information per second. CJIS WAN provides Internet-like connectivity but without the security risk. This network is an intranet with access only permitted to participating laboratories. The National DNA Index System computers are protected by firewalls to maintain a high degree of security. The hub of CJIS WAN is located in Clarksburg, West Virginia.

Each state pays for their end of the system. The computer equipment for a state system costs around $15 000–25 000. The FBI Laboratory provides the CODIS software and maintains the equipment for the national system. SDIS and LDIS laboratories sign memorandums of understandings with the FBI. CODIS users agree to adhere to FBI-issued quality assurance standards (see Appendix IV) and to submit to NDIS audits.

CONVICTED OFFENDER SAMPLES VS. FORENSIC CASEWORK SAMPLES

For a criminal DNA database to be successful convicted offender DNA samples must be entered and crime scene material from cases where there is no suspect must be tested. Because the demand for DNA testing is surpassing the ability of public forensic laboratories to perform the tests, private contract laboratories are now being used to reduce the sample backlogs for convicted offender samples. Much of this work in the United States is being performed with federal government financial assistance through grant programs administered by the National Institute of Justice.

Private DNA typing laboratories can have a higher throughput capacity because the focus is on running samples rather than performing casework and testifying in court. In addition, all of the convicted offender samples are in the same format (i.e., liquid blood or DNA extracted from buccal swabs), which improves the capability for automating the DNA typing process. On the other hand, forensic cases can involve the examination of a dozen or more pieces of biological evidence from a variety of formats (e.g., semen stains, bloodstains, etc.), which makes them much more complex.

Since July 1998, the Virginia Department of Forensic Sciences DNA laboratory system has outsourced many of its convicted offender samples (Pederson 1999). By November 1999, the Bode Technology Group, a contract service laboratory located in Springfield, Virginia, had analyzed more than 100 000 samples. Convicted offender samples were analyzed at a rate of approximately 2000 samples per week (Pederson 1999). This rapid growth in the Virginia DNA database directly led to the hit that solved the rape case discussed in Chapter 1. The Virginia state laboratories have built their capacity over the past few years and are now doing all of their own samples rather than outsourcing them as of early 2004.

To ensure that analysis of convicted offender samples by contract laboratories is performed in a reliable fashion, the DNA Advisory Board issued standards for analysis of convicted offender samples. These guidelines became effective in April 1999. Appendix IV contains a copy of the contract laboratory quality assurance standards in a format that directly compares them with the standards for quality assurance of forensic casework samples.

DATABASES FOR MISSING PERSONS AND MASS DISASTER RECONSTRUCTION

DNA databases can also play an important role in helping identify missing individuals and aiding mass disaster reconstruction following a plane crash or terrorist activity (see Chapter 24). In these cases, DNA samples are often obtained from biological relatives that can be searched against remains recovered from a missing individual or a disaster site. Many states within the United States and nations around the world are beginning to establish missing persons databases to enable matching of recovered remains to their family members (Lorente *et al.* 2002). CODIS/NDIS also has an index for missing person investigations that can store DNA profiles from both recovered remains and family samples that serve as references. Much of the data from missing person investigations is in the form of mitochondrial DNA sequences (see Chapter 10) since this information can be successfully recovered from highly degraded samples. Use of mitochondrial DNA also enables access to a larger number of reference samples from maternal relatives of a victim.

IMPORTANT ISSUES FOR DNA DATABASES

There are a number of important issues for DNA databases. These issues include security of the information contained in them, the ability to perform rapid searches and effective matches from large numbers of entries, maintaining the quality of the inputted data, and handling changes in technology. Both computer and DNA technologies are constantly improving at a rapid rate. DNA databases have to be flexible enough to handle this change. Legacy data must be maintained or the value of the database will be diminished (see Gill *et al.* 2004).

PRIVACY ISSUES

One of the major challenges for maintaining a DNA database is the issue of privacy and security of the information stored in the database. Blood samples contain genetic information that could be used against an individual or their family if not handled properly. The issue of privacy is approached in two ways. First, the DNA markers, such as the 13 CODIS core STR loci, are in non-coding regions of the DNA and are not known to have any association with a genetic disease or any other genetic predisposition. Thus, the information in the database is only useful for human identity testing.

Second, no names of individuals or other characterizing data is stored with the DNA profiles. The National DNA Index System of CODIS only references the sources of the DNA profiles, such as Orange County Sheriff's Office or Palm Beach County Crime Laboratory. Specific case data is secured and controlled by local law enforcement agencies (Spalding 1995). Thus, only the crime laboratory that submitted the DNA profile has the capability to link the DNA results with a known individual.

Another important facet to the privacy and security of the information in DNA databases is the fact that access to CODIS is solely for law enforcement purposes. There are strict penalties for any one using the information or samples for any purpose other than law enforcement including a $100 000 fine for unauthorized disclosures of information on any sample (McEwen 1995).

MAINTAINING QUALITY CONTROL OF DATA

The old adage of 'garbage in, garbage out' applies with any database containing information that will be probed regularly. If the DNA profiles entered into a DNA database are not accurate, then they will be of little value for making a meaningful match. The high quality of data going into a DNA database is ensured by requiring laboratories to follow quality assurance guidelines (Appendix IV), to submit to audits of their procedures, and by conducting regular proficiency tests of analysts as described in Chapter 16.

SEARCH AND MATCH ALGORITHMS

As DNA databases grow in size, they become more valuable as an intelligence tool, but they also become more of a challenge to search rapidly. In addition, because the STR kits used by the various manufacturers may have different primer binding regions for the same loci, allele dropout could result with one primer set and not the other as noted previously in Chapter 6 (Walsh 1998, Budowle *et al.* 2001b). This would result in an apparent discrepancy between results obtained with one STR kit versus another. Lower stringency search algorithms may be used to address this issue. For example, the CODIS search algorithm and match criteria can be loosened on a search using 26 possible alleles from the 13 STRs by only requiring a match at 25 out of 26 possible alleles.

Differences in measurement capabilities of laboratories, particularly in their ability to detect microvariant (off-ladder) alleles, make it important to have allele equivalency capabilities in the search algorithm. Thus, a TH01 allele 8.3 measured in one laboratory can be matched with an allele 8.x or allele 9 measured in another laboratory. However, this aspect of STR typing is not as problematic now that separation and detection technologies have become more refined and precise (see Butler *et al.* 2004).

SAMPLE COLLECTION FROM CONVICTED OFFENDERS

One of the facts about DNA databases that often gets overlooked is the sample collection process. Law enforcement personnel have to extract blood or obtain a saliva sample from incarcerated felons that are not always cooperative. In some cases, extraordinary efforts including force are required to persuade felons to submit to a blood draw (Spalding 1995). Collecting the actual samples can be a challenge considering the fact that the convicted offender knows his blood or saliva could be used to catch him committing another crime in the future or match him to a previous unsolved crime he committed.

WORKING UNKNOWN SUSPECT CASES

Crime laboratories must work cases that have no suspect in order to take full advantage of DNA databases. Convicted offender samples can be typed in large batches because large numbers come into the laboratory together and they are in the same format, such as liquid blood. Casework samples, on the other hand, present a different kind of challenge. Each case requires significant up-front work including evidence handling, locating DNA within the submitted evidence, and extraction of DNA from different types of substrates. Often sample mixtures must be dealt with and interpreted. Multiple pieces of evidence may also be involved in a case. In addition, significant work is required after analysis of the samples. Lab reports must be written and court testimony may be required.

In spite of the time and effort required to obtain results on crime scene samples, it is working these cases that make DNA databases effective. Law enforcement agencies must be encouraged to collect and submit evidence to the nation's crime laboratories especially if the statute of limitations is about to expire on a case (D.N.A. Box 18.2). In some cases, thousands of rape kits are sitting in police evidence rooms that are not submitted to crime laboratories (Lovrich *et al.* 2004).

MEASURING THE SUCCESS OF A DATABASE

The purpose of DNA databases is to solve crimes that would otherwise be unsolvable. A common method of measuring the effectiveness of CODIS or any

D.N.A. Box 18.2

The 'John Doe' warrant

The capabilities of forensic DNA testing have generated new legal issues for prosecutors. The sensitivity of the polymerase chain reaction enables DNA profiles to be obtained from previously intractable evidence. Furthermore, the existence of DNA databases now permits matches between perpetrators of crimes spanning jurisdictions and 'cold hits' on unsolved crimes many years after they occurred.

Many states have *statutes of limitations* meaning that after a certain period of time a crime cannot be prosecuted. If DNA evidence exists from a crime scene yet no suspect has been located to be charged with the crime, a 'John Doe' warrant may be issued based solely on the assailant's genetic code. In September 1999 Norman Gahn, Assistant District Attorney from Milwaukee County, Wisconsin, filed the first warrant for the arrest of 'John Doe', an unknown male who could be identified by his 13 locus STR profile. This approach has been successful in stopping the ticking clock of a crime's statute of limitations making it possible to prosecute the crime when the assailant is identified through a DNA database cold hit in the future. Several of these 'John Doe's' have been subsequently identified with DNA database cold hits and successfully prosecuted for the crimes they committed.

Wisconsin law governing the statute of limitations was amended in September 2001 to provide for the use of DNA profiles from individuals unknown to the prosecution at the time the warrant for arrest is issued. The new legislation creates an exception to the time limits for prosecuting sexual assault crimes if the State has DNA evidence related to the crime. John Doe warrants have also been issued in other states.

Source:
Silent Witness, Volume 7, Number 1, 2002, American Prosecutors Research Institute (see www.ndaa-apri.org)

other DNA database is in what is referred to as a 'hit.' A hit is a confirmed match between two or more DNA profiles discovered by the database search. Within CODIS, hits may occur at a local (LDIS), state (SDIS), or national (NDIS) level.

Hits fall into two different categories. A forensic hit occurs when two or more forensic casework samples are linked at LDIS, SDIS, or NDIS. These types of hits are sometimes called case-to-case hits and are especially important to solving serial crimes. An offender hit occurs when one or more forensic samples are linked to a convicted offender sample. These types of hits are sometimes referred to as case-to-offender hits. Either type of hit contributes to the bottom-line performance metric of a DNA database – the number of criminal investigations aided. In the first five years of operation (1998–2003), the CODIS system aided more than 15 000 investigations in the United States.

DNA DATABASE LAWS

DNA databases work because most criminals are repeat offenders (McEwen and Reilly 1994, Langan and Levin 2002). If their DNA profile can be entered into the system early in their criminal career, then they can be identified when future crimes are committed. Serial crimes can also be linked effectively with a computer database. Ultimately, the value of the DNA database is in its ability to apprehend criminals that are not direct suspects in a case and to prevent further victims from crimes committed by those individuals.

CRIMES FOR INCLUSION IN A STATE DNA DATABASE

As of June 1998, all 50 states in the United States had passed legislation requiring convicted offenders to provide samples for DNA databasing. Each state though has different requirements as to what types of offenses are considered for DNA sample collection. In many states these requirements are changing over time to include more and more criminal offenses. The requirements for having to donate a blood sample range from all felons to strictly sex offenses. The trend is for laws that require a DNA sample submission for any felony crime. Table 18.2 includes a summary list of the qualifying offenses for entry into a state's DNA database and the number of states within the U.S. that fall into each category as of June 2003.

Some state DNA database statutes specify exactly how the sample will be taken while others simply require any biological sample containing DNA. California, for example, requires two specimens of blood, a saliva sample and right thumb

Offenses	Number of States
Sex crimes	50
Murder	50
All violent crimes	47
Burglary	44
Drug crimes	35
All felons	30
Juveniles	32
Some misdemeanors	23
Arrestees/suspects	4

Table 18.2

Summary of U.S. state DNA database laws and qualifying offenses for DNA collection as of June 2003 (from http://www.dnaresource.com).

and full palm print impression for verifying identity of the submitting convicted offender (Herkenham 1999). The law for South Carolina, on the other hand, asks only for a suitable sample from which DNA may be obtained.

The ability of state and local forensic DNA laboratories to improve their capabilities for DNA analysis, especially with the STR technology described in this book, has been greatly aided by federal funding. The DNA Identification Act of 1994 provided approximately $40 million in federal matching grants to aid states in DNA analysis activities. This funding has been a great benefit to forensic DNA laboratories, which are typically understaffed and underfunded. While a convicted offender backlog of several hundred thousand samples exists in the United States as of early 2004, efforts are underway to alleviate this sample backlog within the next few years. In March 2003, Attorney General John Ashcroft announced an initiative from President George W. Bush to put $1 billion into forensic DNA typing over the next five years. Only time will tell if political promises become financial facts. However, society is bound to benefit from this investment in forensic DNA technology (see D.N.A. Box 18.1). Progress on legislation regarding the use of DNA is available through the web site: http://www.dnaresource.com.

NATIONAL DNA DATABASES AROUND THE WORLD

National DNA databases are being used in many countries around the world (Martin 2004, Walsh 2004). The same STR markers are being used in many instances. There are eight STR loci (FGA, TH01, VWA, D3S1358, D8S1179, D16S539, D18S51, and D21S11) that overlap between European and United States DNA database collection efforts. This fact will permit international collaboration on cases that warrant them.

The pioneering national DNA database was formed on 10 April 1995 in the United Kingdom and commonly referred to as the National DNA Database (NDNAD). Since the debut of this database, more than 2.5 million convicted felon DNA profiles have been processed by the United Kingdom's Forensic Science Service (Werrett and Sparkes 1998, Forensic Science Service 2003). Their original database involved six STR loci and the amelogenin gender identification marker with a random match probability of approximately 1 in 50 million. In 1999, the set of STR markers was expanded with the availability of the AmpFℓSTR® SGM Plus™ kit to include 10 STRs and amelogenin with a random match probability of approximately 1 in 3 trillion (see Chapters 5 and 21). The NDNAD delivers over 1700 crime scene-to-crime scene or suspect-to-crime scene hits per week (Asplen 2004) and has definitely demonstrated its value as an important tool for law enforcement. England's government has invested more than £182 million into NDNAD in the first 10 years of its existence, which equates to approximately $5 per citizen invested in DNA databasing (Asplen 2004). England has also had good success with mass screens, where

As described by Joseph Wambaugh's *The Blooding*, the first use of forensic DNA testing involved a genetic dragnet of over 4000 adult males in the Narborough, England area. Samples that failed to be excluded from the crime scene sample with traditional blood typing were subjected to 'DNA fingerprinting' or multi-locus RFLP testing. Colin Pitchfork was eventually apprehended based on this mass DNA intelligence screen (see D.N.A. Box 1.1).

DNA intelligence or 'mass' screens to aid identification of a perpetrator and exclusion of innocent individuals in no-suspect cases have been successfully used many times by the Forensic Science Service (FSS) and other law enforcement agencies. The largest mass screen conducted to date by the FSS was in conjunction with the investigation of the murder of Louise Smith, whose body was found near Chipping Sodbury, England in 1996. Over 4500 samples were analyzed from local volunteers at an expense of over one million pounds. Eventually police realized that one of the potential suspects had since moved to South Africa. He was tracked down and his DNA sample taken, which was found to match a crime scene STR profile recovered from the scene. David Frost is now serving time for the crime he committed and fled from hoping to escape justice.

Of course this type of effort and expense is not conducted in every case but it has proven useful in some situations. However, collecting samples from every individual fitting a particular description or living in a particular geographical region is not always greeted fondly by the general public. Questions about genetic privacy and civil liberties are often raised particularly in the United States when mass screens are initiated. In April 2004, a DNA dragnet was conducted in Charlottesville, Virginia to try and stop a rapist that had attacked at least six women between 1997 and 2003. Community concerns that black men were being targeted led police in Charlottesville to eventually suspend the mass screen after only collecting and analyzing about 200 men. Hopefully the proper balance can be found in the future to fully utilize the power of DNA testing and yet preserve the privacy and civil liberties of innocent citizens.

Source:
http://www.forensic.gov.uk; http://www.washingtonpost.com
(14, 15 and 17 April 2004)

DNA samples are collected from local volunteers in a particular region to help solve a crime without a suspect (D.N.A. Box 18.3).

Other countries in Europe besides England have also developed successful DNA databases (Martin *et al.* 2001, Schneider and Martin 2001). Each country has different laws regarding reasons for obtaining a DNA profile, when a profile would be expunged from the database, whether or not a DNA sample will be stored following analysis, and which STR loci are included. Most countries within the European Union have standardized on use of the AmpF*l*STR® SGM Plus™ kit, which will enable fruitful collaboration of criminal DNA information in the future.

REFERENCES AND ADDITIONAL READING

Asplen, C.H. (2004) The application of DNA technology in England and Wales. Available at: http://www.ojp.usdoj.gov/nij/pdf/uk_finaldraft.pdf.

Budowle, B., Moretti, T.R., Niezgoda, S.J. and Brown, B.L. (1998) *Proceedings of the Second European Symposium on Human Identification*, pp. 73–88. Madison, Wisconsin: Promega Corporation.

Budowle, B. and Moretti, T.R. (1998) *Proceedings of the Ninth International Symposium on Human Identification*, pp. 64–73. Madison, Wisconsin: Promega Corporation.

Budowle, B., Shea, B., Niezgoda, S. and Chakraborty, R. (2001a) *Journal of Forensic Sciences*, 46, 453–489.

Budowle, B., Masibay, A., Anderson, S.J., Barna, C., Biega, L., Brenneke, S., Brown, B.L., Cramer, J., DeGroot, G.A., Douglas, D., Duceman, B., Eastman, A., Giles, R., Hamill, J., Haase, D.J., Janssen, D.W., Kupferschmid, T.D., Lawton, T., Lemire, C., Llewellyn, B., Moretti, T., Neves, J., Palaski, C., Schueler, S., Sgueglia, J., Sprecher, C., Tomsey, C. and Yet, D. (2001b) *Forensic Science International*, 124, 47–54.

Butler, J.M., Buel, E., Crivellente, F. and McCord, B.R. (2004) *Electrophoresis*, 25, 1397–1412.

Coffman, D. (1998) *Proceedings of the Ninth International Symposium on Human Identification*, p. 63. Madison, Wisconsin: Promega Corporation.

Ferrara, P.B. and Li, G.C. (2004) *Profiles in DNA*, 7 (1), 3–5. Madison, Wisconsin: Promega Corporation. Available at: http://www.promega.com/profiles.

Forensic Science Service (2003) The National DNA Database Annual Report 2002–03. Available at: http://www.forensic.gov.uk/forensic/news/press_releases/2003/NDNAD_Annual_Report_02-03.pdf.

Gill, P., Urquhart, A., Millican, E.S., Oldroyd, N.J., Watson, S., Sparkes, R. and Kimpton, C.P. (1996) *International Journal of Legal Medicine*, 109, 14–22.

Gill, P., Werrett, D.J., Budowle, B. and Guerreri, R. (2004) *Science and Justice,* 44, 51–53.

Herkenham, M.D. (1999). *State DNA Database Statues: Summary of Provisions*, U.S. Department of Justice.

Herkenham, D. (2002) *Profiles in DNA*, 5 (1), 6–7. Madison, Wisconsin: Promega Corporation. Available at: http://www.promega.com/profiles.

Kimmelman, J. (2000) *Nature Biotechnology*, 18, 695–696.

Langan, P.A. and Levin, D.J. (2002) Recidivism of prisoners released in 1994. Bureau of Justice Statistics, U.S. Department of Justice, Washington, DC. Available at: http://www.ojp.usdoj.gov/bjs/abstract/rpr94.htm.

Langan, P.A., Schmitt, E.L. and Durose, M.R. (2003) Recidivism of sex offenders released from prison in 1994. Bureau of Justice Statistics, U.S. Department of Justice, Washington, DC. Available at: http://www.ojp.usdoj.gov/bjs/abstract/rsorp94.htm.

Lorente, J.A., Entrala, C., Alvarez, J.C., Lorente, M., Arce, B., Heinrich, B., Carrasco, F., Budowle, B. and Villanueva, E. (2002) *International Journal of Legal Medicine*, 116, 187–190.

Lovirich, N.P., Gaffney, M.J., Pratt, T.C., Johnson, C.L., Asplen, C.H., Hurst L.H. and Schellberg, T.M. (2004) National Forensic DNA Study Report. Available at: http://www.ojp.usdoj.gov/nij/pdf/dna_studyreport_final.pdf.

Martin, P.D., Schmitter, H. and Schneider, P.M. (2001) *Forensic Science International*, 119, 225–231.

Martin, P.D. (2004) National DNA databases – practice and practicability. A forum for discussion. *Progress in Forensic Genetics 10*, ICS 1261, 1–8.

McEwen, J.E. and Reilly, P.R. (1994) *American Journal of Human Genetics*, 54, 941–958.

McEwen, J.E. (1995) *American Journal of Human Genetics*, 56, 1487–1492.

National Commission on the Future of DNA Evidence (2000) *The Future of Forensic DNA Testing: Predictions of the Research and Development Working Group*. Washington, D.C.: National Institute of Justice. Available at: http://www.ojp.usdoj.gov/nij/pubs-sum/183697.htm.

Niezgoda, S.J. and Brown, B. (1995) *Proceedings of the Sixth International Symposium on Human Identification*, pp.149–153. Madison, Wisconsin: Promega Corporation.

Niezgoda, S.J. (1997) *Proceedings of the Eighth International Symposium on Human Identification*, pp. 48–49. Madison, Wisconsin: Promega Corporation.

Pederson, J. (1999) *Profiles in DNA*, 3, 3–7. Madison, Wisconsin: Promega Corporation. Available at: http://www.promega.com/profiles.

Royal Canadian Mounted Police (2003) National DNA Data Bank Advisory Committee 2002–2003 Annual Report. Available at: http://www.rcmp-grc.gc.ca/dna_ac/2002_2003_annualreport_e.htm.

Scheck, B. (1994) *American Journal of Human Genetics*, 54, 931–933.

Schneider, P.M. and Martin, P.D. (2001) *Forensic Science International*, 119, 232–238.

Spalding, V.B. (1995) *Proceedings of the Sixth International Symposium on Human Identification*, pp. 137–148. Madison, Wisconsin: Promega Corporation.

Walsh, P.S. (1998) *Journal of Forensic Sciences*, 43, 1103–1104.

Walsh, S.J. (2004) *Expert Reviews in Molecular Diagnostics*, 4 (1), 31–40.

Werrett, D.J. and Sparkes, R. (1998) *Proceedings of the Ninth International Symposium on Human Identification*, pp. 55–62. Madison, Wisconsin: Promega Corporation.

GENETICS

CHAPTER 19

BASIC GENETIC PRINCIPLES, STATISTICS, AND PROBABILITY

We balance probabilities and choose the most likely. It is the scientific use of the imagination.

(Sherlock Holmes, *The Hound of the Baskervilles*)

You can… never foretell what any one man will do, but you can say with precision what an average number will be up to. Individuals vary, but percentages remain constant. So says the statistician.

(Sherlock Holmes, *The Sign of Four*)

The genetics section of this book (Chapters 19–23) discusses basic principles that are important when considering forensic DNA profiles. These principles are based on the DNA Advisory Board Recommendations on Statistics (see Appendix V). Example equations are examined and discussed using population allele frequency information from Appendix II. The goal is not to perform an in-depth examination of each genetic and statistical principle but rather to keep things in an understandable format for beginners in the field. Those readers desiring more extensive information on the topics discussed within this book may refer to additional references listed at the end of this chapter.

Statistical genetic information is often more difficult for DNA analysts to grasp than the technology and biology issues addressed earlier in this book because of its heavy use of mathematics particularly algebra. The concepts of probabilities can be challenging to forensic scientists schooled in biology rather than mathematics. In the course of the next few chapters, we endeavor to explain and expand on equations used to calculate random match probabilities.

In this chapter we introduce the basic concepts of probability, statistics, and population genetics. Chapter 20 covers the analysis of DNA population data for allele and locus independence using tests for Hardy–Weinberg equilibrium and linkage equilibrium. Chapter 21 describes calculations for random match probability estimates and introduces the concept of likelihood ratios. Chapter 22 delves into various approaches for interpreting mixtures and results from degraded DNA profiles including probabilities of exclusion. Finally, Chapter 23 concludes the genetics section with the unique challenges of kinship analysis and paternity testing.

PURPOSE OF STATISTICS AND POPULATION CONSIDERATIONS

In Chapter 15 on short tandem repeat (STR) interpretation, we concluded with a section entitled 'to match or not to match – that is the question.' If a DNA profile from a suspect does not match the evidence from a crime scene (and the testing has been performed properly), then we can reliably conclude that the individual in question did not contribute the biological sample recovered from the crime scene.

However, the more interesting outcome of a DNA profile comparison is what to conclude when the profiles between suspect and evidence match. Are they from the same individual or is there someone else out there who might just happen to match the evidence in question? Since we do not have the luxury of access to DNA profiles of everyone living on planet Earth, we must use smaller population data sets to extrapolate the possibility of a random match. As will be described in more detail in Chapter 20, allele frequencies are collected from various ethnic/racial sample sets, such as contained in Appendix II. Based on their allele frequencies from validated databases, population genetic principles are applied to infer how reasonable it is that a random, unrelated individual could have contributed the DNA profile in question (see Chapter 21).

It is important to distinguish between unrelated and related individuals in assumptions being made for the calculations that follow. Obviously related individuals have DNA profiles that are more similar than unrelated individuals who are compared. In most equations that will be used in Chapters 19–22, we will be assuming that unrelated individuals are involved. In Chapter 23, we consider paternity testing and other kinship scenarios where closely related individuals are being studied.

Of the three possible outcomes of a DNA test – 'no match', 'inconclusive', or 'match' between samples examined – only the third requires statistics. Statistics attempt to provide meaning to the match. These match statistics are usually provided in the form of an estimate of the random match probability or in other words, the frequency for the particular genotype (DNA profile) in a population. However, different laboratories may use different methods for calculating the statistical topics discussed in this book.

IS THERE MORE THAN ONE STATISTICAL SOLUTION?

It is important to recognize that not all approaches are universally accepted and discussion/debate still exists regarding the application of some statistics to forensic DNA typing results (e.g., Bayesian approaches, see Chapter 21).

Models are used in statistics to help interpret data. Yet there are usually assumptions involved so these models are simplified versions of true genetic processes and are attempts to model the real world. The examples provided in this text will be those approaches that are most widely used today largely due to the acceptance of the National Research Council's report on *The Evaluation of Forensic DNA Evidence*, which was published in 1996 and is commonly referred to as the NRCII. Both the NRCII report (1996) and the DNA Advisory Board (DAB) recommendations on statistics (DAB 2000, see Appendix V) recognize that rarely is there only one statistical approach to interpret and explain evidence. In fact the DAB recommendations state, 'The choice of approach is affected by the philosophy and experience of the user, the legal system, the practicality of the approach, the question(s) posed, available data, and/or assumptions' (DAB 2000). The DAB further states that simplistic and less rigorous approaches can be employed, as long as false inferences are not conveyed (DAB 2000).

With the caveat in mind that multiple approaches may exist in the literature, we will attempt to build a foundation for the reader in the next few sections on probability, basic statistics, and population genetics.

PROBABILITY

In the case of a rape or murder, there may be no witnesses available to assist in verification of who was the actual perpetrator of the crime. Therefore, DNA evidence developed as part of a criminal investigation of necessity has to be made in the face of uncertainty. While a crime scene sample may match the DNA profile of a suspect, the result is typically cast in the language of probabilities rather than certainty. Probability statements are designed to attach numerical values to issues of uncertainty.

Probability is the number of times an event happens divided by the number of opportunities for it to happen (i.e., the number of trials). The concepts of probabilities can be difficult to grasp because we are often in the mindset of thinking simply that something either happened or it did not. Probability is usually viewed on a continuum between zero and one. At the lower extreme of zero, it is not possible for the event to occur (or to have occurred). In other words, there is a certainty of non-occurrence. At the upper end, where the probability is equal to one, the event being measured or calculated did in fact occur. Quite often in scientific determinations, the probability of an event occurring is understood to never be completely zero or completely one. Thus, decisions in science, as in life, often need to be made in the face of uncertainty.

If a weather bureau predicts a 60% chance of rain, then this probability was arrived at because experience has shown that under similar meteorological conditions it has rained six out of ten times. If one of two events is equally possible, such as heads or tails when flipping a coin, then the probability is considered 50% or 0.5 for either one of the events. Probabilities are mathematically described with symbols, such as P. The probability that an event can occur is given by the notation or formula: $P(H|E)$ or $Pr(H|E) = \ldots$ This notation is shorthand for stating 'the probability of event H occurring given evidence E is equal to …'. Every probability is conditional on knowing something or on something else occurring.

LAWS OF PROBABILITY

The three laws of probability can be summarized as follows (see Evett and Weir 1998). First, as stated earlier, probabilities can take place in the range zero to one. Events that are certain have a probability of one while those that are not possible have a probability of zero. Thus, if a proposition or possibility is false, it has a probability of zero.

Second, events can be *mutually exclusive* meaning that if any one of a particular set of events has occurred then none of the others has occurred. If two events are mutually exclusive and we wish to know the probability that one or other of them is true then we can simply add their probabilities. This concept can be written out in the form:

$$P(G \text{ or } H|E) = P(G|E) + P(H|E)$$

or verbally, the probability of events G or H occurring given evidence E is equal to the probability of event G occurring given evidence E plus the probability of event H occurring given evidence E. In this example, all possibilities are captured by events G or H. Thus, if event G occurred then event H did not and *visa versa*. Another way to write this concept is that $P(G|E) + P(H|E) = 1$ and therefore upon rearranging the equation $P(H|E) = 1 - P(G|E)$. This then means that the probability that H is false is equal to one minus the probability that H is true.

The third law of probability centers on the fact that when two events are independent of one another their probabilities can be multiplied with one another.

$$P(G \text{ and } H|E) = P(G|E) \times P(H|G,E)$$

or verbally, the probability of events G and H occurring given evidence E is equal to the probability of event G given evidence E multiplied by the probability of event H given event G and evidence E.

If the conditioning information (evidence E) is clearly specified and consistent for all possible events, then we can drop the '$|E$' or 'given evidence E' portion of the equation to arrive at:

$$P(G \text{ and } H) = P(G) \times P(H|G)$$

And if G and H are statistically independent or unassociated events then:

$$P(G \text{ and } H) = P(G) \times P(H)$$

To summarize, probabilities fall in the range of 0 to 1. When considering the possibilities of two events occurring, if either one of two mutually exclusive events can occur, their individual probabilities are added (sum rule). Alternatively, if we wish to consider the probability of two independent events occurring simultaneously, then the individual probabilities can be multiplied (product rule).

LIKELIHOOD RATIOS AND BAYESIAN STATISTICS

A likelihood ratio (LR) involves a comparison of the probabilities of the evidence under two alternative propositions. As will be described later in this chapter, these two propositions are often referred to as the null hypothesis and the alterative hypothesis. In forensic DNA settings, these mutually exclusive hypotheses represent the position of the prosecution – namely that the DNA from the crime scene originated from the suspect – and the position of the defense – that the DNA just happens to coincidently match the defendant and is instead from an unknown person out in the population at large (see Chapter 21). In mathematical terms, the likelihood ratio is written as:

$$LR = H_p / H_d$$

where H_p represents the hypothesis of the prosecution and H_d represents the hypothesis of the defense. The likelihood ratio is used in Bayes' theorem to relate the probabilities of the propositions after the evidence to the probabilities prior to the evidence (Weir 2003):

$$\frac{Pr(H_p | E)}{Pr(E | H_d)} = \frac{Pr(E | H_p)}{Pr(E | H_d)} \times \underbrace{\frac{Pr(H_p)}{Pr(H_d)}}_{\substack{Likelihood \\ Ratio}}$$

which can be stated as the posterior odds on H_p are equal to the LR times the prior odds on H_p.

Prior odds relates to the relative guilt or innocence of the suspect. Thus, in order to perform this calculation, one must make assumptions about the prior odds of guilt or innocent. As you might imagine, this approach has not caught on in the United States where the judicial system tries to maintain 'innocent until proven guilty.' However, there is nothing wrong with using the likelihood ratio by itself and have the judge and jury decide on the prior and post odds of guilt or innocence.

A good DNA typing system should provide large likelihood ratios when the defendant and the perpetrator of a crime is the same person. Likewise, if they are different people, then the likelihood ratio will be less than 1. Relative levels of likelihood ratios will be discussed in more detail in Chapter 21.

STATISTICS

Statistics is the science of uncertainty and its measurement. It also provides a sense of how reliable a measurement is when the measurement is made multiple times. Statistics involves using samples to make inferences about populations. A *population* is considered in this context to be a set of objects of interest, which may be infinite or otherwise unmeasurable in their entirety. An observable subset of a population can be referred to as a *sample* with a *statistic* being some observable property of the sample. In the context of DNA testing, the 'population' would be the entire group of individuals that could be considered (e.g., billions of people around the world or those living within a particular country or region). The 'sample' would be a set of individuals from the population at large (e.g., 100 males) that were selected at random and tested at particular genetic markers to try and establish a reliable representation of the entire population. The 'statistic' examined might be the observed allele or genotype frequencies for the tested genetic markers.

HYPOTHESIS TESTING FOR STATISTICAL SIGNIFICANCE

One of the most important things to understand about statistics is the concept of hypothesis testing. Hypothesis testing is the formal procedure for using statistical concepts and measures in performing decision-making (Ayyub and McCuen 2003, p. 290). This concept forms the basis for likelihood ratios that were mentioned briefly in the previous section and will be described in more detail in Chapter 21.

Six steps are typically involved in making a statistical analysis of a hypothesis (Figure 19.1): (1) formulate two competing hypotheses; (2) select the appropriate statistical model (theorem) that identifies the test statistic; (3) specify the level of significance, which is a measure of risk; (4) collect a sample of data and compute an estimate of the test statistic; (5) define the region of rejection for the test statistic; and (6) select the appropriate hypothesis (Ayyub and McCuen 2003, p. 290).

The first step is to formulate usually two hypotheses for testing. The first hypothesis is called the *null hypothesis*, and is denoted by H_0. The null hypothesis is formulated as an equality and indicates that a difference does not exist. The second hypothesis is usually referred to as the *alternative hypothesis* and is

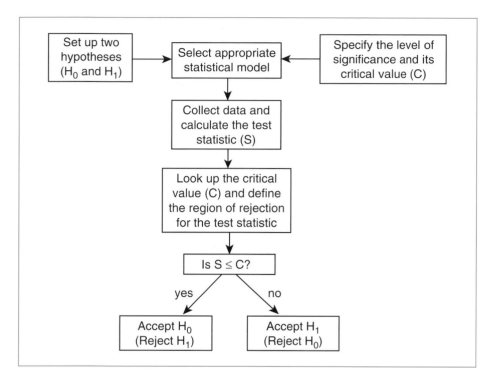

Figure 19.1

Flow chart illustrating the steps in hypothesis testing. The null hypothesis (H_0) is mutually exclusive of the alternative hypothesis (H_1). Adapted from Graham (2003).

denoted by H_1 or H_A. The null and alternative hypotheses are set up to represent mutually exclusive conditions so that when a statistical analysis of the sampled data suggests that the null hypothesis should be rejected, the alternative hypothesis must be accepted. Thus, the data collected (evidence gathered) should tip the scales towards either the null hypothesis or the alternative hypothesis.

In the context of a forensic DNA evidence examination, the null hypothesis put forward by the prosecution is that the defendant contributed the crime scene DNA profile while the alternative hypothesis championed by the defense is that someone else other than the defendant contributed the crime scene DNA profile in question. These two hypotheses are then expressed in the form of a likelihood ratio with H_0 or H_p (hypothesis of the prosecution) in the numerator and H_1 or H_d (hypothesis of the defense) in the denominator.

The available situations and potential decisions/outcomes of a hypothesis test are shown in Figure 19.2. There are two types of errors that can be made with hypothesis testing. A type I error involves rejecting the null hypothesis when in fact it is really true. This might be considered a 'false negative.' A type II error on the other hand involves accepting the null hypothesis when in fact it is really false. A type II error is a 'false positive.'

The level of significance, which is a primary element of the decision-making process in hypothesis testing, represents the probability of making a type I error and is denoted by α (D.N.A. Box 19.1). The value chosen for α is typically based on convention and the historical custom, with values for α of 0.05 and 0.01 being

Figure 19.2

(a) Comparison of decisions based on hypothesis testing and the relationship of type I and type II errors. (b) Example demonstrating how type I and type II errors correlate to false-positive and false-negative results. Adapted from Graham (2003).

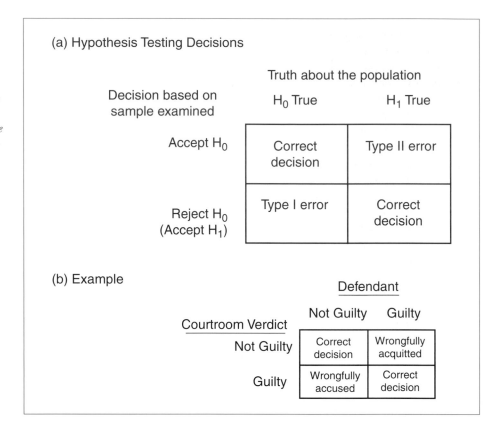

selected frequently. The value of 0.05 is equivalent to a 95% 'confidence limit' while α of 0.01 represents a 99% confidence limit. When we select the significance level (α) of a test, we are setting the probability of a type I error ('wrongfully accused'). Small p-values lead to rejection of the null hypothesis (and acceptance of the alternative hypothesis) while large p-values favor the null hypothesis.

The null hypothesis is rejected when the computed p-value lies in the region of rejection, which is usually $p < 0.05$. Rejection of the null hypothesis implies acceptance of the alternative hypothesis. When multiple analyses are being performed simultaneously for significance testing, the Bonferroni correction may be applied (D.N.A. Box 19.2).

It is important to recognize that by chance 5% of tests will have p-values below $p = 0.05$. Thus, just because a test of significance is below a certain value, the null hypothesis should not necessarily be rejected.

CHI SQUARE TEST

Chi square test is a 'goodness-to-fit' test. In other words, how close do observations of an event come to the expected results? The chi square (χ^2) is determined by summing the squared value of the difference between the observed results (obs) and expected results (exp) divided by the expected results.

D.N.A. Box 19.1

What does it mean to be 'statistically significant'?

Connected to the process of hypothesis testing is the concept of a statistically significant result, which involves a probability value or 'p-value'. A p-value reflects the probability that a variable being measured would assume a value greater than or equal to the observed value strictly by chance. In mathematical terms this can be described as $P(z \geq z_{observed})$. The threshold whereby a p-value is considered significant is set by an 'alpha value' (α). With the commonly used 95% confidence limit, $\alpha = 0.05$ (since 100–95% is 5% or 0.05).

A variety of alpha values are used in different fields, but probably the most common is 0.05 for a 95% confidence interval around a measurement. Thus, if a p-value is >0.05, then the test statistic and comparison are considered 'not significant'. If the p-value is computed to be between 1% and 5%, then it is generally considered 'significant' in which case the value can be denoted with an asterisk (e.g., 0.0435*). When the computed p-value is less than 1%, it is thought to be 'highly significant' and can be marked with a double asterisk (e.g., 0.00273**).

Thus, in cases where the p-value, which is the probability of obtaining an observed result or a more extreme result, is less than the conventional 0.05, we conclude that there is a 'significant relationship' between the two classification factors. However, it is important to keep in mind that the outcome of the significance testing is very much dependant on how the question is framed as part of the hypothesis testing.

Source:
http://mathworld.wolfram.com/P-value.html;
  http://mathworld.wolfram.com/Significance.html;
  http://mathworld.wolfram.com/AlphaValue.html; See also Graham (2003).

$$\chi^2 = \sum_{i=1}^{k} \frac{(\text{obs}_i - \text{exp}_i)^2}{\text{exp}_i}$$

The resultant chi square value is then compared against a table of numbers to see if there is a significant deviation from the 'normal' values expected. High chi-square values indicate discrepancies between observed and expected results. Different 'degrees of freedom' may be applied to data depending on the situation. Paul Lewis at the University of Connecticut has created a nice little freeware program that can quickly relate user inputted chi-square values and degrees of freedom to their p-value. This program is available at: http://lewis.eeb.uconn.edu/lewishome/software.html.

CONFIDENCE INTERVALS

Another important statistical concept is that of confidence intervals. Confidence intervals are useful for determining the precision of a point estimate. Typically a

D.N.A. Box 19.2

The Bonferroni correction

Carlo Emilio Bonferroni was an Italian mathematician that lived from 1892–1960 who developed theories for simultaneous statistical analysis. The Bonferroni correction is a multiple-comparison correction used when several independent statistical tests are being performed simultaneously. While a given alpha value (e.g., 0.05) may be appropriate for each individual comparison, it is probably not sufficient for the set of all comparisons. Thus, the alpha value needs to be lowered to account for the number of comparisons being performed.

The Bonferroni correction lowers the significance level for the entire set of n comparisons by dividing n into the alpha value for each comparison. The adjusted significance level becomes:

$$1 - (1 - \alpha)^{1/n} \approx \alpha/n$$

Thus, a set of 10 comparisons would lower the alpha value from 0.05 to 0.005 (0.05/10) so only p-values below 0.005 would be considered statistically significant rather than the conventional $p < 0.05$.

In the analysis of genetic data for Hardy–Weinberg equilibrium, application of the Bonferroni correction almost always removes the stigma of a locus being below the 5% threshold level. Essentially applying the Bonferroni correction means that the more sources of data being tested (e.g., more STR loci), the less sensitive the testing regime becomes since the p-value threshold has been lowered.

Sources:
http://mathworld.wolfram.com/BonferroniCorrection.html
Perneger, T.V. (1998) What's wrong with Bonferroni adjustments. *British Medical Journal*, 316, 1236–1238.
Weir, B.S. (1996) *Genetic Data Analysis II*, Multiple Tests, pp. 133–135.

95% confidence interval is computed reflecting the probability that in 95% of the samples tested, the interval should contain the actual value measured. A 95% confidence interval effectively is the sample average plus or minus two standard deviations. A confidence interval around some value π is a function of the frequency of the observation (p) and the number of individuals or items sampled in a population (n).

$$p - z_{\alpha/2}\sqrt{\frac{p(1-p)}{n}} \leq \pi \leq p + z_{\alpha/2}\sqrt{\frac{p(1-p)}{n}}$$

For 90% confidence intervals, $z_{0.05} = 1.645$ and for 95% confidence intervals, $z_{0.025} = 1.96$.

$$p - 1.96\sqrt{\frac{p(1-p)}{n}} \leq \pi \leq p + 1.96\sqrt{\frac{p(1-p)}{n}}$$

Lower bound   *Upper bound*

Note that the upper bound of 95% confidence interval is what is used in D.N.A. Box 10.3 for mitochondrial DNA frequency estimates with the counting method.

RANDOMIZATION TESTS

To confirm validity of data sets, randomization tests are often performed usually with the aid of computer programs. These randomization tests permit an investigator to ask the question if the data were collected differently, could the overall results be significantly different. Permutation tests, such as the 'exact test' shuffle the original set of genotypes obtained in a population database to examine how unusual the original sampling of genotypes is. This shuffling generates a new genotypic distribution that can be compared to the original one.

Re-sampling tests referred to as 'bootstrapping' or 'jack-knifing' can be performed on data sets as well. Bootstrapping is a computer simulation where the original n observations are re-sampled with replacement. Jack-knifing on the other hand involves re-sampling by leaving one observation out of the original n observations to create n samples each of size $n-1$. Most papers in the literature describing population data sets utilize the exact test with 2000 shuffles although some have reported shuffling as many as 100 000 times.

Since we are only sampling a DNA profile one time in most cases we perform statistical tests to estimate expected variability if the test were performed again. In the end, a number of statistical tests are performed on genetic data to estimate genotype frequencies since many genotypes are rare and may not be seen in population samples tested (see Chapters 20 and 21).

PRINCIPLES OF POPULATION GENETICS

Population genetics is the study of inherited variation and its modulation in time and space. It is an attempt to quantify the variation observed within a population group or among different population groups in terms of allele and genotype frequencies. Great genetic variation exists within species at the individual nucleotide level. For example in humans, approximately 10 million nucleotides can differ between individuals (see Chapter 2). The genetic difference between individuals within human racial groups is much greater than the average difference between races (Barbujani *et al.* 1997).

As discussed in Chapter 2, diploid individuals have two copies of each autosomal gene (or DNA marker): one of paternal origin (sperm) and one of maternal origin (egg). If the alleles obtained from the sperm and the egg differ, then they are termed *heterozygous* for that locus whereas if the individual received two identical alleles from both parents they are described as being *homozygous*.

Variability within a locus has to be stable enough to accurately pass the allele to the next generation (i.e., possess a low mutation rate) yet not be too stable or else only a few alleles would exist over time and the locus would not be as informative (i.e., useful in human identity testing applications). The simplest description of variation is the frequency distribution of genotypes. A measure of this variation is the number of heterozygote individuals present in a population.

Population genetic forces including mutation, migration (gene flow), natural selection, and random genetic drift all affect gene frequency of alleles present in a population. Over time, isolated populations diverge from one another, each losing heterozygosity due to inbreeding. The gene selection pool is smaller in isolated groups and therefore, not as much shuffling of genes exists.

LAWS OF MENDELIAN GENETICS

There are several basic laws or principles of genetics first described by Gregory Mendel (1822–1884) that form the foundation for interpretation of DNA evidence. One of Mendel's laws is that the two members of a gene pair segregate (separate) from each other during sex-cell formation (meiosis), so that one-half of the sex cells carry one member of the pair and the other one half of the sex cells carry the other member of the gene pair. In other words, chromosome pairs separate during meiosis so that the sex cells (gametes) become haploid and possess only a single copy of a chromosome. Another of Mendel's laws is that different segregating gene pairs behave independently. This is sometimes called the law of independent assortment and is due to recombination (see Chapter 2). The law of segregation and the law of independent assortment are the basis for linkage equilibrium and Hardy–Weinberg equilibrium that are tested for when examining DNA population databases (see Chapter 20).

For a genetic marker with two alleles A and a in a random-mating population, the expected genotype frequencies of AA, Aa, and aa are given by p^2, $2pq$, and q^2, where p and q are the allele frequencies of A and a, respectively, with $p + q = 1$ (Hartl and Clark 1997). Figure 19.3 illustrates these principles, which constitute Hardy–Weinberg equilibrium (HWE). This graphical representation of the

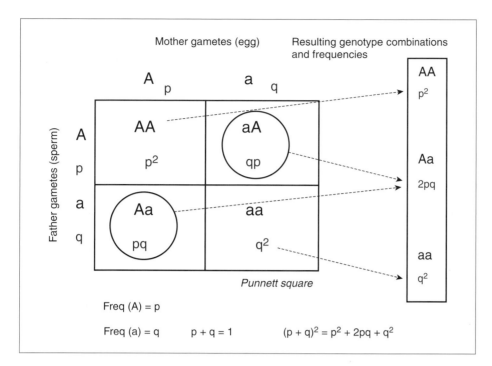

Mother gametes (egg)   Resulting genotype combinations and frequencies

Father gametes (sperm)

	A p	a q
A p	AA p^2	aA qp
a q	Aa pq	aa q^2

AA p^2

Aa $2pq$

aa q^2

Punnett square

Freq (A) = p

Freq (a) = q   p + q = 1 $(p + q)^2 = p^2 + 2pq + q^2$

Figure 19.3

A cross-multiplication (Punnett) square showing Hardy–Weinberg frequencies resulting from combining two alleles A and a with frequencies p and q, respectively. Note that p+ q= 1 and that the Hardy–Weinberg genotype proportions are simply a binomial expansion of $(p+ q)^2$, or $p^2 + 2pq + q^2$.

cross between alleles *A* and *a* from both parents is referred to as a Punnett square. Godfrey Hardy (1877–1947) and Wilhelm Weinberg (1862–1937) both independently discovered the mathematics for independent assortment that is now associated with their names as the Hardy–Weinberg principle (Crow 1999). HWE proportions of genotype frequencies can be reached in a single generation of random mating. HWE is simply a way to relate allele frequencies to genotype frequencies.

Checking for HWE is performed by taking the observed allele frequencies and calculating the expected genotype frequencies based on the allele frequencies. If the observed genotype frequencies are close to the expected genotype frequencies calculated from the observed allele frequencies, then the population is in Hardy–Weinberg equilibrium and allele combinations are assumed to be independent of one another.

One of the principal implications of HWE is that the allele and genotype frequencies remain constant from generation to generation (Hartl and Clark 1997). Another implication is that when an allele is rare, the population contains more heterozygotes for the allele than it contains homozygotes for the same allele (e.g., allele 15 in Table 20.2).

Genes (or genetic markers like STR loci) that are in random association are said to be in a state of linkage equilibrium while those genes that are not in random association are said to be in linkage disequilibrium (Hartl and Clark 1997). Computer programs and tests for checking linkage equilibrium will be discussed in more detail in Chapter 20.

RELATIONSHIP BETWEEN NUMBER OF ALLELES AND NUMBER OF POSSIBLE GENOTYPES

Allele frequency refers to the number of copies of an allele in a population divided by the total number of all alleles in this population. Genotype frequency refers to the number of individuals with a particular genotype divided by the total number of individuals. Figure 19.4 graphically depicts the relationships between allele frequencies and genotype frequencies. In terms of the total number of alleles and genotypes, if n alleles exist for a genetic marker, then $n(n+1)/2$ genotypes are possible.

Table 19.1 includes the number of theoretically possible genotypes for the 13 Combined DNA Index System (CODIS) core STR loci based on the number of reported alleles that are listed in Appendix I. For example, there have been 20 different CSF1PO alleles reported in the literature. Thus, $20(20+1)/2$ or 210 genotypes are theoretically possible for CSF1PO based on the alleles reported so far. However, many alleles are rather rare and thus not likely to give rise to a particular genotype containing them. Note that in a U.S. population study of

Figure 19.4

Graph depicting genotype frequencies for AA, Aa, and aa when Hardy–Weinberg equilibrium conditions are met. The highest amount of heterozygotes Aa are observed when alleles frequencies for both A and a are 0.5. Adapted from Hartl and Clark (1997).

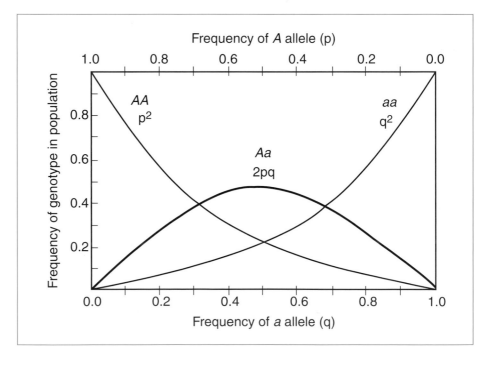

(a) Theoretical

STR Locus	Number of Reported Alleles	Number of Possible Genotypes
	n	$n(n+1)/2$
CSF1PO	20	210
FGA	80	3240
TH01	20	210
TPOX	15	120
VWA	29	435
D3S1358	25	325
D5S818	15	120
D7S820	30	465
D8S1179	15	120
D13S317	17	153
D16S539	19	190
D18S51	51	1326
D21S11	89	4005
	Product	2.51×10^{33}

(b) Observed

STR Locus	Number of Genotypes	Number of Alleles	Number of Samples Examined
CSF1PO	31	8	728
FGA	79	23	731
TH01	21	8	733
TPOX	28	9	732
VWA	34	11	732
D3S1358	24	10	733
D5S818	33	9	731
D7S820	30	12	731
D8S1179	44	11	732
D13S317	29	8	733
D16S539	27	8	732
D18S51	84	21	728
D21S11	79	23	732

Table 19.1

(a) Number of theoretically possible genotypes for the 13 CODIS STR loci based on reported alleles listed in Appendix I. The combination of all possible genotypes is 2.5×10^{33}. (b) Observed number of alleles and genotypes for the same loci in a U.S. population study (Butler et al. 2003). Allele frequencies from this study are found in Appendix II.

728 individuals there were only eight CSF1PO alleles observed with 31 different genotypes. Based on the total number of alleles seen across all 13 STR loci, there are over 2.5 decillion ($\times 10^{33}$) possible genotypes (Table 19.1). However, the vast majority of these genotypes will never be observed. The use of STR alleles and genotypes to form DNA population databases and predict STR profile frequencies will be covered in the next two chapters.

REFERENCES AND ADDITIONAL READING

Ayyub, B.M. and McCuen, R.H. (2003) *Probability, Statistics, and Reliability for Engineers and Scientists*, 2nd Edition. Washington, DC: Chapman & Hall/CRC.

Barbujani, G., Magagni, A., Minch, E. and Cavalli-Sforza, L.L. (1997) *Proceedings of the National Academy of Sciences U.S.A.*, 94, 4516–4519.

Brenner, C.H. (2003) Forensic genetics: mathematics. In Cooper, D.N. (ed) *Nature Encyclopedia of the Human Genome*, Volume 2, pp. 513–519. New York: Macmillan Publishers Ltd, Nature Publishing Group.

Cavilli-Sforza, L.L. and Bodmer, W.F. (1971) *The Genetics of Human Populations*. Mineola, NY: Dover Publications.

Chakraborty, R. (1992) *Human Biology*, 64, 141–159.

Crow, J.F. and Kimura, M. (1970) *An Introduction to Population Genetics Theory*. New York: Harper and Row.

Crow, J.F. (1999) *Genetics*, 152, 821–825.

DNA Advisory Board (2000) Statistical and population genetic issues affecting the evaluation of the frequency of occurrence of DNA profiles calculated from pertinent population database(s) (approved 23 February 2000). *Forensic Science Communications*, July 2000. Available at: http://www.fbi.gov/programs/lab/fsc/backissu/july2000/dnastat.htm; see also Appendix V.

Evett, I.W., Werrett, D.J., Pinchin, R. and Gill, P. (1989) *Proceedings of the International Symposium on Human Identification*, pp. 77–101. Madison, WI: Promega Corporation.

Evett, I.W. and Weir, B.S. (1998) *Interpreting DNA Evidence: Statistical Genetics for Forensic Scientists*. Sunderland, MA: Sinauer Associates.

Gonick, L. and Wheelis, M. (1983) *The Cartoon Guide to Genetics, Updated Edition*. New York: HarperCollins Publishers.

Graham, A. (2003) *Teach Yourself Statistics*. Blacklick, Ohio: McGraw-Hill.

Hardy, G.H. (1908) Mendelian proportions in a mixed population. *Science*, 17, 49–50.

Hartl, D.L. and Clark, A.G. (1997) *Principles of Population Genetics, Third Edition*. Sunderland, Massachusetts: Sinauer Associates.

Hartl, D.L. and Jones, E.W. (1998) *Genetics: Principles and Analysis, Fourth Edition*. Sudbury, Massachusetts: Jones and Bartlett Publishers.

National Research Council Committee on DNA Forensic Science (1996) *The Evaluation of Forensic DNA Evidence*. National Academy Press: Washington, DC; usually referred to as NRCII; recommendations listed in Appendix VI.

Planz, J. (2003) Introduction to Forensic Statistics: Probability and Statistics. Workshop presented at the 14th International Symposium on Human Identification. Presentation available at: http://www.promega.com/geneticidproc/ussymp14proc/stats_workshop.htm.

Rumsey, D. (2003) *Statistics for Dummies*. Indianapolis, Indiana: Wiley Publishing, Inc.

Weir, B.S. (1996) *Genetic Data Analysis II: Methods for Discrete Population Genetic Data*. Sunderland, Massachusetts: Sinauer Associates.

Weir, B.S. (2003) Forensics. In Balding, D. J., Bishop, M. and Cannings, C. (eds) *Handbook of Statistical Genetics, 2nd Edition*. Hoboken, New Jersey: John Wiley & Sons, pp. 830–852.

Weir, B.S. (2003) DNA evidence: inferring identity. In Cooper, D. N. (ed) *Nature Encyclopedia of the Human Genome*, Volume 2, pp. 85–88. New York City: Macmillan Publishers Ltd, Nature Publishing Group.

CHAPTER 20

STR POPULATION DATABASE ANALYSES

As population databases increase in numbers, virtually all populations will show some statistically significant departures from random mating proportions. Although statistically significant, many of the differences will be small enough to be practically unimportant.

(NRC II, p. 58)

Population DNA databases are important for comparison purposes to understand how frequent or how rare a crime scene DNA profile may be in a particular population. Hundreds of publications in the literature contain information on DNA profiles generated with genotype information from common short tandem repeat (STR) loci across tens of thousands of individuals collected from various populations around the world (see reference listing available on the NIST STRBase web site: http://www.cstl.nist.gov/biotech/strbase). Table 20.1 summarizes several sets of population data that have been published in the literature.

Population comparison DNA databases are often generated by individual forensic laboratories to assess variation in common local populations. This is particularly important to locales that may have an isolated and inbred population within its jurisdiction (see discussion on population substructure later in this chapter). For example, in Arizona it would be helpful to have a population database involving Native Americans such as Apaches and Navajos since they live in fairly close-knit communities within Arizona and would be expected to have different genotype frequencies compared to Caucasians or African-Americans living in Arizona (see Chakraborty *et al.* 1999).

The primary goal of generating a population database is to find all 'common' alleles and sample these alleles multiple times in order to reliably estimate the alleles present in the population under consideration. A listing of observed alleles for the commonly used STR markers that have been reported in numerous population studies from around the world is contained in Appendix I. However, it is worth noting that some of these alleles, particularly the microvariant alleles (see Chapter 6) have only been observed a few times and are therefore rather rare (e.g., TH01 allele 6.3). Allele frequencies for common alleles at 15 different STR loci in the three major U.S. population groups may be found in Appendix II (Butler *et al.* 2003).

Source	Population	Samples Typed (N)	Loci Typed	Forensic Parameters Evaluated	Reference
Arab	Yemenites	100	Profiler Plus kit (9 STRs)	MP, PD, PIC, PE, PI, H, h	Klintschar *et al.* (2001)
Australian	Caucasian	2645	Profiler Plus kit (9 STRs)	H, MP, PE	Bagdonavicius *et al.* (2002)
Austria	Caucasian	204	SGM Plus kit (10 STRs)	h, PD, PE	Steinlechner *et al.* (2001)
Guam	Filipinos	99	Profiler Plus kit (9 STRs)	h, PD, PE	Budowle *et al.* (2000)
Israel	Jewish	124	Profiler Plus kit (9 STRs), DYS19, DYS389I/II, DYS390, DYS391, DYS393, DYS287, D4S243, F13A1, D18S535, D12S391	H, PD, CE, CE2	Picornell *et al.* (2002)
United States	Caucasians	302	Identifiler kit (15 STRs)	H	Butler *et al.* (2003)

Table 20.1

Summary of a few published population data sets. Forensic parameters evaluated include heterozygosity (H), homozygosity (h), power of discrimination (PD), power of exclusion (PE), chance of exclusion (CE), chance of exclusion if only one parent and child are typed (CE2), polymorphic information content (PIC), matching probability (MP), paternity index (PI).

This chapter will cover the primary aspects of generating and testing population DNA databases prior to their use in estimating STR profile frequencies. Calculations for determining the rarity of a particular STR profile will be discussed in Chapter 21 and will be based on allele frequencies found in Appendix II.

GENERATING A POPULATION DNA DATABASE

The primary steps in generating a population database, such as found in Appendix II, are illustrated in Figure 20.1. A laboratory must first decide on the number of samples that will be tested and what particular ethnic/racial groups are relevant to estimating DNA profile frequencies that might be encountered by the laboratory. Population databases are often generated by gathering a set of biological samples in the form of liquid blood from a local hospital or blood bank. Usually the individuals selected are healthy and hopefully unrelated to one another so that they reliably represent the population of interest. These 'convenience' samples are deemed reliable since they are similar with other data sets (NRC II, p. 58) (see Table 20.4). Usually the individual samples are devoid of identifiers that could be used to link the DNA typing results back to the donor (see below).

After the samples have been gathered, they are extracted, polymerase chain reaction (PCR) amplified, and genotyped at the STR loci of interest, such as the 13 core loci used in the FBI's Combined DNA Index System (CODIS). These single-source samples are typically processed in the same manner as the convicted felon samples mentioned in Chapter 18 using commercial STR kits (see Chapter 5) and standard interpretation guidelines to designate alleles (see Chapter 15).

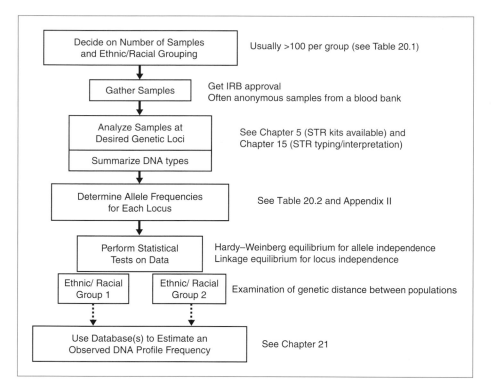

Figure 20.1

Steps in generating and validating a population database that can then be used to estimate the frequency of an observed DNA profile in the population.

Following the gathering of the genotype data, the information is converted into allele frequencies by counting the number of times each allele is observed. Table 20.2 shows an example of allele counting for the locus D13S317 used to determine the Caucasian data in Appendix II. The observed alleles, ranging from 8–15 repeats, are listed across the top and down the left side. At the intersection of the rows and columns, the numbers of observed genotypes are listed. For example, starting in the top left hand corner, the genotype 8,8 is seen nine times in the set of 302 individuals examined while the genotype 11,14 is seen 12 times. On the right side of Table 20.2, the numbers of observed alleles are counted by summing the row and column containing the allele of interest. Thus, the number of chromosomes containing allele 8 is equal to 68 from $9 + 9 + 1 + 17 + 13 + 10 + 0 + 0$ for the row containing allele 8 plus nine for the column with the 8,8 genotype. The number of 8,8 genotypes is counted twice since both chromosomes contain an allele 8 at the D13S317 marker. The frequency for allele 8 is determined by dividing 68 into the total number of chromosomes, which are 604 since there are two chromosomes for each of the 302 individuals typed. Note that there is only one allele 15 observed in this study, which comes from a 10,15 genotype. A little later we will discuss the concept of minimum allele frequencies in order to reliably estimate population allele frequencies.

Allele frequency information allows for more compact data storage and enables independence testing, such as exact tests for Hardy–Weinberg

Table 20.2

Genotype array and allele count for the STR locus D13S317 from unrelated U.S. Caucasian samples (n = 302 or 604 chromosomes measured). Only the observed allele frequency is reported in Appendix II. Note that allele 15 is only observed once and therefore is below the minimum recommended allele count of five. Thus, the value of 0.00828 (5/2N) should be used instead of 0.00166 for accurate estimation of genotype frequencies according to the recommendations of the National Research Council report (1996).

Genotype Array	8	9	10	11	12	13	14	15		Allele Count	Observed Frequency
8	8,8 9	8,9 9	8,10 1	8,11 17	8,12 13	8,13 10	8,14 0	8,15 0	8	68	0.11258
9		9,9 1	9,10 2	9,11 15	9,12 10	9,13 4	9,14 3	9,15 0	9	45	0.07450
10			10,10 2	10,11 12	10,12 6	10,13 3	10,14 2	10,15 1	10	31	0.05132
11				11,11 37	11,12 54	11,13 21	11,14 12	11,15 0	11	205	0.33940
12					12,12 21	12,13 18	12,14 7	12,15 0	12	150	0.24834
13						13,13 7	13,14 5	13,15 0	13	75	0.12417
14							14,14 0	14,15 0	14	29	0.04801
15								15,15 0	15	1	**0.00166**
										604	

equilibrium (see below). Typically the sample genotypes and allele frequencies associated with a particular ethnic/racial group are segregated to enable both intra- and inter-group comparisons. Several important issues will be considered in the next few pages including adequate sample sizes and sample selection for population databases.

ADEQUATE SAMPLE SIZES IN DNA DATABASES USED FOR ALLELE FREQUENCY ESTIMATION

In an ideal world, DNA databases would include STR genotypes from every individual in a particular population to permit extremely accurate evaluations of DNA profile frequencies. However, practical considerations of cost and time necessitate a smaller database. Fortunately, it is possible to run a small subset of the population and reliably predict allele and genotype frequencies in the entire population – much like a telephone survey of several hundred individuals is used to try and predict the outcome of a political election. The key is collecting information from enough individuals to reliably estimate the frequency of the major alleles for a genetic locus.

Most published population data includes on the order of 100–200 STR types per locus per population examined (see Table 20.1). In a key paper in 1992 entitled 'sample size requirements for addressing the population genetic issues of forensic use of DNA typing', Ranajit Chakraborty concluded that 100–150 individuals per population could provide an adequate sampling for a genetic locus provided that allele frequencies below 1% were not used in forensic calculations (Chakraborty 1992). The concept of minimum allele frequencies will be discussed below. Evett and Gill (1991) arrived at a similar conclusion with multilocus matches of DNA profiles, namely that 100–120 individuals per locus per population were sufficient for robust likelihood calculations. Collecting information from more samples usually only improves the precision of a result rather than the accuracy of the allele count (see Foreman and Evett 2001).

MINIMUM ALLELE FREQUENCY

In order to make a reliable estimation of an allele frequency, it is important to collect more than one data point for that allele. Recall that in Table 20.2 we only observed a single allele 15. As noted by Chakraborty (1992), a conservative minimum allele frequency is used to insure that an allele has been sampled sufficiently to be used reliably in statistical tests. Furthermore, the National Research Council report (1996) states that an estimate of an allele frequency can be very inaccurate if the allele is so rare that it is represented only once or a few times in a database; and some rare alleles might not be represented at all (NRC II, p. 148). Thus, it is recommended that each allele should be observed at least five times to be included in reliable statistical calculations. The minimum allele frequency is therefore $5/2N$, where N is the number of individuals sampled from a population and 2N is the number of chromosomes counted because autosomes are in pairs due to inheritance of one allele from one's mother and one from one's father.

Going back to the example of allele 15 in Table 20.2, we find that the minimum allele frequency is $5/2N$ or $5/604$, which equals 0.00828. Therefore, 0.00828 should be used in calculations involving a D13S317 allele 15 rather than the empirically observed 0.00166. In other words, because allele 15 was not observed enough times to reliably estimate its true value in the population, its frequency is inflated to five times its observed value in order to be conservative. The impact of database size on minimum allele frequency can be seen in Table 20.3.

A conservative minimum allele frequency is used to insure that an allele has been sampled sufficiently to be used reliably in statistical tests. On the basis of these criteria, usually 100–150 DNA samples from unrelated individuals is sufficient for STR loci that possess between five and 15 alleles (Chakraborty 1992).

SAMPLING OF INDIVIDUALS FOR DNA DATABASE

If possible, population databases for use in forensic DNA testing should contain unrelated individuals of known ethnicity. However, this may not be completely possible in a practical sense, as many laboratories are required to use samples

Number of Individuals (N)	Chromosomes Sampled (2N)	Minimum Allele Frequency (5/2N)
100	200	0.025
200	400	0.0125
500	1000	0.005
1000	2000	0.0025
10 000	20 000	0.00025

that have been made anonymous prior to study. In addition, categories of ethnicity or race are often subjective and may be based on perceived phenotype or cultural classification. Sampled individuals may have more than one easily definable racial background and may prefer to be grouped differently from a cultural standpoint than they might otherwise biologically. Finally, people who have been adopted or conceived through *in vitro* fertilization may not know their true genetic heritage.

While relatedness may not always be completely prevented, inclusion of some relatives in the database will not invalidate allele frequency estimates. Systematic bias in allele frequency estimates should only occur with inclusion of relatives if particular alleles substantially affect fertility or viability of offspring. However, markers used in human identity testing applications are selected to avoid this type of sampling bias. The primary purpose in trying to obtain unrelated individuals is to improve the precision of allele frequency estimates by increasing the number of independent genes sampled (i.e., to observe more alleles that would represent the population being studied).

Individuals included in a DNA database should be selected without prior knowledge of genotypes at the loci under examination to insure randomness of the samples. A frequent practice is to collect samples from blood donors or hospital volunteers. For example, the samples used to generate the STR typing data in Appendix II are from anonymous blood donors with self-identified ethnicities purchased from two different blood banks (Butler *et al.* 2003). Well-characterized population samples with anthropological descriptions would be desirable in many cases to carefully define population groups but are not necessary to obtain valid information in forensic DNA population databases. Self-declaration of ethnicity can be a suitable method of categorizing samples on the basis of ethnicity (see Walsh *et al.* 2003).

Broad racial/ethnic categories are usually adequate for most forensic databases, unless an isolated population is of interest, such as Native American Apache Indians. An examination of allele frequencies observed with different sample sets from around the United States demonstrates that the differences between sample sets within a racial/ethnic group are small (Table 20.4). Note for example, the D13S317 allele frequencies for African-Americans seen with Holt *et al.* (2002) versus the data from Appendix II (Butler *et al.* 2003). The most common allele in both data sets is allele 12, which has a frequency of 0.424 in Appendix II and 0.444 in Holt *et al.* (2002). Likewise, allele frequencies for Florida and Virginia Caucasians are very similar with the most common allele

Reference	Population (number typed)	D13S317 Allele Frequencies								
	African American	< 8	8	9	10	11	12	13	14	15
Appendix II	NIST U.S. Samples (*n* = 258)	—	0.033	0.033	0.023	0.306	0.424	0.145	0.035	—
Holt *et al.* 2000	ABI U.S. Samples (*n* = 195)	—	0.036	0.023	0.023	0.272	0.444	0.141	0.062	—
Budowle *et al.* 2001	FBI U.S. Samples (*n* = 179)	—	0.036	0.028	0.050	0.237	0.483	0.126	0.036	**0.003**
Budowle *et al.* 2001	California (*n* = 200)	—	0.045	0.033	0.023	0.270	0.405	0.153	0.073	—
Budowle *et al.* 2001	Alabama (*n* = 124)	—	0.032	**0.012**	**0.008**	0.367	0.379	0.169	0.032	—
Budowle *et al.* 2001	Florida (*n* = 100)	—	**0.015**	0.030	**0.010**	0.310	0.380	0.180	0.075	—
Budowle *et al.* 2001	Virginia (*n* = 199)	—	0.030	0.025	0.013	0.329	0.405	0.148	0.048	**0.003**
Budowle *et al.* 2001	New York (*n* = 150)	—	0.017	0.040	**0.003**	0.340	0.420	0.143	0.037	—
Budowle *et al.* 2001	Illinois (*n* = 155)	**0.003**	0.016	0.042	**0.013**	0.326	0.361	0.171	0.068	—
Budowle *et al.* 2001	Minnesota (*n* = 150)	—	0.047	0.033	0.030	0.270	0.410	0.157	0.053	—
	Minimum value		0.015	0.012	0.003	0.237	0.361	0.126	0.032	
	Maximum value		0.047	0.042	0.050	0.367	0.483	0.180	0.075	
	Caucasian	< 8	8	9	10	11	12	13	14	15
Appendix II	NIST U.S. Samples (*n* = 302)	—	0.113	0.075	0.051	0.339	0.248	0.124	0.048	**0.002**
Holt *et al.* 2000	ABI U.S. Samples (*n* = 200)	**0.003**	0.115	0.078	0.068	0.313	0.283	0.098	0.043	**0.003**
Budowle *et al.* 2001	FBI U.S. Samples (*n* = 196)	—	0.100	0.077	0.051	0.319	0.309	0.110	0.036	—
Budowle *et al.* 2001	California (*n* = 150)	**0.003**	0.130	0.057	0.073	0.313	0.277	0.100	0.040	**0.007**
Budowle *et al.* 2001	Alabama (*n* = 150)	—	0.117	0.080	0.067	0.320	0.267	0.110	0.037	**0.003**
Budowle *et al.* 2001	Florida (*n* = 246)	**0.002**	0.120	0.087	0.059	0.323	0.254	0.124	0.031	—
Budowle *et al.* 2001	Virginia (*n* = 197)	**0.003**	0.137	0.064	0.048	0.317	0.277	0.114	0.038	**0.003**
Budowle *et al.* 2001	New York (*n* = 141)	—	0.121	0.082	0.053	0.348	0.277	0.085	0.036	—
Budowle *et al.* 2001	Minnesota (*n* = 150)	—	0.130	0.070	0.063	0.297	0.307	0.090	0.043	—
Budowle *et al.* 2001	Canada (*n* = 166)	—	0.093	0.093	0.045	0.301	0.301	0.124	0.042	—
	Minimum value		0.093	0.057	0.045	0.297	0.248	0.085	0.031	
	Maximum value		0.137	0.093	0.073	0.348	0.309	0.124	0.048	

Reference	Population (number typed)	D13S317 Allele Frequencies								
	Hispanic	< 8	8	9	10	11	12	13	14	15
Appendix II	NIST U.S. Samples (n = 140)	—	0.121	0.154	0.100	0.236	0.221	0.118	0.046	**0.004**
Budowle et al. 2001	FBI U.S. Samples (n = 203)	—	0.067	0.219	0.101	0.202	0.217	0.138	0.057	—
Budowle et al. 2001	California (n = 200)	—	0.073	0.217	0.100	0.203	0.235	0.130	0.043	—
Budowle et al. 2001	Florida (n = 240)	—	0.115	0.115	0.077	0.306	0.229	0.108	0.050	—
Budowle et al. 2001	New York (n = 152)	—	0.092	0.112	0.066	0.240	0.306	0.141	0.043	—
Budowle et al. 2001	Michigan (n = 150)	0.003	0.143	0.120	0.083	0.227	0.250	0.120	0.050	**0.003**
Budowle et al. 2001	Minnesota (n = 149)	—	0.121	0.195	0.064	0.262	0.228	0.084	0.047	—
Budowle et al. 2001	Arizona (n = 234)	—	0.109	0.135	0.098	0.235	0.239	0.126	0.056	**0.002**
Budowle et al. 2001	Mexico (n = 143)	—	0.091	0.231	0.070	0.196	0.252	0.122	0.039	—
	Minimum value		0.067	0.112	0.064	0.196	0.217	0.084	0.039	
	Maximum value		0.143	0.231	0.101	0.306	0.306	0.141	0.057	

Table 20.4
(Continued)

being allele 11 at 0.323 and 0.317, respectively. Therefore, estimates of DNA profile frequencies will likely not vary significantly if a Caucasian population data set from Florida was used versus data from Virginia.

LEGAL REQUIREMENTS WHEN GATHERING SAMPLES FOR DNA TESTING

Many institutions now require institutional review board (IRB) approval when performing tests on human subjects. These IRB forms may request that informed consent be provided by those who provide the DNA samples or require that DNA samples be made completely anonymous to the testing laboratory so that genetic information cannot be reconnected with the tested subject. Government agencies will often not provide funding for research unless a DNA laboratory has an IRB in place.

STATISTICAL TESTS ON DNA DATABASES

COMPUTER PROGRAMS AVAILABLE

Once STR genotypes have been generated from population samples the data are typically evaluated through statistical tests to insure that the database will be a useful one when applied to human identity testing. Statistical tests on genetic data have been greatly aided by the availability of computer processing power and a number of computer programs are now available to perform the various tests for independence that will be described below.

Ranajit Chakraborty's group, formerly at the University of Texas and now at the University of Cincinnati, has prepared a software program named DNATYPE that has been used to evaluate a wide set of population data with a number of statistical tests (Figure 20.2). These tests are primarily designed to evaluate independence of alleles within a locus (program 'H') and between multiple loci (program 'D'). Pair-wise comparisons of loci for independence can be done between any two loci in any database (program 'K') and two databases can be compared with one another for genetic similarity (program 'N').

These programs are written in BASIC and run in DOS windows that will pop up when programs are initiated. The individual programs request specific input information within the DOS window (such as name of database file and names of loci to be examined) before they run. The most challenging part about using the program is getting the genotype data into a uniform format that the program can recognize. The program checks a database file for entry errors and accuracy in format and can also search for duplicates.

Program 'H' within DNATYPE examines the genotypes inputted from a data set and outputs the distribution of genotypes and allele frequencies along with several tests that are designed to check whether or not the genotype frequencies are in Hardy–Weinberg equilibrium (HWE) proportions based on the observed allele frequencies. Four HWE tests are performed: the exact test

Figure 20.2

Screenshot of the DNATYPE computer program developed by Ranajit Chakraborty, David Stivers and Yixi Zhong in the 1990's and widely used for original analysis of much of the restriction fragment length polymorphism (RFLP) and STR data generated by the North American forensic community. It is a DOS based program that has a Microsoft Windows user interface added by Snehit Cherian and Robert Gaensslen through funding by the National Institute of Justice. Tests performed with this program include checking for duplicate sample types in databases, single-locus tests for Hardy–Weinberg equilibrium, multiple locus tests for linkage equilibrium, and genetic distances between pairs of population databases.

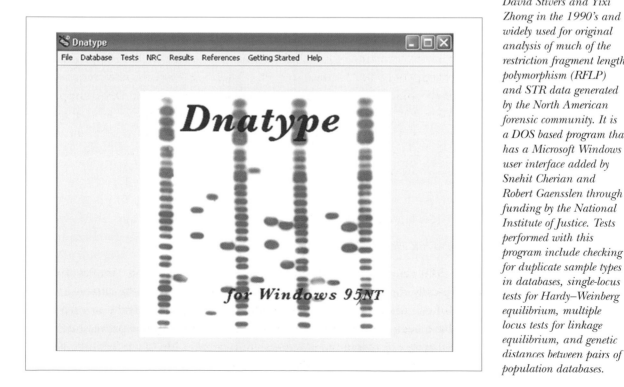

(Guo and Thomson 1992), heterozygosity-biased, heterozygosity-unbiased, and likelihood ratio. If all empirical p-values are above the significance level of 5% (i.e., $p > 0.05$) (see Chapter 19), then the observed genotypes suggest no significant departure from HWE.

Program 'D' evaluates independence of all loci with data where individuals were typed at all loci. This test is based on the distribution of the number of heterozygous loci observed across individuals (Chakraborty 1984). Locus-specific heterozygosities can be used to compute the expected variance of the number of heterozygous loci if linkage equilibrium exists (i.e., independence exists across all loci examined). Global independence of alleles across loci is inferred if the observed variance in the number of heterozygous loci falls within the 95% confidence interval of the expected frequency for the number of heterozygous loci calculated under the assumption of independence. When this occurs, use of the product rule is deemed appropriate for multi-locus genotype probability calculations involving the tested loci.

A number of other computer programs are also available to perform statistical tests on DNA markers and population databases (Table 20.5). Most of these programs are available as free downloads or can be run over the Internet to conduct the statistical tests for Hardy–Weinberg and linkage equilibrium. It is important to note that the output from these programs (e.g., p-values) may not always be the same due to different algorithms used for analyzing the data.

TESTING FOR INDEPENDENCE OF ALLELES WITHIN A GENETIC LOCUS

Hardy–Weinberg equilibrium (HWE) predicts the stability of allele and genotype frequencies from one generation to the next. The primary purpose in testing for HWE is to determine if alleles within a locus are independent of each other. Frequencies should not change over the course of many generations if the locus is genetically stable. However, natural populations usually violate HWE to some degree and thereby cause allele frequencies to change over time.

Another purpose of performing a HWE test is to look for any indications of excess homozygosity. The primary biological explanation for excess homozygosity is due to allelic dropout or 'null alleles' (see Chapter 6) where only one allele is observed from a true heterozygous individual. Thus, allele dropout causes this heterozygote to be interpreted as a homozygote. A good example of how an extreme departure from HWE was helpful in detecting null alleles is D8S1179 in samples from Guam (Budowle *et al.* 2000; D.N.A. Box 20.1).

Table 20.6 describes a number of assumptions that are made when testing for Hardy–Weinberg equilibrium. HWE assumes a random mating population of infinite size with no migration or mutation to introduce new alleles, which of course does not exist in real human populations. Yet allele frequencies are

PowerStats (version 1.2) from Promega Corporation (see Tereba 1999); ~3 Mb
Can be freely downloaded: http://www.promega.com/geneticidtools/default.htm.
Uses a Microsoft Excel workbook template to obtain statistics on allele distributions
within populations examined. Summary statistics include frequency of each allele,
polymorphism information content, probability of a match, power of discrimination,
power of exclusion, and the paternity index.

GDA (Genetic Data Analysis) version 1.0; ~750 kb as zipped file
Free program distributed by the authors Paul Lewis and Dmitri Zaykin via download:
http://lewis.eeb.uconn.edu/lewishome/software.html. Can perform a variety of
genetic tests described in Weir's *Genetic Data Analysis II* (1996).

GENEPOP (version 3.4); see (Raymond and Rousset 1995)
The original program runs under the DOS operating system and can be downloaded
via: ftp://ftp.cefe.cnrs-mop.fr/PC/MSDOS/GENEPOP/. A web-based browser version is
accessible: http://wbiomed.curtin.edu.au/genepop/index.html. This program can do a
variety of statistical tests necessary for validating STR population frequency database
including Hardy–Weinberg exact tests (http://wbiomed.curtin.edu.au/genepop/
genepop_op1.html) and a linkage disequilibrium test (http://wbiomed.curtin.edu.au/
genepop/genepop_op2.html).

DNA-VIEW; see http://www.dna-view.com/; costs $7500
Program written by Charles Brenner to enable a wide variety of DNA tests including
kinship analysis and parentage testing.

DNATYPE (Windows 95/NT version); *Contact authors for availability* (see Figure 20.2)
A collection of computer programs developed in the 1990's by Ranajit Chakraborty,
David Stivers and Yixi Zhong (then at University of Texas Health Science Center in
Houston). The programs run in DOS windows that pop up when they are executed
through a Windows interface added more recently by Snehit Cherian and Robert
Gaensslen (University of Chicago-Illinois). This suite of programs has been used
extensively by Bruce Budowle (FBI Laboratory) and others to analyze forensic
population databases for both RFLP and STR data.

ARLEQUIN (version 2.0); available for download: http://lgb.unige.ch/arlequin/; ~7 Mb
This software suite is written in Java and is widely used in population genetics
because of its flexibility in handling large samples of DNA sequences, STR, or
SNP information.

PowerMarker (version 3.07); available for download: http://www.powermarker.net
Developed by Jack Liu at North Carolina State University and is capable of
performing a wide range of summary and population statistics.

PopStats (version 5.3); standalone version is not publicly available
Used as part of the FBI's CODIS system to perform population statistical analyses and
DNA profile frequency estimates.

TFPGA (Tools for Population Genetic Analyses) version 1.3; ~350 kb as zipped file
Available for free download from author Mark Miller at: http://bioweb.usu.edu/
mpmbio/tfpga.asp. Performs a variety of tests including Hardy–Weinberg and
F-statistic calculations. Inter-population comparisons can be performed using
http://bioweb.usu.edu/mpmbio/rxc.asp to perform analysis of R × C contingency tables.

Table 20.5

Computer programs available for performing statistical tests on genetic data.

An example of a significant difference between the observed and the expected number of homozygotes was seen at the D8S1179 locus when testing some Chamorros and Filipinos from Guam (Budowle *et al.* 2000). A total of 38.4% homozygotes were observed rather than the 22.8% expected in 97 Chamorros (*p*-value < 0.0001). Likewise, in 99 Filipinos the observed homozygosity was 25.8% rather than the expected 15.8% (*p*-value = 0.007). This excess homozygosity later led to discovery of a primer binding site polymorphism in the D8S1179 reverse primer for the original Profiler Plus kit (Leibelt *et al.* 2003). A second unlabeled (degenerate) reverse primer has been added to the Identifiler and Profiler Plus ID kit that matches this mutation and enables recovery of the null allele (Leibelt *et al.* 2003).

Sources:
Budowle, B., Defenbaugh, D.A. and Keys, K.M. (2000) *Legal Medicine*, 2, 26–30.
Leibelt, C., Budowle, B., Collins, P., Daoudi, Y., Moretti, T., Nunn, G., Reeder, D. and Roby, R. (2003) *Forensic Science International*, 133, 220–227.

inherited in a Mendelian fashion and frequencies of occurrence follow a predictable pattern of probability. If two alleles *A* and *a* occur with frequencies p and q in the population, then the genotype *AA* (a homozygote) should occur p^2 and the genotype *Aa* (heterozygote) should occur with frequency 2pq. Allele frequencies are used to generate expected genotype frequencies that are then compared to the observed genotype frequencies. If observed and expected values are similar then it is assumed that alleles within the genetic locus are stable or in other words 'in equilibrium.'

DEPARTURES FROM HARDY–WEINBERG EQUILIBRIUM

The null hypothesis (see Chapter 19) with Hardy–Weinberg equilibrium testing is that the genotype frequencies are identical to the proportions expected

Table 20.6

Assumptions with Hardy–Weinberg equilibrium.

The Assumption	The Reason
Large population	Lots of possible allele combinations
No natural selection	No restriction on mating so all alleles have equal chance of becoming part of next generation
No mutation	No new alleles being introduced
No immigration/emigration	No new alleles being introduced or leaving
Random mating	Any allele combination is possible

based on the allele frequencies. In other words, perfect independence of the alleles at the measured locus exists in the population under examination. Thus, small *p*-values cast doubt on the validity of the null hypothesis. The Bonferroni correction is often applied to lower the *p*-value and thus make results with multiple comparisons not statistically significant (see D.N.A. Box 19.2). It is important to keep in mind that since the perfect conditions for HWE and linkage equilibrium (LE) cannot exist in real human populations (random mating, no migration, etc.), therefore the null hypothesis cannot be true by definition.

If minor departures are seen from HWE, there is generally no major cause for concern with using a particular database. Some authors will do little more than note that there is a statistically significant departure from HWE for a particular locus in their population data set. It is important to keep in mind that there are three principal reasons for observations of major differences (departures) from Hardy–Weinberg equilibrium: (1) parents might be related leading to inbreeding and a higher than expected number of homozygotes, (2) population substructure, and (3) selection because persons with different genotypes might survive and reproduce at different rates (NRCII 1996, p. 98).

TESTING FOR INDEPENDENCE OF ALLELES BETWEEN LOCI

The ability to combine information from multiple loci strengthens the statistics of a match and lends real power to STR typing. For random match probabilities to be combined from multiple loci, it is required that those loci are independent of one another (i.e., that recombination occurs between them). When STR loci or any DNA sequence is transferred independently of another DNA segment during meiosis, the two DNA regions are said to be in *linkage equilibrium*.

Independence of frequencies for all genotypes can be tested with exact tests. These tests use the probability of the observed genotypic array, conditional on the observed allelic array. Exact tests are calculated for the database being considered as well as for a series of new databases obtained by permuting the alleles among individuals. Typically 2000 shuffles or permutations are performed with computer programs. Shuffling through permutation testing is performed to enable a look at a virtual population database that is much larger than the one actually measured (Guo and Thomson 1992).

Multiplication of allele frequencies in the form of the product rule benefits demonstration that a DNA profile is extremely rare. If the product rule cannot be used, then the power of a genetic test is vastly reduced. Therefore, the assumption that loci and alleles are inherited independently is an important one so that match probabilities can be multiplied.

FREQUENCY ESTIMATE CALCULATIONS FOR RANDOM MATCH PROBABILITIES

The random match probability is usually equated with the population frequency of the DNA profile of interest. This population frequency is estimated based on allele frequencies from various population databases. Statistical approaches for determining profile frequency estimates will be presented in more detail in Chapter 21.

Significant deviations from independence are rare with the 13 CODIS STRs commonly used for forensic DNA typing because all are located on separate chromosomes with the exception of CSF1PO and D5S818, which occur on the long arm of chromosome 5 (see Chapter 5). If independence can be demonstrated between alleles and between loci, then the product rule can be invoked to produce the extremely rare random match probabilities reported in forensic cases, such as 1 in 5 trillion or greater.

THE FALLACY OF INDEPENDENCE TESTING

Several authors have concluded that independence testing for HWE and LE do not help validate the product rule (Evett and Buckleton 1996, Buckleton *et al.* 2001, Forman and Evett 2001). The requirements for Hardy–Weinberg equilibrium (infinite, randomly mating population with no migration or mutation) cannot be met in real human populations (Evett and Buckleton 1996). Thus, obtaining a *p*-value of >0.05 and demonstrating that the STR alleles in our population database are not statistically significantly different from HWE does not mean that the samples are in HWE.

In order to use the product rule and multiple genotype frequencies across all tested loci, an assumption of within-locus and between loci independence is made. According to Buckleton *et al.* (2001), independence testing using Fisher's exact test (Guo and Thomson 1992) does not validate the product rule. Nevertheless, Evett and Buckleton (1996) maintain that the model of perfect independence performs adequately in the context of estimating DNA profile frequencies. However, they argue that classical hypothesis testing has only a small part to play in assessing reliability of a DNA population database. Evett and Buckleton (1996) do not advocate the abandonment of testing for disequilibria but they do contend that the results of independence testing (e.g., HWE) do not address the questions of practical impact and provide no guidance on how to proceed after testing. Thus, if a *p*-value of <0.05 is observed with a set of alleles measured at a particular STR locus, it does not mean that a laboratory should avoid using this data because it 'failed' a test for Hardy–Weinberg equilibrium (e.g., see Gorman *et al.* 2004).

Despite their harsh position on the lack of value for testing of independence assumptions, Foreman and Evett (2001) recognize that there is merit in using within-locus significance tests as part of quality control of the data to enable detection of null alleles (see D.N.A. Box 20.1). In short, critics of conventional independence testing question the relevance of HWE and LE evaluation since these tests are not capable of proving the assumption of independence true. However, these tests are useful for quality assurance of databases and can help detect the presence of null alleles.

COMPARISON TO OTHER POPULATION DATA SETS

Once allele frequencies have been evaluated within a locus and between loci it is also informative to test whether two or more populations differ in their allele frequencies at a given locus. If two sets of population data are similar, then perhaps they can be combined to yield a larger data set. Comparisons between two or more sets of alleles may be performed with $R \times C$ contingency tables, which involve showing the responses of one variable as a function of another variable.

When examining 24 populations from European databases run with SGM Plus loci, Gill *et al.* (2003) utilize the following function to determine if a genotype match probability determined from the combined allele frequencies of all population data ($P_{combined}$) is conservative relative to a genotype match probability calculated from allele frequencies with just a single population data set (P_{origin}):

$$d = \log_{10}\left(\frac{P_{origin}}{P_{combined}}\right)$$

If the *d* value is negative, then P_{origin} is less than $P_{combined}$ suggesting that $P_{combined}$ is conservative (Gill *et al.* 2003).

In order to get a good feel for whether or not a particular sample set is similar to another population data set, comparisons of allele frequencies may be made to information from the same or different population groups (see Table 20.4). A visual comparison of allele frequencies between different population groups may be conducted with histogram plots (Figure 20.3).

In the end, many laboratories, particularly forensic DNA typing laboratories in the United States using the FBI's Combined DNA Index System (see Chapter 18) will utilize the PopStats database that is part of the CODIS software. PopStats allele frequencies for the 13 CODIS core loci were determined from multiple populations around the U.S. (Budowle *et al.* 2001). These population data sets have been extensively examined prior to their routine use in determining frequency estimates for DNA profiles.

POPULATION SUBSTRUCTURE

Genetic mixing of alleles is not completely random because parents often share some common ancestry. The consequence of this non-random mating is that

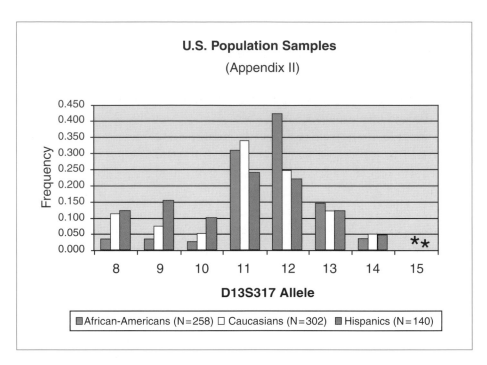

there is usually a decrease in heterozygotes and an increase in homozygotes. This population substructure can be adjusted for with the use of a correction factor referred to as θ (NRC II, p. 102). The NRC II report recommends the use of $\theta = 0.01$ for general populations and $\theta = 0.03$ for small, isolated populations where inbreeding is more likely (see NRC II recommendations in Appendix VI). This correction factor is applied to genotype frequency estimates as described in Chapter 21 (e.g., Table 21.4). Data supports the NRC II recommendation that a value for Fst (θ) of 0.01 is conservative for U.S. populations (Budowle and Chakraborty 2001, Budowle *et al.* 2001).

Sewall Wright defined the inbreeding coefficient, F, to compute any degree of relationship between parents (Wright 1951). The kinship coefficient is also designated by F and used to measure degree of relationship between two people (NRC II, p. 99). Fst is Wright's measure of population subdivision (Wright 1951) and is the same as theta (θ) when mating within subpopulations is random. F-statistics can be thought of as a measure of the correlation of alleles within individuals and are related to inbreeding coefficients (NRC II, p. 102). F-statistics describe the amount of inbreeding-like effects within subpopulations (Fst or θ), among populations (Fis or f), and within the entire population (Fit or F). Weir and Cockerham (1984) and Weir and Hill (2002) contain more information on calculations behind θ and the theory of population substructure.

PRACTICAL CONSIDERATIONS

FORENSIC JOURNAL PUBLICATION POLICIES REGARDING POPULATION DATA

Several years ago the *Journal of Forensic Science* (JFS) (Gaensslen 1999) and *Forensic Science International* (FSI) (Lincoln and Carracedo 2000) decided to reduce the amount of space taken up by published DNA population data information since STR typing has become fairly routine with the use of commercial STR kits. JFS population data is reported in 'For the Record' articles while FSI population information is recorded in 'Announcements of Population Data.' The goal of this policy is to enable researchers and forensic scientists to report population genetic data for common loci in an efficient manner and to encourage formation of web-based repositories of population data.

UNDERSTANDING THE NUMBERS USED IN POPULATION DATA PUBLICATIONS

At the end of allele frequency tables in forensic population data publications is usually a list of values that can be used to evaluate the relative usefulness of each STR marker that has been typed. These measures of the evaluated DNA markers include the power of discrimination (PD), the power of exclusion (PE), the *a priori* chance of exclusion (CE), the polymorphism information content (PIC), and a marker's heterozygosity (H).

Unfortunately, authors publishing such data fail to describe what the various measures of DNA markers mean or how the values are related to one another either because of lack of space or lack of understanding. Typically the numbers for each of these statistical measures are generated by a computer program and therefore the user does not need to think about what the values indicate. Table 20.7 attempts to put a number of calculated genetic functions into context with one another.

Remember that the number of homozygotes (h) plus the number of heterozygotes (H) equals 100% of the samples tested. Thus, since $h + H = 1$, then $H = 1 - h$ and $h = 1 - H$.

Heterozygosity (H) is simply the proportion of heterozygous individuals in the population. It is calculated by dividing the number of samples containing heterozygous alleles into the total number of samples (Weir 1996, pp. 141–150). A higher heterozygosity means that more allele diversity exists and therefore there is less chance of a random sample matching. Edwards *et al.* (1992) described the following formula for calculating an unbiased estimate of the expected heterozygosity:

$$H = \frac{n}{(n-1)}\left[1 - \sum_{j=1}^{k}\left(n_j/n\right)^2\right] = \frac{n}{(n-1)}\left[1 - \sum_{j=1}^{k}\left(p_j\right)^2\right]$$

Table 20.7

Summary of formulas and calculations used to compute various parameters for population data analyses. The worked example information was generated from the D13S317 allele and genotype frequencies contained in Table 20.2. The symbols below are according to the following key: P_i is frequency of ith allele in a population of n samples; x_i is frequency of ith genotype; h = homozygosity; H = heterozygosity.

Genetic Function and Formula	Worked example (D13S317 data from Table 20.2)
Homozygosity (h) $= \sum_{i=1}^{n} P_i^2$	0.2154
Heterozygosity (H) $= 1 - $ Homozygosity $= 1 - \sum_{i}^{n} P_i^2$	0.7845
Effective number of alleles $(n_e) = \dfrac{1}{\text{Homozygosity}} = \dfrac{1}{\sum_{i=1}^{n} P_i^2}$	4.6413
PIC $= 1 - \sum_{1}^{n} P_i^2 - \left(\sum_{1}^{n} P_i^2 \right)^2 + \sum_{1}^{n} P_i^4$	0.7556
PD $= 1 - 2 \left(\sum_{1}^{n} P_i^2 \right)^2 - \sum_{1}^{n} P_i^4$	0.8896
Power of Exclusion (PE) $= H^2(1-(1-H)H^2)$	
PE $= 1 - 2\sum_{1}^{n} P_i^2 - 2 \left(\sum_{1}^{n} P_i^2 \right)^2 + 2\sum_{1}^{n} P_i^4 + \sum_{1}^{n} P_i^3 - 3\sum_{1}^{n} P_i^5 + 3\sum_{1}^{n} P_i^2 \sum_{1}^{n} P_i^3$	0.5910
Probability of Identity (Matching Probability) $= P_I = \sum_{i=1}^{n} x_i^2$	0.0771
Paternity Index (PI) $= \dfrac{H+h}{2h} = \dfrac{(1-h)+h}{2h} = \dfrac{1}{2h} = \dfrac{1}{2\sum_{i=1}^{n} P_i^2}$	2.3207

where n_1, n_2, ..., n_k are the allele counts of K alleles at a locus in a sample of n genes drawn from the population and p_j is the allele frequency.

Gene diversity, often referred to as expected heterozygosity, is defined as the probability that two randomly chosen alleles from the population are different (Weir 1996, pp. 150–156).

Power of discrimination (PD), or probability of discrimination, was first described by Fisher (1951). PD is equal to 1 minus the sum of the square of the genotype frequencies. PD is equal to $1-P_I$ (see below).

Power of exclusion (PE) or probability of exclusion was first described by Fisher (1951) and may be determined by the formula: $PE = H^2(1-(1-H)H^2)$, where H = heterozygosity.

The *probability of identity* (P_I) value is the probability that two individuals selected at random will have an identical genotype at the tested locus

(Sensabaugh 1982). It is calculated by summing the square of the genotype frequencies.

Polymorphism information content or power of information content (PIC) reflects the probability that a given offspring of a parent carrying a rare allele at a locus will allow deduction of the parental genotype at the locus and is determined by summing the mating frequencies multiplied by the probability that an offspring will be informative (Botstein *et al.* 1980).

Probability of a match (PM or *p*M) is sometimes referred to as the probability of a random match and is the inverse of the genotype frequency for a marker (or full profile).

Paternity index (PI) is the likelihood that the genetic alleles obtained by the child support the assumption that the tested man is the true biological father rather than an untested randomly selected unrelated man. The *combined paternity index* (CPI) is determined by multiplying the individual PIs for each locus tested.

Probability of paternity exclusion (PPE) is the probability, averaged over all possible mother-child pairs, that a random alleged father will be excluded from paternity (Chakraborty and Stivers 1996).

WHAT ARE THE 'BEST' MARKERS FOR USE?

Having learned now how to judge measured parameters such as the probability of exclusion for a marker, it is perhaps worthwhile to step back and consider the STR markers under use in human identity applications. Are some markers better than others and if so is it possible to get results from the best markers when partial profiles or mixtures arise (see Chapter 22)?

In human identity testing, a primary goal is to reliably distinguish unrelated individuals from one another (see Chapter 21) or to match related ones through kinship analysis (see Chapter 23). This is done by using polymorphic markers so that we have less of a chance of a type from another individual randomly matching when it should not. In order to have this high discrimination power we typically use approximately a dozen unlinked STR markers. European laboratories are largely using the SGM Plus kit that amplifies 10 STR loci and North American laboratories use the 13 core CODIS loci with sometimes the addition of two more loci depending on the commercial megaplex (see Chapter 5). There are eight STR loci in common between the European and North American STR testing sets: FGA, TH01, VWA, D3S1358, D8S1179, D16S539, D18S51, and D21S11.

The probability of obtaining a match between two distinct and unrelated individuals (random match probability, RMP) provides a useful measure for evaluating the discriminating power of the DNA profiling system (Foreman and Evett 2001). Table 20.8 examines the average random match probabilities for

Table 20.8

Random match probabilities (RMPs) for the commonly used STR markers determined from Caucasian population databases. Power of discrimination (PD) was calculated for the available kits by multiplying the RMPs for the loci included in the kits (see Table 5.3). The CODIS core STR loci and the most commonly used kits are listed in bold font. The 13 CODIS loci data are from FBI data reported in NIJ's The Future of Forensic DNA Testing (2000). The product of these RMPs is 1.7 × 10⁻¹⁵ for the 13 CODIS loci. D2S1138 and D19S433 RMP data are from the Identifiler Kit User's Manual from Applied Biosystems while the SE33 information is from the SEfiler Kit User's Manual. The Penta D and Penta E information comes from PowerPlex 16 population data available from Promega Corporation.

Locus	RMP	STR Kit	1/PD	PD
CSF1PO	0.112	PowerPlex 1.1	8.5×10^{-9}	1.2×10^{8}
FGA	0.036	PowerPlex 2.1	5.8×10^{-12}	1.7×10^{11}
TH01	0.081	**PowerPlex 16**	3.0×10^{-18}	3.4×10^{17}
TPOX	0.195	PowerPlex ES	2.0×10^{-11}	5.0×10^{10}
VWA	0.062	**Profiler Plus**	1.1×10^{-11}	9.4×10^{10}
D3S1358	0.075	**COfiler**	7.7×10^{-7}	1.3×10^{6}
D5S818	0.158	**SGM Plus**	2.1×10^{-13}	4.8×10^{12}
D7S820	0.065	Profiler	2.6×10^{-10}	3.9×10^{9}
D8S1179	0.067	**Identifiler**	4.0×10^{-18}	2.5×10^{17}
D13S317	0.085	SEfiler	4.1×10^{-15}	2.4×10^{14}
D16S539	0.089	CTT	1.8×10^{-3}	5.7×10^{2}
D18S51	0.028	CTTv	1.1×10^{-4}	9.1×10^{3}
D21S11	0.039	GammaSTR	7.8×10^{-5}	1.3×10^{4}
D2S1338	0.027	FFFL	3.8×10^{-4}	2.7×10^{3}
D19S433	0.087			
Penta D	0.059			
Penta E	0.03			
SE33	0.02			
F13A1	0.098			
F13B	0.144			
FES/FPS	0.172			
LPL	0.155			

the commonly used STR loci and the various combinations that are available with the use of different commercial kits.

A higher gene diversity (heterozygosity) level will result from having more alleles at a locus and having alleles that are fairly equal in terms of their observed frequency. Note that the RMP value of TPOX is poorer than D16S539 even though they have a similar number of alleles because D16S539 has a better balance of allele frequencies.

ESTIMATING THE MOST COMMON AND RAREST GENOTYPES POSSIBLE

Another measure to reflect the usefulness of a particular set of DNA markers is to examine the frequencies of the most common genotypes, which would therefore be the least powerful in terms of being able to differentiate between two unrelated individuals (Edwards *et al.* 1992). The theoretically most common type can be calculated by considering a sample type that is heterozygous at all loci possessing the two most common alleles at each locus (Foreman and Evett 2001). For example, with D13S317 the two most common alleles in U.S. Caucasians are 11 (frequency of 0.33940) and 12 (frequency of 0.24834) (see Appendix II).

In Table 20.9 frequencies from the two most common alleles at each of the 13 CODIS loci are used to estimate a theoretical most common STR profile, which occurs with a frequency of approximately 6.26×10^{-12} or 1 in 160 billion. The rarest theoretical 13-locus profile, using the 5/2N minimum allele frequency

Locus	A1	A2	Allele 1 Freq (p)	Allele 2 Freq (q)		Most Common Genotype Frequency
D13S317	11	12	0.33940	0.24834	2pq	0.1686
D16S539	12	11	0.32616	0.32119	2pq	0.2095
D18S51	15	16	0.15894	0.13907	2pq	0.0442
D21S11	30	29	0.27815	0.19536	2pq	0.1087
D3S1358	15	16	0.26159	0.25331	2pq	0.1325
D5S818	12	11	0.38411	0.36093	2pq	0.2773
D7S820	10	11	0.24338	0.20695	2pq	0.1007
D8S1179	13	12	0.30464	0.18543	2pq	0.1130
CSF1PO	12	11	0.36093	0.30132	2pq	0.2175
FGA	22	21	0.21854	0.18543	2pq	0.0810
TH01	9.3	6	0.36755	0.23179	2pq	0.1704
TPOX	8	11	0.53477	0.24338	2pq	0.2603
VWA	17	18	0.28146	0.20033	2pq	0.1128
						6.26×10^{-12}
					1 in...	1.60×10^{11} (160 billion)

Table 20.9

Calculations for theoretically most common genotype frequencies and profile frequency based on two most common alleles found in a U.S. Caucasian allele frequency database (Appendix II).

rule or 0.00828 in the case of the Caucasian database in Appendix II, and calculated with heterozygous rare alleles is as follows:

$$(2pq)^{13} = [2(0.00828)(0.00828)]^{13}$$
$$= 6.06 \times 10^{-51} \ (1.65 \times 10^{50}).$$

Thus, based on the allele frequencies present in a U.S. Caucasian database and independence assumptions between alleles and loci, the theoretical 13-locus STR profile estimates can range from 6.26×10^{-12} to 6.06×10^{-51}.

CONCLUSIONS

In order to determine the probability of a random match between the DNA profile obtained from a crime scene and that of an accused suspect, a reliable estimation of allele and genotype frequencies in relevant population(s) are needed (Chakraborty 1992). Hence population data have been gathered in numerous studies and examined for Hardy–Weinberg equilibrium and linkage equilibrium. Once a specific database has been deemed 'reliable' following allele and locus independence tests the allele frequencies are then used to estimate a specific DNA profile's frequency as will be described in Chapter 21.

POINTS FOR DISCUSSION

1. Should population data be removed from use in genotype frequency estimations if samples are not found to be in HWE based on independence testing?
2. How far will results be off if the wrong ethnic/racial database is used or a database where allele frequencies are not in HWE?
3. Can population databases be combined? If so, what tests are needed before data can be deemed reliable to combine with another set?

REFERENCES AND ADDITIONAL READING

Bagdonavicius, A., Turbett, G.R., Buckleton, J.S. and Walsh, S.J. (2002) *Journal of Forensic Sciences*, 47, 1149–1153.

Botstein, D., White, R.L., Skolnick, M. and Davis, R.W. (1980) *American Journal of Human Genetics*, 32, 314–331.

Buckleton, J.S., Walsh, S. and Harbison, S.A. (2001) *Science & Justice*, 41, 81–84.

Budowle, B., Defenbaugh, D.A. and Keys, K.M. (2000) *Legal Medicine*, 2, 26–30.

Budowle, B. and Chakraborty, R. (2001) *Legal Medicine*, 3, 29–33.

Budowle, B., Shea, B., Niezgoda, S. and Chakraborty, R. (2001) *Journal of Forensic Sciences*, 46, 453–489.

Butler, J.M., Schoske, R., Vallone, P.M., Redman, J.W. and Kline, M.C. (2003) *Journal of Forensic Sciences*, 48, 908–911.

Chakraborty, R. (1984) *Genetics*, 108, 719–731.

Chakraborty, R. (1992) *Human Biology*, 64, 141–159.

Chakraborty, R. and Stivers, D.N. (1996) *Journal of Forensic Sciences*, 41, 671–677.

Chakraborty, R., Stivers, D.N., Su, Y. and Budowle, B. (1999) *Electrophoresis*, 20, 1682–1696.

Crow, J.F. and Kimura, M. (1970) *An Introduction to Population Genetics Theory*. New York: Harper & Row.

Edwards, A., Hammond, H.A., Jin, L., Caskey, C.T. and Chakraborty, R. (1992) *Genomics*, 12, 241–253.

Evett, I.W. and Gill, P. (1991) *Electrophoresis*, 12, 226–230.

Evett, I.W. and Buckleton, J.S. (1996) Statistical analysis of STR data. In Carracedo, A., Brinkmann, B., and Bar, W. (eds) *Advances in Forensic Haemogenetics 6*, pp. 79–86. New York: Springer-Verlag.

Fisher, R.A. (1951) *Heredity*, 5, 95–102.

Foreman, L.A. and Evett, I.W. (2001) *International Journal of Legal Medicine*, 114, 147–155.

Gaensslen, R.E. (1999) *Journal of Forensic Sciences*, 44(4), 671–674.

Gill, P., Foreman, L., Buckleton, J.S., Triggs, C.M. and Allen, H. (2003) *Forensic Science International*, 131, 184–196.

Gorman, K., Slinker, Z., Johnson, S. and Beckwith, M. (2004) *Profiles in DNA*. Madison, Wisconsin: Promega Corporation, Volume 7, Number 1, pp. 9–11. see http://www.promega.com/profiles.

Guo, S.W. and Thompson, E.A. (1992) *Biometrics*, 48, 361–372.

Hardy, G. (1908) *Science*, 28, 49–50.

Holt, C.L., Stauffer, C., Wallin, J.M., Lazaruk, K.D., Nguyen, T., Budowle, B. and Walsh, P.S. (2000) *Forensic Science International*, 112, 91–109.

Hosking, L., Lumsden, S., Lewis, K., Yeo, A., McCarthy, L., Bansal, A., Riley, J., Purvis, I. and Xu, C.-F. (2004) *European Journal of Human Genetics*, 12, 395–399.

Jones, D.A. (1972) *Journal of the Forensic Science Society*, 12, 355–359.

Klintschar, M., Al Hammadi, N. and Reichenpfader, B. (2001) *International Journal of Legal Medicine*, 114, 211–214.

Leibelt, C., Budowle, B., Collins, P., Daoudi, Y., Moretti, T., Nunn, G., Reeder, D. and Roby, R. (2003) *Forensic Science International*, 133, 220–227.

Lincoln, P. and Carracedo, A. (2000) *Forensic Science International*, 110, 3–5.

National Research Council Committee on DNA Forensic Science (1996) *The Evaluation of Forensic DNA Evidence*. National Academy Press: Washington, DC; usually referred to as NRCII; recommendations listed in Appendix VI.

Nei, M. (1978) *Genetics*, 89, 583–590.

Picornell, A., Tomas, C., Jimenez, G., Castro, J.A. and Ramon, M.M. (2002) *Forensic Science International*, 125, 52–58.

Raymond, M. and Rousset, F. (1995) *Journal of Heredity*, 86, 248–249.

Sensabaugh, G. (1982) Biochemical markers of individuality. In Saferstein, R. (ed) *Forensic Science Handbook,* pp. 338–415. New York: Prentice-Hall, Inc.

Steinlechner, M., Berger, B., Scheithauer, R. and Parson, W. (2001) *International Journal of Legal Medicine*, 114, 288–290.

Tereba, A. (1999) Tools for analysis of population statistics. *Profiles in DNA*, 2 (3), 14–16. PowerStats can be downloaded from http://www.promega.com/geneticidtools.

Walsh, S.J., Triggs, C.M., Curran, J.M., Cullen, J.R. and Buckleton, J.S. (2003) *Journal of Forensic Sciences*, 48, 1091–1093.

Weinberg, W. (1908) *Naturkd Wurttemberg*, 64, 368–382.

Weir, B.S. and Cockerham, C.C. (1984) *Evolution*, 38, 1358–1370.

Weir, B.S. and Hill, W.G. (2002) *Annual Review of Genetics*, 36, 721–750.

Wright, S. (1951) *Annuals of Eugenetics*, 15, 159–171.

CHAPTER 21

PROFILE FREQUENCY ESTIMATES, LIKELIHOOD RATIOS, AND SOURCE ATTRIBUTION

DNA evidence can be infuriating, particularly if you're a criminal defendant.

(Henry Lee and Frank Tirnady, *Blood Evidence*, 2003)

It would not be scientifically justifiable to speak of a match as proof of identity in the absence of underlying data that permit some reasonable estimate of how rare the matching characteristics actually are.

(NRC II, p. 192)

When a match is observed between an evidence sample (the 'unknown' or question sample – Q) and a reference sample (the 'known' – K), then statistical methods are typically invoked to provide information regarding the relevance of this match. The prosecution advocates to the court that the Q and K samples have a common source while the defense typically argues that the samples happen to match by chance. The possibility of another *unrelated* individual pulled at random from the population possessing an identical genotype can be determined by calculating the frequency with which the observed genotype occurs in a representative population database (see Chapter 20). When a DNA profile is fairly common, then it is easier to imagine that the suspect might not be connected to the crime scene. If on the other hand, the genotype is found to be extremely rare, then the evidence is stronger that the suspect contributed the crime scene sample in question.

As described in Chapter 20, a number of population databases have been generated in recent years to which a DNA profile may be compared (e.g., Budowle *et al.* 2001). The U.S. population data shown in Appendix II (Butler *et al.* 2003) will be used throughout this chapter to illustrate the values for various allele frequencies. For calculations performed in one's own laboratory, a relevant population database, usually specific to possible populations in one's local area, would be used instead.

It is important to keep in mind that methods for reporting DNA evidence vary between laboratories. Some laboratories present random match probabilities that are based on genotype frequency estimates. Another approach is to report likelihood ratios to convey relative support for the weight of DNA evidence under the hypothesis that the defendant is the source of the DNA profile versus an unrelated individual from the population at large. The Federal Bureau of Investigation (FBI) Laboratory has opted for a source attribution approach when random match probabilities are sufficiently rare. In this chapter, we will discuss the issues surrounding each approach and go through the statistical calculations performed with each method.

FREQUENCY ESTIMATE CALCULATIONS

DNA profile frequency estimates are calculated by first considering the genotype frequency for each locus and then multiplying the frequencies across all loci. The most effective method to understand how the probability of a random match is calculated is to work through an example. The frequency for any DNA profile can be calculated with knowledge of the alleles from the DNA profile and allele frequencies seen in a population database. Of course a different size database or one with different allele frequencies can result in a different expected genotype frequency for each tested locus and hence a different DNA profile frequency. It is therefore important that the database used is large enough and representative of the population of the suspect(s).

In Table 21.1 the DNA profile frequencies for three short tandem repeat (STR) loci are determined using allele frequencies from two different databases. One database contains 302 U.S. Caucasians or 604 measured alleles (Table 21.1a) and the other contains 140 U.S. Hispanics or 280 measured alleles (Table 21.1b). Both a paternal and a maternal allele are counted when considering autosomal markers (see Chapter 2). The DNA profile in question contains the following alleles: 11 and 14 at D13S317 (heterozygous), 6 at TH01 (homozygous), and 14 and 16 at D18S51 (heterozygous).

In the population sample of 604 alleles (302 U.S. Caucasian individuals), allele 11 for D13S317 was observed 205 times or 34% of the time (Table 21.1a). (Note that in Appendix II, allele 11 for D13S317 is listed as 0.33940. Thus, the number of significant figures has been reduced for this example.) The frequency of allele 11 for D13S317 can therefore be recorded as $p = 0.34$. In other words, we can assume that there is a 34% chance that any particular D13S317 allele selected at random from an unrelated individual will be an 11. In the same manner, the chance for observing an allele 14 is $q = 0.05$ since this allele was seen 29 times in 604 allele measurements (in Appendix II this value is listed as 0.04801).

If the individual with the 11,14 D13S317 genotype received these alleles at random from each of his parents, then the chance to receive an 11 from his mother and a 14 from his father is pq and to receive the 14 from his mother and the 11 from his father is another pq. With either combination possible, the probability to be 11,14 by chance is pq + pq or 2pq. Plugging the frequency

(a)

DNA Profile		Allele Frequency from Database			Genotype Frequency for Locus	
Locus	Alleles	Times Allele Observed	Size of Database	Frequency	Formula	Number
D13S317	11	205	604	p = 0.34	2pq	0.03
	14	29		q = 0.05		
TH01	6	140	604	p = 0.23	p^2	0.05
	6					
D18S51	14	83	604	p = 0.14	2pq	0.04
	16	84		q = 0.14		

Profile Frequency = 0.000060
1 in 17 000

(b)

DNA Profile		Allele Frequency from Database			Genotype Frequency for Locus	
Locus	Alleles	Times Allele Observed	Size of Database	Frequency	Formula	Number
D13S317	11	66	280	p = 0.24	2pq	0.02
	14	13		q = 0.05		
TH01	6	60	280	p = 0.21	p^2	0.04
	6					
D18S51	14	39	280	p = 0.14	2pq	0.04
	16	38		q = 0.14		

Profile Frequency = 0.000032
1 in 31 000

Table 21.1

Example calculation of the DNA profile frequency or random match probability using alleles from three STR loci.(a) The database used in this case involves 302 U.S. Caucasian individuals or 604 measured alleles (Appendix II). (b) The database used in this case involves 140 U.S. Hispanic individuals or 280 measured alleles (Appendix II).

values of p = 0.34 and q = 0.05 into the formula 2pq ($2 \times 0.34 \times 0.05$) results in an estimated genotype frequency of 0.034 or in other words approximately 3% of people from a Caucasian population are expected to have an 11,14 genotype at the D13S317 locus. Conducting the same analysis with a U.S. Hispanic database will result in a similar genotype frequency of 0.02 or 2% (Table 21.1b).

With the TH01 locus, a homozygous allele 6 was observed (Table 21.1). The same comparison of the profile's observed allele to a measured allele frequency in a population database is performed with TH01 but in this case the combined probability of inheriting allele 6 from both parents is pp or p^2 (see Chapter 19). Since allele 6 was observed 140 times out of 604 allele measurements in U.S. Caucasians, p = 0.23 and p^2 = 0.05 (Table 21.1a). Likewise, in the U.S. Hispanic population, p = 0.21 and p^2 = 0.04 (Table 21.1b).

Since these two STR loci are on separate chromosomes (e.g., chromosome 13 for D13S317 and chromosome 11 for TH01, see Table 5.2), they will segregate independently during meiosis allowing the genotype frequencies to be multiplied. In the case of a U.S. Caucasian population, the chance of a person having the combined genotype of 11,14 at D13S317 and 6,6 at TH01 is 5% of 3% (i.e., 0.05×0.03) or 0.15%. Similar calculations for the D18S51 locus with alleles 14 and 16 result in an estimated genotype frequency of 4% (Table 21.1a). The combined profile frequency with these three loci thus becomes 0.000060 – the product of the three individual genotype frequencies ($0.03 \times 0.05 \times 0.04$) or about 1 in 17 000. Note that when using the Hispanic database, the DNA profile frequency in this example drops to 0.000032 ($0.02 \times 0.04 \times 0.04$) or about 1 in 31 000 individuals (Table 21.1b).

As more and more loci match during a Q and K sample comparison, it becomes less and less likely that an *unrelated*, random person in the population contributed the crime scene sample. Thus, either the suspect contributed the evidence or a very unlikely coincidence happened. Later in this chapter and also in Chapter 23 we will consider the impact of a *related* individual on the DNA profile frequency estimate calculations.

WHAT A RANDOM MATCH PROBABILITY IS NOT

It is important to realize what a random match probability is not. It is not the chance that someone else is guilty or that someone else left the biological material at the crime scene. Likewise, it is not the chance of the defendant not being guilty or the chance that someone else in reality would have that same genotype. Rather a random match probability is simply the estimated frequency at which a particular STR profile would be expected to occur in a population. This random match probability may also be thought of as the theoretical chance that if you sample one person at random from the population they will have the particular DNA profile in question.

Switching the language and meaning of a random match probability is something referred to as the *Prosecutor's Fallacy* or the fallacy of the transposed conditional (Thompson and Schumann 1987, Balding and Donnelly 1994). Statements such as 'there is only a 1 in 17 000 chance that the DNA profile came from someone else' or 'there is only a 1 in 17 000 chance that the defendant is not guilty' are examples of the prosecutor's fallacy. A correct statement would be 'the probability of selecting the observed profile from a population of random unrelated individuals is expected to be 1 in 17 000 based on the alleles present in this sample…' Note that with a 13 STR locus-match instead of just the three used in the above example the random match probabilities are in the range of trillions, quadrillions, and beyond (see Table 21.2 and D.N.A. Box 21.1).

The *Defense Attorney's Fallacy* is equally problematic where the assumption is made that everyone else with the same genotype has an equal chance of being guilty or that every possible genotype in a mixture (see Chapter 22) has an equal chance of having committed the crime (NRC II, p. 133). Access to the crime scene, motive, and legitimate alibis all play a role in an investigation suggesting that it is unwise to consider DNA evidence and corresponding frequency estimates in a vacuum devoid of other information. A suspect is usually under suspicion and investigation prior to knowing his DNA profile, and thus the DNA results are most often used to corroborate and connect a criminal perpetrator to his crime scene rather than as the sole evidentiary material.

TREATMENT OF RARE ALLELES AND TRI-ALLELIC PATTERNS

In Chapter 6 we discussed rare variant or 'off-ladder' alleles and tri-allelic patterns. Since these anomalies do exist, they must be accounted for in statistical calculations and STR profile frequency estimates. Because they are rare, it is expected that the minimum allele rule discussed in Chapter 20 would be applied and therefore they would be treated as 5/2N (see NRC II, p. 148). In the case of the Caucasian population from Appendix II where N = 302, the minimum allele threshold of 5/2N is 0.00828 or 0.828%. On the variant allele listing of STRBase (http://www.cstl.nist.gov/biotech/strbase/var_tab.htm) some frequency estimates for particular variant alleles and tri-allelic patterns are provided as these alleles are discovered.

THE PRODUCT RULE

The lack of population structuring with allele frequencies in Hardy–Weinberg equilibrium and linkage equilibrium (see Chapter 20) justifies the assumption that genotypes are independent at unlinked loci. With the assumption of independence, it then becomes possible to equate the overall match probability with the product of the locus-specific match probabilities. This combination of locus-specific match probabilities is referred to as the *product rule*. In other words, the match probability for the STR locus D13S317 can be combined with additional STR loci such as TH01 and D18S51 to decrease the odds of a random match to an unrelated individual.

Table 21.2 calculates the match probabilities for all 13 CODIS loci with the same STR profile using the three different U.S. population databases contained in Appendix II. Since the STR profile used for these examples is from a Caucasian and is present in the Appendix II allele frequency database, the Caucasian population is expected to provide the most common frequency estimate. This turns out to be the case where the calculated STR profile frequency estimates are 1.20×10^{-15} using the Caucasian allele frequencies and 6.04×10^{-17} and 5.57×10^{-17} using the African-American and Hispanics allele frequencies, respectively (Table 21.2). As noted by Weir (2003), focusing on a suspect's racial group in the calculations has an element of conservativeness.

Often the rarity of a calculated DNA profile goes beyond one in billions (10^9) or trillions (10^{12}) to numbers that are not frequently used because they are so large. A list of some big number names is contained in Table 21.3 to aid in verbal descriptions of rare DNA profiles. For example, the inverted value of 1.20×10^{-15} is 1 in 8.37×10^{14} or 0.84×10^{15} (one in 0.84 quadrillion).

Table 21.2

Random match probabilities with all 13 CODIS loci

	A1	A2	Allele 1 freq (p)	Allele 2 freq (q)		Expected genotype freq	
From U.S. Caucasian (N = 302); Appendix II – sample in database							
D13S317	11	14	0.33940	0.04801	2pq	**0.0326**	
TH01	6	6	0.23179		p²	0.0537	
D18S51	14	16	0.13742	0.13907	2pq	0.0382	
D21S11	28	30	0.15894	0.27815	2pq	0.0884	
D3S1358	16	17	0.25331	0.21523	2pq	0.1090	
D5S818	12	13	0.38411	0.14073	2pq	0.1081	
D7S820	9	9	0.17715		p²	0.0314	
D8S1179	12	14	0.18543	0.16556	2pq	0.0614	
CSF1PO	10	10	0.21689		p²	0.0470	
FGA	21	22	0.18543	0.21854	2pq	0.0810	
D16S539	9	11	0.11258	0.32119	2pq	0.0723	
TPOX	8	8	0.53477		p²	0.2860	
VWA	17	18	0.28146	0.20033	2pq	0.1128	
AMEL	X	Y					Product Rule
						1.20E-15	Combined Frequency
					1 in	8.37E + 14	
From African-American (N = 258); Appendix II – sample not in database; wrong population							
D13S317	11	14	0.30620	0.03488	2pq	**0.0214**	
TH01	6	6	0.12403		p²	0.0154	
D18S51	14	16	0.07198	0.15759	2pq	0.0227	
D21S11	28	30	0.25775	0.17442	2pq	0.0899	
D3S1358	16	17	0.33527	0.20543	2pq	0.1377	

Continued

	A1	A2	Allele 1 freq (p)	Allele 2 freq (q)		Expected genotype freq	
From African-American (N = 258); Appendix II – sample not in database; wrong population							
D5S818	12	13	0.35271	0.23837	2pq	0.1682	
D7S820	9	9	0.10853		p²	0.0118	
D8S1179	12	14	0.14147	0.30039	2pq	0.0850	
CSF1PO	10	10	0.25681		p²	0.0660	
FGA	21	22	0.11628	0.19574	2pq	0.0455	
D16S539	9	11	0.19574	0.31783	2pq	0.1244	
TPOX	8	8	0.37209		p²	0.1385	
VWA	17	18	0.24225	0.15504	2pq	0.0751	
AMEL	X	Y					
						6.04E-17	Product Rule Combined Frequency
					1 in	1.66E + 16	

	A1	A2	Allele 1 freq (p)	Allele 2 freq (q)		Expected genotype freq	
From Hispanic (N = 140); Appendix II – sample not in database; wrong population							
D13S317	11	14	0.23571	0.04643	2pq	**0.0219**	
TH01	6	6	0.21429		p²	0.0459	
D18S51	14	16	0.13929	0.13571	2pq	0.0378	
D21S11	28	30	0.09643	0.26071	2pq	0.0503	
D3S1358	16	17	0.28571	0.20357	2pq	0.1163	
D5S818	12	13	0.35000	0.12500	2pq	0.0875	
D7S820	9	9	0.11071		p²	0.0123	
D8S1179	12	14	0.14286	0.25000	2pq	0.0714	
CSF1PO	10	10	0.23214		p²	0.0539	
FGA	21	22	0.16786	0.15000	2pq	0.0504	
D16S539	9	11	0.13929	0.26071	2pq	0.0726	
TPOX	8	8	0.47143		p²	0.2222	
VWA	17	18	0.21786	0.17143	2pq	0.0747	
AMEL	X	Y					
						5.57E-17	Product Rule Combined Frequency
					1 in	1.80E + 16	

Table 21.3

Names of big numbers with their corresponding scientific notation. From http://www.gomath.com/ htdocs/ToGoSheet/Algebra /bignumber.html.

10^1	Ten
10^2	Hundred
10^3	Thousand
10^6	Million
10^9	Billion
10^{12}	Trillion
10^{15}	Quadrillion
10^{18}	Quintillion
10^{21}	Sextillion
10^{24}	Septillion
10^{27}	Octillion
10^{30}	Nonillion
10^{33}	Decillion
10^{36}	Undecillion
10^{39}	Duodecillion
10^{42}	Tredecillion
10^{45}	Quattuordecillion
10^{48}	Quindecillion
10^{51}	Sexdecillion
10^{54}	Septendecillion
10^{57}	Octodecillion
10^{60}	Novemdecillion
10^{63}	Vigintillion
10^{100}	Google

IMPACT OF VARIOUS POPULATION DATABASES

From the three combined STR profile frequencies calculated in Table 21.2 it is apparent that different populations can yield different frequency estimates due to variations in allele frequencies in these populations. A calculation of the same STR profile as used in the previous examples against 97 different published population databases (including the three present in Appendix II) found that the cumulative profile frequency ranged from 1 in 3.43×10^{14} to 1 in 2.65×10^{21} (D.N.A. Box 21.1).

D.N.A. Box 21.1

STR profile frequency calculations against multiple population databases

The ability to determine simultaneously the frequency for a particular STR profile in multiple population databases was recently made easier with the development of a Microsoft Excel macro called OmniPop. Below the cumulative profile frequency range is calculated for the particular STR profile listed against 166 published population studies involving Profiler Plus kit loci and 97 published reports containing all 13 CODIS core loci. The cumulative profile frequency obtained with U.S. Caucasian alleles present in the Appendix II data set are listed as well.

These profile frequencies were all calculated with a theta value of 0.01. When using a theta value of 0.03 as recommended by NRC II for native (more inbred) populations, the range for the computed profile with all 13 STR loci across the 97 published population data sets is 2.77×10^{14} to 1.27×10^{21}.

It is worth noting that the computed profile is part of the U.S. Caucasian data set used to generate the allele frequencies described in Appendix II and thus this database would be expected to compute fairly conservative values for this particular 13-locus STR profile as demonstrated below.

STR Locus	Profile Computed	Number of Populations Used	Cumulative Profile Frequency Range (1 in ...)	Cumulative Profile Frequency against U.S. Caucasians (Appendix II)
D3S1358	16,17	166	5.24 to 62.6	9.19
VWA	17,18	166	37.6 to 1080	81.8
FGA	21,22	166	737 to 119 000	1010
D8S1179	12,14	166	8980 to 5 430 000	16 400
D21S11	28,30	166	165 000 to 248 000 000	186 000
D18S51	14,16	166	3.85×10^6 to 2.68×10^{10}	4.88×10^6
D5S818	12,13	166	2.28×10^7 to 4.22×10^{11}	4.51×10^7
D13S317	11,14	166	4.32×10^8 to 1.69×10^{13}	1.38×10^9
D7S820	9,9	166	1.17×10^{10} to 2.98×10^{16}	4.22×10^{10}
D16S539	9,11	97	4.06×10^{11} to 1.11×10^{18}	5.82×10^{11}
TH01	6,6	97	9.30×10^{12} to 1.45×10^{19}	1.05×10^{13}
TPOX	8,8	97	3.33×10^{13} to 1.54×10^{20}	3.63×10^{13}
CSF1PO	10,10	97	3.43×10^{14} to 2.65×10^{21}	7.43×10^{14}

Source:

OmniPop 150.4.2 was used for these calculations. This Excel macro is available from Brian Burritt of the San Diego Police Department; telephone: +1 (619) 531 2215; email: bburritt@pd.sandiego.gov.

It is probably worth noting that the final calculated value in the far right column of D.N.A. Box 21.1 (1 in 7.43 x 10¹⁴) differs slightly from that determined in Table 21.2 (1 in 8.37 x 10¹⁴) due to the number of significant figures carried throughout the calculations. Thus, in order to obtain consistent frequency estimates with the same allele frequency information it is essential to maintain the same significant figures between calculations.

Another source of population databases that enables an online search is the European Network of Forensic Science Institutes (ENFSI) DNA Working Group STR Population Database located at http://www.str-base.org/index2.php. An estimated random match probability for a DNA profile of interest can be calculated using allele frequencies produced from 5700 profiles covering 24 European populations that have been generated with the SGM Plus loci (see Chapter 5) (see Gill *et al.* 2003).

IMPACT OF POPULATION STRUCTURE CORRECTIONS

The National Research Council (NRC II) report entitled *The Evaluation of Forensic DNA Evidence* discusses issues that surround population structure. The NRC II report made several recommendations for taking population structure into account (see Appendix VI). The impacts of these recommendations on homozygote and heterozygote frequency calculations are illustrated in Table 21.4. Again examples are given with a TH01 homozygote 6,6 and a D13S317 heterozygote 11,14.

Table 21.4

Comparison of statistical treatment for homozygotes and heterozygotes under different assumptions. Allele frequency values (p_i, p_j) for the TH01 and D13S317 example data are from Appendix II (U.S. Caucasians). Note that if θ is zero then unconditional and conditional formulas collapse to their Hardy–Weinberg equilibrium (HWE) functions.

	Under HWE	Unconditional (NRCII Recommendation 4.1)	Conditional with Substructure Adjustment
			(NRCII recommendation 4.10a)
Homozygote	p_i^2	$p_i^2 + p_i(1-p_i)\theta$	$\dfrac{[p_i(1-\theta)+2\theta][p_i(1-\theta)+3\theta]}{(1+\theta)(1+2\theta)}$
TH01 **6,6** $p_i = 0.23$ θ = 0.01	$(0.23)^2$ = **0.053**	$(0.23)^2 + (0.23)(1-0.23)(0.01)$ = 0.053 + 0.0018 = **0.055**	$\dfrac{[(0.23)(1-0.01)+2(0.01)][(0.23)(1-0.01)+3(0.01)]}{(1+0.01)(1+2(0.01))}$ = (0.2477)(0.2577)/(1.01)(1.02) = **0.062**
			(NRCII recommendation 4.10b)
Heterozygote	$2p_ip_j$	$2p_ip_j(1-\theta)$	$\dfrac{2[p_i(1-\theta)+2\theta][p_j(1-\theta)+\theta]}{(1+\theta)(1+2\theta)}$
D13S317 **11,14** $p_i = 0.34$ $p_j = 0.05$ θ = 0.01	$2(0.34)(0.05)$ = **0.0340**	$2(0.34)(0.05)(1-0.01)$ = **0.0337**	$\dfrac{2[(0.34)(1-0.01)+0.01][(0.05)(1-0.01)+0.01]}{(1+0.01)(1+2(0.01))}$ = 2(0.3466)(0.0595)/(1.01)(1.02) = **0.0400**

The NRCII recommendation 4.1 substructure adjustments replace p^2 for homozygote calculations with $p^2+p(1-p)\theta$, where θ is an empirically determined measure of population subdivision (see Chapter 20). A conservative value for θ is 0.01 for typical at-large populations and 0.03 with smaller, isolated, and more inbred groups of people. Budowle *et al.* (2001) have demonstrated that $\theta = 0.01$ is a reliable and conservative estimate of population substructure with extensive population data.

Further calculations with these population substructure adjustment equations are illustrated in Table 21.5. Note that NRC II recommendation 4.1 adjustments only change the overall profile frequency estimate from 1.20×10^{-15} to 1.35×10^{-15} with $\theta = 0.01$ and to 1.70×10^{-15} with $\theta = 0.03$. NRC II recommendation 4.10a and 4.10b generate further minor differences as well.

IMPACT OF RELATIVES ON STR PROFILE FREQUENCY ESTIMATES

If the suspect and the true perpetrator of a crime are related, then their genotype frequencies are not independent and a different calculation is required (Weir 1994, Weir 2003). Since STR profiles from relatives are expected to be more similar to the individual in question than a random, unrelated individual, NRC II recommendation 4.4 covers probability calculations from various scenarios of individuals related to the suspect. Table 21.6 works through the same DNA profile according to NRC II recommendations 4.4 and equations 4.8a and 4.8b (see Appendix VI).

More recently, Weir (2003) has developed match probability formulas to calculate the effects of family relatedness that incorporate population substructure functions into them. These formula are listed in Table 21.7 along with a worked example using a homozygote TH01 6,6 Caucasian allele frequency and a heterozygote D13S317 11,14 Caucasian allele frequencies. Note that in the case of the TH01 6,6 example the match probability increases from 0.06283 for an unrelated individual to 0.38921 for full siblings – an increase of more than sixfold. Likewise, the random match probability of the heterozygote D13S317 11,14 changes 9.3-fold from 0.03864 to 0.35955 when moving from an unrelated individual to a full sibling.

GENERAL MATCH PROBABILITY

As noted in this entire section, profile probabilities need to be calculated for a variety of scenarios. Balding (1999) points out that there are five different sets of people and possible relationships to a suspect: (1) the suspect's siblings, (2) his other relatives, (3) other members of his sub-population, (4) other members of his racial group, and (5) anyone else outside of his population (e.g., racial) group (see also Foreman and Evett 2001, Weir 2003).

Table 21.5

Example calculations with NRC II recommendations for population substructure adjustments (see Appendix VI). Scenarios with theta equal to 0.01 and 0.03 are examined.

From U.S. Caucasian (N =302); Appendix II - sample in database

	A1	A2	Allele 1 freq (p)	Allele 2 freq (q)		Under HWE Calc freq	NRCII Recommendation 4.1	θ=0.01	θ=0.03	NRCII Recommendation 4.10	θ=0.01	θ=0.03
D13S317	11	14	0.33940	0.04801	2pq	0.0326	2pq	0.0326	0.0326	eq. 4.10b	0.0386	0.0504
TH01	6	6	0.23179	—	p^2	0.0537	$p^2 + p(1-p)\theta$	0.0555	0.0591	eq. 4.10a	0.0628	0.0821
D18S51	14	16	0.13742	0.13907	2pq	0.0382	2pq	0.0382	0.0382	eq. 4.10b	0.0419	0.0493
D21S11	28	30	0.15894	0.27815	2pq	0.0884	2pq	0.0884	0.0884	eq. 4.10b	0.0927	0.1011
D3S1358	16	17	0.25331	0.21523	2pq	0.1090	2pq	0.1090	0.1090	eq. 4.10b	0.1129	0.1206
D5S818	12	13	0.38411	0.14073	2pq	0.1081	2pq	0.1081	0.1081	eq. 4.10b	0.1131	0.1228
D7S820	9	9	0.17715	—	p^2	0.0314	$p^2 + p(1-p)\theta$	0.0328	0.0358	eq. 4.10a	0.0390	0.0556
D8S1179	12	14	0.18543	0.16556	2pq	0.0614	2pq	0.0614	0.0614	eq. 4.10b	0.0654	0.0733
CSF1PO	10	10	0.21689	—	p^2	0.0470	$p^2 + p(1-p)\theta$	0.0487	0.0521	eq. 4.10a	0.0558	0.0744
FGA	21	22	0.18543	0.21854	2pq	0.0810	2pq	0.0810	0.0810	eq. 4.10b	0.0851	0.0930
D16S539	9	11	0.11258	0.32119	2pq	0.0723	2pq	0.0723	0.0723	eq. 4.10b	0.0773	0.0871
TPOX	8	8	0.53477	—	p^2	0.2860	$p^2 + p(1-p)\theta$	0.2885	0.2934	eq. 4.10a	0.2983	0.3227
VWA	17	18	0.28146	0.20033	2pq	0.1128	2pq	0.1128	0.1128	eq. 4.10b	0.1167	0.1245
AMEL	X	Y										
						1.20E-15		1.35E-15	1.70E-15		3.92E-15	3.02E-14

Table 21.6

Example calculations with corrections for relatives using the NRC II recommended formula.

From U.S. Caucasian (N = 302): Appendix II – sample in database					Under HWE		NRCII Recommendation 4.4				
	A1	A2	Allele 1 freq (p)	Allele 2 freq (q)		Calc freq		F = 1/4 (parent)	F = 1/8 (half sib)	F = 1/16 (1st cousin)		Full sib
D13S317	11	14	0.33940	0.04801	$2pq$	**0.0326**	eq. 4.8b	0.1937	0.1131	0.0729	eq. 4.9b	0.3550
TH01	6	6	0.23179	—	p^2	0.0537	eq. 4.8a	0.2318	0.1428	0.0982	eq. 4.9a	0.3793
D16S539	9	11	0.11258	0.32119	$2pq$	0.0723	eq. 4.8b	0.2169	0.1446	0.1085	eq. 4.9b	0.3765
D18S51	14	16	0.13742	0.13907	$2pq$	0.0382	eq. 4.8b	0.1382	0.0882	0.0632	eq. 4.9b	0.3287
D21S11	28	30	0.15894	0.27815	$2pq$	0.0884	eq. 4.8b	0.2185	0.1535	0.1209	eq. 4.9b	0.3814
D3S1358	16	17	0.25331	0.21523	$2pq$	0.1090	eq. 4.8b	0.2343	0.1717	0.1403	eq. 4.9b	0.3944
D5S818	12	13	0.38411	0.14073	$2pq$	0.1081	eq. 4.8b	0.2624	0.1853	0.1467	eq. 4.9b	0.4082
D7S820	9	9	0.17715	—	p^2	0.0314	eq. 4.8a	0.1772	0.1043	0.0678	eq. 4.9a	0.3464
D8S1179	12	14	0.18543	0.16556	$2pq$	0.0614	eq. 4.8b	0.1755	0.1184	0.0899	eq. 4.9b	0.3531
CSF1PO	10	10	0.21689	—	p^2	0.0470	eq. 4.8a	0.2169	0.1320	0.0895	eq. 4.9a	0.3702
FGA	21	22	0.18543	0.21854	$2pq$	0.0810	eq. 4.8b	0.2020	0.1415	0.1113	eq. 4.9b	0.3713
TPOX	8	8	0.53477	—	p^2	0.2860	eq. 4.8a	0.5348	0.4104	0.3482	eq. 4.9a	0.5889
VWA	17	18	0.28146	0.20033	$2pq$	0.1128	eq. 4.8b	0.2409	0.1768	0.1448	eq. 4.9b	0.3986
AMEL	X	Y										
						1.20E-15		3.17E-09	1.68E-11	3.74E-13		4.04E-06
												1 in 247 616

Relationship	Match probability formula		
	Homozygotes (A$_i$A$_i$)		Result with TH01 6,6
Full siblings	$\dfrac{(1+p_i)^2+(7+7p_i-2p_i^2)\theta+(16-9p_i+p_i^2)\theta^2}{4(1+\theta)(1+2\theta)}$		0.38921
Parent and child	$\dfrac{2\theta+(1-\theta)p_i}{(1+\theta)}$		0.24700
Half siblings	$\dfrac{[2\theta+(1-\theta)p_i][2+4\theta+(1-\theta)p_i]}{2(1+\theta)(1+2\theta)}$		0.27479
First cousins	$\dfrac{[2\theta+(1-\theta)p_i][2+11\theta+3(1-\theta)p_i]}{4(1+\theta)(1+2\theta)}$		0.10888
Unrelated	$\dfrac{[2\theta+(1-\theta)p_i][3\theta+(1-\theta)p_i]}{4(1+\theta)(1+2\theta)}$	(NRC II, 4.10a)	0.06283
			$p^2=0.05373$
	Heterozygotes (A$_i$A$_j$)		Result with D13 11,14
Full siblings	$\dfrac{(1+p_i+p_j+2p_ip_j)+(5+3p_i+3p_j-4p_ip_j)\theta+2(4-2p_i-2p_j+p_ip_j)\theta^2}{4(1+\theta)(1+2\theta)}$		0.35955
Parent and child	$\dfrac{2\theta+(1-\theta)(p_i+p_j)}{2(1+\theta)}$		0.19977
Half siblings	$\dfrac{(p_i+p_j+4p_ip_j)+(2+5p_i+5p_j+8p_ip_j)\theta+(8-6p_i-6p_j+4p_ip_j)\theta^2}{4(1+\theta)(1+2\theta)}$		0.11921
First cousins	$\dfrac{(p_i+p_j+12p_ip_j)+(2+13p_i+13p_j-24p_ip_j)\theta+2(8-7p_i-7p_j+6p_ip_j)\theta^2}{8(1+\theta)(1+2\theta)}$		0.07893
Unrelated	$\dfrac{2\theta[+(1-\theta)p_i][\theta+(1-\theta)p_j]}{(1+\theta)(1+2\theta)}$	(NRC II, 4.10b)	0.03864 $2pq=0.03259$

Table 21.7

Effects of family relatedness on match probabilities (adapted from Weir 2003, p.839). Notice that the unrelated formulas are the same as those for NRC II recommendations 4.10a and 4.10b (see Appendix VI). Worked examples are using θ = 0.01 and the allele frequencies found in Appendix II for Caucasians: p(TH01 allele 6) = 0.23179; p(D13 allele 11) = 0.33940; p(D13 allele 14) = 0.04801.

One solution to this is the use of general match probabilities that have been calculated from the theoretically most conservative method involving the most two common alleles for each locus (D.N.A. Box 21.2) (Foreman and Evett 2001). The primary advantage of this approach is that repeated calculations are not required for each profile observed. Another reason that Foreman and Evett (2001) advocate a general match probability that avoids case-specific calculations is that it is difficult to provide any sound statistical support for probabilities of such a small magnitude (e.g., 10^{-21}).

LIKELIHOOD RATIO

As noted previously, when matching STR profiles are obtained between a suspect (who then becomes the defendant in a court case; in other words the known reference sample, K) and the crime scene evidence (questioned sample, Q), it is necessary to quantify the evidentiary value of this match.

In a paper performing statistical analyses to support forensic interpretation of the 10-loci present in the SGM Plus kit, Foreman and Evett (2001) advocate the use of general probability values when reporting full matching STR profiles. With the 10 STR loci present in the SGM Plus kit used in the UK and Europe, the probabilities are as follows (see Foreman and Evett 2001, Table 4):

Relationship with suspect	Match probability
Sibling	1 in 10 000
Parent/child	1 in 1 million
Half-sibling or uncle/nephew	1 in 10 million
First cousin	1 in 100 million
Unrelated	1 in 1 billion

They argue that adoption of such figures would eliminate the need to per-form case-specific match probabilities making it much easier to present infor-mation to the court. The match probabilities for specific STR profiles are typically several orders of magnitude smaller than those given above, which were calculated from the theoretically most common SGM Plus profile. Thus, these probabilities should provide a fair and reasonable assessment of the weight of DNA evidence for each category and in the end would probably be favorable to the suspect (defendant).

A similar calculation for a full match with the 13 CODIS loci using the most common alleles observed in U.S. population databases, such as Appendix II, would result in even higher general match probability values since more STR loci are being examined.

Sources:

Foreman, L.A. and Evett, I.W. (2001) Statistical analyses to support forensic interpretation for a new 10-locus STR profiling system. *International Journal of Legal Medicine*, 114, 147–155.

Balding, D.J. (1999) When can a DNA profile be regarded as unique? *Science & Justice*, 39, 257–260.

Another approach besides the match probability profile frequency estimate just described is the use of likelihood ratios (LR). LRs involve a comparison of the probabilities of the evidence under two alternative propositions. These mutu-ally exclusive hypotheses represent the position of the prosecution – namely that the DNA from the crime scene originated from the suspect – and the position of the defense – that the DNA just happens to coincidently match the defendant and is instead from an unknown person out in the population at large.

A likelihood ratio is a ratio of two probabilities of the same evidence under different hypotheses. For example, if a DNA profile generated from a crime scene evidence sample matches a suspect's DNA profile, then there are generally two possible hypotheses for why the profiles match each other: (1) the suspect matches because he left his biological sample at the crime scene or (2) the true perpetrator is still at large and just happens to match the suspect at the DNA markers examined.

Typically the first hypothesis (and that championed by the prosecution) is placed in the numerator of the likelihood ratio while the second hypothesis – that someone else other than the defendant committed the crime (which is of course the defense's position) – is placed in the denominator.

Thus, in mathematical terms:

$$LR = H_p/H_d$$

or verbally the likelihood ratio equals the hypothesis of the prosecution divided by the hypothesis of the defense. Since the hypothesis of the prosecution is that the defendant committed the crime, then $H_p = 1$ (assumes 100% probability). On the other hand the hypothesis of the defense that the profile originated from someone else can be calculated from the genotype frequency of the particular STR profile. If the STR typing result is heterozygous, then this probability would be $2pq$, where p is the frequency of allele 1 and q is the frequency of allele 2 in the relevant population for the locus in question. Alternatively, for a homozygous STR type the H_d would be p^2.

Therefore,

$$LR = \frac{H_p}{H_d} = \frac{1}{2pq}$$

If the STR type in question was D13S317 alleles 11 and 14, then p is 0.3394 and q is 0.04801 for the Caucasian population (Appendix II). The likelihood ratio for the D13S317 genotype match then becomes

$$LR = \frac{H_p}{H_d} = \frac{1}{2pq} = \frac{1}{2(0.33940)(0.04801)} = \frac{1}{0.03259} = 30.7$$

Note that the rarer the particular STR genotype is, the higher the likelihood ratio will be since there is a reciprocal relationship. In its simplest form, a LR is the inverse of the estimated genotype frequency for each locus and if discrete alleles and independent marker systems are utilized, then the LR is simply the inverse of the relative frequency of the observed genotype in the relevant population. Of course, LRs can become much more complicated if mixtures or alternative scenarios for the evidence are possible (see Chapter 22, Table 22.1). The product of all locus-specific LRs results in the full profile LR, which in the example of the Caucasian data shown in Table 21.2 comes to 8.37×10^{14} (the inverse of 1.20×10^{-15}).

If the value for a likelihood ratio is greater than one, then it provides support to the prosecution's case. If on the other hand, the LR is less than one, then the defense's case is supported. In the example shown here, if there is a match between a crime stain possessing D13S317 alleles 11 and 14 and the suspect who also possesses a D13S317 genotype of 11,14, then it is 30.7 times more likely if the suspect left the evidence than if it came from some unknown person out of the general Caucasian population.

When considering the strength of a likelihood ratio in terms of supporting the prosecution's position, the following guidelines have been suggested (Evett and Weir 1998, p. 226):

If likelihood ratio is…	Then the evidence provides…
1 to 10	limited support…
10 to 100	moderate support…
100 to 1000	strong support…
1000 and greater	very strong support…

With a 13-locus STR match likelihood ratio of 8.37×10^{14} based on a full profile with unambiguous results (e.g., no mixture present), the evidence has extremely strong support from the proposition that the suspect supplied the evidentiary sample.

SOURCE ATTRIBUTION

Given that DNA evidence can provide strong likelihood ratios and random match probabilities from forensic samples that exceed the world population many fold, the Federal Bureau of Investigation (FBI) Laboratory has adopted a source attribution policy (Budowle *et al.* 2000, DAB 2000). With average random match probabilities of less than one in a trillion using the 13 core STR loci (Chakraborty *et al.* 1999), there comes within the context of a particular case a high degree of confidence that an individual is the source of an evidentiary DNA sample with reasonable scientific certainty. The logic behind this source attribution policy is provided below.

If p_x is the random match probability for a given evidentiary profile X, then $(1-p_x)^N$ is the probability of not observing the particular profile in a sample of N *unrelated* individuals.

When this probability is greater than or equal to a $1-\alpha$ confidence level (with α being 0.01 for 99%), then $(1-p_x)^N \geq 1-\alpha$ or $p_x \leq 1-(1-\alpha)^{1/N}$, which enables the calculation that if N is approximately the size of the U.S. population (N = 300 000 000), then a random match probably of less than 3.35×10^{-11} will confer at least 99% confidence that the evidentiary profile is unique in the population (Budowle *et al.* 2000). Table 21.8 lists the random match probability thresholds for various population sizes and confidence levels.

Table 21.8

Random match probability thresholds for source attribution at various population sizes and confidence levels (adapted from Budowle et al. 2000). With a random match probability of 1.20×10^{-15} in U.S. Caucasians (see Tables 21.2 and 21.5), the example STR profile would be considered 'unique.'

	Sample Size (N)	Confidence Levels $(1 - \alpha)$			
		0.90	0.95	0.99	0.999
	2	5.13×10^{-2}	2.53×10^{-2}	5.01×10^{-3}	5.00×10^{-4}
	3	3.45×10^{-2}	1.70×10^{-2}	3.34×10^{-3}	3.33×10^{-4}
	4	2.60×10^{-2}	1.27×10^{-2}	2.51×10^{-3}	2.50×10^{-4}
	5	2.09×10^{-2}	1.02×10^{-2}	2.01×10^{-3}	2.00×10^{-4}
	6	1.74×10^{-2}	8.51×10^{-3}	1.67×10^{-3}	1.67×10^{-4}
	7	1.49×10^{-2}	7.30×10^{-3}	1.43×10^{-3}	1.43×10^{-4}
	8	1.31×10^{-2}	6.39×10^{-3}	1.26×10^{-3}	1.25×10^{-4}
	9	1.16×10^{-2}	5.68×10^{-3}	1.12×10^{-3}	1.11×10^{-4}
	10	1.05×10^{-2}	5.12×10^{-3}	1.00×10^{-3}	1.00×10^{-4}
	25	4.21×10^{-3}	2.05×10^{-3}	4.02×10^{-4}	4.00×10^{-5}
	50	2.10×10^{-3}	1.03×10^{-3}	2.01×10^{-4}	2.00×10^{-5}
	100	1.05×10^{-3}	5.13×10^{-4}	1.00×10^{-4}	1.00×10^{-5}
	1000	1.05×10^{-4}	5.13×10^{-5}	1.01×10^{-5}	1.00×10^{-6}
	100 000	1.05×10^{-6}	5.13×10^{-7}	1.01×10^{-7}	1.00×10^{-8}
	1 000 000	1.05×10^{-7}	5.13×10^{-8}	1.01×10^{-8}	1.00×10^{-9}
	10 000 000	1.05×10^{-8}	5.13×10^{-9}	1.01×10^{-9}	1.00×10^{-10}
	50 000 000	2.11×10^{-9}	1.03×10^{-9}	2.01×10^{-10}	2.00×10^{-11}
U.S. (1999)	260 000 000	4.05×10^{-10}	1.97×10^{-10}	3.87×10^{-11}	3.85×10^{-12}
U.S. (2005)	300 000 000	3.51×10^{-10}	1.71×10^{-10}	$\mathbf{3.35 \times 10^{-11}}$	3.33×10^{-12}
	1 000 000 000	1.05×10^{-10}	5.13×10^{-11}	1.01×10^{-11}	1.00×10^{-12}
World pop	6 000 000 000	1.76×10^{-11}	8.55×10^{-12}	1.68×10^{-12}	1.67×10^{-13}

A statement provided with a report involving a source attribution might include the following words: 'In the absence of identical twins or close relatives, it can be concluded to a reasonable scientific certainty that the DNA from (x) and from (y) came from the same individual' or 'Reasonable scientific certainty means that you are ($x\%$) certain that you would not see this profile in a sample of (y) unrelated individuals.'

It should be pointed out that if the possibility exists that a close relative of the accused had access to the crime scene and may have been a contributor of the evidence, then the best action is to obtain a reference sample from the relative (DAB 2000). This scenario should be sufficient probable cause for obtaining a reference sample, typing it with the same STR markers as the evidence, and using this information to resolve the question of whether or not the relative carries the same DNA profile as the accused.

OTHER TOPICS OF INTEREST

MATCH OBTAINED FROM A DNA DATABASE SEARCH

The development of national DNA databases filled with both convicted offenders and unsolved casework samples (see Chapter 18) permits searches for matches between evidentiary and database profiles. The NRC II report recommendation 5.1 advocates that the random match probability should be multiplied by N, the number of persons in the database (see Appendix VI, Stockmarr 1999). The DNA Advisory Board endorsed this NRC II report recommendation (DAB 2000, see Appendix V).

USING THE COUNTING METHOD WITH LINEAGE MARKERS

Lineage markers include mitochondrial DNA (see Chapter 10) and Y chromosome haplotypes (see Chapter 9) that are transferred directly from generation to generation either from mother to child in the case of mitochondrial DNA or from father to son in the case of the Y chromosome. As described in the earlier chapters, the counting method in conjunction with an upper bound confidence limit is typically used with mtDNA or Y chromosome haplotypes. The counting method relies on the size of the database and involves counting the number of times the profile (haplotype) has been observed within the database. A frequency estimate with a confidence interval is then made based on this count (see D.N.A. Box 10.3).

REFERENCES AND ADDITIONAL READING

Balding, D.J. (1999) When can a DNA profile be regarded as unique? *Science and Justice*, 39, 257–260.

Balding, D.J. and Donnelly, P. (1994) The prosecutor's fallacy and DNA evidence. *Criminal Law Review*, 711–721.

Balding, D. and Donnelly, P. (1995) Inferring identity from DNA profile evidence. *Proceedings of the National Academy of Sciences U.S.A.*, 92, 11741–11745.

Balding, D.J. and Donnelly, P. (1996) Evaluating DNA profile evidence when the suspect is identified through a database search. *Journal of Forensic Sciences*, 41, 603–607.

Brenner, C.H. (2003) Forensic genetics: mathematics. In Cooper, D.N. (ed) *Nature Encyclopedia of the Human Genome,* Volume 2, pp. 513–519. New York: Macmillan Publishers Ltd, Nature Publishing Group.

Budowle, B., Chakraborty, R., Carmody, G. and Monson, K.L. (2000) Source attribution of a forensic DNA profile. *Forensic Science Communications*, 2 (3). Available online at: http://www.fbi.gov/hq/lab/fsc/backissu/july2000/source.htm.

Budowle, B., Shea, B., Niezgoda, S. and Chakraborty, R. (2001) *Journal of Forensic Sciences*, 46, 453–489.

Chakraborty, R. (1992) *Human Biology*, 64, 141–159.

Chakraborty, R., Stivers, D.N., Su, Y. and Budowle, B. (1999) *Electrophoresis*, 20, 1682–1696.

Collins, A. and Morton, N.E. (1994) Likelihood ratios for DNA identification. *Proceedings of the National Academy of Sciences U.S.A.*, 91, 6007–6011.

DNA Advisory Board (2000) Statistical and population genetic issues affecting the evaluation of the frequency of occurrence of DNA profiles calculated from pertinent population database(s) (approved 23 February 2000). *Forensic Science Communications*, July 2000. Available at: http://www.fbi.gov/programs/lab/fsc/backissu/july2000/dnastat.htm; see also Appendix V.

Evett, I.W. and Buckleton, J.S. (1996) Statistical analysis of STR data. In Carracedo, A., Brinkmann, B. and Bar, W. (eds) *Advances in Forensic Haemogenetics 6,* pp. 79–86. New York: Springer-Verlag.

Evett, I.W. and Weir, B.S. (1998) *Interpreting DNA Evidence: Statistical Genetics for Forensic Scientists.* Sunderland, MA: Sinauer Associates.

Foreman, L.A. and Evett, I.W. (2001) *International Journal of Legal Medicine*, 114, 147–155.

Foreman, L.A., Champod, C., Evett, I.W., Lambert, J.A. and Pope, S. (2003) Interpreting DNA evidence: a review. *International Statistical Review*, 71, 473–495.

Gill, P., Foreman, L., Buckleton, J.S., Triggs, C.M. and Allen, H. (2003) *Forensic Science International*, 131, 184–196.

Henderson, J.P. (2002) *Forensic Science International*, 128, 183–186.

Krawczak, M. (1999) Statistical inference from DNA evidence. In Epplen, J.T. and Lubjuhn, T. (eds) *DNA Profiling and DNA Fingerprinting*, pp. 229–244. Basel, Switzerland: Birkhauser Verlag.

National Research Council Committee on DNA Forensic Science (1996) *The Evaluation of Forensic DNA Evidence.* National Academy Press: Washington, DC; usually referred to as NRCII; recommendations listed in Appendix VI.

Stockmarr, A. (1999) *Biometrics*, 55, 671–677.

Taroni, F., Lambert, J.A., Fereday, L. and Werrett, D.J. (2002) *Science and Justice*, 42, 21–28.

Thompson, W.C. and Schumann, E.L. (1987) *Law Human Behavior*, 11, 167–187.

Tracey, M. (2001) *Croatian Medical Journal*, 42, 233–238.

Weir, B.S. (1994) *Annual Reviews of Genetics*, 28, 597–621.

Weir, B.S. (1996) *Genetic Data Analysis II: Methods for Discrete Population Genetic Data.* Sunderland, MA: Sinauer Associates.

Weir, B.S. (1997) The co-ancestry coefficient in forensic science. *Proceedings from the Eighth International Symposium on Human Identification,* pp. 87–91. Madison, Wisconsin: Promega Corporation.

Weir, B.S. (2003) Forensics. In Balding, D.J., Bishop, M. and Cannings, C. (eds) *Handbook of Statistical Genetics,* 2nd Edition, pp. 830–852. Hoboken, NJ: John Wiley & Sons.

Weir, B.S. (2003b) DNA evidence: inferring identity. In Cooper, D.N. (ed) *Nature Encyclopedia of the Human Genome,* Volume 2, pp. 85–88. New York: Macmillan Publishers Ltd, Nature Publishing Group.

CHAPTER 22

APPROACHES TO STATISTICAL ANALYSIS OF MIXTURES AND DEGRADED DNA

Any truth is better than indefinite doubt.

(Sherlock Holmes, *The Yellow Face*)

For every complex problem, there is a simple solution... and it is wrong.

(Henry Louis Mencken, Mencken's Law)

The points that have been addressed in Chapters 19–21 thus far involve obtaining a 'clean' full DNA profile from a single source. However, crime scene evidence can produce mixed DNA profiles from more than one individual (see Chapter 7). These mixtures can be challenging to interpret and depending on the alleles present in the profile, the individual components may never be unambiguously separated. In addition, partial profiles where entire loci have dropped out sometimes occur due to the presence of degraded DNA or polymerase chain reaction (PCR) inhibitors (see Chapter 7).

As pointed out in Chapter 7, many forensic cases involve multiple pieces of DNA evidence and not all of these will be mixtures. Thus, if additional samples can be tested that are easier to interpret they should be sought after versus complicated mixtures. Fortunately, mixtures do not represent a majority of cases especially if a good differential extraction is performed in a sexual assault case and the sperm fraction can be fully separated from the victim's DNA (see Chapter 3). As an example, over a four-year period, one forensic laboratory worked 1547 criminal cases that involved a total of 2424 samples, yet only 163 showed a mixed profile or 6.7% (Torres *et al.* 2003).

MIXTURE INTERPRETATION

Mixtures are DNA samples containing two or more contributors. As described in Chapter 7, the presence of a mixture may be ascertained when three or more alleles are observed at multiple loci or notable differences in allele intensities are detected in a short tandem repeat (STR) profile. Sometimes it is relatively straightforward to determine a contributor profile, such as when DNA analysis from an intimate swab reveals a mixture consistent with the victim and perpetrator.

As stated in earlier chapters, there is no one universal formula that fits all situations when interpreting DNA evidence. This reality certainly applies for mixture interpretation and reporting. Various laboratories have adopted different approaches to these challenging situations. Ladd *et al.* (2001) discuss four primary approaches to statistical evaluation of DNA mixtures: (1) issuing a qualitative statement involving no calculations and simply stating that the suspect is included or excluded from being a possible contributor to the mixture observed; (2) performing match probability estimations after deducing the possible genotypes of the contributors; (3) using exclusion probabilities; and (4) performing likelihood ratio calculations. These four approaches will be discussed below in the context of example data (Figure 22.1).

QUALITATIVE ASSESSMENT OF A MIXTURE

By far the most conservative approach to dealing with a DNA mixture result is to simply make a qualitative assessment that the suspect cannot be excluded as having contributed to the crime scene evidence. In the example shown in Figure 22.1, suspects possessing genotypes 11,14 or 11,12 or 11,11 could all have contributed to the evidence possessing alleles 11, 12 and 14 at the STR locus D13S317. Alternatively, if a suspect's DNA profile contains alleles not

Figure 22.1

Hypothetical example of a mixture with STR data at the D13S317 locus. The crime scene evidence sample possesses alleles 11, 12 and 14.

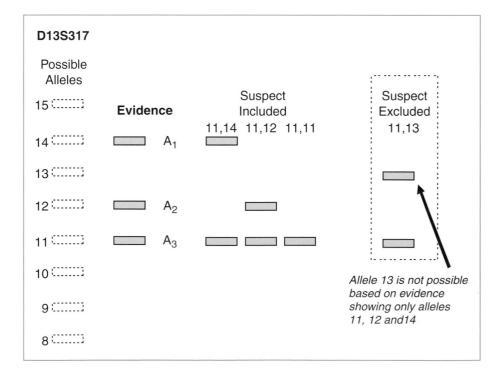

present in the mixture profile (e.g., allele 13 in Figure 22.1), then the suspect can be excluded (provided there is no evidence of low-copy number conditions that might cause allele dropout). Of course, this type of inclusion/exclusion evaluation would be conducted at all loci examined in DNA profiles from the evidentiary mixture and the suspects.

After reviewing DNA results across all profiles, a casework report with a qualitative assessment might state that 'the defendant cannot be excluded as a contributor to the mixed DNA profile.' In other words, the defendant's genotype is completely represented in the alleles observed in the mixed DNA profile. However, many other people not considered as suspects in the case may also possess DNA profiles that are not excluded particularly if only a few genetic loci are examined. This is not as much of a problem now with examination of multi-allelic polymorphic STR markers as it was with previous DNA tests such as HLA-DQA1 and PolyMarker that only had a few possible alleles.

While presenting this simple qualitative statement of inclusion or exclusion rather than performing any calculations is definitely a conservative approach, it is often not satisfactory in a court of law where a judge typically requires some kind of numerical estimate to give statistical weight to the evidence. As pointed out in *The Evaluation of Forensic DNA Evidence*, 'to make appropriate use of DNA technology in the courtroom, the trier of fact must give the DNA evidence appropriate weight' (NRC II, p. 203). Approaches to providing a statistic to the DNA mixture are provided below.

MATCH PROBABILITY ESTIMATIONS

In some cases it may be possible to confidently pull apart the alleles from individual contributors to a mixture. In cases of sexual assault, the victim's DNA profile is typically compared to the mixed profile and this comparison may be helpful in identifying the STR alleles present in the perpetrator's DNA profile.

The DNA Advisory Board recommendations on statistics issued in February 2000 state that 'when intensity differences are sufficient to identify the major contributor in the mixed profile, it can be treated statistically as a single source sample' (DAB 2000; see Appendix V). In such a situation, after deciphering the individual components for the major and minor contributors, the statistical treatment of their profiles could be conducted as described in Chapter 21 for single-source samples.

Interpretation of genotypes present in a mixture is much more complicated when the contributions of the donors is approximately equal and thus a major contributor cannot be definitively determined or when true alleles for a contributor are masked by stutter products (see Chapter 6) or other alleles in the mixture (DAB 2000). It is not always possible to unambiguously determine all of the alleles present in a mixture especially with a partial profile from a degraded DNA sample. Likewise, it is not always possible to infer the complete genotypes of all contributors with a high degree of confidence because the mixture combination may be too complex to easily decipher. Perhaps in the future computer programs will exist that will be able to pull apart mixture components with a high degree of reliability (see Perlin and Szabady 2001, Wang *et al.* 2002). Until that time, conservative approaches such as the calculation of exclusion probabilities and likelihood ratios will likely be used.

EXCLUSION PROBABILITIES

Complex DNA mixtures can be conservatively interpreted using the combined probability of exclusion (CPE) (Devlin 1992, DAB 2000, Ladd *et al.* 2001). The probability of exclusion provides an estimate of the portion of the population that has a genotype composed of at least one allele not observed in the mixed profile (DAB 2000). If evidence includes three alleles at a locus (A_1, A_2, A_3), then:

$$p = A_1 + A_2 + A_3$$
$$\text{and } q = 1 - p$$
$$\text{so PE} = 2\,pq + q^2.$$

The probability of exclusion (PE) at each locus is first determined and then PEs from multiple loci are combined through the equation listed below:

$$\text{Combined PE} = 1 - [(1 - \text{PE1}) \times (1 - \text{PE2}) \times \ldots \times (1 - \text{PEn})]$$

In the example shown in Figure 22.2, determination of CPE would involve calculation of the frequency of genotypes that do not possess the alleles 11, 12 or 14 present in the evidentiary mixed profile. The CPE approach avoids the potential pitfalls associated with extrapolating the genotypes of mixture contributors (e.g., overlooking an allele because it is in a stutter position). No prior knowledge or assumptions regarding the number of possible contributors to the mixture are needed and results can still be reported without knowledge of a known profile, such as the victim's profile.

While calculating the combined probability of exclusion is not as powerful of a technique as the likelihood ratio method that is discussed in the next section, a major advantage with the CPE approach is that the number of contributors to the crime scene DNA profile does not need to be taken into account. Simply all other alleles not observed in the stain are considered. This approach is rather conservative because an individual can be excluded if he has any allele at any locus that is not detected in the stain.

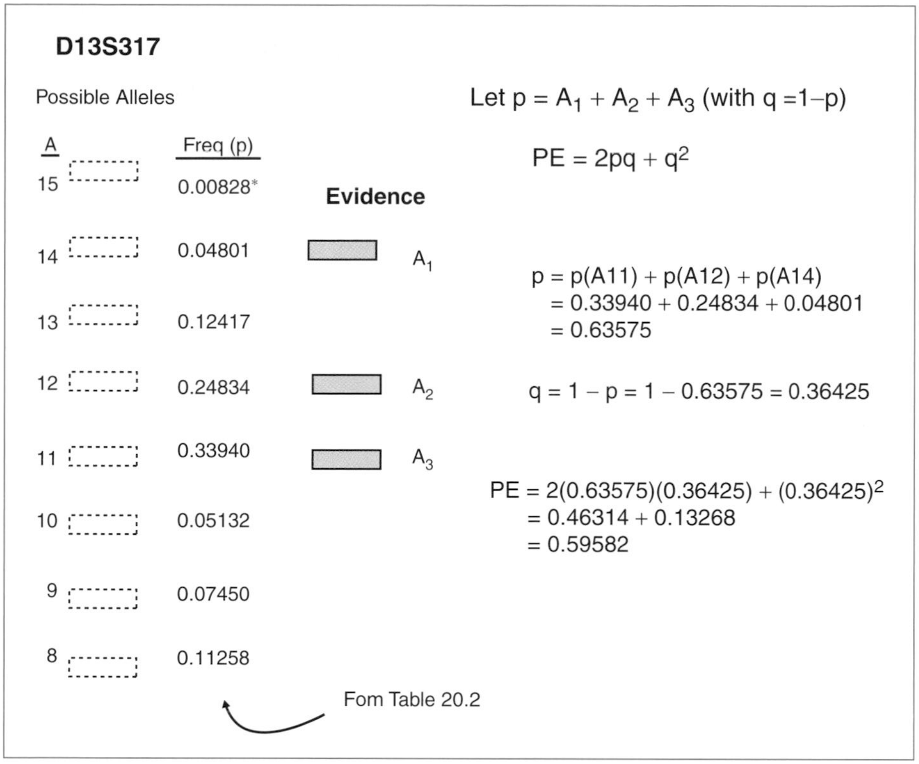

Figure 22.2

Calculation of probability of exclusion with a mixture example.

LIKELIHOOD RATIO APPROACH

The likelihood ratio (LR) is the ratio of possibilities under alternative propositions and provides a reliable method that is able to make full use of available genetic data (Evett and Weir 1998). Two competing hypotheses are setup, the hypothesis of the prosecution (H_p), which is that the defendant committed the crime, and the hypothesis of the defense (H_d) that some unknown individual committed the crime. Thus, the likelihood ratio involves a ratio describing the probability of the evidence given the prosecution's hypothesis over the probability of the evidence given the defense's hypothesis:

$$LR = \frac{P(E \mid H_p)}{P(E \mid H_d)}$$

Unfortunately, determination of which hypotheses to consider is not necessarily straightforward (Ladd *et al.* 2001). Interpretation of a mixture depends on the circumstances of the case and involves assumptions about the identity and number of contributors to the mixture in question (DAB 2000). LR calculations are more widely used in Europe than the United States for forensic applications. Paternity testing on the other hand routinely uses

LR calculations. In the end though, the LR method makes better use of the available genetic data than does the probability of exclusion method discussed previously.

If evidence contains four alleles at a locus (A_1, A_2, A_3, A_4) and the victim possesses A_3 and A_4 while the suspect exhibits A_1 and A_2, then the prosecution's hypothesis would be that the DNA evidence is from the victim and the suspect. On the other hand, the defense's hypothesis would be that the DNA evidence is from the victim and an unknown person. The probability of the prosecution's hypothesis is one because their position is that they are 100% confident (probability = 1) that the defendant committed the crime, which is why the trial is occurring in the first place. The defense's hypothesis can vary depending on the circumstances of the case, such as the number of other possible contributors under consideration and the alleles present in the evidentiary DNA profile. In the end, the likelihood ratio describes the relative chance of observing a specific mixture and combination of STR alleles. Some LR examples for various scenarios are listed in Table 22.1.

Using the example shown in Figure 22.1 of evidence containing alleles 11, 12 and 14 at the STR locus D13S317 and a suspect with alleles 11 and 14, we can use the second formula in Table 22.1 to deduce the likelihood ratio for the scenario where three peaks are seen in the crime scene evidence and two of those alleles are present in the suspect. In this case the LR is 8.23 with $p_1 = 0.24834$ (allele frequency of D13S317 allele 12 from Caucasians in Appendix II), $p_2 = 0.33940$ (allele frequency of D13S317 allele 11 from Caucasians in Appendix II), and $p_3 = 0.04801$ (allele frequency of D13S317 allele 14 from Caucasians in Appendix II).

Table 22.1

Likelihood ratios for mixed stains using p^2 rule for homozygotes (from Table 5.1 of NRC II, p. 163). Allele frequencies (p) for observed (A) can be obtained from population databases, such as Appendix II.

Crime Scene	Suspect	Likelihood Ratio
$A_1A_2A_3A_4$	A_1A_2	$\dfrac{1}{12p_1p_2}$
$A_1A_2A_3$	A_2A_3	$\dfrac{p_1+2p_2+2p_3}{12p_2p_3(p_1+p_2+p_3)}$
$A_1A_2A_3$	A_1	$\dfrac{1}{6p_1(p_1+p_2+p_3)}$
A_1A_2	A_1A_2	$\dfrac{(p_1+p_2)^2}{2p_1p_2(3p_1p_2+2p_1^2+2p_2^2)}$
A_1A_2	A_1	$\dfrac{2p_1+p_2}{2p_1(3p_1p_2+2p_1^2+2p_2^2)}$

ADDITIONAL THOUGHTS ON MIXTURES

Mixtures can be quite complicated to interpret, such as in the case of People of the State of California versus Orenthal James Simpson (Weir and Buckleton 1996). However, that particular case would probably have been simplified if STR testing had been available at the time since STRs have more possible alleles than the dot blot methods used to recover DNA evidence in the O.J. Simpson trial.

Evett and Weir (1998) note that the essence of mixture interpretation is to first identify the alleles in the crime scene evidence sample and alleles carried by the known contributor(s) to the sample, such as the victim. Then any alleles present in the evidence sample that are not provided by the known contributor(s) must be carried by one or more unknown contributors, which may or may not include the suspect.

The DNA Advisory Board recommends that either or both CPE and LR calculations be performed whenever feasible in the event of a mixture (DAB 2000). However, there will be mixture results that due to low-copy number stochastic limits, DNA template degradation or PCR inhibition, no interpretation of the profile can be made. In the end, as pointed out in Chapter 15, the interpretation of results in forensic casework, whether arising from single-source samples or mixtures, is a matter of professional judgment and expertise.

Mixtures will be complicated by the fact that some loci will possess intensity differences that permit contributors to be deciphered while other loci may not be fully interpretable due to overlapping allele combinations (see Figure 7.5). With STRs and peak intensity differences, some loci may be interpretable so that contributors can be statistically treated as single sources, while other loci may be too complex to confidently attribute alleles to their sources. Thus, when performing mixture interpretation do everything possible to first eliminate artifacts such as stutter products from consideration and then interpret remaining alleles to determine how many contributors are present.

Other challenges with interpreting mixtures involve taking into account the ethnicity of contributors (Fung and Hu 2000, 2001, 2002a, 2002b) and assessment of DNA mixtures with the presence of relatives (Hu and Fung 2003a). A computer program has been written to help in evaluating forensic DNA mixtures involving contributors from different ethnic origins (Hu and Fung 2003b). Probabilistic expert systems are also in development for aid in DNA mixture resolution (Mortera *et al.* 2003).

DETERMINING THE NUMBER OF CONTRIBUTORS TO A MIXTURE

Typically assumptions are made as to the number of contributors possible in a mixture. The vast majority of cases most laboratories face are two-person

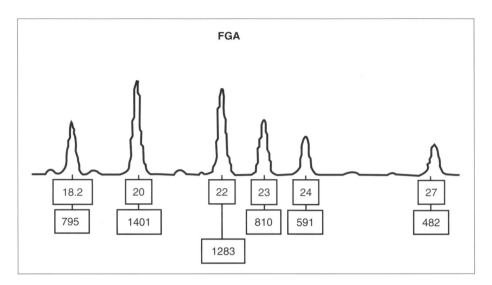

mixtures where one of the contributors is known (e.g., the victim in a sexual assault case). However, situations with gang rape become much more complicated. Examining results in the context of the entire DNA profile is important in terms of assessing the number of contributors. Frequently, the more polymorphic loci with more allele possibilities, such as D18S51 and FGA, will provide the best chance of determining the number of contributors to a mixture (Figure 22.3). Evett and Weir (1998) point out that when only a few of the possible alleles are present in a mixture profile then it becomes less probable that a large number of contributors will have that set of alleles in common between them (pp. 204–205).

PARTIAL DNA PROFILES

Interpretation of a DNA profile can only be performed on loci that have results. Unfortunately, with degraded DNA specimens or low-copy number samples (see Chapter 7) the PCR amplification may fail to generate signals above the detection threshold of the instrument and individual alleles and entire loci may be lost from the final DNA profile. Foreman and Evett (2001) note that partial profiles occur in approximately 20% of cases seen by the Forensic Science Service. Given that it is often not possible to know what alleles should have been present had the sample not been degraded, the standard practice is to interpret only the detected alleles as described in Chapter 21.

Of course, obtaining matching alleles between a full-profile suspect and a partial-profile evidentiary sample is not as powerful as a full-profile to full-profile match. However, any data is better than none. Even if results are obtained on only a few STR loci, this information can provide ample assistance to either include or exclude the suspect and therefore aid in resolving the case.

Occasionally results from additional loci may be recovered from degraded DNA samples through the use of validated miniSTR primer sets (see Chapter 7) or other genetic systems such as single nucleotide polymorphisms (see Chapter 8) that amplify smaller regions of the DNA template. Finally, in most cases, the forensic sample has been divided into two or more parts so that unused portions are retained to permit additional tests as desired by the court according to NRC II recommendation 3.3 (see Appendix VI). These retained samples can be tested as occasion warrants in order to verify previous test results.

REFERENCES AND ADDITIONAL READING

Brenner, C.H. (1997) Proof of a mixed stain formula of Weir. *Journal of Forensic Sciences*, 42, 221–222.

Curran, J.M., Triggs, C.M., Buckleton, J. and Weir, B.S. (1999) *Journal of Forensic Sciences*, 44, 987–995.

Devlin, B. (1992) *Statistical Methods in Medical Research*, 2, 241–262.

DNA Advisory Board (2000) Statistical and population genetic issues affecting the evaluation of the frequency of occurrence of DNA profiles calculated from pertinent population database(s) (approved 23 February 2000). *Forensic Science Communications*, July 2000. Available at: http://www.fbi.gov/programs/lab/fsc/backissu/july2000/dnastat.htm; printed in Appendix V.

Evett, I.W. and Weir, B.S. (1998) *Interpreting DNA Evidence: Statistical Genetics for Forensic Scientists.* Sunderland, MA: Sinauer Associates.

Foreman, L.A. and Evett, I.W. (2001) *International Journal of Legal Medicine*, 114, 147–155.

Foreman, L.A., Champod, C., Evett, I.W., Lambert, J.A. and Pope, S. (2003) *International Statistical Review*, 71, 473–495.

Fung, W.K. and Hu, Y.-Q. (2000) *Forensic Science Communications*, 2 (4). Available online at: http://www.fbi.gov/hq/lab/fsc/backissu/oct2000/fung.htm.

Fung, W.K. and Hu, Y.Q. (2001) *International Journal of Legal Medicine*, 115, 48–53.

Fung, W.K. and Hu, Y.-Q. (2002a) *Statistics in Medicine*, 21, 3583–3593.

Fung, W.K. and Hu, Y.Q. (2002a) *International Journal of Legal Medicine*, 116, 79–86.

Hu, Y.Q. and Fung, W.K. (2003a) *International Journal of Legal Medicine*, 117, 39–45.

Hu, Y.Q. and Fung, W.K. (2003b) *International Journal of Legal Medicine*, 117, 248–249.

Ladd, C., Lee, H.C., Yang, N. and Bieber, F.R. (2001) *Croatian Medical Journal*, 42 (3), 244–246.

Lauritzen, S.L. and Mortera, J. (2002) *Forensic Science International*, 130, 125–126.

Mortera, J., Dawid, A.P. and Lauritzen, S.L. (2003) *Theoretical Population Biology*, 63, 191–205.

National Research Council Committee on DNA Forensic Science (1996) *The Evaluation of Forensic DNA Evidence.* National Academy Press: Washington, DC; usually referred to as NRCII; recommendations listed in Appendix VI.

Perlin, M.W. and Szabady, B. (2001) *Journal of Forensic Sciences*, 46, 1372–1378.

Tomsey, C.S., Kurtz, M., Flowers, B., Fumea, J., Giles, B. and Kucherer, S. (2001) *Croatian Medical Journal*, 42, 276–280.

Torres, Y., Flores, I., Prieto, V., Lopez-Soto, M., Farfan, M.J., Carracedo, A. and Sanz, P. (2003) *Forensic Science International,* 134, 180–186.

Wang, T., Xue, N. and Wickenheiser, R. (2002) Least-square deconvolution (LSD): a new way of resolving STR/DNA mixture samples. *Proceedings of the Thirteenth International Symposium on Human Identification.* Available at: http://www.promega.com/ geneticidproc/ussymp13proc/contents/wang.pdf.

Weir, B.S. (1996) *Genetic Data Analysis II: Methods for Discrete Population Genetic Data.* Sunderland, MA: Sinauer Associates, pp. 221–225.

Weir, B.S. and Buckleton, J.S. (1996) Statistical issues in DNA profiling. In Carracedo, A., Brinkmann, B., and Bar, W. (eds) *Advances in Forensic Haemogenetics 6,* pp. 457–464. New York: Springer-Verlag.

Weir, B.S. (2003) Forensics. In Balding, D.J., Bishop, M. and Cannings, C. (eds) *Handbook of Statistical Genetics,* 2nd Edition, pp. 830–852. Hoboken, New Jersey: John Wiley & Sons.

Weir, B.S., Triggs, C.M., Starling, L., Stowell, L.I., Walsh, K.A. and Buckleton, J. (1997) *Journal of Forensic Sciences*, 42, 213–222.

CHAPTER 23

KINSHIP AND PARENTAGE TESTING

Recommendation 4.4: If the possible contributors of the evidence sample include relatives of the suspect, DNA profiles of those relatives should be obtained. If these profiles cannot be obtained, the probability of finding the evidence profile in those relatives should be calculated...

(NRCII 1996)

In this chapter we will examine the situation when DNA samples being compared are from related individuals. Since many alleles will be shared when samples are from related individuals, different statistical equations must be applied. In Chapter 21 equations were introduced to correct for relatedness in determining frequency estimates of genotypes and DNA profiles. Tables 21.6 and 21.7 worked through example data to determine the frequency estimates from full siblings, parents, half siblings, and cousins based on NRC II recommendation 4.4 (see Appendix VI).

Besides its use in criminal investigations, DNA data plays an important role in parentage and kinship testing. Different questions are usually being asked in parentage testing rather than in criminal casework where a direct match is being considered between evidence and suspect. We will consider several applications involving DNA evidence from related individuals including traditional parentage testing that usually involves addressing questions of paternity (i.e., who is the father?) and missing persons and mass disaster investigations that involve reverse parentage analysis (i.e., could these sets of remains have come from a child of these reference samples?). Immigration cases also involve kinship testing to determine if an individual could have a proposed relationship to reference samples. Figure 23.1 illustrates the questions posed with parentage and reverse parentage analysis.

In the case of people who are close relatives, such as parents and offspring, full siblings, half siblings, and cousins, we can use the model of Mendelian segregation (see Chapter 19) to calculate the amount of shared genetic information. Once we establish the expected conditional probabilities between two relatives we can either calculate an exclusion probability, or more precisely a likelihood ratio expressing how much more likely it is that we would see the

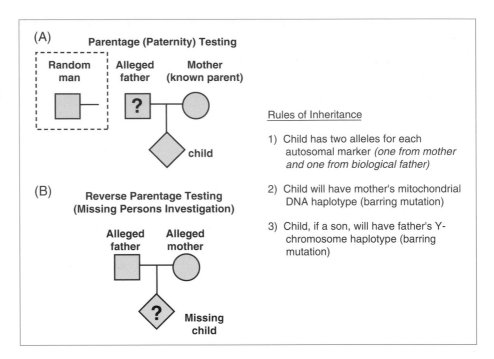

DNA evidence under the hypothesis that they came from people with a specific relationship as opposed to the hypothesis that they came from two ostensibly unrelated people.

This approach has a long-standing precedent in the calculations for paternity determination where an exclusion probability can be calculated to express how rare it would be to find a random man who could not be excluded as the biological father of the child. In 1938 the first publication (Essen-Möller 1938) appeared that developed the theoretical framework for calculating what we now refer to as the paternity index (PI). Since that time it has become common to calculate a likelihood ratio that quantifies the DNA evidence under two competing hypotheses.

PATERNITY (PARENTAGE) TESTING

Every year in the United States more than 300 000 paternity cases are performed where the identity of the father of a child is in dispute (The American Association of Blood Banks [AABB] 2003). These cases typically involve the mother, the child, and one or more alleged fathers. In 2002, almost one million samples were run for this purpose in the United States alone (AABB 2003). A number of different laboratories perform parentage testing (see Appendix III).

The determination of parentage is made based on whether or not alleles are shared between the child and the alleged father when a number of genetic markers are examined. Thus, the outcome of parentage testing is simply inclusion or exclusion. Paternity testing laboratories often utilize the same short tandem repeat (STR) multiplexes and commercial kits as employed by forensic testing laboratories (see Chapter 5). However, rather than looking for a complete one-to-one match in a DNA profile, the source of the non-maternal or 'obligate paternal allele' at each genetic locus is under investigation.

The basis of paternity comes down to the fact that in the absence of mutation a child receives one allele matching each parent at every genetic locus examined (Figure 23.2). Thus, parents with genotypes 11,14 (father) and 8,12 (mother) may produce offspring with the following types: 8,11 8,14 11,12 and 12,14. Inversely, if the mother's genotype is known to be 8,12 and the children possess alleles 8, 11, 12, and 14, then we may deduce that the father contributed alleles 11 and 14 – provided of course that the same individual fathered all of the children.

The obligate paternal allele for each child in this example is shown in Figure 23.2. In this particular example, the parents had non-overlapping alleles. Paternity testing becomes more complicated when mother and father share

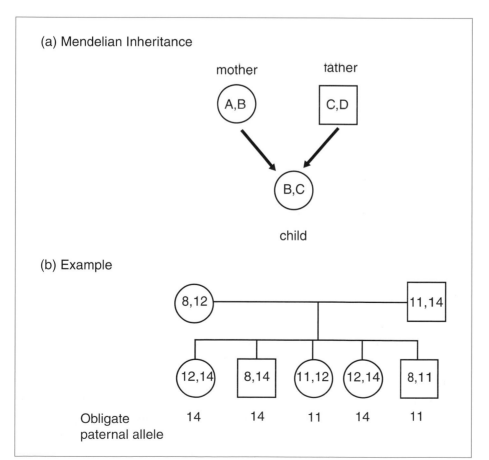

Figure 23.2

(a) Mendelian inheritance patterns with a mother possessing alleles A and B contributing one of them to the child while the father who possesses alleles C and D also contributes one of his alleles to the child. (b) An example pedigree for a family where the parents possess different alleles enabling identification of the obligate paternal allele in each of the children. This scenario can become more complicated to interpret if mutations occur or maternal and paternal alleles are shared.

alleles, but the logic remains the same in calculating exclusion probability and the paternity index likelihood ratio described below.

STATISTICAL CALCULATIONS

If the man tested cannot be excluded as the biological father of the child in question, then statistical calculations are performed to aid in understanding the strength of the match. The most commonly applied test in this regard is the paternity index.

The *paternity index* (PI) is the ratio of two conditional probabilities where the numerator assumes paternity and the denominator assumes a random man of similar ethnic background was the father. The numerator is the probability of observed genotypes, given the tested man is the father, while the denominator is the probability of the observed genotypes, given that a random man is the father. The paternity index then is a likelihood ratio of two probabilities conditional upon different competing hypotheses. This likelihood ratio reflects how many times more likely it is to see the evidence (e.g., a particular set of alleles) under the first hypothesis compared to the second hypothesis. When mating is random, the probability that the untested alternative father will transmit a specific allele to his child is equal to the allele frequency in his race (see Eisenberg 2003).

The PI is generally represented in the formula X/Y, where X is the chance that the alleged father (AF) could transmit the obligate allele and Y is the chance that some other man of the same race could have transmitted the allele. Typically, X is assigned the value of 1 if the AF is homozygous for the allele of interest and 0.5 if the AF is heterozygous. A population database containing frequency distributions for the various alleles at the tested genetic markers, such as Appendix II, is used to calculate the potential of a randomly selected man passing the obligate allele to the child.

The PI is calculated for each locus and then individual PI values are multiplied together to obtain the combined paternity index (CPI) for the entire set of genetic loci examined. The generally accepted minimum standard for an inclusion of paternity is a PI of 100 or greater (Coleman and Swenson 2000). A PI of 100 correlates to the probability that the alleged father has a 99 to 1 better chance of being the father than a random man.

Another statistical test performed in paternity testing is the *exclusion probability*, which is the combined frequency of all genotypes that would be excluded if the pedigree relationships were true assuming Hardy–Weinberg equilibrium. Computer programs used for statistical calculations in parentage testing include DNA View (http://www.dna-view.com/dnaview.htm), Familias (http://www.math.chalmers.se/~mostad/familias/), and EasyDNA (Fung 2003, Fung *et al.* 2004).

IMPACT OF MUTATIONAL EVENTS

Mutations happen and must be accounted for in parentage investigations. In Chapter 6, we discussed the fact that mutation rates for STR loci are on the order of 1–4 per thousand meioses or 0.1–0.4% (see Table 6.3). The American Association of Blood Banks (AABB) has issued standards for parentage testing laboratories regarding mutations. The AABB standards recognize that mutations are naturally occurring genetic events and standard 5.4.2 states that the mutation frequency at a given locus shall be documented. Furthermore, standard 6.4.1 emphasizes that an opinion of non-paternity shall not be rendered on the basis of an exclusion at a single DNA locus (single inconsistency).

The 'two exclusion' rule is commonly accepted in parentage testing laboratories. In other words, if two genetic loci do not match between an alleged father and a child, the alleged father cannot be excluded as being the true biological father. It is important to keep in mind that the more genetic systems examined the greater the chance of a random mutation to be observed. With STR analysis often examining a battery of a dozen or more loci, it is not uncommon to see two inconsistencies between a child and the true biological father (Gunn *et al.* 1997, Nutini *et al.* 2003).

REFERENCE SAMPLES

A variety of reference samples can be used in kinship and parentage testing. Figure 23.1 mentions several rules of inheritance that are valuable in parentage testing. Mitochondrial DNA (see Chapter 10) and Y chromosome loci (see Chapter 9) permit evaluation of samples from different genetic angles and also extend the range of reference samples that are possible to more distantly related relatives (see Figures 9.3 and 10.2).

Unfortunately, complete parentage trios are not always available. Sometimes the mother's DNA sample may not be included in a case or the father is not available for testing (e.g., the Thomas Jefferson case described in Chapter 9). Even the child may not be available in some cases. While more statistical uncertainty can arise in these cases, they can still be brought to a reasonable degree of resolution. In addition, mitochondrial DNA or Y chromosome results can help with confirming relationships. A combination of samples from more than one close relative can help provide greater confidence in this kinship analysis (see D.N.A. Box 23.1).

REVERSE PARENTAGE TESTING

In identification of remains as part of missing persons investigations or mass disaster victim identification work (see Chapter 24), the question under

'We've got him!' were the words of Paul Bremer, U.S. governor in Iraq, at a press conference on 14 December 2003. From the beginning of the war in Iraq in March 2003, one of the stated missions of the United States military was to kill or capture Saddam Hussein, the dictator of Iraq, in order to remove him from power after more than two decades of threatening the world and terrorizing his own people. However, Saddam was known to have many 'stunt doubles' to protect his life from assassins. Therefore, the ability to verify his identity through genetic testing was essential to knowing that the United States in fact 'had their man.' Forensic DNA testing using short tandem repeat (STR) markers played an important role in the identification effort behind the words of Paul Bremer.

Validated reference samples are required in any missing persons investigation or paternity testing when kinship is being verified through similarities in nuclear or mitochondrial DNA profiles. In this case, DNA from Saddam's two sons provided the family reference samples. Saddam's sons Uday and Qusay were killed in a gunfight near Mosul, Iraq on 22 July 2003. DNA samples were collected from their remains shortly after they were killed for use as reference samples in verifying the identity of their father if he was ever located. Both autosomal and Y chromosome STR profiles were generated from Uday and Qusay's biological samples.

Shortly after Saddam was captured in a small hole underneath a farmhouse near Tiqrit, Iraq, in December 2003, blood and hair samples were flown to the United States where it was immediately examined by DNA scientists at the Armed Forces DNA Identification Laboratory (AFDIL) located near Washington, DC. Working through the night, several scientists carefully extracted and then amplified the DNA sample using the autosomal STR kit Profiler Plus (see Chapter 5) to obtain a full 13 locus STR profile. These STR profiles possessed alleles in common with the previously generated DNA profiles of Saddam's two sons. Additionally, the Y chromosome STR kit Y-PLEX 6 (see Chapter 9) also showed full allele sharing between Saddam and his two sons indicating that the sample in question was from their same paternal lineage. Saddam's capture is a great relief to many who feared his re-emergence and continued terror.

Source:
Personal communication from AFDIL; *The Scientist* 19 December 2003 news article.

consideration may be whether or not a child belongs to the mother and father tested or other biological references available. This is essentially the opposite question as that asked in parentage testing, namely given a child's genotype, who are the parents? The samples examined may be the same family trio as studied in parentage testing: alleged mother, alleged father, and child. Unfortunately, it is normally a luxury to have samples from both parents available. Typically only a single parent or sibling samples are available making the reverse parentage analysis more challenging.

REFERENCES AND ADDITIONAL READING

Alford, R.L., Hammond, H.A., Coto, I. and Caskey, C.T. (1994) *American Journal of Human Genetics*, 55, 190–195.

AABB (2003) American Association of Blood Banks Annual Report Summary for Testing in 2002 (see http://www.aabb.org/About_the_AABB/Stds_and_Accred/ptannrpt02.pdf).

Birus, I., Marcikic, M., Lauc, D., Dzijan, S. and Lauc, G. (2003) *Croatian Medical Journal*, 44, 322–326.

Brenner and Morris (1989) Paternity index calculations in single locus hypervariable DNA probes: validation and other studies. *Proceedings of the First International Symposium on Human Identification*, pp. 21–53.

Brenner, C.H. (1997) Symbolic kinship program. *Genetics*, 145, 535–542.

Brenner, C.H. and Weir, B.S. (2003) *Theoretical Population Biology*, 63, 173–178.

Brinkmann, B., Pfeiffer, H., Schurenkamp, M. and Hohoff, C. (2001) *International Journal of Legal Medicine*, 114, 173–177.

Cavilli-Sforza, L.L. and Bodmer, W.F. (1971, 1999) *The Genetics of Human Populations*. Mineola, New York: Dover Publications.

Coleman, H. and Swenson, E. (2000) *DNA in the Courtroom: A Trial Watcher's Guide*. Chapter 4: DNA in parentage testing. Available at: http://www.genelex.com/paternitytesting/paternitybook.html.

Donnelly, P. (1995) *Heredity*, 75, 26–34.

Egeland, T. and Mostad, P.F. (2002) *Scandinavian Journal of Statistics*, 29, 1–11.

Eisenberg, A. (2003a) Popstats Parentage Statistics Strength of Genetic Evidence In Parentage Testing. Workshop presented at 14th International Symposium on Human Identification. Available at: http://www.promega.com/geneticidproc/ussymp14proc/stats_workshop.htm.

Eisenberg, A. (2003b) Popstats Relatedness Statistics. Workshop presented at 14th International Symposium on Human Identification. Available at: http://www.promega.com/geneticidproc/ussymp14proc/stats_workshop.htm.

Essen-Möller, E. (1938) Die Beweiskraft der Aehnlichkeit im Vaterschaftsnachweis; theoretische Grundlagen. *Mitt. Anthrop. Ges. (Wein)*, 68, 9–53.

Evett, I.W. and Weir, B.S. (1998) *Interpreting DNA Evidence: Statistical Genetics for Forensic Scientists*. Sunderland, MA: Sinauer Associates.

Fung, W.K., Chung, Y.-K. and Wong, D.-M. (2002) *International Journal of Legal Medicine*, 116, 64–67.

Fung, W.K. (2003) *Forensic Science International*, 136, 22–34.

Fung, W.K., Yang, C.T. and Guo, W. (2004) EasyDNA: user-friendly paternity and kinship testing programs. *Progress in Forensic Genetics 10 (International Congress Series)*, 1261, 628–630.

Gaytmenn, R., Hildebrand, D.P., Sweet, D. and Pretty, I.A. (2002) *International Journal of Legal Medicine*, 116, 161–164.

Gonick, L. and Wheelis, M. (1983) *The Cartoon Guide to Genetics, Updated Edition*. New York: HarperCollins Publishers.

Gunn, P.R., Trueman, K., Stapleton, P. and Klarkowski, D.B. (1997) *Electrophoresis*, 18, 1650–1652.

Hartl, D.L. and Clark, A.G. (1997) *Principles of Population Genetics, Third Edition*. Sunderland, MA: Sinauer Associates.

Hartl, D.L. and Jones, E.W. (1998) *Genetics: Principles and Analysis, Fourth Edition*. Sunderland, MA: Jones and Bartlett Publishers.

Li, C.C. and Sachs, L. (1954) *Biometrics*, 10, 347–360.

Macan, M., Uvodic, P. and Botica, V. (2003) *Croatian Medical Journal*, 44, 347–349.

National Research Council Committee on DNA Forensic Science (1996) *The Evaluation of Forensic DNA Evidence*. Washington, DC: National Academy Press.

Nutini, A.L., Mariottini, A., Giunti, L., Torricelli, F. and Ricci, U. (2003) *Croatian Medical Journal*, 44, 342–346.

Pena, S.D.J. and Chakraborty, R. (1994) *Trends in Genetics*, 10 (6), 204–209.

Presciuttini, S., Ciampini, F., Alu, M., Cerri, N., Dobosz, M., Domenici, R., Peloso, G., Pelotti, S., Piccinini, A., Ponzano, E., Ricci, U., Tagliabracci, A., Baley-Wilson, J.E., De Stefano, F. and Pascali, V. (2003) *Forensic Science International*, 131, 85–89.

Sjerps, M. and Kloostermann, A.D. (1999) *International Journal of Legal Medicine*, 112, 176–180.

Thomson, J.A., Ayres, K.L., Pilotti, V., Barrett, M.N., Walker, J.I.H. and Debenham, P.G. (2001) *International Journal of Legal Medicine*, 115, 128–134.

Weir, B.S. (1996) *Genetic Data Analysis II: Methods for Discrete Population Genetic Data.*
    Sunderland, MA: Sinauer Associates.

Weir, B.S. (2003) Forensics. In Balding, D. J., Bishop, M. and Cannings, C. (eds)
    Handbook of Statistical Genetics, Second Edition, pp. 830–852. Hoboken, NJ:
    John Wiley & Sons.

BIOLOGY, TECHNOLOGY, AND GENETICS

CHAPTER 24

MASS DISASTER DNA VICTIM IDENTIFICATION

On Sept. 11, New York City suffered the darkest day in our history. It's now up to us to make it its finest hour...

(Mayor Rudolph Giuliani)

Identification of the victims of mass disasters, such as the terrorist attacks of 11 September 2001 on the World Trade Center twin towers, can require application of innovative biology, technology, and genetics. Indeed, as will be described in this chapter, the New York City Office of the Chief Medical Examiner brought to bear new DNA markers (autosomal single nucleotide polymorphisms [SNPs]), new technology (mini short tandem repeat [STR] assays, high-throughput mitochondrial DNA[mtDNA] sequencing, TrueAllele), and new genetic analysis (software for kinship analysis and remains association) to help in identifying many of the remains of the World Trade Center victims that would have otherwise been unidentifiable. The power of forensic DNA typing is at its finest hour when applied to identification of victims of mass disasters.

ISSUES FACED DURING DISASTER VICTIM IDENTIFICATION EFFORTS

In the United States, DNA testing has now become routine and expected in disaster victim identification in the event of a plane crash, large fire, or terrorist attack. Military casualties are also identified through STR typing or mitochondrial DNA sequencing by the Armed Forces DNA Identification Laboratory (AFDIL). All airplane crashes within the United States are examined by the National Transportation Safety Board (http://www.ntsb.gov), which often contracts with AFDIL to identify the air crash victims through DNA testing as part of the investigation.

Often mass disasters leave human remains that are literally in pieces or burned beyond recognition. In some cases it is possible to visually identify a victim, but unfortunately body parts can be separated from one another and the remains co-mingled making identification without DNA techniques virtually impossible. The use of fingerprints and dental records (odontology) still plays an important role in victim identification but these modalities obviously require a finger or an intact skull or jawbone along with previously archived fingerprint and dental records.

DNA testing has a major advantage in that it can be used to identify *each and every portion of the remains* recovered from the disaster site, provided (1) that there is sufficient intact DNA present to obtain a DNA type and (2) a reference sample is available for comparison purposes from a surviving family member or some verifiable personal item containing biological material. Personal items from the deceased including toothbrushes, combs, razors, or even dirty laundry can provide biological material to generate a reference DNA type for the victim. The direct comparison of DNA results from disaster victim remains, to DNA recovered from personal items (Figure 24.1a) represents the easiest way to obtain a match and hence an identification provided it is possible to verify the source (e.g., the toothbrush was not used by some other household member). The use of DNA from biological relatives (Figure 24.1b) necessitates the added complexity of kinship analysis similar to that employed for paternity or reverse parentage testing (see Chapter 23).

There are several important aspects of mass fatality incidents that will be discussed prior to moving into examples of victim identification efforts through DNA testing in recent mass disasters. The areas include collection of reference samples and federal assistance programs. The National Institute of Justice plans to publish two documents in the near future which will provide additional information on mass disaster investigations: *Mass Fatality Incidents: A Guide for*

Figure 24.1

Example demonstrating the use of reference samples in mass disaster victim identification using DNA typing. (a) Direct comparison involves analysis of a direct reference sample from some kind of personal effect of the victim. (b) Kinship analysis utilizes close biological relatives, such as those illustrated in Figure 24.2, to reconstruct a victim's DNA profile.

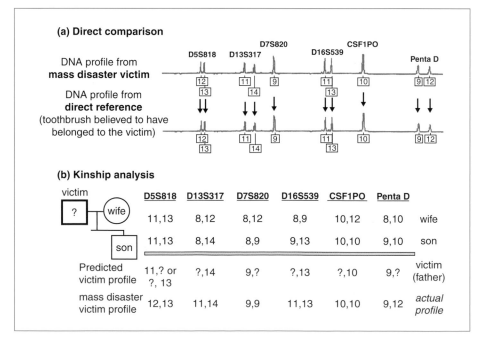

Human Forensic Identification (prepared by the National Center for Forensic Science) and *DNA Identification in Mass Fatality Incidents* (prepared by the Kinship and Data Analysis Panel, see Table 24.2).

COLLECTION OF REFERENCE SAMPLES

In order to be able to identify victims of mass fatality incidents, reference samples are needed in order to sort out DNA profiles obtained from recovered remains. If possible, it is preferable to obtain personal effects that enable a direct match to a victim (see Figure 24.1a). These personal effects may be in the form of a used razor, hairbrush, toothbrush, unwashed dirty laundry or other items that were handled solely by the victim and from which usable biological material may be recovered to generate a DNA profile.

Living biological relatives can also provide needed and valuable reference samples. Immediate relatives including siblings, parents, and children are the most effective indirect reference samples (Figure 24.2). More extended family members can provide helpful samples though if mitochondrial DNA or Y chromosome testing is performed.

Kinship samples can also help confirm the validity of personal effects received for a missing individual. Often a lengthy and complicated administrative review of DNA results and reference sample chain-of-custody is needed to verify both direct references samples from personal effects and indirect references from kinship samples to enable confidence in reporting DNA identifications (see Hennessey 2002).

FEDERAL ASSISTANCE IN DISASTER SITUATIONS

The Disaster Mortuary Operational Response Team (DMORT) is a federally funded group of professionals with experience in disaster victim identification

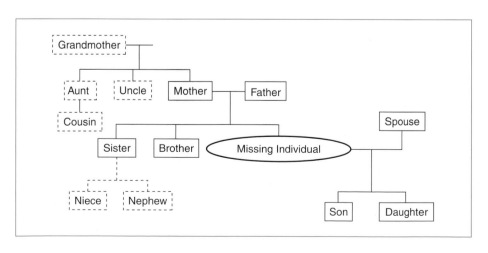

Figure 24.2

Direct biological relatives that can provide valuable reference samples for aiding identification of the missing individual. Ideally samples are available from multiple relatives to help establish robust kinship. A sample from a spouse is only valuable in connection with a child's sample in order to determine the expected alleles coming from the missing individual. Of course, more extended family members, with direct maternal or paternal linkage, can provide helpful samples when mitochondrial DNA or Y chromosome testing is performed. Samples that would be valuable for the maternally transmitted mitochondrial DNA are indicated in dashed boxes.

that becomes activated in response to a major disaster in the United States (see http://www.dmort.org). DMORT is part of the National Disaster Medical System, U.S. Department of Health and Human Services and can be activated by one of four methods: (1) a request for assistance from a local official through the Federal Emergency Management Agency (FEMA) when a federal disaster has been declared, (2) a request for assistance from the National Transportation Safety Board when a passenger aircraft accident occurs, (3) under the U.S. Public Health Act to support a state or locality that cannot provide the necessary response to a disaster, and (4) when requested by a federal agency, such as the FBI, to provide disaster victim identification.

Usually within 24–48 hours after the disaster, the DMORT team is set up to provide professional personnel and technical support and assistance to the local medical examiner or coroner in forensic services and victim identification and is composed of forensic pathologists, forensic anthropologists, forensic dentists, medical investigators, funeral directors and other technical support staff.

Dental records and X-rays along with fingerprints are normally the primary methods used in victim identification. DNA will be used as a last resort and only after all conventional means of identification are exhausted. DMORT does not perform DNA testing but rather gathers samples for future DNA testing, often by a laboratory such as AFDIL.

FEMA provides funding for the DNA identification effort along with other aspects of a disaster investigation after an official declaration of a major disaster or state of emergency has been declared by the President of the United States following a request for such a declaration from the governor of the affected state. This 'state of emergency' declaration then activates a number of federal programs to assist in the response and recovery effort (see http://www.fema.gov).

WHY IDENTIFY REMAINS OF VICTIMS?

Perhaps some wonder why the limited and precious resources of law enforcement and forensic laboratories are devoted to the challenging effort of identifying the remains of disaster victims. Although a variety of reasons could probably be cited, there are two primary reasons that impact remaining family members. First, in many jurisdictions some form of identification is required before a death certificate can be issued that enables remaining family members to collect on life insurance policies. Second, the remaining family members may want the remains of their loved one returned to bring closure to the tragedy and to provide a proper burial and memorial service for their relative. As noted by Ballantyne (1997), there seems to be an overwhelming desire by many relatives to retrieve even the most miniscule tissue sample of a loved one.

CHALLENGES WITH DNA IDENTIFICATION OF VICTIMS

Occasionally there are no known or living biological relatives or every immediate member of a family is among the victims making it difficult to associate the remains or to connect them to a valid reference sample. An additional challenge can come with simply locating surviving family members or communicating with them in the case of an international disaster.

Unfortunately, there can be challenges in dealing with family members of disaster victims such as family disputes (e.g., feuding family members fighting over who is entitled to the recovered remains) or discovery of illegitimate relationships when biological non-paternity is demonstrated for someone who previously thought that they were the father or the child of a victim. Care must be taken by the laboratory director or other laboratory personnel that may interact directly with the families of victims to be sensitive to their grieving process.

Collection of biological material from a disaster site is sometimes anything but trivial. For example, if a plane goes down at sea (e.g., Swissair Flight 111) especially in deep water, then recovery of the remains can be quite challenging. By the very nature of the disaster, there is typically damage done to the biological samples and hence the DNA molecules contained therein. Extreme environmental conditions both during and after the disaster impact the quality of the recovered remains so that the DNA may be degraded when it is analyzed in the laboratory. As described in Chapter 7, degraded DNA gives rise to partial profiles, which then lower the statistics of a match because not all loci tested can be reported (see Chapter 22).

TRAUMA TO LABORATORY PERSONNEL

Being exposed to large numbers of badly damaged human remains can have a psychological impact on laboratory personnel. In addition, loved ones may be among those who died making it difficult to cope with the rigors of careful laboratory work. Strain can be placed on the laboratory by political officials and the news media to produce results rapidly and to give regular updates of progress.

EARLY EFFORTS WITH APPLYING DNA ANALYSIS TO MASS DISASTERS

In the following pages, several mass disaster situations are highlighted where STR typing proved to be a valuable means for identifying human remains from burn victims, airplane crashes, terrorist acts, and mass grave sites (D.N.A. Box 24.1). In these cases, STR typing was successfully performed in spite of heavy damage inflicted by a high-temperature fire (Branch Davidian compound in Waco, Texas, April 1993) or severe water damage (airplane crash of Swissair Flight 111 near Nova Scotia, September 1998). The more recent use of DNA typing

D.N.A. Box 24.1

DNA identification from mass graves

DNA can play an important role in identifying remains in unmarked graves. Unfortunately, there are regions of the world that have suffered severely under the hands of ruthless dictators who do not value human life. Mass graves are often the tragic trademark of such tyrants.

The former Yugoslavia contains an estimated 40 000 unidentified bodies in mass graves (Huffine *et al.* 2001, Williams and Crews 2003). The International Commission on Missing Persons (ICMP) was created in 1996 to help with identifying human remains in these mass graves. In effect, the ICMP is using DNA technology to map human genocide (Huffine *et al.* 2001).

One of the major challenges for performing DNA identification from mass graves is obtaining biological reference samples from relatives. Family outreach centers take information and blood samples from living relatives such as parents or a spouse and a child of a missing loved one. These reference samples are then typed with STRs and mitochondrial DNA sequencing to provide points of comparison to results obtained from the mass grave remains.

Mitochondrial DNA often is the only source of successful DNA recovery from bones that have been in the ground for many years although improved DNA extraction methods have enabled successful STR typing results to be obtained in many cases. DNA often provides the only way to confirm the death of a missing person, enable return of the remains to a living relative, and help bring justice to the criminals who initiated the massacres that led to the mass grave sites.

Sources:

Huffine, E., Crews, J., Kennedy, B., Bomberger, K. and Zinbo, A. (2001) *Croatian Medical Journal*, 42, 271–275.
Williams, E.D. and Crews, J.D. (2003) *Croatian Medical Journal*, 44, 251–258.

methods to identify victims of the 11 September 2001 terrorist attacks against the United States is reviewed in the final section of this chapter.

WACO BRANCH DAVIDIAN FIRE

On 19 April 1993, the Mount Carmel Branch Davidian compound in Waco, Texas burned during a raid by FBI agents. Over 80 individuals died, and their remains were severely damaged following a high-temperature fire. While approximately half of these individuals could be identified by dental or fingerprint comparison and anthropological and pathological findings, the rest had to be identified based on information that could only be provided by DNA analysis. This was the first mass disaster investigation where DNA analysis with STR markers was used.

The Armed Forces DNA Identification Laboratory (AFDIL) and the Forensic Science Service (FSS) in England analyzed these samples by examining a variety of DNA markers including HLA DQ alpha, AmpliType PM, D1S80, amelogenin sex-typing, mitochondrial DNA sequencing, and four STR loci (TH01, F13A1, FES/FPS, VWA). Without the use of polymerase chain reaction (PCR)-based DNA typing procedures, specifically STR markers, approximately half of the individuals who perished in the Mount Carmel Compound of Branch Davidians would not have been identified.

AFDIL received 242 samples from the Mount Carmel Branch Davidian compound representing 82 sets of human remains (DiZinno *et al.* 1994). Blood stained cards from living relatives were also tested to serve as reference samples for the unknowns. When usable tissue from the badly burned bodies was not available, then portions of rib bones were removed and the DNA extracted.

Body identifications were made by matching observed sample genotypes with predicted possible genotypes obtained from using results of relatives' reference blood samples and information gathered from family trees. This approach is basically a reverse parentage analysis where the parent genotypes are used to predict the child's genotype. A total of 26 positive identifications were made using the family tree matching approach (Clayton *et al.* 1995). A shortage of relatives prevented the identification of the other bodies. These results highlight the need for reference samples in order to take full advantage of DNA testing in mass disaster situations (Ballantyne 1997).

Two airplane crashes in 1996 helped demonstrate the value of DNA analysis in victim identification. On 26 August 1996, an airplane carrying 64 Russian and 77 Ukrainian passengers crashed into a mountain near Spitsbergen killing all 128 passengers and 13 crewmembers. The establishment of victim identity took only 20 days due to rapid analysis of samples in the University of Oslo (Norway) Institute of Forensic Medicine.

DNA profiles comprised of three STRs and five minisatellite loci enabled sorting of 257 body parts into 141 individuals of which 139 were identified based on available reference samples submitted by 154 relatives (Olaisen *et al.* 1997). In this particular case, the rapid recovery of samples from a frigid environment (~0°C) led to 100% success in producing DNA profiles from the remains. The cost of the DNA typing portion of the investigation was approximately 3–5% of the total cost of the entire operation.

Shortly after take-off on 17 July 1996, TWA Flight 800, a Boeing 747 on route to Paris, blew up in the sky above Long Island, New York killing all 230 passengers and crew. Their fragmented remains were also identified with STR typing results (Ballantyne 1997). These early successes with DNA testing in aiding airline disaster victim identification paved the way for more recent uses.

SWISSAIR FLIGHT 111

On the evening of 2 September 1998, while en route to Geneva, Switzerland, from New York City, Swissair Flight 111 crashed into the Atlantic Ocean not far from Halifax, Nova Scotia. All 229 people on board (214 passengers and 15 crewmembers) were killed. The plane went down about 10 kilometers from land requiring the wreckage to be raised from a depth of more than 60 meters (~180 feet) of water.

Over the next few weeks, a large task force of investigators collected human remains from the crash scene. These remains were carefully collected and subsequently identified for two important reasons. First, without a reason for the plane falling from the sky, criminal activity was a possibility. The plane's manifest listed 229 people on board. But were they all who they claimed to be? Any discrepancy from that number could be a sign of terrorist activity. A missing individual or an extra passenger that could not be accounted for might have been a terrorist with a bomb.

The second important reason for identifying the victims of any mass disaster is to bring closure to the living relatives. If the remains of their loved ones can be identified, then something can be given back to the living relatives for burial and memorial purposes. However, one of the challenges for identifying the remains of airline crashes is that entire families often travel together. In this case, closely related individuals have to be distinguished from each other and sometimes without the benefit of a living relative to act as a reference sample.

In the case of the 229 victims of the Swissair Flight 111 tragedy, a number of methods were used to identify the victims. These methods included fingerprints, dental records, X-ray evaluation, and DNA testing. Only one body was intact enough for visual identification. DNA testing played an important role in this investigation because of the lack of fingerprint and dental records on many of the victims.

A total of 147 victims could be identified by means other than DNA. For example, 1020 fingers were recovered from the crash site. However, these fingers allowed only 43 victims to be identified based on their fingerprints because only a small percentage of the victims had fingerprint records that could be located. Police visited the homes of the victims and tried to recover latent fingerprints from objects that they may have handled. These efforts led to the recovery of over 200 latent prints that were used to identify 33 of 43 victims mentioned above. An effective method of initial identification involved dental records, which were used to positively ID 102 of the victims and enable a certificate of death to be issued. Dental comparisons provided the fastest identification when reference samples were available. However, DNA analysis was the most effective method of identification overall especially when crash scene victims were not intact.

Concurrent to other efforts to identify the crash victims, DNA testing was performed by the Royal Canadian Mounted Police (RCMP). DNA analysis was performed by four RCMP laboratories from across Canada and the Ontario Provincial forensic laboratory, each contributing a vital and specific subset of data.

The DNA identification process was coordinated by the DNA Methods and Database section in Ottawa. This team of more than 50 DNA scientists consisted of members of the Biology sections of the RCMP Forensic Laboratories located in Halifax, Regina, Vancouver and Ottawa and the Centre for Forensic Sciences in Toronto. DNA typing with the STR markers described in this book was used to help identify all 229 people on board Swissair Flight 111. In every case where other forms of identification were performed, DNA analysis helped confirm and support those results.

Two separate identification issues were addressed by DNA testing. First, recovered human remains showing identical STR genotypes were associated, and second, each passenger was identified through comparisons of human remains with the reference samples isolated from personal effects of the victims or reference blood samples submitted by relatives of the victims. In many cases, DNA analysis was the only means by which the samples could be positively identified.

Over 2400 human remains were recovered from the crash site of which 1277 were analyzed by DNA testing. These samples were analyzed along with 310 reference samples from relatives that were submitted on FTA paper blood cards (see Chapter 3) to be genotyped and used in a relational database (Figure 24.3). In addition, 89 personal effects, such as toothbrushes and hair from combs were taken from the homes of 17 victims because no relatives were available to serve as a reference. One of the challenges of collecting the reference samples was the fact that the living relatives were from 21 different countries. The FTA kit enabled rapid blood collection and room temperature delivery of the reference samples and aided the successful completion of the investigation.

The AmpF*l*STR Profiler Plus™ STR markers D3S1358, VWA, FGA, D8S1179, D21S11, D18S51, D5S818, D13S317, and D7S820 were the primary means of identifying the remains although additional STR loci TH01, TPOX, CSF1PO, and D16S539 from the AmpF*l*STR COfiler™ kit were used to gain a higher power of discrimination (Fregeau *et al.* 2000). COfiler amplifications were performed on 118 crash scene remains and 129 'known' reference comparison samples. The crash scene samples were analyzed in approximately one week using either the nine STRs from Profiler Plus™ or all 13 STRs from Profiler Plus™ and COfiler™. The challenge of making appropriate associations between the samples took a little longer. Genotypes of all 229 victims were compared to genotypes of all other victims and family relatives, which represented 71 490 genotype comparisons (Leclair *et al.* 2000). The comparison of reference sample genotypes to questions involved over 180 000 comparisons because more known sample genotypes existed for cross-comparison purposes.

Traditional parentage trios with both living parents were encountered for only 25% of the 229 victims. Even more challenging was the fact that 43 families of two to five individuals were present among the victims. A pair of identical twins were also present on the plane and could not be individually identified

Figure 24.3

Strategy for disaster victim identification in the Swissair Flight 111 crash scene investigation. Most samples were tested with nine STRs and amelogenin using the AmpFlSTR Profiler Plus kit and then four additional STRs were added with the COfiler kit as needed to obtain a higher power of discrimination between closely related individuals. The relational database compared genotypes of 1277 crash scene samples to genotypes of all 229 victims and family relatives, a total of 71 490 genotype comparisons (Leclair et al. 1999).

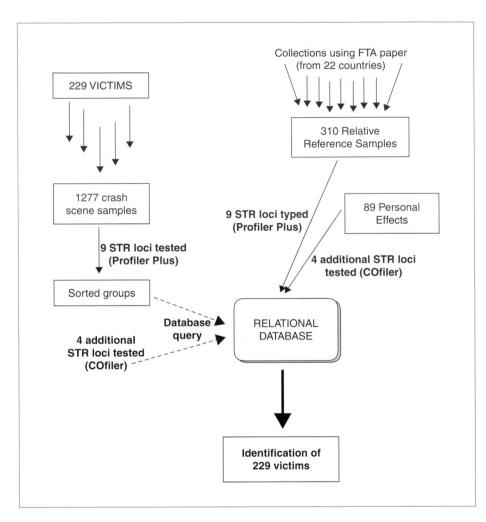

with DNA testing. Nevertheless, the DNA testing led to confident kinship analysis in the case of 218 victims for whom reference samples from close relatives or personal effects were submitted (Leclair *et al.* 2000).

The efforts of the RCMP demonstrated a successful model for how a mass disaster investigation should be conducted. Tremendous cooperation is required from forensic laboratories, law enforcement personnel and family members of victims, often from a number of countries, in order to successfully identify the victims of mass fatality incidents. The need for readily available reference samples was also highlighted by this investigation. In fact, a formal recommendation was made from the RCMP to the Canadian Transport Safety Board for all airline personnel and any private citizens who are frequent fliers to have fingerprints and DNA samples made available for identification purposes if ever the need arises. These records cannot be stored by the police but rather could be maintained by the airline or stored in an individual's safety deposit box. Of the 229 victims of Swissair Flight 111, all 229 were positively identified, an astounding success compared to previously used conventional identification techniques and one that would not have been possible without the power of DNA typing with STR markers.

DNA IDENTIFICATION WORK WITH 11 SEPTEMBER 2001 VICTIMS

The terrorist attacks against the United States on 11 September 2001 left over 3000 victims in three different locations: the Pentagon in Washington, DC, a field near Shanksville (Somerset County), Pennsylvania, and the twin towers of the World Trade Center in New York City. Several teams of forensic scientists were involved in the DNA analysis of these mass disasters.

THE PENTAGON SITE

Hijacked by five terrorists on the morning of 11 September 2001, American Airlines Flight 77 crashed into the Pentagon shortly after 9:40 a.m. killing all 59 passengers and crew on board, the five terrorists, and 124 building occupants at the Pentagon. The remains, ranging in size from whole bodies to small fragments, were taken to Dover Air Force Base in Delaware where they were sampled for DNA testing. From the 2000 containers of evidence recovered, 938 samples were collected for DNA analysis. STR typing was then performed at the Armed Forces DNA Identification Laboratory in Rockville, Maryland using Profiler Plus. Reference samples included 49 direct reference bloodstain cards from the Armed Forces DNA Repository and 348 family references.

   Over the course of the next two months (17 Sept 2001 to 15 Nov 2001), 177 identifications were made using DNA only or a combination of DNA, dental records, and fingerprints. In addition, one victim was identified solely with dental records. Unfortunately, five missing individuals could not be identified because no biological material was recovered from the crash site (Edson *et al.* 2004). In addition, there were five male STR profiles that did not match reference samples from any of the victims. These samples were classified as belonging to the terrorists who hijacked the plane and were further confirmed using mitochondrial DNA testing and comparison to Near Eastern mtDNA haplotypes (Edson *et al.* 2004).

THE SOMERSET COUNTY (SHANKSVILLE, PENNSYLVANIA) SITE

Shortly after departing Newark, New Jersey on the morning of 11 September 2001, United Airlines Flight 93 was hijacked by four terrorists. The plane crashed near Shanksville, Pennsylvania at 10:10 a.m. killing all 40 passengers and crew along with the four terrorists. The 1319 total remains recovered from the Somerset County crash site were all highly fragmented because the plane went straight into the ground going more than 500 miles per hour. Scientists from the Armed Forces DNA Identification Laboratory collected 592 samples for DNA analysis including 423 bones, 141 tissues, 23 teeth, two hairs, and three fingernails. Reference samples used included 55 family references and 50 direct references.

All 40 passengers and crewmembers were able to be identified through DNA alone or a combination of DNA, dental records, and fingerprints. In addition, four male STR profiles that did not match family references were observed and ascribed to the four terrorists. These samples were tested with mtDNA and found to be associated with Near Eastern mtDNA haplotypes (Edson *et al.* 2004).

THE WORLD TRADE CENTER DNA IDENTIFICATION EFFORT

The DNA identification efforts for the World Trade Center (WTC) victims have become arguably the world's largest forensic case to date. More than 19 917 pieces of human remains were collected from a pile of rubble weighing over a million tons and extending more than 70 feet in height following the crushing collapse of the twin towers. The initial removal and sorting of human remains took place between September 2001 and May 2002. However, the DNA identification efforts went on for more than three years – almost two-and-a-half years after the last piece of debris had been removed from the WTC site. In the end, more than 1558 victims of the 2749 present when the twin towers collapsed were identified (see Brenner and Weir 2003). Without the capabilities of DNA testing, there would have been only a fraction of the victims identified based on other modalities such as fingerprints and dental records (Table 24.1). As of 15 June 2004, a total of 8705 remains from the World Trade Center site have been identified and associated with one of the 1558 victims identified so far.

Biological samples recovered from the WTC site had been subjected to extreme pressures with the building collapse and then subterranean fires of 1500°F or more for the three months following the terrorist attack. The jet fuel from both planes that rammed the WTC towers burned intensely enough to melt the steel support beams and bring down the buildings. Thus, human remains in this pressure cooker were often co-mingled, very fragmented, and in many cases likely vaporized.

Dr. Robert Shaler, the director of Forensic Biology at the New York City's Office of Chief Medical Examiner (NYC OCME) stood at the helm of the WTC DNA efforts throughout the many months involved in this investigation. Dr. Shaler and his dedicated staff coordinated the efforts and assistance of outside companies and consultants. They worked tirelessly to collate every piece of possible information in the complex process of matching a genetic match between a victim's DNA profile and that of a reference sample in the form

Modality	Victims Identified			Remains Identified		
	Single Modality	Multiple Modalities	Total ID's	Single Modality	Multiple Modalities	Total ID's
DNA	817	465	1282	4231	3685	7916
Photo	11	14	25	11	14	25
Viewed	12	2	14	12	2	14
Body X-ray	0	3	3	0	4	4
Dental	102	424	526	117	497	614
Prints	53	215	268	56	240	296
Tattoos	0	6	6	0	6	6
Personal effects	16	59	75	18	61	79
Other	7	34	41	7	101	108

Table 24.1

Summary of World Trade Center victim and remains identification efforts completed as of June 2004. The number of victims stands at 2749 of which 1558 have been identified. The number of remains is 19 917 with some form of identification completed on 8705 of these pieces as of 15 June 2004. Although DNA far outweighs other methods in terms of success at recovering information in this disaster, other modalities were useful in identifying victims or sets of remains. Information courtesy of Dr. Robert Shaler, New York City Office of Chief Medical Examiner.

of a personal effect or a victim's biological relative (see Figure 24.2). A computer program named DNA View written by Dr. Charles Brenner played an important part in determining many of these kinship identifications.

Several innovations came out of the 9/11 tragedy (see Vastag 2002). These included new extraction methods from bone (Holland *et al.* 2003), reduced size amplicons or miniSTRs (Butler *et al.* 2003, Holland *et al.* 2003), panels of single nucleotide polymorphisms (SNPs), and high-throughput mitochondrial DNA sequencing. In addition, new software was developed to aid in matching reference samples and recovered remains as well as associating remains with the same DNA profile.

Three different software programs were used extensively in the WTC victim identification efforts: M-FISys (Gene Codes Forensics; Hennessey 2002), DNA View (Brenner and Weir 2003), and MDKAP (Leclair *et al.* 2002). M-FISys worked through a direct match algorithm and helped in collapsing and sorting data sets to obtain identifications. DNA View deduced kinship by pedigree analyses while MDKAP performed kinship analyses through pair-wise comparisons. A variation of the MDKAP (Mass Disaster Kinship Analysis Program) is now commercially available as the Bloodhound program (see http://www.ananomouse.com).

One of the largest challenges from this investigation was review of the massive amounts of data produced by contracting laboratories. The flow of DNA samples and data between the various laboratories involved in this tremendous effort is illustrated in Figure 24.4. In the end, more than 52 528 STR profiles, 16 938 SNP profiles, and 31 155 mtDNA sequences were generated in an effort to identify

Figure 24.4

Illustration of (a) material and (b) data flow between laboratories involved in processing World Trade Center samples. Laboratories included Office of Chief Medical Examiner (OCME, New York City), New York State Police (NYSP, Albany, NY), Myriad Genetics (Salt Lake City, UT), Orchid Cellmark (Dallas, TX), Celera Genomics (Rockville, MD), and Bode Technology Group (Springfield, VA). Physical materials shipped between laboratories included DNA extracts (solid red line), buccal swabs from biological relative reference samples (dashed red line), personal effects (dotted red line), recovered bones (solid blue line), and recovered tissue (dashed blue line). Note that most of the DNA samples were extracted at the NYSP and OCME laboratories although some tissue and bone were shipped directly to Myriad and Bode. All bones were extracted at Bode. Figure courtesy of National Institute of Justice World Trade Center Kinship and Data Analysis Panel and New York City Office of Chief Medical Examiner.

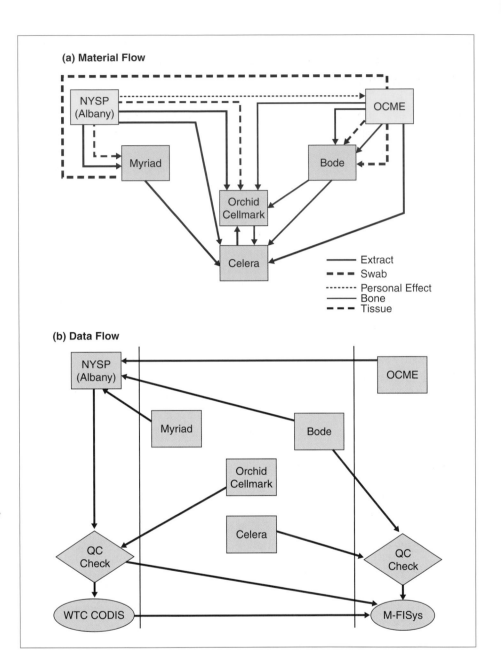

the 2749 victims of the World Trade Center collapse based on 19 917 recovered remains (Bob Shaler, personnel communication) – truly a heroic effort. Most of the data from the recovered remains contained only partial DNA profiles making it even more difficult to sort through and piece together sufficient information to make a reliable identification. Brenner and Weir (2003) describe the major issues and strategies for identifying victims of mass disasters. The process for assigning identities to specific remains has three stages: collapsing, screening, and testing. Clustering algorithms can be used to help reconstruct likely pedigree information when multiple family members are part of the sample set being investigated (Cowell and Mostad 2003).

As noted by Gill *et al.* (2004), mass disaster investigations can more easily implement new genetic markers like SNPs than criminal casework because no national DNA database is involved that only contains information on a core set of loci and is used for open investigations over a long period of time. Rather in a mass disaster victim identification effort, both reference samples and recovered remains are processed with the same genetic markers and assays.

KINSHIP AND DATA ANALYSIS PANEL (KADAP)

The National Institute of Justice (NIJ) organized a panel of experts that convened every other month in the two years following 11 September 2001 to aid in reviewing data and providing guidance and recommendations to NYC OCME regarding statistical thresholds for reporting DNA matches with direct matches and kinship associations. The names and affiliations of the advisory WTC Kinship and Data Analysis Panel (KADAP) are listed in Table 24.2. KADAP gathered in NYC, Albany, Baltimore, and Washington, DC for two-day meetings during Oct 2001, Dec 2001, Feb 2002, Apr 2002, July 2002, Sept 2002, Jan 2003, and July 2003.

Member	Affiliation
Lisa Forman, Ph.D.	National Institute of Justice (U.S. Department of Justice)
Steve Niezgoda, MBA	NIJ Contractor
Amanda Sozer, Ph.D.	NIJ Contractor
Joan Bailey-Wilson, Ph.D.	National Institutes of Health-National Human Genome Research Institute (U.S. Department of Health and Human Services)
Jack Ballantyne, Ph.D.	University of Central Florida
Fred Bieber, M.D, Ph.D.	Harvard Medical School
Les Biesecker, Ph.D.	National Institutes of Health-National Human Genome Research Institute (U.S. Department of Health and Human Services)
Charles Brenner, Ph.D.	DNA-View
Bruce Budowle, Ph.D.	FBI Laboratory (U.S. Department of Justice)
John Butler, Ph.D.	National Institute of Standards and Technology (U.S. Department of Commerce)

Table 24.2

World Trade Center Kinship and Data Analysis Panel (WTC KADAP) members convened to aid New York City with WTC identifications. National Institute of Justice funded-meetings were held bimonthly from October 2001 to July 2003 in New York City, Albany, Baltimore, or Washington, DC (not all members attended every meeting) to discuss progress with the investigation and to aid in reviewing technology validation and statistical genetic issues.

Table 24.2
(Continued)

Member	Affiliation
George Carmody, Ph.D.	Carleton University
Maureen Casey	New York City Police Department
Ken Kidd, Ph.D.	Yale University School of Medicine
Cheryl Conley	Orchid Genescreen Ohio
Michael Conneally, Ph.D.	Indiana University School of Medicine
Art Eisenberg, Ph.D.	University of North Texas Health Science Center
Mike Hennessy, MBA	Gene Codes Forensics
Benoit Leclair, Ph.D.	Myriad Genetics
Judy Nolan, Ph.D.	Gene Codes Forensics
Tom Parsons, Ph.D.	Armed Forces DNA Identification Laboratory (U.S. Department of Defense)
Elizabeth Pugh, Ph.D.	John Hopkins University/Center for Inherited Disease Research
Steve Sherry, Ph.D.	National Institutes of Health-National Center for Biotechnology Information (U.S. Department of Health and Human Services)
Amy Sutton	Gene Codes Forensics
Lois Tully, Ph.D.	National Institute of Justice
Anne Walsh, Ph.D.	New York State Department of Health
OCME	
Robert Shaler, Ph.D.	New York City Office of Chief Medical Examiner (OCME)
Howard Baum, Ph.D.	New York City Office of Chief Medical Examiner (OCME)
Erik Bieschke	New York City Office of Chief Medical Examiner (OCME)
Elaine Mar	New York City Office of Chief Medical Examiner (OCME)
Amy Mundorff	New York City Office of Chief Medical Examiner (OCME)
Mecki Prinz, Ph.D.	New York City Office of Chief Medical Examiner (OCME)
Noelle Umback	New York City Office of Chief Medical Examiner (OCME)
NYSP	
Mark Dale	New York State Police (NYSP); New York City Police Department Laboratory
Barry Duceman, Ph.D.	New York State Police
Dennis Gaige	New York State Police

Member	Affiliation
Steven Hogan, J.D.	New York State Police
John Snyder	New York State Police Contractor
Steve Swinton	New York State Police
Peter Wistort	New York State Police
Guest Presenters	
Jeanine Baisch, Ph.D.	Orchid Cellmark (SNP assays)
Todd Bille	Bode Technology Group (miniSTRs, new DNA extractions)
Howard Cash	Gene Codes Forensics (MFISYS software)
Joyce deJong	Disaster Mortuary Operational Response Team (DMORT)
Bob Giles, Ph.D.	Orchid Cellmark (SNP assays)
Rhonda Roby, MPM	Applied Biosystems/Celera (mtDNA sequencing)
James Ross	Armed Forces DNA Identification Laboratory (LIMS software)

Table 24.2
(Continued)

While we hope to never see the likes of another 11 September 2001 terrorist attack, forensic DNA typing laboratories should be prepared to aid in victim identification efforts in future mass fatality incidences. Lessons learned from the WTC DNA identification efforts by the KADAP will be available from a National Institute of Justice report entitled 'DNA Identification in Mass Fatality Incidents' that provides information on project management, sample tracking and flow of information, suggestions on structuring communications between victim family members and the DNA laboratory, and vital issues to consider immediately after an incident occurs. It is safe to conclude that DNA analysis will continue to play a valuable support role in future mass fatality incidents, just as it does with law enforcement and the criminal justice system.

REFERENCES AND ADDITIONAL READING

Alonso, A., Andelinovic, S., Martin, P., Sutlovic, D., Erceg, I., Huffine, E., de Simon, L.F., Albarran, C., Definis-Gojanovic, M., Fernandez-Rodriguez, A., Garcia, P., Drmic, I., Rezic, B., Kuret, S., Sancho, M. and Primorac, D. (2001) *Croatian Medical Journal*, 42, 260–266.

Ballantyne, J. (1997) *Nature Genetics*, 15, 329–331.

Boles, T.C., Snow, C.C. and Stover, E. (1995) *Journal of Forensic Sciences*, 40, 349–355.

Brenner, C.H. (1997) *Genetics*, 145, 535–542.

Brenner, C.H. (2000) Kinship analysis by DNA when there are many possibilities. In Sensabaugh, G.F., Lincoln, P.J. and Olaisen, B. (eds) *Progress in Forensic Genetics 8,* pp. 94–96. New York: Elsevier.

Brenner, C.H. and Weir, B.S. (2003) *Theoretical Population Biology*, 63, 173–178.

Budimlija, Z.M., Prinz, M.K., Zelson-Mundorff, A., Wiersema, J., Bartelink, E., MacKinnon, G., Nazzaruolo, B.L., Estacio, S.M., Hennessey, M.J. and Shaler, R.C. (2003) *Croatian Medical Journal*, 44, 259–263.

Butler, J.M., Shen, Y. and McCord, B.R. (2003) *Journal of Forensic Sciences*, 48, 1054–1064.

Cowell, R.G. and Mostad, P. (2003) *Journal of Forensic Sciences*, 48, 1239–1248.

DiZinno, J., Fisher, D., Barritt, S., Clayton, T., Gill, P., Holland, M., Lee, D., McGuire, C., Raskin, J., Roby, R., Ruderman, J. and Weedn, V. (1994) *Proceedings from the Fifth International Symposium on Human Identification,* pp. 129–135. Madison, Wisconsin: Promega Corporation.

Edson, S.M., Ross, J.P., Coble, M.D., Parsons, T.J. and Barritt, S.M. (2004) *Forensic Science Review*, 16, 63–90.

Egeland, T., Mostad, P.F. and Olaisen, B. (1997) *Science and Justice*, 37, 269–274.

Fregeau, C.J., Leclair, B., Bowen, K., Elliot, J., Borys, S. and Fourney, R. (2000) The Swissair Flight 111 disaster: short tandem repeat mutations observed. In Sensabaugh, G.F., Lincoln, P.J. and Olaisen, B. (eds) *Progress in Forensic Genetics 8,* pp. 40–42. New York: Elsevier.

Fregeau, C.J., Vanstone, H., Borys, S., McLean, D., Maroun, J.A., Birnboin, H.C. and Fourney, R.M. (2001) *Journal of Forensic Sciences*, 46, 1180–1190.

Gill, P., Werrett, D.J., Budowle, B. and Guerreri, R. (2004) *Science and Justice,* 44, 51–53.

Goodwin, W., Linacre, A. and Vanezis, P. (1999) *Electrophoresis*, 20, 1707–1711.

Gornik, I., Marcikic, M., Kubat, M., Primorac, D. and Lauc, G. (2002) *International Journal of Legal Medicine*, 116, 255–257.

Hennessey, M. (2002) *Proceedings of the Thirteenth International Symposium on Human Identification*. Available online at: http://www.promega.com/geneticidproc/ussymp13proc/contents/hennesseyrev1.pdf.

Hoff-Olsen, P., Mevag, B., Staalstrom, E., Hovde, B., Egeland, T. and Olaisen, B. (1999) *Forensic Science International*, 105, 171–183.

Hoff-Olsen, P., Jacobsen, S., Mevag, B. and Olaisen, B. (2001) *Forensic Science International*, 119, 273–278.

Holland, M.M. and Parsons, T.J. (1999) *Forensic Science Review*, 11, 21–50.

Holland, M.M., Cave, C.A., Holland, C.A. and Bille, T.W. (2003) *Croatian Medical Journal*, 44, 264–272.

Huffine, E., Crews, J., Kennedy, B., Bomberger, K. and Zinbo, A. (2001) *Croatian Medical Journal*, 42, 271–275.

Leclair, B., Fregeau, C.J., Bowen, K.L., Borys, S.B., Elliot, J. and Fourney, R.M. (2000) Enhanced kinship analysis and STR-based DNA typing for human identification in mass disasters. In Sensabaugh, G.F., Lincoln, P.J. and Olaisen, B. (eds) *Progress in Forensic Genetics 8,* pp. 91–93. New York: Elsevier.

Leclair, B., Niezgoda, S., Carmody, G.R. and Shaler, R.C. (2002) *Proceedings of the Thirteenth International Symposium on Human Identification*. Available online at: http://www.promega.com/geneticidproc/ussymp13proc/contents/leclair.pdf.

Olaisen, B., Stenersen, M. and Mevag, B. (1997) *Nature Genetics*, 15, 402–405.

Primorac, D., Andelinovic, S., Definis-Gojanovic, M., Drmic, I., Rezic, B., Baden, M.M., Kennedy, M.A., Schanfield, M.S., Skakel, S.B. and Lee, H.C. (1996) *Journal of Forensic Sciences*, 41, 891–894.

Primorac, D. and Schanfield, M.S. (2000) *Croatian Medical Journal*, 41, 32–46.

Prinz, M., Caragine, T., Kamnick, C., Cheswick, D. and Shaler, R. (2002) *Proceedings of the Thirteenth International Symposium on Human Identification*. Available online at: http://www.promega.com/geneticidproc/ussymp13proc/contents/prinz.pdf.

Vastag, B. (2002) *Journal of the American Medical Association*, 288, 1221–1223.

Whitaker, J.P., Clayton, T.M., Urquhart, A.J., Millican, E.S., Downes, T.J., Kimpton, C.P. and Gill, P. (1995) *BioTechniques*, 18, 670–677.

Williams, E.D. and Crews, J.D. (2003) *Croatian Medical Journal*, 44, 251–258.

APPENDIX I

REPORTED STR ALLELES:
SIZES AND SEQUENCES

EXPLANATION OF INFORMATION INCLUDED IN THE FOLLOWING TABLES:

This appendix material describes the reported alleles for the 13 short tandem repeat (STR) loci most commonly used in the United States and around the world. Note that the number of alleles present for a particular locus is an indication of the polymorphic nature of that marker and its value for use in human identity testing. Thus, FGA is more useful than TPOX because it possesses more alleles and there is a greater chance that two individuals selected at random would have different genotypes at FGA than at TPOX.

STR alleles are named based on the number of full repeat units that they contain with partial repeats (i.e., microvariants) being designated by the number of full repeats, a decimal, and the number of nucleotides present in the partial repeat in accordance with International Society of Forensic Genetics (ISFG) recommendations (1994). The polymerase chain reaction (PCR) product size for each of the possible alleles is listed based on the commercial STR kit used to amplify the particular locus. Different commercially available STR kits produce different DNA fragment sizes due to the fact that their primers hybridize to different positions in the flanking regions of the STR sequence. Hence, we have calculated the expected DNA fragment sizes (based on their actual sequence) for all reported alleles if one uses the designated Promega kit or the Applied Biosystems AmpF/STR® kit. These PCR product sizes are listed without any non-template addition, i.e., they are in the '−A' rather than the '+A' form. STR allele sizes measured in a laboratory may also vary from the actual sequence-based size listed here due to the internal sizing standard used and the particular electrophoretic conditions.

The common repeat sequence motif for each STR locus is listed according to ISFG recommendations (1997). In most cases, the sequence changes in the repeat region are the only variation occurring and the flanking sequences remain constant. However, variation in the flanking sequence is also a possibility. Finally, we have listed the reference where each new allele (and its sequence if published) has been described. As more and more samples are analyzed using these STR loci, we recognize that new (rare) alleles will be discovered and that this listing will quickly become outdated. We encourage the reader to consult the STRBase variant allele listing (http://www.cstl.nist.gov/biotech/strbase/var_tab.htm) and to contribute newly discovered alleles so that they may be categorized for fellow workers in this field. The complete sequence for one of the alleles listed for each locus may be found by using the GenBank accession number listed for that locus or by checking the reference sequence in STRBase (http://www.cstl.nist.gov/biotech/strbase/seq_info.htm).

CSF1PO

GenBank Accession: X14720 (allele 12). PCR product sizes of observed alleles

Allele (Repeat #)	Promega PowerPlex 1.1	ABI COfiler (w/o + A)	Repeat Structure $[AGAT]_n$	Reference
5	287 bp	276 bp	$[AGAT]_5$	STRBase
6	291 bp	280 bp	$[AGAT]_6$	COfiler
7	295 bp	284 bp	$[AGAT]_7$	Huang *et al.* (1995)
7.3	298 bp	287 bp	Not published	STRBase
8	299 bp	288 bp	$[AGAT]_8$	Puers *et al.* (1993)
8.3	302 bp	291 bp	Not published	STRBase
9	303 bp	292 bp	$[AGAT]_9$	Puers *et al.* (1993)
9.1	304 bp	293 bp	Not published	STRBase
10	307 bp	296 bp	$[AGAT]_{10}$	Puers *et al.* (1993)
10.1	308 bp	297 bp	Not published	STRBase
10.2	309 bp	298 bp	Not published	STRBase
10.3	310 bp	299 bp	Not published	Lazaruk *et al.* (1998)
11	311 bp	300 bp	$[AGAT]_{11}$	Puers *et al.* (1993)
11.1	312 bp	301 bp	Not published	Scherczinger *et al.* (2000)
12	315 bp	304 bp	$[AGAT]_{12}$	Puers *et al.* (1993)
12.1	316 bp	305 bp	Not published	Budowle and Moretti (1998)
13	319 bp	308 bp	$[AGAT]_{13}$	Puers *et al.* (1993)
14	323 bp	312 bp	$[AGAT]_{14}$	Puers *et al.* (1993)
15	327 bp	316 bp	$[AGAT]_{15}$	COfiler
16	331 bp	320 bp	$[AGAT]_{16}$	Margolis-Nunno *et al.* (2001)

20 observed alleles.

FGA

GenBank Accession: M64982 (allele 21). PCR product sizes of observed alleles

Allele (Repeat #)	Promega PowerPlex 2.1	ABI Profiler Plus (w/o + A)	Repeat Structure $[TTTC]_3$ TTTT TTCT$[CTTT]_n$ CTCC $[TTCC]_2$	Reference
12.2	308 bp	196 bp	Not published	STRBase
13	310 bp	198 bp	Not published	SGM Plus
13.2	312 bp	200 bp	Not published	STRBase
14.3	317 bp	205 bp	Not published	STRBase
15	318 bp	206 bp	$[TTTC]_3$TTTT TTCT$[CTTT]_7$CTCC$[TTCC]_2$	Barber *et al.* (1996)
15.3	321 bp	209 bp	Not published	STRBase
16	322 bp	210 bp	Not published	STRBase

Allele (Repeat #)	Promega PowerPlex 2.1	ABI Profiler Plus (w/o + A)	Repeat Structure $[TTTC]_3$ TTTT TTCT$[CTTT]_n$ CTCC $[TTCC]_2$	Reference
16.1	323 bp	211 bp	$[TTTC]_3$TTTT TTCT$[CTTT]_5$T $[CTTT]_3$CTCC$[TTCC]_2$	Griffiths *et al.* (1998)
16.2	324 bp	212 bp	Not published	SGM Plus
17	326 bp	214 bp	$[TTTC]_3$TTTT TTCT$[CTTT]_9$CTCC$[TTCC]_2$	Barber *et al.* (1996)
17.1	327 bp	215 bp	Not published	SGM Plus
17.2	328 bp	216 bp	Not published	STRBase
18	330 bp	218 bp	$[TTTC]_3$TTTT TTCT$[CTTT]_{10}$CTCC$[TTCC]_2$	Barber *et al.* (1996)
18.1	331 bp	219 bp	Not published	SGM Plus
18.2	332 bp	220 bp	$[TTTC]_3$TTTT TT $[CTTT]_{11}$CTCC$[TTCC]_2$	Barber *et al.* (1996)
19	334 bp	222 bp	$[TTTC]_3$TTTT TTCT$[CTTT]_{11}$CTCC$[TTCC]_2$	Barber *et al.* (1996)
19.1	335 bp	223 bp	Not published	STRBase
19.2	336 bp	224 bp	$[TTTC]_3$TTTT TT $[CTTT]_{12}$CTCC$[TTCC]_2$	STRBase
19.3	337 bp	225 bp	Not published	STRBase
20	338 bp	226 bp	$[TTTC]_3$TTTT TTCT$[CTTT]_{12}$CTCC$[TTCC]_2$	Barber *et al.* (1996)
20.1	339 bp	227 bp	Not published	STRBase
20.2	340 bp	228 bp	$[TTTC]_3$TTTT TT $[CTTT]_{13}$CTCC$[TTCC]_2$	Barber *et al.* (1996)
20.3	341 bp	229 bp	Not published	STRBase
21	342 bp	230 bp	$[TTTC]_3$TTTT TTCT$[CTTT]_{13}$CTCC$[TTCC]_2$	Barber *et al.* (1996)
21.1	343 bp	231 bp	Not published	STRBase
21.2	344 bp	232 bp	$[TTTC]_3$TTTT TT $[CTTT]_{14}$CTCC$[TTCC]_2$	STRBase
21.3	345 bp	233 bp	Not published	STRBase
22	346 bp	234 bp	$[TTTC]_3$TTTT TTCT$[CTTT]_{14}$CTCC$[TTCC]_2$	Barber *et al.* (1996)
22.1	347 bp	235 bp	Not published	STRBase
22.2	348 bp	236 bp	$[TTTC]_3$TTTT TT $[CTTT]_{15}$CTCC$[TTCC]_2$	Barber *et al.* (1996)
22.3	349 bp	237 bp	Not published	Gill *et al.* (1996)
23	350 bp	238 bp	$[TTTC]_3$TTTT TTCT$[CTTT]_{15}$CTCC$[TTCC]_2$	Barber *et al.* (1996)
23.1	351 bp	239 bp	Not published	STRBase
23.2	352 bp	240 bp	$[TTTC]_3$TTTT TT $[CTTT]_{16}$CTCC$[TTCC]_2$	Barber *et al.* (1996)
23.3	353 bp	241 bp	Not published	STRBase
24	354 bp	242 bp	$[TTTC]_3$TTTT TTCT$[CTTT]_{16}$CTCC$[TTCC]_2$	Barber *et al.* (1996)
24.1	355 bp	243 bp	Not published	STRBase
24.2	356 bp	244 bp	$[TTTC]_3$TTTT TT $[CTTT]_{17}$CTCC$[TTCC]_2$	Barber *et al.* (1996)
24.3	357 bp	245 bp	Not published	STRBase
25	358 bp	246 bp	$[TTTC]_3$TTTT TTCT$[CTTT]_{17}$CTCC$[TTCC]_2$	Barber *et al.* (1996)
25.1	359 bp	247 bp	Not published	STRBase
25.2	360 bp	248 bp	$[TTTC]_3$TTTT TT $[CTTT]_{18}$CTCC$[TTCC]_2$	STRBase
25.3	361 bp	249 bp	Not published	STRBase
26	362 bp	250 bp	$[TTTC]_3$TTTT TTCT$[CTTT]_{18}$CTCC$[TTCC]_2$	Barber *et al.* (1996)
26.1	363 bp	251 bp	Not published	STRBase
26.2	364 bp	252 bp	$[TTTC]_3$TTTT TT $[CTTT]_{19}$CTCC$[TTCC]_2$	STRBase
26.3	365 bp	253 bp	Not published	STRBase
27	366 bp	254 bp	$[TTTC]_3$TTTT TTCT$[CTTT]_{19}$CTCC$[TTCC]_2$	Barber *et al.* (1996)
27'	366 bp	254 bp	$[TTTC]_3$TTTT TTCT$[CTTT]_{13}$CCTT$[CTTT]_5$CTCC $[TTCC]_2$	Griffiths *et al.* (1998)
27.1	367 bp	255 bp	Not published	SGM Plus
27.2	368 bp	256 bp	$[TTTC]_3$TTTT TT $[CTTT]_{20}$CTCC$[TTCC]_2$	SGM Plus
27.3	369 bp	257 bp	Not published	SGM Plus
28	370 bp	258 bp	$[TTTC]_3$TTTT TTCT$[CTTT]_{20}$CTCC$[TTCC]_2$	Barber *et al.* (1996)
28.1	371 bp	259 bp	Not published	SGM Plus
28.2	372 bp	260 bp	$[TTTC]_3$TTTT TT $[CTTT]_{21}$CTCC$[TTCC]_2$	SGM Plus
29	374 bp	262 bp	$[TTTC]_3$TTTT TTCT$[CTTT]_{15}$CCTT$[CTTT]_5$CTCC $[TTCC]_2$	Barber *et al.* (1996)
29.2	376 bp	264 bp	Not published	STRBase
30	378 bp	266 bp	$[TTTC]_3$TTTT TTCT$[CTTT]_{16}$CCTT$[CTTT]_5$CTCC $[TTCC]_2$	Griffiths *et al.* (1998)

Allele (Repeat #)	Promega PowerPlex 2.1	ABI Profiler Plus (w/o + A)	Repeat Structure [TTTC]$_3$ TTTT TTCT[CTTT]$_n$ CTCC [TTCC]$_2$	Reference
30.2	380 bp	268 bp	[TTTC]$_4$TTTT TT [CTTT]$_{14}$[CTTC]$_3$[CTTT]$_3$CTCC [TTCC]$_4$	Barber et al. (1996)
31	382 bp	270 bp	Not published	STRBase
31.2	384 bp	272 bp	[TTTC]$_4$TTTT TT [CTTT]$_{15}$[CTTC]$_3$[CTTT]$_3$CTCC [TTCC]$_4$	Griffiths et al. (1998)
32	386 bp	274 bp	Not published	SGM Plus
32.1	387 bp	275 bp	Not published	STRBase
32.2	388 bp	276 bp	[TTTC]$_4$TTTT TT [CTTT]$_{16}$[CTTC]$_3$[CTTT]$_3$CTCC [TTCC]$_4$	Griffiths et al. (1998)
33.1	391 bp	279 bp	Not published	STRBase
33.2	392 bp	280 bp	[TTTC]$_4$TTTT TT [CTTT]$_{17}$[CTTC]$_3$[CTTT]$_3$CTCC [TTCC]$_4$	Griffiths et al. (1998)
34.1	395 bp	283 bp	Not published	STRBase
34.2	396 bp	284 bp	[TTTC]$_4$TTTT TT [CTTT]$_{18}$[CTTC]$_3$[CTTT]$_3$CTCC [TTCC]$_4$	Barber et al. (1996)
35.2	400 bp	288 bp	Not published	STRBase
42.2	428 bp	316 bp	[TTTC]$_4$TTTT TT [CTTT]$_8$ [CTGT]$_4$[CTTT]$_{13}$[CTTC]$_3$[CTTT]$_3$CTCC[TTCC]$_4$	Griffiths et al. (1998)
43.2	432 bp	320 bp	[TTTC]$_4$TTTT TT [CTTT]$_8$ [CTGT]$_5$[CTTT]$_{13}$[CTTC]$_4$[CTTT]$_3$CTCC[TTCC]$_4$	Griffiths et al. (1998)
44	434 bp	322 bp	Not published	Steinlechner et al. (2002)
44.2	436 bp	324 bp	[TTTC]$_4$TTTT TT [CTTT]$_{11}$[CTGT]$_3$[CTTT]$_{14}$ [CTTC]$_3$[CTTT]$_3$CTCC[TTCC]$_4$	Griffiths et al. (1998)
45.2	440 bp	328 bp	[TTTC]$_4$TTTT TT [CTTT]$_{10}$[CTGT]$_5$[CTTT]$_{13}$ [CTTC]$_4$[CTTT]$_3$CTCC[TTCC]$_4$	Griffiths et al. (1998)
46.2	444 bp	332 bp	[TTTC]$_4$TTTT TT [CTTT]$_{12}$[CTGT]$_5$[CTTT]$_{13}$ [CTTC]$_3$[CTTT]$_3$CTCC[TTCC]$_4$	Barber et al. (1996)
47.2	448 bp	336 bp	[TTTC]$_4$TTTT TT [CTTT]$_{12}$[CTGT]$_5$[CTTT]$_{14}$ [CTTC]$_3$[CTTT]$_3$CTCC[TTCC]$_4$	Griffiths et al. (1998)
48.2	452 bp	340 bp	[TTTC]$_4$TTTT TT [CTTT]$_{14}$[CTGT]$_3$[CTTT]$_{14}$ [CTTC]$_4$[CTTT]$_3$CTCC[TTCC]$_4$	Griffiths et al. (1998)
49.2	456 bp	344 bp	Not published	SGM Plus
50.2	460 bp	344 bp	[TTTC]$_4$TTTT TT [CTTT]$_{14}$[CTGT]$_4$[CTTT]$_{15}$ [CTTC]$_4$[CTTT]$_3$CTCC[TTCC]$_4$	Griffiths et al. (1998)
51.2	464 bp	348 bp	Not published	SGM Plus

80 observed alleles.

TH01

GenBank Accession: D00269 (allele 9). PCR product sizes of observed alleles

Allele (Repeat #)	Promega PowerPlex 1.1	Promega PowerPlex 2.1	ABI COfiler (w/o + A)	Repeat Structure [AATG]$_n$ other strand: [TCAT]$_n$	Reference
3	171 bp	152 bp	160 bp	[AATG]$_3$	Espinheira et al. (1996)
4	175 bp	156 bp	164 bp	[AATG]$_4$	Griffiths et al. (1998)
5	179 bp	160 bp	168 bp	[AATG]$_5$	Brinkmann et al. (1996b)
5.3	182 bp	163 bp	171 bp	Not published	SGM Plus
6	183 bp	164 bp	172 bp	[AATG]$_6$	Brinkmann et al. (1996b)

Allele (Repeat #)	Promega PowerPlex 1.1	Promega PowerPlex 2.1	ABI COfiler (w/o + A)	Repeat Structure $[AATG]_n$ other strand: $[TCAT]_n$	Reference
6.1	184 bp	165 bp	173 bp	Not published	SGM Plus
6.3	186 bp	167 bp	175 bp	$[AATG]_3ATG[AATG]_3$	Klintschar (1998)
7	187 bp	168 bp	176 bp	$[AATG]_7$	Brinkmann et al. (1996b)
7.1	188 bp	169 bp	177 bp	Not published	SGM Plus
7.3	190 bp	171 bp	179 bp	Not published	STRBase
8	191 bp	172 bp	180 bp	$[AATG]_8$	Brinkmann et al. (1996b)
8.3	194 bp	175 bp	183 bp	$[AATG]_5ATG[AATG]_3$	Brinkmann et al. (1996b)
9	195 bp	176 bp	184 bp	$[AATG]_9$	Brinkmann et al. (1996b)
9.3	198 bp	179 bp	187 bp	$[AATG]_6ATG[AATG]_3$	Brinkmann et al. (1996b)
10	199 bp	180 bp	188 bp	$[AATG]_{10}$	Brinkmann et al. (1996b)
10.3	202 bp	183 bp	191 bp	$[AATG]_6ATG[AATG]_4$	Brinkmann et al. (1996b)
11	203 bp	184 bp	192 bp	$[AATG]_{11}$	Brinkmann et al. (1996b)
12	207 bp	188 bp	196 bp	$[AATG]_{12}$	Van Oorschot et al. (1994)
13.3	214 bp	195 bp	203 bp	$[AATG][AACG][AATG]_8$ $ATG[AATG]_3$	Gene et al. (1996), Griffiths et al. (1998)
14	215 bp	196 bp	204 bp	Not published	SGM Plus

20 observed alleles.

TPOX

GenBank Accession: M68651 (allele 11). PCR product sizes of observed alleles

Allele (Repeat #)	Promega PowerPlex 1.1	Promega PowerPlex 2.1	ABI COfiler (w/o + A)	Repeat Structure $[AATG]_n$	Reference
4	216 bp	254 bp	209 bp	$[AATG]_4$	STRBase
5	220 bp	258 bp	213 bp	$[AATG]_5$	STRBase
6	224 bp	262 bp	217 bp	$[AATG]_6$	COfiler
7	228 bp	266 bp	221 bp	$[AATG]_7$	Amorim et al. (1996)
7.3	231 bp	269 bp	224 bp	Not published	STRBase
8	232 bp	270 bp	225 bp	$[AATG]_8$	Puers et al. (1993)
9	236 bp	274 bp	229 bp	$[AATG]_9$	Puers et al. (1993)
10	240 bp	278 bp	233 bp	$[AATG]_{10}$	Puers et al. (1993)
11	244 bp	282 bp	237 bp	$[AATG]_{11}$	Puers et al. (1993)
12	248 bp	286 bp	241 bp	$[AATG]_{12}$	Puers et al. (1993)
13	252 bp	290 bp	245 bp	$[AATG]_{13}$	Amorim et al. (1996)
13.1	253 bp	291 bp	246 bp	Not published	STRBase
14	256 bp	294 bp	249 bp	$[AATG]_{14}$	Huang et al. (1995)
15	260 bp	298 bp	253 bp	Not published	STRBase
16	264 bp	302 bp	257 bp	Not published	STRBase

15 observed alleles.

VWA

GenBank Accession: M25858 (allele 18). PCR product sizes of observed alleles

Allele (Repeat #)	Promega PowerPlex 1.1	ABI Profiler Plus (w/o + A)	Repeat Structure TCTA[TCTG]$_{3-4}$[TCTA]$_n$	Reference
10	123 bp	152 bp	TCTA TCTG TCTA [TCTG]$_4$[TCTA]$_3$	Griffiths *et al.* (1998)
11 (13')*	127 bp	156 bp	TCTA[TCTG]$_3$[TCTA]$_7$	Brinkmann *et al.* (1996b)
12	131 bp	160 bp	TCTA[TCTG]$_4$[TCTA]$_7$	Griffiths *et al.* (1998)
13	135 bp	164 bp	[TCTA]$_2$[TCTG]$_4$[TCTA]$_3$TCCA[TCTA]$_3$	Griffiths *et al.* (1998)
13 (15)	135 bp	164 bp	TCTA[TCTA]$_4$[TCTA]$_8$TCCATCTA	Brinkmann *et al.* (1996b)
13 (15″)	135 bp	164 bp	TCTA[TCTA]$_4$[TCTA]$_{10}$	Brinkmann *et al.* (1996b)
14 (16″)	139 bp	168 bp	TCTA[TCTG]$_4$[TCTA]$_{11}$	Brinkmann *et al.* (1996b)
14' (16‴)	139 bp	168 bp	TCTA TCTG TCTA[TCTG]$_4$[TCTA]$_3$TCCA[TCTA]$_3$	Brinkmann *et al.* (1996b)
14″	139 bp	168 bp	TCTA [TCTG]$_5$[TCTA]$_3$TCCA[TCTA]$_3$	Lins (1998)
15 (17)	143 bp	172 bp	TCTA[TCTG]$_4$[TCTA]$_{10}$TCCATCTA	Brinkmann *et al.* (1996b)
15 (17')	143 bp	172 bp	TCTA[TCTG]$_3$[TCTA]$_{11}$TCCATCTA	Brinkmann *et al.* (1996b)
15.2	145 bp	174 bp	[TCTA]$_2$[TCTG]$_4$[TCTA]$_5$T—A[TCTA]$_4$	Gill (1995)
16 (18)	147 bp	176 bp	TCTA[TCTG]$_4$[TCTA]$_{11}$TCCATCTA	Brinkmann *et al.* (1996b)
16 (18')	147 bp	176 bp	TCTA[TCTG]$_3$[TCTA]$_{12}$TCCATCTA	Brinkmann *et al.* (1996b)
16.1	148 bp	177 bp	Not published	STRBase
17 (19)	151 bp	180 bp	TCTA[TCTG]$_4$[TCTA]$_{12}$TCCATCTA	Brinkmann *et al.* (1996b)
18 (20)	155 bp	184 bp	TCTA[TCTG]$_4$[TCTA]$_{13}$TCCATCTA	Brinkmann *et al.* (1996b)
18' (20')	155 bp	184 bp	TCTA[TCTG]$_5$[TCTA]$_{12}$TCCATCTA	Brinkmann *et al.* (1996b)
18.1	156 bp	185 bp	TCTA[TCTG]$_4$[TCTA]$_{12}$A(TCTA)TCCA TCTA	Kido *et al.* (2003)
18.2	157 bp	186 bp	Not published	SGM Plus
18.3	158 bp	187 bp	Not published	STRBase
19 (21)	159 bp	188 bp	TCTA[TCTG]$_4$[TCTA]$_{14}$TCCATCTA	Brinkmann *et al.* (1996b)
19.2	161 bp	190 bp	Not published	SGM Plus
20 (22)	163 bp	192 bp	TCTA[TCTG]$_4$[TCTA]$_{15}$TCCATCTA	Brinkmann *et al.* (1996b)
21 (23)	167 bp	196 bp	TCTA[TCTG]$_4$[TCTA]$_{16}$TCCATCTA	Brinkmann *et al.* (1996b)
22 (24)	171 bp	200 bp	TCTA[TCTG]$_4$[TCTA]$_{17}$TCCATCTA	Brinkmann *et al.* (1996b)
23	175 bp	204 bp	Not published	SGM Plus
24	179 bp	208 bp	Not published	STRBase
25	183 bp	212 bp	Not published	STRBase

*Allele designations shown in parentheses come from Brinkmann *et al.* (1996b)
29 observed alleles.

D3S1358

GenBank Accession: NT_005997 (allele 18). PCR product sizes of observed alleles

Allele (Repeat #)	Promega PowerPlex 2.1	ABI Profiler Plus (w/o + A)	Repeat Structure TCTA[TCTG]$_{2-3}$[TCTA]$_n$	Reference
8	99 bp	97 bp	Not published	STRBase
9	103 bp	101 bp	Not published	STRBase
10	107 bp	105 bp	Not published	STRBase
11	111 bp	109 bp	Not published	STRBase
12	115 bp	113 bp	Not published	SGM Plus
13	119 bp	117 bp	TCTA[TCTG]$_2$[TCTA]$_{10}$	Mornhinweg *et al.* (1998)
14	123 bp	121 bp	TCTA[TCTG]$_2$[TCTA]$_{11}$	Szibor *et al.* (1998)

Allele (Repeat #)	Promega PowerPlex 2.1	ABI Profiler Plus (w/o + A)	Repeat Structure TCTA[TCTG]$_{2-3}$[TCTA]$_n$	Reference
15	127 bp	125 bp	TCTA[TCTG]$_3$[TCTA]$_{11}$	Szibor *et al.* (1998)
15′	127 bp	125 bp	TCTA[TCTG]$_2$[TCTA]$_{12}$	Szibor *et al.* (1998)
15.1	128 bp	126 bp	Not published	STRBase
15.2	129 bp	127 bp	Not published	STRBase
15.3	130 bp	128 bp	Not published	STRBase
16	131 bp	129 bp	TCTA[TCTG]$_3$[TCTA]$_{12}$	Szibor *et al.* (1998)
16′	131 bp	129 bp	TCTA[TCTG]$_2$[TCTA]$_{13}$	Mornhinweg *et al.* (1998)
16.2	133 bp	131 bp	Not published	Budowle *et al.* (1997)
17	135 bp	133 bp	TCTA[TCTG]$_3$[TCTA]$_{13}$	Szibor *et al.* (1998)
17′	135 bp	133 bp	TCTA[TCTG]$_2$[TCTA]$_{14}$	Mornhinweg *et al.* (1998)
17.1	136 bp	134 bp	Not published	STRBase
17.2	137 bp	135 bp	Not published	STRBase
18	139 bp	137 bp	TCTA[TCTG]$_3$[TCTA]$_{14}$	Szibor *et al.* (1998)
18.2	141 bp	139 bp	Not published	STRBase
18.3	142 bp	140 bp	Not published	STRBase
19	143 bp	141 bp	TCTA[TCTG]$_3$[TCTA]$_{15}$	Mornhinweg *et al.* (1998)
20	147 bp	145 bp	TCTA[TCTG]$_3$[TCTA]$_{16}$	Mornhinweg *et al.* (1998)
21	151 bp	149 bp	Not published	Ayres *et al.* (2002)

25 observed alleles.

D5S818

GenBank Accession: AC008512 (allele 11) or G08446 (allele 11). PCR product sizes of observed alleles

Allele (Repeat #)	Promega PowerPlex 1.1	ABI Profiler Plus (w/o + A)	Repeat Structure [AGAT]$_n$	Reference
7	119 bp	134 bp	[AGAT]$_7$	Lins *et al.* (1998)
8	123 bp	138 bp	[AGAT]$_8$	Lins *et al.* (1998)
9	127 bp	142 bp	[AGAT]$_9$	Lins *et al.* (1998)
10	131 bp	146 bp	[AGAT]$_{10}$	Lins *et al.* (1998)
10.1	132 bp	147 bp	Not published	STRBase
11	135 bp	150 bp	[AGAT]$_{11}$	Lins *et al.* (1998)
11.1	136 bp	151 bp	Not published	STRBase
12	139 bp	154 bp	[AGAT]$_{12}$	Lins *et al.* (1998)
12.3	142 bp	157 bp	Not published	STRBase
13	143 bp	158 bp	[AGAT]$_{13}$	Lins *et al.* (1998)
14	147 bp	162 bp	[AGAT]$_{14}$	Lins *et al.* (1998)
15	151 bp	166 bp	[AGAT]$_{15}$	Lins *et al.* (1998)
16	155 bp	170 bp	[AGAT]$_{16}$	Profiler Plus
17	159 bp	174 bp	[AGAT]$_{17}$	STRBase
18	163 bp	178 bp	[AGAT]$_{18}$	STRBase

15 observed alleles.

D7S820

GenBank Accession: AC004848 (allele 13) or G08616 (allele 12). PCR product sizes of observed alleles

Allele (Repeat #)	Promega PowerPlex 1.1	ABI Profiler Plus (w/o + A)	Repeat Structure [GATA]$_n$	Reference
5	211 bp	253 bp	Not published	STRBase
5.2	213 bp	255 bp	Not published	STRBase
6	215 bp	257 bp	[GATA]$_6$	Lins *et al.* (1998)
6.3	218 bp	260 bp	Not published	Amorim *et al.* (2001)
7	219 bp	261 bp	[GATA]$_7$	Lins *et al.* (1998)
7.1	220 bp	262 bp	Not published	STRBase
7.3	222 bp	264 bp	Not published	Ayres *et al.* (2002)
8	223 bp	265 bp	[GATA]$_8$	Lins *et al.* (1998)
8.1	224 bp	266 bp	Not published	Balamurugan *et al.* (2000)
8.2	225 bp	267 bp	Not published	STRBase
8.3	226 bp	268 bp	Not published	STRBase
9	227 bp	269 bp	[GATA]$_9$	Lins *et al.* (1998)
9.1	228 bp	270 bp	Not published	Ayres *et al.* (2002)
9.2	229 bp	271 bp	Not published	STRBase
9.3	230 bp	272 bp	Not published	STRBase
10	231 bp	273 bp	[GATA]$_{10}$	Lins *et al.* (1998)
10.1	232 bp	274 bp	Not published	STRBase
10.3	234 bp	276 bp	Not published	STRBase
11	235 bp	277 bp	[GATA]$_{11}$	Lins *et al.* (1998)
11.1	236 bp	278 bp	Not published	STRBase
11.3	238 bp	280 bp	Not published	STRBase
12	239 bp	281 bp	[GATA]$_{12}$	Lins *et al.* (1998)
12.1	240 bp	282 bp	Not published	STRBase
12.3	242 bp	284 bp	Not published	STRBase
13	243 bp	285 bp	[GATA]$_{13}$	Lins *et al.* (1998)
13.1	244 bp	286 bp	Not published	STRBase
14	247 bp	289 bp	[GATA]$_{14}$	Lins *et al.* (1998)
14.1	248 bp	290 bp	Not published	STRBase
15	251 bp	293 bp	Not published	Profiler Plus
16	255 bp	297 bp	Not published	Ayres *et al.* (2002)

30 observed alleles.

D8S1179 (listed as D6S502 in early papers)

GenBank Accession: AF216671 (allele 13) or G08710 (allele 12). PCR product sizes of observed alleles

Allele (Repeat #)	Promega PowerPlex 2.1	ABI Profiler Plus (w/o + A)	Repeat Structure [TCTR]$_n$	Reference
7	203 bp	123 bp	[TCTA]$_7$	Griffiths *et al.* (1998)
8	207 bp	127 bp	[TCTA]$_8$	Barber and Parkin (1996)
9	211 bp	131 bp	[TCTA]$_9$	Barber and Parkin (1996)
10	215 bp	135 bp	[TCTA]$_{10}$	Barber and Parkin (1996)
11	219 bp	139 bp	[TCTA]$_{11}$	Barber and Parkin (1996)
12	223 bp	143 bp	[TCTA]$_{12}$	Barber and Parkin (1996)
13	227 bp	147 bp	[TCTA]$_1$[TCTG]$_1$[TCTA]$_{11}$	Barber and Parkin (1996)
14	231 bp	151 bp	[TCTA]$_2$[TCTG]$_1$[TCTA]$_{11}$	Barber and Parkin (1996)
15	235 bp	155 bp	[TCTA]$_2$[TCTG]$_1$[TCTA]$_{12}$	Barber and Parkin (1996)
15.3	238 bp	158 bp	Not published	STRBase

Allele (Repeat #)	Promega PowerPlex 2.1	ABI Profiler Plus (w/o + A)	Repeat Structure [TCTR]$_n$	Reference
16	239 bp	159 bp	[TCTA]$_2$[TCTG]$_1$[TCTA]$_{13}$	Barber and Parkin (1996)
17	243 bp	163 bp	[TCTA]$_2$[TCTG]$_2$[TCTA]$_{13}$	Barber and Parkin (1996)
18	247 bp	167 bp	[TCTA]$_2$[TCTG]$_1$[TCTA]$_{15}$	Barber and Parkin (1996)
19	251 bp	171 bp	[TCTA]$_2$[TCTG]$_2$[TCTA]$_{15}$	Griffiths *et al.* (1998)
20	255 bp	175 bp	Not published	STRBase

15 observed alleles.

D13S317

GenBank Accession: AL353628 (allele 11) or G09017 (allele 13). PCR product sizes of observed alleles

Allele (Repeat #)	Promega PowerPlex 1.1	ABI Profiler Plus (w/o + A)	Repeat Structure [TATC]$_n$	Reference
5	157 bp	193 bp	Not published	STRBase
6	161 bp	197 bp	Not published	STRBase
7	165 bp	201 bp	[TATC]$_7$	Lins *et al.* (1998)
7.1	166 bp	202 bp	Not published	STRBase
8	169 bp	205 bp	[TATC]$_8$	Lins *et al.* (1998)
8.1	170 bp	206 bp	Not published	STRBase
9	173 bp	209 bp	[TATC]$_9$	Lins *et al.* (1998)
10	177 bp	213 bp	[TATC]$_{10}$	Lins *et al.* (1998)
10'	177 bp	213 bp	[TATC]$_{10}$ AATC	Lins *et al.* (1998)
11	181 bp	217 bp	[TATC]$_{11}$	Lins *et al.* (1998)
12	185 bp	221 bp	[TATC]$_{12}$	Lins *et al.* (1998)
13	189 bp	225 bp	[TATC]$_{13}$	Lins *et al.* (1998)
13.3	192 bp	228 bp	Not published	STRBase
14	193 bp	229 bp	[TATC]$_{14}$	Lins *et al.* (1998)
14.3	196 bp	232 bp	Not published	STRBase
15	197 bp	233 bp	[TATC]$_{15}$	Lins *et al.* (1998)
16	201 bp	237 bp	Not published	STRBase

17 observed alleles.

D16S539

GenBank Accession: AC024591 (allele 11) or G07925 (allele 11). PCR product sizes of observed alleles

Allele (Repeat #)	Promega PowerPlex 1.1	ABI COfiler (w/o + A)	Repeat Structure [GATA]$_n$	Reference
5	264 bp	233 bp	[GATA]$_5$	Lins *et al.* (1998)
6	268 bp	237 bp	[GATA]$_6$	STRBase
7	272 bp	241 bp	[GATA]$_7$	STRBase
8	276 bp	245 bp	[GATA]$_8$	Lins *et al.* (1998)
9	280 bp	249 bp	[GATA]$_9$	Lins *et al.* (1998)
9.3	283 bp	252 bp	Not published	STRBase
10	284 bp	253 bp	[GATA]$_{10}$	Lins *et al.* (1998)

Allele (Repeat #)	Promega PowerPlex 1.1	ABI COfiler (w/o + A)	Repeat Structure [GATA]$_n$	Reference
11	288 bp	257 bp	[GATA]$_{11}$	Lins et al. (1998)
11.3	291 bp	260 bp	Not published	STRBase
12	292 bp	261 bp	[GATA]$_{12}$	Lins et al. (1998)
12.1	293 bp	262 bp	Not published	STRBase
12.2	294 bp	263 bp	Not published	STRBase
13	296 bp	265 bp	[GATA]$_{13}$	Lins et al. (1998)
13.1	297 bp	266 bp	Not published	STRBase
13.3	299 bp	268 bp	Not published	STRBase
14	300 bp	269 bp	[GATA]$_{14}$	Lins et al. (1998)
14.3	303 bp	272 bp	Not published	STRBase
15	304 bp	273 bp	[GATA]$_{15}$	Lins et al. (1998)
16	308 bp	277 bp	Not published	STRBase

19 observed alleles.

D18S51

GenBank Accession: AP001534 (allele 18) or L18333 (allele 13). PCR product sizes of observed alleles

Allele (Repeat #)	Promega PowerPlex 2.1	ABI Profiler Plus (w/o + A)	Repeat Structure [AGAA]$_n$	Reference
7	286 bp	264 bp	Not published	STRBase
8	290 bp	268 bp	[AGAA]$_8$	Griffiths et al. (1998)
9	294 bp	272 bp	[AGAA]$_9$	Barber and Parkin (1996)
9.2	296 bp	274 bp	Not published	SGM Plus
10	298 bp	276 bp	[AGAA]$_{10}$	Barber and Parkin (1996)
10.2	300 bp	278 bp	Not published	SGM Plus
11	302 bp	280 bp	[AGAA]$_{11}$	Barber and Parkin (1996)
11.2	304 bp	282 bp	Not published	STRBase
12	306 bp	284 bp	[AGAA]$_{12}$	Barber and Parkin (1996)
12.2	308 bp	286 bp	Not published	STRBase
12.3	309 bp	287 bp	Not published	STRBase
13	310 bp	288 bp	[AGAA]$_{13}$	Barber and Parkin (1996)
13.1	311 bp	289 bp	Not published	STRBase
13.2	312 bp	290 bp	[AGAA]$_{13}$ AG	Barber and Parkin (1996)
13.3	313 bp	291 bp	Not published	STRBase
14	314 bp	292 bp	[AGAA]$_{14}$	Barber and Parkin (1996)
14.2	316 bp	294 bp	[AGAA]$_{14}$ AG	Barber and Parkin (1996)
15	318 bp	296 bp	[AGAA]$_{15}$	Barber and Parkin (1996)
15.1	319 bp	297 bp	Not published	SGM Plus
15.2	320 bp	298 bp	[AGAA]$_{15}$ AG	Barber and Parkin (1996)
15.3	321 bp	299 bp	Not published	SGM Plus
16	322 bp	300 bp	[AGAA]$_{16}$	Barber and Parkin (1996)
16.1	323 bp	301 bp	Not published	STRBase
16.2	324 bp	302 bp	Not published	STRBase
16.3	325 bp	303 bp	Not published	STRBase
17	326 bp	304 bp	[AGAA]$_{17}$	Barber and Parkin (1996)
17.1	327 bp	305 bp	Not published	SGM Plus
17.2	328 bp	306 bp	[AGAA]$_{17}$ AG	Gill et al. (1996)
17.3	329 bp	307 bp	Not published	STRBase
18	330 bp	308 bp	[AGAA]$_{18}$	Barber and Parkin (1996)
18.1	331 bp	309 bp	Not published	STRBase
18.2	332 bp	311 bp	Not published	STRBase

Allele (Repeat #)	Promega PowerPlex 2.1	ABI Profiler Plus (w/o + A)	Repeat Structure [AGAA]$_n$	Reference
19	334 bp	312 bp	[AGAA]$_{19}$	Barber and Parkin (1996)
19.2	336 bp	314 bp	[AGAA]$_{19}$ AG	Gill et al. (1996)
20	338 bp	316 bp	[AGAA]$_{20}$	Barber and Parkin (1996)
20.1	339 bp	317 bp	Not published	STRBase
20.2	340 bp	318 bp	Not published	STRBase
21	342 bp	320 bp	[AGAA]$_{21}$	Barber and Parkin (1996)
21.2	344 bp	322 bp	Not published	STRBase
22	346 bp	324 bp	[AGAA]$_{22}$	Barber and Parkin (1996)
22.1	347 bp	325 bp	Not published	STRBase
22.2	348 bp	327 bp	Not published	STRBase
23	350 bp	328 bp	[AGAA]$_{23}$	Barber and Parkin (1996)
23.1	351 bp	329 bp	Not published	SGM Plus
23.2	352 bp	330 bp	Not published	STRBase
24	354 bp	332 bp	[AGAA]$_{24}$	Barber and Parkin (1996)
25	358 bp	336 bp	[AGAA]$_{25}$	Barber and Parkin (1996)
26	362 bp	340 bp	[AGAA]$_{26}$	Barber and Parkin (1996)
27	366 bp	344 bp	[AGAA]$_{27}$	Barber and Parkin (1996)
28.1	371 bp	349 bp	Not published	STRBase
39.2	416 bp	394 bp	Not published	STRBase

51 observed alleles.

D21S11

GenBank Accession: AP000433 (allele 29). PCR product sizes of observed alleles

Allele (Repeat #)	Promega PowerPlex 2.1	ABI Profiler Plus (w/o + A)	Repeat Structure [TCTA]$_n$[TCTG]$_n${[TCTA]$_3$TA[TCTA]$_3$TCA [TCTA]$_2$TCCATA} [TCTA]$_n$ TA TCTA	Reference
12	155 bp	138 bp	Not published	Ayres et al. (2002)
24 (53)*	203 bp	186 bp	[TCTA]$_4$[TCTG]$_6${43 bp}[TCTA]$_6$	Griffiths et al. (1998)
24.2 (54)	205 bp	188 bp	[TCTA]$_5$[TCTG]$_6$ {[TCTA]$_3$ TCA [TCTA]$_2$ TCCA TA}[TCTA]$_9$	Griffiths et al. (1998)
24.3	206 bp	189 bp	Not published	STRBase
25 (55)	207 bp	190 bp	[TCTA]$_4$[TCTG]$_3${43 bp}[TCTA]$_{10}$	Schwartz et al. (1996)
25.2 (56)	209 bp	192 bp	[TCTA]$_5$[TCTG]$_6$ {[TCTA]$_3$ TCA [TCTA]$_2$ TCCA TA][TCTA]$_{10}$	Griffiths et al. (1998)
25.3	210 bp	193 bp	Not published	STRBase
26 (57)	211 bp	194 bp	[TCTA]$_4$[TCTG]$_6${43 bp}[TCTA]$_8$	Moller et al. (1994)
26.1	212 bp	195 bp	Not published	SGM Plus
26.2	213 bp	196 bp	Not published	STRBase
27 (59)	215 bp	198 bp	[TCTA]$_4$[TCTG]$_6${43 bp}[TCTA]$_9$	Moller et al. (1994)
27' (59)	215 bp	198 bp	[TCTA]$_6$[TCTG]$_5${43 bp}[TCTA]$_8$	Schwartz et al. (1996)
27" (59)	215 bp	198 bp	[TCTA]$_5$[TCTG]$_5${43 bp}[TCTA]$_9$	Griffiths et al. (1998)
27.1	216 bp	199 bp	Not published	STRBase
27.2	217 bp	200 bp	Not published	STRBase
27.3	218 bp	201 bp	Not published	SGM Plus
28 (61)	219 bp	202 bp	[TCTA]$_4$[TCTG]$_6${43 bp}[TCTA]$_{10}$	Moller et al. (1994)
28'	219 bp	202 bp	[TCTA]$_5$[TCTG]$_6${43 bp}[TCTA]$_9$	Zhou et al. (1997)
28.2 (62)	221 bp	204 bp	[TCTA]$_4$[TCTG]$_6${43 bp}[TCTA]$_{10}$	Griffiths et al. (1998)
28.2'	221 bp	204 bp	[TCTA]$_5$[TCTG]$_6${43 bp}[TCTA]$_8$ TA TCTA	Zhou et al. (1997)
28.3	222 bp	205 bp	Not published	SGM Plus

Allele (Repeat #)	Promega PowerPlex 2.1	ABI Profiler Plus (w/o + A)	Repeat Structure $[TCTA]_n[TCTG]_n\{[TCTA]_3TA[TCTA]_3TCA$ $[TCTA]_2TCCATA\}$ $[TCTA]_n$ TA TCTA	Reference
29 (63)	223 bp	206 bp	$[TCTA]_4[TCTG]_6\{43\,bp\}[TCTA]_{11}$	Griffiths *et al.* (1998)
29' (63)	223 bp	206 bp	$[TCTA]_6[TCTG]_5\{43\,bp\}[TCTA]_{10}$	Zhou *et al.* (1997)
29.1	224 bp	207 bp	Not published	STRBase
29.2 (64)	225 bp	208 bp	$[TCTA]_5[TCTG]_5\{43\,bp\}[TCTA]_{10}$ TATCTA	Zhou *et al.* (1997)
29.3	226 bp	209 bp	Not published	Amorim *et al.* (2001)
30 (65)	227 bp	210 bp	$[TCTA]_4[TCTG]_6\{43\,bp\}[TCTA]_{12}$	Schwartz *et al.* (1996)
30' (65)	227 bp	210 bp	$[TCTA]_5[TCTG]_6\{43\,bp\}[TCTA]_{11}$	Zhou *et al.* (1997)
30" (65)	227 bp	210 bp	$[TCTA]_6[TCTG]_5\{43\,bp\}[TCTA]_{11}$	Griffiths *et al.* (1998)
30''' (65)	227 bp	210 bp	$[TCTA]_6[TCTG]_6\{43\,bp\}[TCTA]_{10}$	Brinkmann *et al.* (1996a)
30.1	228 bp	211 bp	Not published	SGM Plus
30.2 (66)	229 bp	212 bp	$[TCTA]_5[TCTG]_6\{43\,bp\}[TCTA]_{10}$ TATCTA	Griffiths *et al.* (1998)
30.2' (66)	229 bp	212 bp	$[TCTA]_5[TCTG]_5\{43\,bp\}[TCTA]_{11}$ TATCTA	Schwartz *et al.* (1996)
30.3	230 bp	213 bp	Not published	STRBase
31 (67)	231 bp	214 bp	$[TCTA]_5[TCTG]_6\{43\,bp\}[TCTA]_{12}$	Griffiths *et al.* (1998)
31' (67)	231 bp	214 bp	$[TCTA]_6[TCTG]_5\{43\,bp\}[TCTA]_{12}$	Moller *et al.* (1994)
31" (67)	231 bp	214 bp	$[TCTA]_6[TCTG]_5\{43\,bp\}[TCTA]_{11}$	Zhou *et al.* (1997)
31''' (67)	231 bp	214 bp	$[TCTA]_7[TCTG]_5\{43\,bp\}[TCTA]_{11}$	Schwartz *et al.* (1996)
31.1	232 bp	215 bp	Not published	STRBase
31.2 (68)	233 bp	216 bp	$[TCTA]_5[TCTG]_6\{43\,bp\}[TCTA]_{11}$ TATCTA	Griffiths *et al.* (1998)
31.3	234 bp	217 bp	Not published	SGM Plus
32 (69)	235 bp	218 bp	$[TCTA]_6[TCTG]_5\{43\,bp\}[TCTA]_{13}$	Griffiths *et al.* (1998)
32' (69)	235 bp	218 bp	$[TCTA]_5[TCTG]_6\{43\,bp\}[TCTA]_{13}$	Zhou *et al.* (1997)
32.1	236 bp	219 bp	Not published	Steinlechner *et al.* (2002)
32.2 (70)	237 bp	220 bp	$[TCTA]_5[TCTG]_6\{43\,bp\}[TCTA]_{12}$ TATCTA	Griffiths *et al.* (1998)
32.2' (70)	237 bp	220 bp	$[TCTA]_4[TCTG]_6\{43\,bp\}[TCTA]_{13}$ TATCTA	Brinkmann *et al.* (1996a)
32.2" (70)	237 bp	220 bp	$[TCTA]_5[TCTG]_6\{[TCTA]_2TA[TCTA]_3$ TCA $[TCTA]_2TCCA$ TA$][TCTA]_{13}$ TATCTA	Brinkmann *et al.* (1996a)
32.3	238 bp	221 bp	Not published	SGM Plus
33 (71)	239 bp	222 bp	$[TCTA]_5[TCTG]_6\{43\,bp\}[TCTA]_{14}$	Zhou *et al.* (1997)
33.1	240 bp	223 bp	Not published	Steinlechner *et al.* (2002)
33.2 (72)	241 bp	224 bp	$[TCTA]_5[TCTG]_6\{43\,bp\}[TCTA]_{13}$ TATCTA	Griffiths *et al.* (1998)
33.2' (72)	241 bp	224 bp	$[TCTA]_6[TCTG]_5\{43\,bp\}[TCTA]_{13}$ TATCTA	Brinkmann *et al.* (1996a)
33.2" (72)	241 bp	224 bp	$[TCTA]_6[TCTG]_6\{43\,bp\}[TCTA]_{12}$ TATCTA	Brinkmann *et al.* (1996a)
33.3	242 bp	225 bp	$[TCTA]_5[TCTG]_6\{43\,bp\}[TCTA]_8$ TCA $[TCTA]_3$ TCA $[TCTA]_2$ TA TCTA	Brinkmann *et al.* (1996a)
34 (73)	243 bp	226 bp	$[TCTA]_5[TCTG]_6\{43\,bp\}[TCTA]_{15}$	Zhou *et al.* (1997)
34' (73)	243 bp	226 bp	$[TCTA]_{10}[TCTG]_5\{43\,bp\}[TCTA]_{11}$	Brinkmann *et al.* (1996a)
34.1	244 bp	227 bp	Not published	Ayres *et al.* (2002)
34.2 (74)	245 bp	228 bp	$[TCTA]_5[TCTG]_6\{43\,bp\}[TCTA]_{14}$ TATCTA	Griffiths *et al.* (1998)
34.3	246 bp	229 bp	$[TCTA]_5[TCTG]_6\{43\,bp\}[TCTA]_{10}$ TCA $[TCTA]_4$ TA TCTA	Brinkmann *et al.* (1996a)
35 (75)	247 bp	230 bp	$[TCTA]_{10}[TCTG]_5\{43\,bp\}[TCTA]_{12}$	Griffiths *et al.* (1998)
35' (75)	247 bp	230 bp	$[TCTA]_{11}[TCTG]_5\{43\,bp\}[TCTA]_{11}$	Brinkmann *et al.* (1996a)
35.1	248 bp	231 bp	Not published	Steinlechner *et al.* (2002)
35.2 (76)	249 bp	232 bp	$[TCTA]_5[TCTG]_6\{43\,bp\}[TCTA]_{15}$ TATCTA	Zhou *et al.* (1997)
35.3	250 bp	233 bp	Not published	SGM Plus
36 (77)	251 bp	234 bp	$[TCTA]_{11}[TCTG]_5\{43\,bp\}[TCTA]_{12}$	Griffiths *et al.* (1998)
36' (77)	251 bp	234 bp	$[TCTA]_{10}[TCTG]_5\{43\,bp\}[TCTA]_{13}$	Brinkmann *et al.* (1996a)
36" (77)	251 bp	234 bp	$[TCTA]_{10}[TCTG]_6\{43\,bp\}[TCTA]_{12}$	Brinkmann *et al.* (1996a)
36.1	252 bp	235 bp	Not published	STRBase
36.2	253 bp	236 bp	$[TCTA]_5[TCTG]_6\{43\,bp\}[TCTA]_{16}$ TATCTA	Zhou *et al.* (1997)
36.3	254 bp	237 bp	Not published	SGM Plus
37 (78)	255 bp	238 bp	$[TCTA]_{11}[TCTG]_5\{43\,bp\}[TCTA]_{13}$	Griffiths *et al.* (1998)
37' (78)	255 bp	238 bp	$[TCTA]_9[TCTG]_{11}\{43\,bp\}[TCTA]_{12}$	Brinkmann *et al.* (1996a)

Allele (Repeat #)	Promega PowerPlex 2.1	ABI Profiler Plus (w/o + A)	Repeat Structure [TCTA]$_n$[TCTG]$_n$\{[TCTA]$_3$TA[TCTA]$_3$TCA [TCTA]$_2$TCCATA\} [TCTA]$_n$ TA TCTA	Reference
37.2	257 bp	240 bp	[TCTA]$_7$[TCTG]$_{14}$\{[TCTA]$_3$ TCA [TCTA]$_2$ TCCA TA\}[TCTA]$_{12}$	Walsh et al. (2003)
37.2'	257 bp	240 bp	[TCTA]$_9$[TCTG]$_{12}$\{[TCTA]$_3$ TCA [TCTA]$_2$ TCCA TA\}[TCTA]$_{12}$	Walsh et al. (2003)
37.2"	257 bp	240 bp	[TCTA]$_9$[TCTG]$_{13}$\{[TCTA]$_3$ TCA [TCTA]$_2$ TCCA TA\}[TCTA]$_{11}$	Walsh et al. (2003)
37.2'''	257 bp	240 bp	[TCTA]$_{10}$[TCTG]$_{11}$\{[TCTA]$_3$ TCA [TCTA]$_2$ TCCA TA\}[TCTA]$_{12}$	Walsh et al. (2003)
37.2''''	257 bp	240 bp	[TCTA]$_{11}$[TCTG]$_{11}$\{[TCTA]$_3$ TCA [TCTA]$_2$ TCCA TA\}[TCTA]$_{11}$	Walsh et al. (2003)
38 (79)	259 bp	242 bp	[TCTA]$_{13}$[TCTG]$_5$\{43 bp\}[TCTA]$_{12}$	Griffiths et al. (1998)
38' (79)	259 bp	242 bp	[TCTA]$_9$[TCTG]$_{11}$\{43 bp\}[TCTA]$_{12}$	Brinkmann et al. (1996a)
38" (79)	259 bp	242 bp	[TCTA]$_{10}$[TCTG]$_{11}$\{43 bp\}[TCTA]$_{13}$	Brinkmann et al. (1996a)
38''' (79)	259 bp	242 bp	[TCTA]$_{11}$[TCTG]$_{11}$\{43 bp\}[TCTA]$_{11}$	Brinkmann et al. (1996a)
38.2	261 bp	244 bp	[TCTA]$_9$[TCTG]$_{12}$\{[TCTA]$_3$ TCA [TCTA]$_2$ TCCA TA\}[TCTA]$_{13}$	Walsh et al. (2003)
38.2'	261 bp	244 bp	[TCTA]$_9$[TCTG]$_{13}$\{[TCTA]$_3$ TCA [TCTA]$_2$ TCCA TA\}[TCTA]$_{12}$	Walsh et al. (2003)
38.2"	261 bp	244 bp	[TCTA]$_{10}$[TCTG]$_{11}$\{[TCTA]$_3$ TCA [TCTA]$_2$ TCCA TA\}[TCTA]$_{13}$	Walsh et al. (2003)
39	263 bp	246 bp	Not published	STRBase
39.2	265 bp	248 bp	[TCTA]$_{10}$[TCTG]$_{13}$\{[TCTA]$_3$ TCA [TCTA]$_2$ TCCA TA\}[TCTA]$_{12}$	Bagdonavicius et al. (2002), Walsh et al. (2003)
39.2'	265 bp	248 bp	[TCTA]$_{11}$[TCTG]$_{12}$\{[TCTA]$_3$ TCA [TCTA]$_2$ TCCA TA\}[TCTA]$_{12}$	Bagdonavicius et al. (2002), Walsh et al. (2003)
40.2	269 bp	252 bp	Not published	Ayres et al. (2002)
41.2	273 bp	256 bp	[TCTA]$_{10}$[TCTG]$_{15}$\{[TCTA]$_3$ TCA [TCTA]$_2$ TCCA TA\}[TCTA]$_{12}$	Bagdonavicius et al. (2002), Walsh et al. (2003)

*Allele designations shown in parentheses come from Urquhart et al. (1994).
89 observed alleles.

REFERENCES

Amorim, A., Gusmao, L. and Prata, M. J. (1996) *Advances in Forensic Haemogenetics*, 6, 486–488.

Amorim, A., Gusmao, L. and Alves, C. (2001) *Forensic Science International*, 115, 119–121.

Applied Biosystems (1998) AmpF/STR® COfiler™ PCR Amplification Kit User's Bulletin. Foster City, California: Applied Biosystems.

Applied Biosystems (1998) AmpF/STR® Profiler Plus™ PCR Amplification Kit User's Manual. Foster City, California: Applied Biosystems.

Applied Biosystems (1999) AmpF/STR® SGM Plus™ PCR Amplification Kit User's Manual. Foster City, California: Applied Biosystems.

Ayres, K.L., Chaseling, J. and Balding, D.J. (2002) *Forensic Science International*, 129, 90–98.

Bagdonavicius, A., Turbett, G.R., Buckleton, J.S. and Walsh, S.J. (2002) *Journal of Forensic Sciences*, 47, 1149–1153.

Balamurugan, K., Budowle, B. and Tahir, M.A. (2000) *Journal of Forensic Sciences*, 45, 744–746.

Barber, M.D., McKeown, B.J. and Parkin, B.H. (1996) *International Journal of Legal Medicine*, 108, 180–185.

Barber, M.D. and Parkin, B.H. (1996) *International Journal of Legal Medicine,* 109, 62–65.

Brinkmann, B., Meyer, E. and Junge, A. (1996a) *Human Genetics*, 98, 60–64.

Brinkmann, B., Sajantila, A., Goedde, H.W., Matsumoto, H., Nishi, K. and Wiegand, P. (1996b) *European Journal of Human Genetics*, 4, 175–182.

Budowle, B., Nhari, L.T., Moretti, T.R., Kanoyangwa, S.B., Masuka, E., Defenbaugh, D.A. and Smerick, J.B. (1997) *Forensic Science International*, 90, 215–221.

Budowle, B. and Moretti, T.R. (1998) *Proceedings from the Ninth International Symposium on Human Identification*, pp. 64–73. Madison, Wisconsin: Promega Corporation.

Espinheira, R., Geada, H., Ribeiro, T. and Reys, L. (1996) *Advances in Forensic Haemogenetics,* 6, 528.

Gene, M., Huguet, E., Moreno, P., Sanchez, C., Carracedo, A. and Corbella, J. (1996) *International Journal of Legal Medicine*, 108, 318–320.

Gill, P., Kimpton, C.P., Urquhart, A., Oldroyd, N., Millican, E.S., Watson, S.K. and Downes, T.J. (1995) *Electrophoresis*, 16, 1543–1552.

Gill, P., Urquhart, A., Millican, E.S., Oldroyd, N.J., Watson, S., Sparkes, R. and Kimpton, C.P. (1996) *International Journal of Legal Medicine*, 109, 14–22.

Griffiths, R.A.L., Barber, M.D., Johnson, P.E., Gillbard, S.M., Haywood, M.D., Smith, C.D., Arnold, J., Burke, T., Urquhart, A. and Gill, P. (1998) *International Journal of Legal Medicine,* 111, 267–272.

Huang, N.E., Schumm, J.W. and Budowle, B. (1995) *Forensic Science International,* 71, 131–136.

Kido, A., Hara, M., Yamamoto, Y., Kameyama, H., Susukida, R., Saito, K., Takada, A. and Oya, M. (2003) *Legal Medicine (Tokyo)*, 5, 93–96.

Klintschar, M., Kozma, Z., Al Hammadi, N., Fatah, M.A. and Nohammer, C. (1998) *International Journal of Legal Medicine,* 111, 107–109.

Lazaruk, K., Walsh, P.S., Oaks, F., Gilbert, D., Rosenblum, B.B., Menchen, S., Scheibler, D., Wenz, H.M., Holt, C. and Wallin, J. (1998) *Electrophoresis*, 19, 86–93.

Lins, A.M., Micka, K.A., Sprecher, C.J., Taylor, J.A., Bacher, J.W., Rabbach, D., Bever, R.A., Creacy, S. and Schumm, J.W. (1998) *Journal of Forensic Sciences*, 43, 1178–1190.

Luis, J., Liste, I. and Caeiro, B. (1994) *Advances in Forensic Haemogenetics*, 5, 366–368.

Margolis-Nunno, H., Brenner, L., Cascardi, J. and Kobilinsky, L. (2001) *Journal of Forensic Sciences*, 46, 1480–1483.

Moller, A., Meyer, E. and Brinkmann, B. (1994) *International Journal of Legal Medicine*, 106, 319–323.

Mornhinweg, E., Luckenbach, C., Fimmers, R. and Ritter, H. (1998) *Forensic Science International*, 95, 173–178.

Puers, C., Lins, A. M., Sprecher, C. J., Brinkmann, B. and Schumm, J. W. (1993) *Proceedings from the Fourth International Symposium on Human Identification*, pp. 161–172. Madison, Wisconsin: Promega Corporation.

Scherczinger, C.A., Hintz, J.L., Peck, B.J., Adamowicz, M.S., Bourke, M.T., Coyle, H.M., Ladd, C., Yang, N.C.S., Budowle, B. and Lee, H.C. (2000) *Journal of Forensic Sciences*, 45, 938–940.

Schwartz, D. W. M., Dauber, E. M., Glock, B. and Mayr, W. R. (1996) *Advances in Forensic Haemogenetics*, 5, 622–625.

Steinlechner, M., Schmidt, K., Kraft, H.G., Utermann, G. and Parson, W. (2002) *International Journal of Legal Medicine*, 116, 176–178.

STRBase variant allele report; http://www.cstl.nist.gov/biotech/strbase/var_tab.htm.

Szibor, R., Lautsch, S., Plate, I., Bender, K. and Krause, D. (1998) *International Journal of Legal Medicine*, 111, 160–161.

Urquhart, A., Kimpton, C.P., Downes, T.J. and Gill, P. (1994) *International Journal of Legal Medicine*, 107, 13–20.

van Oorschot, R.A.H., Gutowski, S.J. and Robinson, S.L. (1994) *International Journal of Legal Medicine*, 107, 121–126.

Walsh, S.J., Robinson, S.L., Turbett, G.R., Davies, N.P. and Wilton, A.N. (2003) *Forensic Science International*, 135, 35–41.

Zhou, H.G., Sato, K., Nishimaki, Y., Fang, L. and Hasekura, H. (1997) *Forensic Science International*, 86, 109–188.

APPENDIX II

U.S. POPULATION DATA – STR ALLELE FREQUENCIES

The U.S. population data contained within Appendix II were collected using the Identifiler STR kit and the ABI 3100 and were previously published in Butler *et al.* (2003) *Journal of Forensic Science*, 48 (4), 908–911. For complete genotypes of the individuals used to generate these allele tables, see: http://www.cstl.nist.gov/biotech/strbase/NISTpop.htm.

Allele frequencies from the most common allele in each population are highlighted in bold font. Allele frequencies denoted with an asterisk (*) are below the 5/2N minimum allele threshold recommended by the National Research Council report (NRCII) *The Evaluation of Forensic DNA Evidence* published in 1996. The following minimum allele frequencies should be substituted into genotype frequency calculations for the designated alleles falling below the 5/2N threshold:

Caucasians 0.00828 (N = 302),
African-Americans 0.00969 (N = 258), and
Hispanics 0.01786 (N = 140).

The allele frequencies contained in this Appendix are used throughout the book to illustrate calculations of random match probabilities and STR profile frequencies.

CSF1PO

Allele	Caucasian N = 302	African-American N = 258	Hispanic N = 140
7		0.05253	0.02143
8	0.00497*	0.06031	
9	0.01159	0.03696	0.02143
10	0.21689	0.25681	0.23214
10.3			0.00357
11	0.30132	0.24903	0.29286
12	**0.36093**	**0.29767**	**0.35714**
13	0.09603	0.03696	0.06071
14	0.00828	0.00973	0.00714*
15			0.00357*

FGA

Allele	Caucasian N = 302	African-American N = 258	Hispanic N = 140
16.2		0.00194*	
18	0.02649	0.00194*	0.01786
18.2		0.01163	
19	0.05298	0.06202	0.06429
19.2		0.00388*	
20	0.12748	0.05620	0.08929
21	0.18543	0.11628	**0.16786**
21.2	0.00497*		
22	**0.21854**	**0.19574**	0.15000
22.2	0.01159	0.00388*	
22.3		0.00194*	
23	0.13411	0.17054	0.13571
23.2	0.00331*	0.00194*	0.00357*
24	0.13576	0.12209	0.15000
24.2	0.00166*		
25	0.07119	0.12403	0.12143
26	0.02318	0.08140	0.05357
27	0.00331*	0.02326	0.04286
28		0.01163	
29		0.00388*	
30		0.00194*	0.00357*
30.2		0.00194*	
31.2		0.00194*	

TH01

Allele	Caucasian N = 302	African-American N = 258	Hispanic N = 140
5	0.00166*	0.00388*	
6	0.23179	0.12403	0.21429
7	0.19040	**0.42054**	**0.27857**
8	0.08444	0.19380	0.09643
9	0.11424	0.15116	0.15000
9.3	**0.36755**	0.10465	0.24643
10	0.00828	0.00194*	0.01429*
11	0.00166*		

TPOX

Allele	Caucasian N = 302	African-American N = 258	Hispanic N = 140
5	0.00166*		
6	0.00166*	0.10078	0.00357*
7		0.01744	0.00714*
8	**0.53477**	**0.37209**	**0.47143**
9	0.11921	0.17829	0.10357
10	0.05629	0.08915	0.03214
11	0.24338	0.21899	0.27500
12	0.04139	0.02132	0.10714
13	0.00166*	0.00194*	

VWA

Allele	Caucasian N = 302	African-American N = 258	Hispanic N = 140
12		0.00194*	
13	0.00166*	0.00775*	
14	0.09437	0.07752	0.08571
15	0.11093	0.18605	0.16786
16	0.20033	**0.24806**	**0.26429**
17	**0.28146**	0.24225	0.21786
18	0.20033	0.15504	0.17143
19	0.10430	0.06202	0.07857
20	0.00497*	0.01550	0.01071*
21	0.00166*	0.00388*	0.00357*

D3S1358

Allele	Caucasian N = 302	African-American N = 258	Hispanic N = 140
11	0.00166*		
13		0.00194*	0.00714*
14	0.10265	0.08915	0.07857
15	**0.26159**	0.30233	**0.29286**
15.2		0.00194*	
16	0.25331	**0.33527**	0.28571
17	0.21523	0.20543	0.20357
18	0.15232	0.06008	0.12500
19	0.01159	0.00388*	0.00714*
20	0.00166*		

D5S818

Allele	Caucasian N = 302	African-American N = 258	Hispanic N = 140
7	0.00166*		0.04286
8	0.00331*	0.04845	0.01071*
9	0.04967	0.03876	0.04286
10	0.05132	0.06977	0.06071
11	0.36093	0.23256	**0.35000**
12	**0.38411**	**0.35271**	**0.35000**
13	0.14073	0.23837	0.12500
14	0.00662*	0.01550	0.01429*
15	0.00166*	0.00388*	0.00357*

D7S820

Allele	Caucasian N = 302	African-American N = 258	Hispanic N = 140
6		0.00194*	
7	0.01821	0.01550	0.01429*
8	0.15066	0.23643	0.12143
8.1	0.00166*		
9	0.17715	0.10853	0.11071
9.3		0.00194*	
10	**0.24338**	**0.33140**	**0.29286**
10.3			0.00357*
11	0.20695	0.20349	0.25714
12	0.16556	0.08721	0.16429
13	0.03477	0.01357	0.03571
14	0.00166*		

D8S1179

Allele	Caucasian N = 302	African-American N = 258	Hispanic N = 140
8	0.01159	0.00194*	0.00714*
9	0.00331*	0.00581*	0.01071*
10	0.10099	0.02907	0.10000
11	0.08278	0.04457	0.05714
12	0.18543	0.14147	0.14286
13	**0.30464**	0.21705	**0.26786**
14	0.16556	**0.30039**	0.25000
15	0.11424	0.18411	0.12857
16	0.03146	0.06977	0.02500
17		0.00388*	0.00714*
18		0.00194*	0.00357*

D13S317

Allele	Caucasian N = 302	African-American N = 258	Hispanic N = 140
8	0.11258	0.03295	0.12143
9	0.07450	0.03295	0.15357
10	0.05132	0.02326	0.10000
11	**0.33940**	0.30620	**0.23571**
12	0.24834	**0.42442**	0.22143
13	0.12417	0.14535	0.11786
14	0.04801	0.03488	0.04643
15	0.00166*		0.00357*

D16S539

Allele	Caucasian N = 302	African-American N = 258	Hispanic N = 140
8	0.01821	0.03876	0.02500
9	0.11258	0.19574	0.13929
10	0.05629	0.11628	0.11786
11	0.32119	**0.31783**	**0.26071**
12	**0.32616**	0.19574	0.25357
13	0.14570	0.11822	0.18571
14	0.01987	0.01744	0.01786

D18S51

Allele	Caucasian N = 302	African-American N = 258	Hispanic N = 140
9		0.00389*	
10	0.00828	0.00584*	0.00357*
11	0.01656	0.00195*	0.01071*
12	0.12748	0.07782	0.11786
13	0.13245	0.05253	0.11071
13.2		0.00584*	
14	0.13742	0.07198	0.13929
14.2	0.00166*		
15	**0.15894**	**0.16148**	**0.18929**
15.2		0.00195*	
16	0.13907	0.15759	0.13571
17	0.12583	0.15175	0.12857
18	0.07616	0.12257	0.06786
19	0.03808	0.09922	0.03929
20	0.02152	0.06420	0.03214
21	0.00828	0.00973	0.01071*
21.2		0.00195*	
22	0.00828	0.00584*	0.01429*
23		0.00195*	
24		0.00195*	

D19S433

Allele	Caucasian N = 302	African-American N = 258	Hispanic N = 140
9			0.00357*
10	0.00166*	0.00969	
11	0.00497*	0.06202	0.01429*
12	0.08113	0.11434	0.06429
12.2	0.00166*	0.03488	0.01429*
13	0.25331	**0.24612**	0.25000
13.2	0.00662*	0.05233	0.03214

D19S433 *(continued)*

Allele	Caucasian N = 302	African-American N = 258	Hispanic N = 140
14	**0.36921**	0.22287	**0.37500**
14.2	0.01821	0.07946	0.04286
15	0.15232	0.07752	0.12143
15.2	0.03477	0.06008	0.03571
16	0.04967	0.00388*	0.02143
16.2	0.01490	0.02713	0.02500
17	0.00828		
17.2	0.00166*	0.00581*	
18.2	0.00166*	0.00388*	

D21S11

Allele	Caucasian N = 302	African-American N = 258	Hispanic N = 140
24.2			0.00357*
25.2	0.00166*		
26		0.00194*	
27	0.02649	0.07752	0.03571
28	0.15894	**0.25775**	0.09643
29	0.19536	0.19767	0.20000
29.2	0.00331*		0.00357*
30	**0.27815**	0.17442	**0.26071**
30.2	0.02815	0.00969	0.03929
31	0.08278	0.08140	0.08214
31.2	0.09934	0.04651	0.11071
32	0.00662*	0.00775*	0.00714*
32.2	0.08444	0.05814	0.12857
33	0.00166*	0.00581*	0.00357*
33.1		0.00194*	0.00357*
33.2	0.02649	0.03488	0.02143
34		0.00581*	0.00357*
34.2	0.00497*		
35	0.00166*	0.02326	
36		0.00969	
37		0.00194*	
38		0.00194*	
39		0.00194*	

D2S1338

Allele	Caucasian N = 302	African-American N = 258	Hispanic N = 140
15	0.00166*		
16	0.03311	0.05837	0.03571
17	**0.18212**	0.09922	**0.19643**
18	0.07947	0.03891	0.10000
19	0.11424	**0.14786**	0.17857
20	0.14570	0.10311	0.13571
21	0.04139	0.14397	0.03571
22	0.03808	0.13035	0.06071
23	0.11755	0.11089	0.09643
24	0.12252	0.07977	0.07143
25	0.09272	0.07198	0.07500
26	0.02980	0.01167	0.01429*
27	0.00166*	0.00389*	

APPENDIX III

SUPPLIERS OF DNA ANALYSIS EQUIPMENT, PRODUCTS, OR SERVICES

Company Name	Street Address	Contact Information	Products/Services
Abacus Diagnostics	P.O. Box 4040 West Hills, CA 91308	818-716-4735	ABAcard® HemaTrace®, PSA ABAcard® Tests
AFDIL^{CS} Consultative Services	1413 Research Blvd Building 101, 2nd Floor Rockville, MD 20850	301-319-0210 Fax: 301-295-5932 www.afip.org/oafme/dna	Forensic DNA analysis for mass fatality incidents
Affiliated Genetics	P.O. Box 58535 Salt Lake City, UT 84158	800-362-5559 Fax: 801-582-8460 www.affiliatedgenetics.com	Paternity testing, DNA banking
Affymetrix	3380 Central Expressway Santa Clara, CA 95051	888-362-2447 Fax: 408-731-5441 www.affymetrix.com	GeneChip DNA hybridization products
African Ancestry	5505 Connecticut Ave # 297 Washington, DC 20015	202-439-0641 Fax: 202-318-0742 www.africanancestry.com	Genetic genealogy with mtDNA and Y-STRs
Agilent	395 Page Mill Rd. Palo Alto, CA 94303	650-752-5000 www.agilent.com	Analysis instrumentation; BioAnalyzer 2100
AMRESCO Inc.	30175 Solon Industrial Pkwy Solon, Ohio 44139	800-366-1313 www.amresco-inc.com	Formamide, capillary electrophoresis buffers
American Type Culture Collection (ATCC)	10801 University Boulevard Manassas, VA 20110-2209	703-365-2700 Fax: 703-365-2750 www.atcc.org	Genomic DNA and cell cultures for research purposes
Amersham Biosciences	800 Centennial Avenue P.O. Box 1327 Piscataway, NJ 08855-1327	Tel: 800-526-3593 Fax: 877-295-8102 amershambiosciences.com	Molecular biology supplies; analysis instrumentation; ALF DNA Sequencer
Anjura Technology Corporation	6 Antares Drive Phase II, Suite 100 Ottawa ON K2E 8A9	613-727-1411 Fax: 613-727-1412 www.stacsdna.com	STaCS (sample tracking and control system) LIMS
Applied Biosystems	850 Lincoln Centre Drive Foster City, CA 94404	800-345-5224 Fax: 650-638-5884 www.appliedbiosystems.com	STR typing kits; thermal cyclers; analysis instrumentation; genotyping software; ABI 310 and 377
ARTEL	25 Bradley Drive Westbrook, ME 04092	888-406-3463 www.artel-usa.com	Pipet calibration

Company Name	Street Address	Contact Information	Products/Services
ASCLD/LAB	139 J Technology Drive Garner, NC 27529	919-773-2600 Fax: 919-773-2602 www.ascld-lab.org	Crime laboratory accreditation and audits
Baltimore Rh Typing Laboratory Inc.	400 West Franklin Street Baltimore, MD 21201	800-765-5170 Fax: 410-383-0938 www.rhlab.com	Paternity & immigration testing, forensic casework
Beckman Coulter, Inc.	4300 N. Harbor Blvd. Fullerton, CA 92834	800-742-2345 Fax: 800-643-4366 www.beckmancoulter.com	Analysis instrumentation and robotics for liquid handling
Bio-Rad Laboratories	2000 Alfred Nobel Drive Hercules, California 94547	800-424-6723 Fax: 800-879-2289 www.biorad.com	Chelex beads for DNA extraction
Biosynthesis Inc.	612 East Main St. Lewisville, TX 75057	800-227-0627 Fax: 972-420-0442 www.biosyn.com	Oligo synthesis, molecular biology products, paternity testing
Biotype AG	Moritzburger Weg 67 D-01109 Dresden GERMANY	Fon: +49 351 8838 400 Fax: +49 351 8838 403 www.biotype.de	Autosomal and Y-STR typing kits
BioVentures, Inc.	1435 Kensington Square Ct. Murfreesboro, TN 37130	877-852-7846 Fax: 877-286-0330 www.bioventures.com	DNA size standards
Bode Technology Group	7364 Steel Mill Drive Springfield, VA 22150	866-263-3443 Fax: 703-644-2319 www.bodetech.com	Contract forensic DNA testing and research
Cambrex BioScience Rockland, Inc. (formerly FMC Bioproducts)	191 Thomaston Street Rockland, ME 04841	207-594-3400 Fax: 207-594-3491 www.cambrex.com	LongRanger gels; agarose gel materials
CBR Laboratories, Inc. Parentage Testing	66 Cummings Park Woburn, MA 01801	781-938-3700 www.cbrlabs.com	Paternity testing
Columbia Laboratory Services Paternity Testing	4350 Oakes Road, Suite 522 Davie, FL 33314	800-952-2181 Fax: 954-625-0039 www.columbialab.com	Paternity testing See also www.testingpaternity.com
Commonwealth Biotechnology Inc.	601 Biotech Drive Richmond, VA 23235	800-735-9224 Fax: 804-648-2641 www.cbi-biotech.com	Contract research; oligo synthesis; DNA identity testing including CODIS databasing
Compacbio, Inc.	1216A Rollins Road Burlingame, CA 94010	800-725-1020 www.compacbio.com	DNA extraction
Coriell Institute for Medical Research	403 Haddon Avenue Camden NJ 08103	856-966-7377 Fax: 856-964-0254 arginine.umdnj.edu	Genomic DNA samples and cell cultures for genetic research reagents; CEPH family samples

Company Name	Street Address	Contact Information	Products/Services
Cybergenetics, Inc.	160 N. Craig St., Suite 210 Pittsburgh, PA 15213	888-FAST-MAP Fax: 412-683-3005 www.cybgen.com	Software for automated genotyping
DNA Diagnostics Center Paternity Testing	205-C Corporate Ct. Fairfield, OH 45014	800-613-5768 www.dnacenter.com	Paternity testing
DNA Heritage	40 Preston Road Weymouth, Dorset, DT3 6PZ UK	+44 (0) 1305-834936 www.dnaheritage.com	Genetic genealogy with Y-STRs
DNA Print	900 Cocoanut Ave Sarasota, FL 34236	941-366-3400 Fax: 941-952-9770 www.dnaprint.com	DNA testing for ethnic origin, paternity testing, genetic genealogy; See www.ancestrybydna.com
DNA Reference Laboratory	7434 Louis Pasteur Dr #15 San Antonio, TX 78229	877-362-1362 Fax: 210-615-0100 www.dnareferencelab.com	Forensic and paternity testing, mtDNA testing
DNA Research Innovations Ltd	940 Cornforth Drive Sittingbourne Research Centre Kent ME9 8PX UK	+44 (0) 1795-411-114 Fax: +44 (0) 1795-411-115 www.chargeswitch.com	Nucleic acid purification reagents
Fairfax Identity Laboratories	3025 Hamaker Court #203 Fairfax, VA 22031	800-848-4362 www.fairfaxidlab.com	Paternity testing; DNA data-banking services
Family Tree DNA	1919 North Loop West, #110 Houston, TX 77008	713-868-1438 Fax: 832-201-7147 www.familytreedna.com	Genetic genealogy; Y chromosome and mtDNA testing
Fitzco	4300 Shoreline Drive Spring Park, MN 55384	800-367-8760 Fax: 952-471-0787 www.fitzcoinc.com	FTA paper for DNA storage and extraction
Forensic Science Associates	3053 Research Drive Richmond, CA 94806	510-222-8883 Fax: 510-222-8887 www.fsalab.com	Consulting services for case review
Future Technologies Inc.	3924 Pender Drive, Suite 200 Fairfax, VA 22030	Tel.: 703-278-0199 Fax: 703-385-0886 www.ftechi.com	Develops information management systems such as AFDIL's LISA suite of software programs
The Gel Company	665 Third Street, Suite 240 San Francisco, CA 94107	800-256-8596 www.gelcompany.com	Reagents for gels and capillary electrophoresis
Gene Codes Corporation	640 Avis Drive Suite 300 Ann Arbor, MI 48108	800-497-4939 Fax: 734-769-7074 www.genecodes.com	Sequencher software for DNA sequencing
GeneCode Forensics	640 Avis Drive Suite 300 Ann Arbor, MI 48108 www.genecodesforensics.com	800-497-4939	M-FISys mass disaster reconstruction software (used in WTC victim identification)

Company Name	Street Address	Contact Information	Products/Services
Genelex Corporation	2203 Airport Way South Seattle WA 98134-2027	800-523-6427 Fax: 206-382-6277 www.genelex.com	Forensic and paternity testing, genetic genealogy
GeneTest Corporation	2316 Delaware Ave Buffalo, NY 14216	877-404-4363 www.genetestlabs.com	Paternity testing
Genetica DNA Laboratories, Inc.	8740 Montgomery Road Cincinnati, OH 45236	800-433-6848 Fax: 513-985-9983 www.genetica.com	Paternity testing
Genetic Profiles Corporation	6122 Nancy Ridge Dr., # 205 San Diego, CA 92121	800-551-7763 Fax: 858-623-0842 www.geneticprofiles.com	Paternity testing
GeneTree DNA Testing Center	2495 South West Temple Salt Lake City, UT 84115	888-404-GENE www.genetree.com	Genetic genealogy, paternity testing
GKT Inc. (South Korea)	1304 Century-1 BLD. 1589-5 Seocho 3 dong, Seocho-gu, Seoul 137-073, KOREA	82-2-3472-2258 Fax: 82-2-3472-2259 http://www.genekotech.com	Y-STR silver stain kits
Hamilton Company	4970 Energy Way Reno, NV 89502	800-648-5950 Fax: 775-856-7259 www.hamiltoncomp.com	Robotic pipetting stations
Identigene, Inc.	5615 Kirby Drive, #800 Houston, TX 77005	800-DNA-TYPE Fax: 713-798-9595 www.identigene.com	Forensic and paternity testing
Identity Genetics, Inc.	801 32nd Avenue Brookings, SD 57006	800-861-1054 Fax: 605-697-5306 www.identitygenetics.com	Forensic and paternity testing
Identity Link, Inc.	606 Idol Drive, Suite 2 High Point, NC 27262	800-325-5465 Fax: 336-885-3045 www.identitylink.com	Paternity testing
Interstate Blood Bank	3180 Old Getwell Road Memphis, TN 38118	901-566-2000 Fax: 901-566-2010 interstatebloodbank.com	Liquid blood samples that may be used for population databasing
Invitrogen	1600 Faraday Avenue Carlsbad, California 92008	800-955-6288 www.invitrogen.com	Molecular biology products
Laboratory Corporation of America	1912 Alexander Drive RTP, NC 27709	800-533-0567 (forensics) 800-742-3944 (paternity) www.labcorp.com	Forensic and paternity testing, immigration testing, HLA testing
Lark Technologies, Inc.	9441 West Sam Houston Parkway South, Suite 103 Houston, TX 77099	800-288-3720 Fax: 713-779-1661 www.lark.com	Molecular biology contract research; Sequencing service
LGC	Queens Road, Teddington, Middlesex TW11 OL4 UK	+44 (0) 20-8943-7000 Fax: +44 (0) 20-8943-2767	Forensic casework and databasing

Company Name	Street Address	Contact Information	Products/Services
LICOR	4308 Progressive Ave Lincoln, NE 68504	800-645-4267 bio.licor.com	Instruments for genetic analysis
Long Beach Genetics	2384 E. Pacifica Place Rancho Dominguez, CA 90220	800-824-2699 Fax: 877-307-1454 www.lbgenetics.com	Paternity testing
Memorial Blood Centers of Minnesota Paternity Laboratory	2304 Park Ave., South Minneapolis, MN 55404	612-871-3300 Fax: 612-871-1359 www.mbcm.org	Paternity testing
Midland Certified Reagent Company	3112-A West Cuthbert Avenue Midland, TX 79701	800-247-8766 Fax: 800-359-5789 www.mcrc.com	Oligo synthesis
Millipore Corporation	290 Concord Road Billerica, MA 01821	800-645-5476 www.millipore.com	DNA separation/purification products, Microcon filters
Mirai Bio	1201 Harbor Bay Parkway Alameda, CA 94502	800-624-6176 Fax: 510-337-2099 www.miraibio.com	Genetic analysis instrumentation; FMBIO II, FMBIO III
Misonix Inc.	1938 New Highway Farmingdale, NY 11735	800-645-9846 Fax: 631-694-9412 www.misonix.com	PCR laminar flow hoods
Mitotyping Technologies, LLC	2565 Park Center Blvd, #200 State College, PA 16801	814-861-0676 Fax: 814-861-0576 www.mitotyping.com	Mitochondrial DNA testing
MJ Research	590 Lincoln Street Waltham, MA 02451	888-PELTIER Fax: 617-923-8080 www.mjresearch.com	Thermal cyclers and PCR consumables
Molecular Dynamics (Amersham Biosciences)	928 East Arques Avenue Sunnyvale, CA 94086-4520	800-333-5703 Fax: 408-773-1493 www.mdyn.com	Gel imaging instruments and capillary array systems
MWG Biotech	4170 Mendenhall Oaks Pkwy, High Point, NC 27265	877-MWG-BTEC Fax: 336-812-9983 www.mwgbiotech.com	Oligo synthesis; genetic analysis equipment; thermal cyclers; robotics for liquid handling
Myriad Genetic Laboratories	320 Wakara Way Salt Lake City, UT 84108	800-469-7423 801-584-3600 www.myriad.com	DNA data-banking services; high volume genetic and clinical testing
Nanogen, Inc.	10398 Pacific Center Court San Diego, CA 92121	877-NANOGEN www.nanogen.com	Microchip devices for DNA analysis
National Medical Services	3701 Welch Road Willow Grove, PA 19090	800-522-6671 Fax: 215-657-2972 www.nmslab.com	Forensic testing
NFSTC	7881 114th Avenue North Largo, FL 33773	727-549-6067 Fax: 727-549-6070 www.nfstc.org	Forensic laboratory accreditation and training programs

Company Name	Street Address	Contact Information	Products/Services
National Institute of Standards and Technology (NIST)	100 Bureau Drive Gaithersburg, MD 20899	301-975-NIST www.nist.gov/srm	Standard Reference Materials for confirming DNA testing methodologies
Orchid Cellmark-Germantown	20271 Goldenrod Lane #101 Germantown, MD 20876	800-USA-LABS Fax: 301-428-4877 www.cellmark-labs.com	Forensic and paternity testing; DNA data-banking services
Orchid Cellmark-Dallas	2600 Stemmons Freeway #133 Dallas, TX 75207	800-752-2774 Fax: 214-634-2898 www.orchidcellmark.com	Forensic and paternity testing; DNA data-banking services
Orchid Cellmark-Nashville	1400 Donelson Pike #A-15 Nashville, TN 37217	615-360-5000 Fax: 615-360-5003 www.orchidcellmark.com	Forensic and paternity testing; DNA data-banking services
Orchid Cellmark-UK	PO Box 265, Abingdon, Oxfordshire, OX14 1YX, UK	+44 (0) 1235-528000 www.cellmark.co.uk	Paternity testing
Orchid Diagnostics (formerly Lifecodes Corporation)	550 West Avenue Stamford, CT 06902	800-543-3263 Fax: 203-328-9599 www.lifecodes.com	Forensic and paternity testing; HLA typing www.private paternitytesting.com
Orchid Gene Screen Dayton, Ohio	5698 Springboro Pike Dayton, OH 45449	800-443-2383 Fax: 937-294-0511 www.genescreen.com	Forensic and paternity testing
Orchid Gene Screen Sacramento	7237 East Southgate Drive Sacramento, CA 95823	800-734-3664 Fax: 916-428-3322 www.genescreen.com	Forensic and paternity testing
Orchid Gene Screen East Lansing	2947 Eyde Parkway, Ste. 110 East Lansing, MI 48823	800-DNA-TEST www.genescreen.com	Forensic and paternity testing
Oxford Ancestors Ltd.	P.O. Box 288, Kidlington, Oxfordshire OX5 1WG UK	www.oxfordancestors.com	Genetic genealogy with mtDNA and Y-STR testing
P.A.L.M. Microlaser Technologies AG	Am Neuland 9+12 82347 Bernried, Germany	+49 (0) 8158-9971-0 www.palm-microlaser.com	Laser microdissection equipment
Paternity Testing Corporation	3501 Berrywood Drive Columbia, MO 65201	888-837-8323 Fax: 573-442-9870 www.ptclabs.com	Paternity testing
Porter Lee Corporation	1072 South Roselle Rd. Schaumburg, IL 60193	847-985-2060 www.porterlee.com	Crime Fighter BEAST LIMS system
PRO-DNA Diagnostic Inc.	5345 de l'Assumption #125 Montreal, Quebec HIT4B3	877-236-6444 Fax: 514-899-9669 www.proadn.com	Paternity testing
Promega Corporation	2800 Woods Hollow Road Madison, WI 53711	800-356-9526 Fax: 608-277-2516 www.promega.com	STR typing kits; DNA extraction kits; AluQuant kit
Qiagen, Inc.	28159 Stanford Avenue Valencia, CA 91355	800-426-8157 Fax: 800-718-2056 www.qiagen.com	DNA isolation products; sample preparation robotics

Company Name	Street Address	Contact Information	Products/Services
Qiagen Operon	1000 Atlantic Avenue #108 Alameda, CA 94501	800-426-8157 Fax: 510-865-5255 oligos.qiagen.com	Oligo synthesis
Quality Forensics	4300 Shoreline Drive Spring Park, MN 55384	952-471-1120 www.qualityforensics.com	Proficiency test provider
QuestGen Forensics	1902 East 8th Street Davis, CA 95616	530-758-4254 Fax: 530-750-5758 www.animalforensics.com	Dog and cat DNA testing with STRs and mtDNA
Quest Genetics	P.O. Box 8931 Gulfport, MS 39506	228-896-9536	Paternity testing; www.paternitytesting associates.com
Rainin Instrument Company, Inc.	7500 Edgewater Drive Oakland, CA 94621	800-472-4646 www.rainin.com	Pipetting products and services
Relative Genetics	2495 South West Temple Salt Lake City, UT 84115	801-461-9760 Fax: 801-461-9761 www.relativegenetics.com	Genetic genealogy with Y-STR testing
ReliaGene Technologies, Inc.	5525 Mounes Street #101 New Orleans, LA 70123	800-256-4106 (forensics) 800-728-3764 (paternity) www.reliagene.com	Paternity and forensic testing; DNA data-banking and mtDNA services; Y-PLEX Y-STR kits
Roche Applied Science	9115 Hague Road Indianapolis, IN 46250-0414	800-262-1640 Fax: 800-428-2883 www.biochem.roche.com	Molecular biology supplies, mtDNA LINEAR ARRAYs
Schleicher & Schuell BioScience, Inc.	10 Optical Avenue Keene, N.H. 03431	603-352-3810 Fax: 603-357-3627 www.s-and-s.com	S&S 903 paper, IsoCode for DNA collection
Serac	Bad Homburg Germany		STR kits for autosomal and Y-chromosome markers
Seratec	Ernst-Ruhstrat-Str. 5, 37079 Goettingen Germany	+49 551 50480-0 Fax: +49 551 50480-80 www.seratec.com	PSA kit for presumptive test of semen
Serological Research Institute (SERI)	3053 Research Drive Richmond, CA 94806	510-223-7374 Fax: 510-222-8887 www.serological.com	Proficiency test provider, training services, casework consultation
Sigma-Aldrich	3050 Spruce Street St. Louis, MO 63103	314-771-5765 www.sigmaaldrich.com	Molecular biology supplies, Genosys custom oligos
SpectruMedix	2124 Old Gatesburg Road State College, PA 16803	814-867-8600 www.spectrumedix.com	Instruments for genetic analysis
TECAN U.S., Inc.	4022 Stirrup Creek Dr., #310 Durham, NC 27703	800-338-3226 Fax: 919-361-5201 www.tecan-us.com	Robotics for liquid handling
Transgenomic, Inc.	12325 Emmet Street Omaha, NE 68164	888-233-9283 Fax: 402-452-5453 www.transgenomic.com	Denaturing HPLC instruments for genetic analysis

Company Name	Street Address	Contact Information	Products/Services
University of North Texas Health Sciences Center DNA Identity Lab	3500 Camp Bowie Blvd. Ft. Worth, TX 76107	800-687-5301	Paternity testing
Whatman, Inc.	9 Bridewell Place Clifton, NJ 07014	800-441-6555 www.whatman.com	FTA paper, GeneSpin DNA purification kit

DNA ADVISORY BOARD QUALITY ASSURANCE STANDARDS

The following information has been previously published and is available directly from the U.S. Department of Justice, Federal Bureau of Investigation (FBI). However, we felt it would once again be a useful addition to this book to include these documents in their entirety so that they could be readily accessible to the reader. The DNA Advisory Board (DAB) was established by the Director of the FBI under the DNA Identification Act of 1994 to operate for a period of five years and consists of 13 voting members. One of the primary purposes of the DAB was to recommend standards for quality assurance in conducting analysis of DNA and to provide guidance to forensic analysts performing those DNA analyses. The following national standards supersede the TWGDAM Guidelines for purposes of certifications required to receive federal funding and to participate in the U.S. National DNA Index System. Updates to these DAB Standards can be made by the FBI Director through recommendation of the Scientific Working Group on DNA Analysis Methods (SWGDAM), which meets every January and July at the FBI Academy in Quantico, Virginia.

Two sets of standards have been issued by the DAB. The first set, which became effective on 1 October 1998, was directed towards forensic DNA laboratories conducting casework investigations. The second set, which became effective on 1 April 1999, is directed at government laboratories performing convicted offender DNA databasing. We have merged both documents (as they contain many redundant points) and list any unique items from the convicted offender DNA databasing laboratory standards in italics.

DNA ADVISORY BOARD

QUALITY ASSURANCE STANDARDS FOR FORENSIC DNA TESTING LABORATORIES *FOR CONVICTED OFFENDER DNA DATABASING LABORATORIES*

PREFACE

Throughout its deliberation concerning these quality standards, the DNA Advisory Board recognized the need for a mechanism to ensure compliance with the standards. An underlying premise for these discussions was that accreditation would be required to demonstrate compliance with the standards and therefore assure quality control and a quality program. Accordingly, the Board recommends that forensic laboratories performing DNA analysis seek such accreditation with all deliberate speed. Additionally, the Board strongly encourages the accrediting bodies to begin positioning themselves to accommodate the increasing demand for accreditation.

PROPOSED MECHANISM TO RECOMMEND CHANGES TO STANDARDS

Once the Director of the FBI has issued standards for quality assurance for forensic DNA testing, the DNA Advisory Board may recommend revisions to such standards to the FBI Director, as necessary. In the event that the duration of the DNA Advisory Board is extended beyond March 10, 2000 by the FBI Director, the Board may continue to recommend revisions to such standards to the FBI Director. In the event that the DNA Advisory Board is not extended by the FBI Director after March 10, 2000, the Technical Working Group on DNA Analysis Methods [TWGDAM] may recommend revisions to such standards to the FBI Director, as necessary.

EFFECTIVE DATE

These standards shall take effect October 1, 1998.

These Quality Assurance Standards for Convicted Offender DNA Databasing Laboratories take effect April 1, 1999.

QUALITY ASSURANCE STANDARDS FOR FORENSIC DNA TESTING LABORATORIES

INTRODUCTION

Forensic DNA identification analysis currently involves forensic casework and convicted offender analyses. These complementary functions demand adherence to the highest analytical standards possible to protect both public safety and individual rights. Separate standards have been drafted for laboratories performing these functions. This separation is an acknowledgment of the differences in the nature or type of sample, the typical sample quantity and potential for reanalysis, and specialization that may exist in a laboratory. Standards for convicted offender laboratories, in some instances, are less stringent than those performing forensic casework analyses, but in no case should the two documents be interpreted as conflicting.

This document consists of definitions and standards. The standards are quality assurance measures that place specific requirements on the laboratory.

Equivalent measures not outlined in this document may also meet the standard if determined sufficient through an accreditation process.

REFERENCES

American Society of Crime Laboratory Directors-Laboratory Accreditation Board (ASCLD-LAB), ASCLD-LAB Accreditation Manual, January 1994, and January 1997.

Federal Bureau of Investigation, Quality Assurance Standards for Forensic DNA Testing Laboratories, (1998).

International Standards Organization (ISO)/International Electrotechnical Commission (IEC), ISO/IEC Guide 25-1990, (1990) American National Standards Institute, New York, NY.

Technical Working Group on DNA Analysis Methods, 'Guidelines for a Quality Assurance Program for DNA Analysis,' Crime Laboratory Digest, April 1995, Volume 22, Number 2, pp. 21–43.

42 Code of Federal Regulations, Chapter IV (10-1-95 Edition), Health Care Financing Administration, Health and Human Services.

1. SCOPE

The standards describe the quality assurance requirements that a *government* laboratory, which is defined as a facility in which forensic DNA testing is performed (*convicted offender DNA testing is regularly performed*), should follow to ensure the quality and integrity of the data and competency of the laboratory. These standards do not preclude the participation of a laboratory, by itself or in collaboration with others, in research and development, on procedures that have not yet been validated.

2. DEFINITIONS

As used in these standards, the following terms shall have the meanings specified:

(a)   Administrative review is an evaluation of the report and supporting documentation for consistency with laboratory policies and for editorial correctness.

(b)   Amplification blank control consists of only amplification reagents without the addition of sample DNA. This control is used to detect DNA contamination of the amplification reagents.

(c)   Analytical procedure is an orderly step-by-step procedure designed to ensure operational uniformity and to minimize analytical drift.

(d)   Audit is an inspection used to evaluate, confirm, or verify activity related
to quality.

Batch is a group of samples analyzed at the same time.

(e)   Calibration is the set of operations which establish, under specified conditions,
the relationship between values indicated by a measuring instrument or measuring
system, or values represented by a material, and the corresponding known values of
a measurement.

*CODIS is the Combined DNA Index System administered by the FBI. It houses DNA
profiles from convicted offenders, forensic specimens, population samples and other
specimen types.*

(f)   Critical reagents are determined by empirical studies or routine practice to require
testing on established samples before use on evidentiary samples in order to prevent
unnecessary loss of sample.

(g)   Commercial test kit is a pre-assembled kit that allows the user to conduct a specific
forensic DNA test.

*Convicted offender is an individual who is required by statute to submit a standard sample for
DNA databasing.*

*Convicted offender database (CODIS) manager or custodian (or equivalent role, position,
or title as designated by the laboratory director) is the person responsible for administration
and security of the laboratory's CODIS.*

*Convicted offender standard sample is biological material collected from an individual for
DNA analysis and inclusion into CODIS. See also database sample.*

*Critical equipment or instruments are those requiring calibration prior to use and periodically
thereafter.*

*Database sample is a known blood or standard sample obtained from an individual
whose DNA profile will be included in a computerized database and searched against other
DNA profiles.*

(h)   Examiner/analyst *(or equivalent role, position, or title as designated by the laboratory
director)* is an individual who conducts and/or directs the analysis of forensic
casework samples, interprets data and reaches conclusions.

(i) Forensic DNA testing is the identification and evaluation of biological evidence in criminal matters using DNA technologies.

(j) Known samples are biological material whose identity or type is established.

(k) Laboratory is a *government* facility in which forensic DNA testing (*convicted offender DNA testing*) is performed *or a government facility who contracts with a second entity for such testing.*

(l) Laboratory support personnel *(or equivalent role, position, or title as designated by the laboratory director)* are individual(s) who perform laboratory duties and do not analyze evidence samples.

(m) NIST is the National Institute of Standards and Technology.

(n) Polymerase Chain Reaction (PCR) is an enzymatic process by which a specific region of DNA is replicated during repetitive cycles which consist of (1) denaturation of the template; (2) annealing of primers to complementary sequences at an empirically determined temperature; and (3) extension of the bound primers by a DNA polymerase.

(o) Proficiency test sample is biological material whose DNA type has been previously characterized and which is used to monitor the quality performance of a laboratory or an individual.

(p) Proficiency testing is a quality assurance measure used to monitor performance and identify areas in which improvement may be needed. Proficiency tests may be classified as:

  1. Internal proficiency test is one prepared and administered by the laboratory;
  2. External proficiency test, which may be open or blind, is one which is obtained from a second agency.

(q) Qualifying test measures proficiency in both technical skills and knowledge.

(r) Quality assurance includes the systematic actions necessary to demonstrate that a product or service meets specified requirements for quality.

(s) Quality manual is a document stating the quality policy, quality system and quality practices of an organization.

(t)   Quality system is the organizational structure, responsibilities, procedures, processes and resources for implementing quality management.

(u)   Reagent blank control consists of all reagents used in the test process without any sample. This is to be used to detect DNA contamination of the analytical reagents.

(v)   Reference material (certified or standard) is a material for which values are certified by a technically valid procedure and accompanied by or traceable to a certificate or other documentation, which is issued by a certifying body.

(w)   Restriction Fragment Length Polymorphism (RFLP) is generated by cleavage by a specific restriction enzyme and the variation is due to restriction site polymorphism and/or the number of different repeats contained within the fragments.

(x)   Review is an evaluation of documentation to check for consistency, accuracy, and completeness.

(y)   Second agency is an entity or organization external to and independent of the laboratory and which performs forensic DNA *(DNA identification)* analysis.

(z)   Secure area is a locked space (for example, cabinet, vault or room) with access restricted to authorized personnel.

(aa)   Subcontractor is an individual or entity having a transactional relationship with a laboratory.

(bb)   Technical manager or leader (or equivalent position or title as designated by the laboratory system) is the individual who is accountable for the technical operations of the laboratory.

(cc)   Technical review is an evaluation of reports, notes, data, and other documents to ensure an appropriate and sufficient basis for the scientific conclusions. This review is conducted by a second qualified individual.

(dd)   Technician *(or equivalent role, position, or title as designated by the laboratory director)* is an individual who performs analytical techniques on evidence samples under the supervision of a qualified examiner/analyst and/or performs DNA analysis on samples for inclusion in a database. Technicians do not evaluate or reach conclusions on typing results or prepare final reports.

(ee) Traceability is the property of a result of a measurement whereby it can be related to appropriate standards, generally international or national standards, through an unbroken chain of comparisons.

(ff) Validation is a process by which a procedure is evaluated to determine its efficacy and reliability for forensic casework analysis *(DNA analysis)* and includes:

1. Developmental validation is the acquisition of test data and determination of conditions and limitations of a new or novel DNA methodology for use on forensic samples;

2. Internal validation is an accumulation of test data within the laboratory to demonstrate that established methods and procedures perform as expected in the laboratory.

3. QUALITY ASSURANCE PROGRAM

STANDARD 3.1 The laboratory shall establish and maintain a documented quality system that is appropriate to the testing activities.

3.1.1   The quality manual shall address at a minimum:
   (a)   Goals and objectives
   (b)   Organization and management
   (c)   Personnel qualifications and training
   (d)   Facilities
   (e)   Evidence control *(sample control)*
   (f)   Validation
   (g)   Analytical procedures
   (h)   Calibration and maintenance
   (i)   Proficiency testing
   (j)   Corrective action
   (k)   Reports *(documentation)*
   (l)   Review
   (m)   Safety
   (n)   Audits

4. ORGANIZATION AND MANAGEMENT

STANDARD 4.1 The laboratory shall:

(a) Have a managerial staff with the authority and resources needed to discharge their duties and meet the requirements of the standards in this document.

(b)   Have a technical manager or leader who is accountable for the technical operations. *(Have a CODIS manager or custodian who is accountable for CODIS operations).*

(c)   Specify and document the responsibility, authority, and interrelation of all personnel who manage, perform or verify work affecting the validity of the DNA analysis.

5. PERSONNEL

STANDARD 5.1 Laboratory personnel shall have the education, training and experience commensurate with the examination and testimony provided. The laboratory shall:

5.1.1 Have a written job description for personnel to include responsibilities, duties and skills.

5.1.2 Have a documented training program for qualifying all technical laboratory personnel.

5.1.3 Have a documented program to ensure technical qualifications are maintained through continuing education.

   5.1.3.1 Continuing education – the technical manager or leader, *(CODIS manager or custodian)* and examiner/analyst(s) must stay abreast of developments within the field of DNA typing by reading current scientific literature and by attending seminars, courses, professional meetings or documented training sessions/classes in relevant subject areas at least once a year.

5.1.4 Maintain records on the relevant qualifications, training, skills and experience of the technical personnel.

STANDARD 5.2 The technical manager or leader shall have the following:

5.2.1 <u>Degree requirements:</u> The technical manager or leader of a laboratory shall have at a minimum a Master's degree in biology-, chemistry- or forensic science- related area and successfully completed a minimum of 12 semester or equivalent credit hours of a combination of undergraduate and graduate course work covering the subject areas of biochemistry, genetics and molecular biology (molecular genetics, recombinant DNA technology), or other subjects which provide a basic understanding of the foundation of forensic DNA analysis as well as statistics and/or population genetics as it applies to forensic DNA analysis.

   5.2.1.1 The degree requirements of section 5.2.1 may be waived by the American Society of Crime Laboratory Directors (ASCLD) or other organization designated by the Director of the FBI in accordance with criteria approved by the Director of the FBI. This waiver shall be available for a period of two years from the effective date of these standards. The waiver shall be permanent and portable.

5.2.2 <u>Experience requirements</u>: A technical manager or leader of a laboratory must have a minimum of three years of forensic DNA laboratory experience *(relevant problem solving or related analytical laboratory experience)*.

5.2.3 Duty requirements:

5.2.3.1 <u>General</u>: manages the technical operations of the laboratory.

5.2.3.2 <u>Specific duties</u>

(a) Is responsible for evaluating all methods used by the laboratory and for proposing new or modified analytical procedures to be used by examiners.

(b) Is responsible for technical problem solving of analytical methods and for the oversight of training, quality assurance, safety and proficiency testing in the laboratory.

5.2.3.3 The technical manager or leader shall be accessible to the laboratory to provide onsite, telephone or electronic consultation as needed.

STANDARD 5.3 *CODIS manager or custodian shall have the following:*

5.3.1 *<u>Degree requirements</u>: A CODIS manager or custodian shall have, at a minimum, a Bachelor's degree in a natural science or computer science.*

5.3.2 *<u>Experience requirements</u>: A CODIS manager or custodian shall have a working knowledge of computers, computer networks, and computer database management, with an understanding of DNA profile interpretation.*

5.3.3 *<u>Duty requirements:</u>*

(a) *Is the system administrator of the laboratory's CODIS network and is responsible for the security of DNA profile data stored in CODIS.*

(b) *Is responsible for oversight of CODIS computer training and quality assurance of data.*

(C) *Has the authority to terminate the laboratory's participation in CODIS in the event of a problem until the reliability of the computer data can be assured. The state CODIS manager or custodian has this authority over all CODIS sites under his/her jurisdiction.*

STANDARD 5.3 Examiner/analyst shall have:

5.3.1 *Degree requirements: An examiner/analyst shall have,* at a minimum a BA/BS *(Bachelors)* degree or its equivalent degree in biology-, chemistry- or forensic science- related area and must have successfully completed college course work (graduate or under-graduate level) covering the subject areas of biochemistry, genetics and molecular biology (molecular genetics, recombinant DNA technology) or other subjects which provide a basic understanding of the foundation of forensic DNA analysis, as well as course work and/or training in statistics and population genetics as it applies to forensic DNA analysis.

5.3.2 *Experience requirements: An examiner/analyst shall have* a minimum of six (6) months of forensic DNA laboratory experience, including the successful analysis of a range of samples typically encountered in forensic case work prior to independent case work analysis using DNA technology.

5.3.3 *An examiner/analyst shall have* successfully completed a qualifying test before beginning independent casework responsibilities.

STANDARD 5.4 Technician shall have:

5.4.1 On the job training specific to their job function(s).

5.4.2 Successfully completed a qualifying test before participating in forensic DNA typing responsibilities.

STANDARD 5.5 Laboratory support personnel shall have:

5.5.1 Training, education and experience commensurate with their responsibilities as outlined in their job description.

6. FACILITIES

STANDARD 6.1 The laboratory shall have a facility that is designed to provide adequate security and minimize contamination. The laboratory shall ensure that:

6.1.1 Access to the laboratory is controlled and limited.

6.1.2 Prior to PCR amplification, evidence examinations, *liquid sample examinations,* DNA extractions, and PCR setup are conducted at separate times or in separate spaces.

6.1.3 Amplified DNA product is generated, processed and maintained in a room(s) separate from the evidence examination, *liquid blood examinations,* DNA extractions and PCR setup areas.

6.1.4 A robotic work station may be used to carry out DNA extraction and amplification in a single room, provided it can be demonstrated that contamination is minimized and equivalent to that when performed manually in separation rooms.

6.1.4 The laboratory follows written procedures for monitoring, cleaning and decontaminating facilities and equipment.

7. EVIDENCE CONTROL (SAMPLE CONTROL)

STANDARD 7.1 The laboratory shall have and follow a documented evidence *(sample inventory)* control system to ensure the integrity of physical evidence. This system shall ensure that:

7.1.1 Evidence is *(Offender samples are)* marked for identification.

7.1.2 Chain of custody for all evidence is maintained.

7.1.2 Documentation of sample identity, collection, receipt, storage, and disposition is maintained.

7.1.3 The laboratory follows documented procedures that minimize loss, contamination, and/or deleterious change of evidence.

7.1.4 The laboratory has secure areas for evidence storage *(sample storage including environmental control consistent with the form or nature of the sample).*

STANDARD 7.2 Where possible, the laboratory shall retain or return a portion of the evidence sample or extract.

7.2.1 The laboratory shall have a procedure requiring that evidence sample/extract(s) are stored in a manner that minimizes degradation.

8. VALIDATION

STANDARD 8.1 The laboratory shall use validated methods and procedures for forensic casework analyses *(DNA analyses).*

8.1.1 Developmental validation that is conducted shall be appropriately documented.

8.1.2 Novel forensic DNA methodologies shall undergo developmental validation to ensure the accuracy, precision and reproducibility of the procedure. The developmental validation shall include the following:

8.1.2.1 Documentation exists and is available which defines and characterizes the locus.

8.1.2.2 Species specificity, sensitivity, stability and mixture studies are conducted.

8.1.2.3 Population distribution data are documented and available.

8.1.2.3.1 The population distribution data would include the allele and genotype distributions for the locus or loci obtained from relevant populations. Where appropriate, databases should be tested for independence expectations.

8.1.3 Internal validation shall be performed and documented by the laboratory.

8.1.3.1 The procedure shall be tested using known and non-probative evidence samples *(known samples only)*. The laboratory shall monitor and document the reproducibility and precision of the procedure using human DNA control(s).

8.1.3.2 The laboratory shall establish and document match criteria based on empirical data.

8.1.3.3 Before the introduction of a procedure into forensic casework *(database sample analysis)*, the analyst or examination team shall successfully complete a qualifying test.

8.1.3.4 Material modifications made to analytical procedures shall be documented and subject to validation testing.

8.1.4 Where methods are not specified, the laboratory shall, wherever possible, select methods that have been published by reputable technical organizations or in relevant scientific texts or journals, or have been appropriately evaluated for a specific or unique application.

9. ANALYTICAL PROCEDURES

STANDARD 9.1 The laboratory shall have and follow written analytical procedures approved by the laboratory management/technical manager.

9.1.1 The laboratory shall have a standard operating protocol for each analytical technique used.

9.1.2 The procedures shall include reagents, sample preparation, extraction, equipment, and controls, which are standard for DNA analysis and data interpretation.

9.1.3 The laboratory shall have a procedure for differential extraction of stains that potentially contain semen.

STANDARD 9.2 The laboratory shall use reagents that are suitable for the methods employed.

9.2.1 The laboratory shall have written procedures for documenting commercial supplies and for the formulation of reagents.

9.2.2 Reagents shall be labeled with the identity of the reagent, the date of preparation or expiration, and the identity of the individual preparing the reagent.

9.2.3 The laboratory shall identify critical reagents *(if any)* and evaluate them prior to use in casework. These critical reagents include but are not limited to: *(THIS LAST PORTION NOT IN CONVICTED OFFENDER DATABASING STANDARDS)*

(a) Restriction enzyme
(b) Commercial kits for performing genetic typing
(c) Agarose for analytical RFLP gels
(d) Membranes for Southern blotting
(e) K562 DNA or other human DNA controls
(f) Molecular weight markers used as RFLP sizing standards
(g) Primer sets
(h) Thermostable DNA polymerase

STANDARD 9.3 The laboratory shall have and follow a procedure for evaluating the quantity of the human DNA in the sample where possible. *(NOT IN CONVICTED OFFENDER DATABASING STANDARDS)*

9.3.1 For casework RFLP samples, the presence of high molecular weight DNA should be determined.

STANDARD 9.4 The laboratory shall monitor the analytical procedures using appropriate controls and standards.

9.4.1 The following controls shall be used in RFLP casework analysis:

9.4.1.1 Quantitation standards for estimating the amount of DNA recovered by extraction. *(When required by the analytical procedure, standards for estimating the amount of DNA recovered by extraction shall be used.)*

9.4.1.2 K562 as a human DNA control. (In monitoring sizing data, a statistical quality control method for K562 cell line shall be maintained.)

9.4.1.3 Molecular weight size markers to bracket known and evidence samples. *(Molecular weight size markers to bracket samples on an analytical gel. No more than five lanes shall exist between marker lanes.)*

9.4.1.4 *A Procedure shall be available* to monitor the completeness of restriction enzyme digestion. *(Interpretation of the autorad/lumigraph is the ultimate method of assessment but a test gel or other method may be used as necessary.)*

9.4.2 The following controls shall be used for PCR casework analysis *(database analysis)*:

9.4.2.1 Quantitation standards, which estimate the amount of human nuclear DNA recovered by extraction. *(When required by the analytical procedure, standards which estimate the amount of human nuclear DNA recovered by extraction shall be used.)*

9.4.2.2 Positive and negative amplification controls.

9.4.2.3 Reagent blanks. *(Contamination controls.)*

9.3.2.3.1 *Samples extracted prior to the effective date of these standards without reagent blanks are acceptable as long as other samples analyzed in the batch do not demonstrate contamination.*

9.4.2.4 Allelic ladders and/or internal size makers for variable number tandem repeat sequence PCR based systems.

STANDARD 9.5 The laboratory shall check its DNA procedures annually or whenever substantial changes are made to the protocol(s) against an appropriate and available NIST standard reference material or standard traceable to a NIST standard.

STANDARD 9.6 The laboratory shall have and follow written general guidelines for the interpretation of data.

9.6.1   The laboratory shall verify that all control results are within established tolerance limits.

9.6.2   Where appropriate, visual matches shall be supported by a numerical match criterion. *(NOT IN CONVICTED OFFENDER DATABASING STANDARDS)*

9.6.3   For a given population(s) and/or hypothesis of relatedness, the statistical interpretation shall be made following the recommendations 4.1, 4.2 or 4.3 as deemed applicable of the National Research Council report entitled 'The Evaluation of Forensic DNA Evidence' (1996) and/or court directed method. These calculations shall be derived from a documented population database appropriate for the calculation. *(NOT IN CONVICTED OFFENDER DATABASING STANDARDS)*

10. EQUIPMENT CALIBRATION AND MAINTENANCE

STANDARD 10.1 The laboratory shall use equipment suitable for the methods employed.

STANDARD 10.2 The laboratory *(shall identify critical equipment and)* shall have a documented program for calibration of instruments and equipment.

10.2.1   Where available and appropriate, standards traceable to national or international standards shall be used for the calibration.

10.2.1.1   Where traceability to national standards of measurement is not applicable, the laboratory shall provide satisfactory evidence of correlation of results.

10.2.2   The frequency of the calibration shall be documented for each instrument requiring calibration. Such documentation shall be retained in accordance with applicable Federal or state law.

STANDARD 10.3 The laboratory shall have and follow a documented program to ensure that instruments and equipment are properly maintained.

10.3.1   New *(critical)* instruments and equipment, or *(critical)* instruments and equipment that have undergone repair or maintenance, shall be calibrated before being used in casework analysis.

10.3.2   Written records or logs shall be maintained for maintenance service performed on instruments and equipment. Such documentation shall be retained in accordance with applicable Federal or state law.

11. REPORTS

STANDARD 11.1 The laboratory shall have and follow written procedures for taking and maintaining case notes to support the conclusions drawn in laboratory reports. *(The laboratory shall have and follow written procedures for generating and maintaining documentation for database samples.)*

11.1.1 The laboratory shall maintain, in a case record, all documentation generated by examiners related to case analyses. *(The laboratory shall have written procedures for the release of database sample information.)*

11.1.2 Reports according to written guidelines shall include: *(NOT IN CONVICTED OFFENDER DATABASING STANDARDS)*

  (a)   Case identifier
  (b)   Description of evidence examined
  (c)   A description of the methodology
  (d)   Locus
  (e)   Results and/or conclusions
  (f)   An interpretative statement (either quantitative or qualitative)
  (g)   Date issued
  (h)   Disposition of evidence
  (i)   A signature and title, or equivalent identification, of the person(s) accepting responsibility for the content of the report.

11.1.3 The laboratory shall have written procedures for the release of case report information. *(NOT IN CONVICTED OFFENDER DATABASING STANDARDS)*

12. REVIEW

STANDARD 12.1 The laboratory shall conduct administrative and technical reviews of all case files and reports to ensure conclusions and supporting data are reasonable and within the constraints of scientific knowledge. *(The laboratory shall have and follow written procedures for reviewing database sample information, results, and matches.)*

12.1.1 The laboratory shall have a mechanism in place to address unresolved discrepant conclusions between analysts and reviewer(s).

STANDARD 12.2 The laboratory shall have and follow a program that documents the annual monitoring of the testimony of each examiner *(laboratory personnel).*

13. PROFICIENCY TESTING

STANDARD 13.1 Examiners and other personnel designated by the technical manager or leader who are actively engaged in DNA analysis shall undergo, at regular intervals of not to exceed 180 days, external proficiency testing in accordance with these standards. Such external proficiency testing shall be an open proficiency testing program.

13.1.1 The laboratory shall maintain the following records for proficiency tests:

    (a)   The test set identifier.

    (b)   Identity of the examiner.

    (c)   Date of analysis and completion.

    (d)   Copies of all data and notes supporting the conclusions.

    (e)   The proficiency test results.

    (f)   Any discrepancies noted.

    (g)   Corrective actions taken.

    (h)   Such documentation shall be retained in accordance with applicable Federal or state law.

13.1.2 The laboratory shall establish at a minimum the following criteria for evaluation of proficiency tests:

    (a)   All reported inclusions are correct or incorrect.

    (b)   All reported exclusions are correct or incorrect.

    (c)   All reported genotypes and/or phenotypes are correct or incorrect according to consensus genotypes/phenotypes or within established empirically determined ranges.

    (d)   All results reported as inconclusive or uninterpretable are consistent with written laboratory guidelines. The basis for inconclusive interpretations in proficiency tests must be documented.

    (e)   All discrepancies/errors and subsequent corrective actions must be documented.

    (f)   All final reports are graded as satisfactory or unsatisfactory. A satisfactory grade is attained when there are no analytical errors for the DNA profile typing data. Administrative errors shall be documented and corrective actions taken to minimize the error in the future.

    (g)   All proficiency test participants shall be informed of the final test results.

14. CORRECTIVE ACTION

STANDARD 14.1 The laboratory shall establish and follow procedures for corrective action whenever proficiency testing discrepancies and/or casework *(analytical)* errors are detected.

14.1.1 The laboratory shall maintain documentation for the corrective action. Such documentation shall be retained in accordance with applicable Federal or state law.

15. AUDITS

STANDARD 15.1 The laboratory shall conduct audits annually in accordance with the standards outlined herein.

15.1.1 Audit procedures shall address at a minimum:

(a) Quality assurance program

(b) Organization and management

(c) Personnel

(d) Facilities

(e) Evidence control *(sample control)*

(f) Validation

(g) Analytical procedures

(h) Calibration and maintenance

(i) Proficiency testing

(j) Corrective action

(k) Reports *(documentation)*

(l) Review

(m) Safety

(n) Previous audits

15.1.2 The laboratory shall retain all documentation pertaining to audits in accordance with relevant legal and agency requirements.

STANDARD 15.2 Once every two years, a second agency shall participate in the annual audit.

16. SAFETY

STANDARD 16.1 The laboratory shall have and follow a documented environmental health and safety program.

17. SUBCONTRACTOR OF ANALYTICAL TESTING FOR WHICH VALIDATED PROCEDURES EXIST

STANDARD 17.1 A laboratory operating under the scope of these standards will require certification of compliance with these standards when a subcontractor performs forensic DNA *(convicted offender DNA)* analyses for the laboratory.

17.1.1 The laboratory will establish and use appropriate review procedures to verify the integrity of the data received from the subcontractor *including but not limited to:*

(a) *Random reanalysis of samples.*

(b) *Visual inspection and evaluation of results/data.*

(c) *Inclusion of QC samples.*

(d) *On-site visits.*

APPENDIX V

DAB RECOMMENDATIONS ON STATISTICS

The following document was published in the July 2000 issue of *Forensic Science Communications*, the online FBI journal. It is available at the following web site: http://www.fbi.gov/hq/lab/fsc/backissu/july2000/dnastat.htm.

DNA ADVISORY BOARD

February 23, 2000

Statistical and population genetics issues affecting the evaluation of the frequency of occurrence of DNA profiles calculated from pertinent population database(s)

INTRODUCTION

When a comparison of DNA profiles derived from evidence and reference samples fails to exclude an individual(s) as a contributor(s) of the evidence sample, statistical assessment and/or probabilistic reasoning are used to evaluate the significance of the association. Proper statistical inference requires careful formulation of the question to be answered, including, in this instance, the requirements of the legal system. Inference must take into account how and what data were collected, which, in turn, determine how the data are analyzed and interpreted.

Previously, the DNA Advisory Board (DAB; June 21, 1996, New York) endorsed the recommendations of the National Research Council's Report (1996; henceforth NRC II Report):

> The DAB congratulates Professor Crow and his NRC [National Research Council] Committee for their superb report on the statistical and population genetics issues surrounding forensic DNA profiling. We wholeheartedly endorse the findings of the report in these substantive matters.

As the NRC II Report (1996) describes, there are alternate methods for assessing the probative value of DNA evidence. Rarely is there only one statistical approach to interpret and explain the evidence. The choice of approach is affected by the philosophy and experience of the user, the legal system, the practicality of the approach, the question(s) posed, available data, and/or assumptions. For forensic applications, it is important that the statistical conclusions be conveyed meaningfully. Simplistic or less rigorous approaches are often sought. Frequently, calculations such as the random match probability and probability of exclusion convey to the trier of fact the probative value of the evidence in a straightforward fashion. Simplified approaches are appropriate, as long as the analysis is conservative or does not provide false inferences. Likelihood ratio (LR) approaches compare mutually exclusive hypotheses and can be quite useful for evaluating the data. However, some LR calculations and interpretations can be complicated, and their significance to the case may not be apparent to the practitioner and the trier of fact.

Bayesian inference, which accounts for information other than the DNA evidence, also could be applied. Bayesian approaches sometimes require knowledge of circumstances beyond the domain of the DNA scientist and have not been addressed in U.S. criminal courts for DNA analysis. The DAB believes it is for the courts to decide whether or not Bayesian statistics are solely the responsibility of the trier of fact. The DAB recognizes that these different approaches can be applied, as long as the question to be answered and the assumptions underlying the analyses are clearly conveyed to the trier of fact.

We have been charged with clarifying issues that arise for the following special cases:

- Source attribution or identity;
- Cases where relatives may be involved;
- Interpretation of mixtures; and
- The significance of a match derived through a felon database search.

SOURCE ATTRIBUTION

According to *Webster's Third New International Dictionary* (Merriam-Webster 1961; henceforth Webster's Third), the term unique can convey several meanings, including *the only one, unusual,* and *some [circumstance] that is the only one of its kind.* Those who question the concept of assigning source attribution for DNA evidence often dwell on the former (e.g., Balding 1999). In their argument against source attribution, some critics say that it is difficult to establish, beyond doubt, that a DNA profile is carried by only one individual in the entire world. Within that context, their argument can be compelling, especially if the profile consists of a fairly small number of loci. Their conclusion, however, is problematic because source attribution should be evaluated within the *context defined by the case,* and the world's population rarely would be the appropriate context. Because source attribution can only be meaningful within the context of the instant case, *Webster's Third* definition of uniqueness comes closest to that required by the legal setting: a circumstance that is the only one of its kind.

By contrast to the world's population, examples of limiting, case-specific contexts are more common. Suppose, for example, the presence of a small group of individuals at the crime scene is stipulated. However, the identity of the single individual who sexually assaulted the victim is at issue. DNA evidence on the victim matches a DNA profile from only one of the named defendants. In this instance, it is simple to assign source because all other individuals are excluded. Now suppose the identities of some individuals at the crime scene are unknown, yet the DNA profile matches one of the defendants. Further suppose this defendant has no close relatives aside from parents. Source attribution is not challenging in this setting. While the answer depends on the number and kind of loci examined, in most instances the source can be assigned with a very high degree of scientific certainty. Suppose, instead, the defendant has multiple siblings, one of whom may have been the assailant and whose profiles are not available for some reason. Even then source can be assigned with a high degree of scientific certainty when a sufficient number of highly polymorphic loci are typed.

Inference regarding source attribution should always be based on the facts in the case. Arguments against source attribution based on premises having nothing to do with the case at hand should not be compelling.

Another set of questions arises when commentators fail to distinguish between source attribution and guilt. Some commentators, for example, set up the following scenario: suppose inculpating DNA evidence appears to come from the defendant with high probability, yet all the non-DNA evidence is exculpatory (e.g., Balding 1999). In this instance, they say, source attribution is impossible. We do not agree. If, to a high degree of scientific certainty, the DNA evidence appears to come from the defendant, then the only reasonable conclusion is that the DNA did indeed come from the defendant. The trier of fact, however, has a different question to ponder: What value is source attribution if the preponderance of the evidence suggests the defendant cannot be the perpetrator? The trier of fact should seek other explanations for the data, some or all of which may exculpate the defendant.

As described above, the possible source of the DNA depends on the context of the case, and thus calculations for source attribution must reflect the appropriate reference population. If relatives are potential contributors, the calculations for source attribution must reflect that fact. If relatives are not potential contributors, the calculations for source attribution should be based on a defined population; that population could be as small as two unrelated individuals or an entire town, city, state, or country. The DNA analyst should take great care with evidence presentation, with two important facts in mind:

■ Inference about source attribution is a probabilistic statement, and its degree of uncertainty is governed by the genetic information contained in the profile; and

■ Inference about source attribution is distinct from inference regarding guilt.

One way to develop criteria to assess the question of source attribution is to let p_x equal the random match probability for a given evidentiary profile X. The random match probability is calculated using the NRC II Report (1996) Formulae 4.1b and 4.4a for general population scenarios or Formula 4.10 under the assumption that the contributor and the accused could only come from one subgroup. The value θ is 0.01, except for estimates for isolated subgroups, where 0.03 is used. The rarity of the estimate is decreased by a factor of 10 (NRC II Report 1996).

Then $(1-p_x)^N$ is the probability of not observing the particular profile in a population of N unrelated individuals. We require that this probability be greater than or equal to a $1-\alpha$ confidence level $(1-p_x)^N \geq 1-\alpha$ or $p_x \leq 1-(1-\alpha)^{1/N}$. Specifying a confidence level of 0.95 (0.99; i.e., an α of 0.05 or 0.01) will enable determination of the random match probability threshold to assert with 95% (99%) confidence that the particular evidentiary profile is unique, given a population of N unrelated individuals.

In practice, p_x is calculated for each of the major population groups residing in the geographic area where the crime was committed (i.e., typically African-American, Caucasian, and Hispanic). When there is no reason to believe a smaller population is relevant, the FBI, for example, has set N to 260 million, the approximate size of the U.S. population. For smaller, defined populations, N should be based on census values or other appropriate values determined by the facts of the case. The source attribution formula advocated here is simple and likely to be conservative, especially when N is larger than the size of the population that would inhabit a geographic area where a crime is committed.

RELATIVES

As described previously in the Source Attribution section, the possibility of a close relative (typically a brother) of the accused being in the pool of potential contributors of crime scene evidence should be considered in case-specific context. It is not appropriate to proffer that a close relative is a potential contributor of the evidence when there are no facts in evidence to suggest this instance is relevant. However, if a relative had access to a crime scene and there is reason to believe he/she could have been a contributor of the evidence, then the best action to take is to obtain a reference sample from the relative. After all, this scenario should be sufficient probable cause for obtaining a reference sample. Typing with the same battery of short tandem repeat (STR) loci will resolve the question of whether or not the relative carries the same DNA profile as the accused.

When a legitimate suspected relative cannot be typed, a probability statement can be provided. Given the accused DNA profile, the conditional probability that the relative has the same DNA profile can be calculated. Examples of methods for estimating the probability of the same DNA profile in a close relative are described in the NRC II Report (1996) and Li and Sacks (1954).

MIXTURES

Mixtures, which for our purposes are DNA samples derived from two or more contributors, are sometimes encountered in forensic biological evidence. The presence of a mixture is evident typically by the presence of three or more peaks, bands, dots, and/or notable differences in intensities of the alleles for at least one locus in the profile. In some situations, elucidation of a contributor profile is straightforward. An example would be the analysis of DNA from an intimate swab revealing a mixture consistent with the composition of the perpetrator and the victim. When intensity differences are sufficient to identify the major contributor in the mixed profile, it can be treated statistically as a single source sample. At times, when alleles are not masked, a minor contributor to the mixed profile may be elucidated. Almost always in a mixture interpretation, certain possible genotypes can be excluded. It may be difficult to be confident regarding the number of contributors in some complex mixtures of more than two individuals; however, the number of contributors often can be inferred by reviewing the data at all loci in a profile.

Interpretation of genotypes is complicated when the contributions of the donors are approximately equal (i.e., when a major contributor cannot be determined unequivocally) or when alleles overlap. Also, stochastic fluctuation during polymerase chain reaction (PCR) arising from low quantity of DNA template can make typing of a minor contributor complicated. When the contributors of a DNA mixture profile cannot be distinguished, two calculations convey the probative value of the evidence.

The first calculation is the probability of exclusion (PE; Devlin 1992 and references therein). The PE provides an estimate of the portion of the population that has a genotype composed of at least one allele not observed in the mixed profile. Knowledge of the accused and/or victim profiles is not used (or needed) in the calculation. The calculation is particularly useful in complex mixtures, because it requires no assumptions about the identity or number of contributors to a mixture. The probabilities derived are valid and for all practical purposes are conservative. However, the PE does not make use of all of the available genetic data.

The LR provides the odds ratio of two competing hypotheses, given the evidence (Evett and Weir 1998). For example, consider a case of sexual assault for which the victim reported there were two assailants. A mixture of two profiles is observed in the 'male fraction,' and the victim is excluded as a contributor of the observed mixed profile. Two men are arrested, and their combined profiles are consistent with the mixture evidence. A likelihood calculation logically might compare the probability that the two accused individuals are the source of the DNA in the evidence versus two unknown (random men) are the source of the evidence. Various alternate hypotheses can be entertained as deemed appropriate, given the evidence. Calculation of a LR considers the identity and actual number of contributors to the observed DNA mixture. Certainly, LR makes better use of the available genetic data than does the PE.

Interpretation of DNA mixtures requires careful consideration of factors including, but not limited to, detectable alleles; variation of band, peak, or dot intensity; and the number of alleles. There are a number of references for guidance on calculating the PE or LR (Evett and Weir 1998; NRC II Report 1996; PopStats in CODIS). The DAB finds either one or both PE or LR calculations acceptable and strongly recommends that one or both calculations be carried out whenever feasible and a mixture is indicated.

DATABASE SEARCH

As felon DNA databases develop in all 50 states, searches for matches between evidentiary and database profiles will become increasingly common. Two questions arise when a match is derived from a database search: (1) What is the rarity of the DNA profile? and (2) What is the probability of finding such a DNA profile in the database searched? These two questions address different issues. That the different questions produce different answers should be obvious. The former question addresses the random match probability, which is often of particular interest to the fact finder. Here we address the latter question, which is especially important when a profile found in a database search matches the DNA profile of an evidence sample.

When the DNA profile from a crime scene sample matches a single profile in a felon DNA database, the NRC II Report (1996) recommended the evaluation of question number (2) be based on the size of the database. They argued for this evaluation because the probability of identifying a DNA profile by chance increases with the size of the database. Thus this chance event must be taken into account when evaluating value of the matching profile found by a database search. Those who argue against NRC II's recommended treatment (e.g., Balding and Donnelly 1996; Evett and Weir 1998; Evett, Foreman and Weir 2000) say the NRC II Report's formulation is wrong and undervalues the evidence. In fact, they argue that the weight of the evidence (defined in terms of a likelihood ratio) for a DNA database search exceeds the weight provided by the same evidence in a 'probable cause' case – a case in which other evidence first implicates the suspect and then DNA evidence is developed.

When other evidence first implicates the suspect, the DNA evidence can be evaluated using the probability p_x of randomly drawing the profile X from the (appropriate) population, which expresses the degree of surprise that the suspect and evidentiary profiles match. Equivalently, we can express it as a LR for two competing hypotheses, namely the likelihood of the evidence when the data come from the same individual (H_s) versus the likelihood of the evidence when the data come from two different individuals (H_d). The LR in this instance is:

$$\text{Lik}(\text{Profile} \mid H_s) / \text{Lik}(\text{Profile} \mid H_d) = p_x / (p_x * p_x) = 1/p_x.$$

For the DNA database search, the NRC II Report recommended the calculation (defined in terms of a LR) to be evaluated as $1/(N p_x)$, where N is the size of the database. While justification for this calculation is given in their report, it is often misunderstood. Stockmarr (1999) re-derives this result in a way that should be more comprehensible. As a special case, assume only one profile in the database matches the evidentiary profile; we can consider that individual is a suspect. Now consider two competing hypotheses, namely the source is or is not in the database (H_{in} versus $H_{not\ in}$). These likelihood's are relevant because we wish to identify whether the suspect is likely to be the source of the sample (H_{in}) or if it is more likely he was identified merely by chance ($H_{not\ in}$). What is the LR for these hypotheses?

$$\text{Lik}(\text{Profile} \mid H_{in}) / \text{Lik}(\text{Profile} \mid H_{not\ in}) = 1/(N p_x).$$

Stockmarr (1999) argues this formulation is the appropriate treatment of the data, as did the NRC II Report (1996) before him. Both recognize an intuitive counter example. Suppose we had a DNA database of the entire world's population (size N), except one individual ($N-1$). A DNA profile from a crime scene is found to match one and only one profile in the database, and its frequency is $1/N$. According to critics (e.g., Balding 1997), this example demonstrates the fallacious nature of the NRC II Report's proposed evaluation of the evidence for a database search, because the value of the evidence appears to be nil (the likelihood ratio is essentially one instead of a large number). Both Stockmarr (1999) and the NRC II Report recognize this interesting result; however, by treating the problem from a Bayesian perspective and invoking prior probabilities that are a function of the size of the database, they argue the example is irrelevant. In essence, the prior probability of H_{in} rises as N rises. This approach is coherent, from the statistical perspective, but it may not be particularly helpful for the legal system. Without the use of prior probabilities, it should be apparent that the treatment of the database search recommended by the NRC II Report can be conservative when the database is extremely large.

It is important to consider the treatment proposed by Balding and Donnelly (1996) and recently endorsed by Evett, Foreman, and Weir (2000). By their line of reasoning, the LR is no different whether other evidence first implicates the suspect or the suspect is identified by a database search. In fact, they argue the true weight of the evidence is actually larger for the latter, albeit the increase is small unless N is large. This argument has some intuitive appeal, especially in light of the example given above, and it is true that their LR is unaffected by sampling.

Both camps appear to present rigorous arguments to support their positions. Indeed the proper treatment superficially appears to rest in the details of arcane mathematics (Balding and Donnelly 1996; NRC II Report 1996; Stockmarr 1999). We believe, however, there is a way to see which of the two treatments is better for the legal setting without resorting to mathematical details. Consider the following scenario: a murder occurs, and the only evidence left at the crime scene is a cigarette butt. DNA analysis types five loci from the saliva on the cigarette butt. The probability of drawing the resulting profile X from a randomly selected individual is $p_x = 1/100\,000$. A search of the DNA database, which contains $N = 100\,000$ profiles, reveals a single match. No other evidence can be found to link the 'suspect,' whose profile matches, to the murder.

If we follow Balding and Donnelly (1996), the message for the investigators is that the evidence is $100\,000$ times more likely if the suspect is the source than if he is not. Alternatively, by the NRC II Report (1996) recommendations, the evidence is not compelling because the likelihood the profile, _a priori_, is/is not in the database is the same. In probabilistic terms, it is not surprising to find a matching profile in the database of size $100\,000$ when the profile probability is $1/100\,000$. Curiously, the mathematics underlying both approaches are correct, despite the apparently divergent answers. It is the foundations of the formulations that differ, and they differ substantially.

At present there are about $20\,000$ known, variable STR loci in the human genome. Of these, forensic scientists use a little more than a dozen, which is sufficient for most forensic analyses. Although not strictly accurate, let us think of the selection of STR loci as random and return to our case. The forensic scientists who worked on the cigarette butt could assay only five loci of the dozen they might type. Suppose they were to type five different loci and generate a new profile based on only these additional five loci? If our suspect were the true source of the sample, a match at those loci would be obtained; however, if he were not the source, a match would be highly unlikely. If the new (i.e., second) profile probability were again on the order of $1/100\,000$, someone else may have been selected. If our suspect is not the source, no one else in the database is, and yet we can easily imagine selecting a set of five loci (out of the thousands possible) to single out each individual therein. This seems like an unsatisfactory state in light of the LR espoused by Balding and Donnelly (1996).

Thus we are left with an interesting dilemma. Within a Bayesian context, the NRC II Report's LR and Balding and Donnelly's (1996) LR could be interpreted to yield a coherent evaluation of the evidence. Unfortunately, Bayesian logic has not been considered by the U.S. criminal legal system for DNA analysis. Clearly, what is required is a formulation of the LR that transparently conveys its import without resorting to Bayesian statistics. In this setting, the treatment of the database search recommended by the NRC II Report can be conservative, but only for the unlikely scenario of a very large N is it very conservative. Apparently the treatment of the database search recommended by Balding and Donnelly (1996) is not conservative when the number of loci genotyped is small and remains so until the number of loci becomes large enough to essentially ensure *uniqueness*. To put it another way, without the Bayesian framework, the Balding and Donnelly (1996) formulation is easily misinterpreted in a fashion unfavorable to the suspect. Stockmarr's (1999) formulation, which is a more formal exposition of what originally appeared in the NRC II Report (1996), communicates value of a database search far better, and it is always conservative. Thus, we continue to endorse the recommendation of the NRC II Report for the evaluation of DNA evidence from a database search.

CONCLUSION

Statistical analyses are sometimes thought to yield automatic rules for making a decision either to accept or reject a hypothesis. This attitude is false in any setting and should be especially avoided for forensic inference. One rarely rests his/her decisions wholly on any single statistical test or analysis. To the evidence of the test should be added data accumulated from the scientist's own past work and that of others (Snedecor and Cochran 1967). Thus, in this light, statistical analyses should be thought of as useful guides for interpreting the weight of the DNA evidence.

REFERENCES

Balding, D. J. (1997) Errors and misunderstandings in the second NRC report. *Jurimetrics*, 37, 603–607.

Balding, D. J. (1999) When can a DNA profile be regarded as unique? *Science & Justice*, 39, 257–260.

Balding, D. J. and Donnelly, P. (1996) Evaluating DNA profile evidence when the suspect is identified through a database search. *Journal of Forensic Sciences*, 41, 603–607.

Devlin, B. (1992) Forensic inference from genetic markers. *Statistical Methods in Medical Research*, 2, 241–262.

Evett, I. W., Foreman, L. A. and Weir, B. S. (2000) *Biometrics*, 56 (4), 1247–1276.

Evett, I. W. and Weir, B. S. (1998) *Interpreting DNA Evidence*. Sinaue, Sunderland, Massachusetts.

Li, C. C. and Sacks, L. (1954) The derivation of joint distribution and correlation between relatives by the use of stochastic matrices. *Biometrics*, 10, 347–360.

Merriam-Webster, Incorporated (1961) *Webster's Third New International Dictionary*. Merriam-Webster, Incorporated, Springfield, Massachusetts.

National Research Council Committee on DNA Forensic Science (1996) *An Update: The Evaluation of Forensic DNA Evidence*. Washington, D.C., National Academy Press.

Snedecor, G. W. and Cochran, W. G. (1967) *Statistical Methods* (6th edition), p.28. Ames: Iowa State University Press.

Stockmarr, A. (1999) Likelihood ratios for evaluating DNA evidence when the suspect is found through a database search. *Biometrics*, 55, 671–677.

APPENDIX VI

NRC II RECOMMENDATIONS

The use of DNA for criminal cases within the United States began in the late 1980s. Numerous scientific and legal issues emerged and a National Research Council (NRC) Committee on DNA Technology was convened in 1989. This committee's report entitled *DNA Technology in Forensic Science* was issued in 1992. The report recommended the use of DNA analysis for forensic purposes, but it was strongly criticized and a second committee was established to carefully consider statistical issues surrounding DNA analysis. NRC II made recommendations in 1996 through publication of *The Evaluation of Forensic DNA Evidence.* Today these recommendations, which are listed below, are carefully considered in U.S. courts when reviewing DNA evidence.

RECOMMENDATIONS TO IMPROVE LABORATORY PERFORMANCE

Recommendation 3.1. Laboratories should adhere to high quality standards (such as those defined by TWGDAM and the DNA Advisory Board) and make every effort to be accredited for DNA work (by such organizations as ASCLD-LAB).

Recommendation 3.2. Laboratories should participate regularly in proficiency tests, and the results should be available for court proceedings.

Recommendation 3.3. Whenever feasible, forensic samples should be divided into two or more parts at the earliest practicable stage and the unused parts retained to permit additional tests. The used and saved portions should be stored and handled separately. Any additional tests should be performed independently of the first by personnel not involved in the first test and preferably in a different laboratory.

RECOMMENDATIONS FOR ESTIMATING RANDOM-MATCH PROBABILITIES

Recommendation 4.1. In general, the calculation of a profile frequency should be made with the product rule. If the race of the person who left the evidence-sample

DNA is known, the database for the person's race should be used; if the race is not known, calculations for all the racial groups to which possible suspects belong should be made. For systems such as variable number tandem repeats (VNTRs), in which a heterozygous locus could be mistaken for a homozygous one, if an upper bound on the frequency of the genotype at an apparently homozygous locus (single band) is desired, then twice the allele (bin) frequency, $2p$, should be used instead of p^2. For systems in which exact genotypes can be determined, $p^2 + p(1 - p)\theta$ should be used for the frequency at such a locus instead of p^2. A conservative value of θ for the US population is 0.01; for some small, isolated populations, a value of 0.03 may be more appropriate. For both kinds of systems, $2p_i p_j$ should be used for heterozygotes.

Recommendation 4.2. If the particular subpopulation from which the evidence sample came is known, the allele frequencies for the specific subgroup should be used as described in Recommendation 4.1. If allele frequencies for the subgroup are not available, although data for the full population are, then the calculations should use the population-structure equations 4.10 for each locus, and the resulting values should then be multiplied.

$$\text{Homozygote: } P(A_iA_i|A_iA_i) = \frac{[2\theta + (1 - \theta)p_i][3\theta + (1 - \theta)p_i]}{(1 + \theta)(1 + 2\theta)} \tag{4.10a}$$

$$\text{Heterozygote: } P(A_iA_j|A_iA_j) = \frac{2[\theta + (1 - \theta)p_i][\theta + (1 - \theta)p_j]}{(1 + \theta)(1 + 2\theta)} \tag{4.10b}$$

Recommendation 4.3. If the person who contributed the evidence sample is from a group or tribe for which no adequate database exists, data from several other groups or tribes thought to be closely related to it should be used. The profile frequency should be calculated as described in Recommendation 4.1 for each group or tribe.

Recommendation 4.4. If the possible contributors of the evidence sample include relatives of the suspect, DNA profiles of those relatives should be obtained. If these profiles cannot be obtained, the probability of finding the evidence profile in those relatives should be calculated with Formulae 4.8 or 4.9.

Genotype of suspect	Probability of same genotype in a relative	
Homozygote: A_iA_i	$p_i^2 + 4p_i(1 - p_i)F$	(4.8a)
Heterozygote: A_iA_j	$2p_ip_j + 2(p_i + p_j - 4p_ip_j)F$	(4.8b)

For parent and offspring, $F = 1/4$; for half-siblings, $1/8$; for uncle and nephew, $1/8$; for first cousins, $1/16$.

Full siblings, being bilineal rather than unilineal, require different formulae:

$$A_iA_i: \quad (1 + 2\,p_i + p_i^2)/4 \tag{4.9a}$$

$$A_iA_j: \quad (1 + p_i + p_j + 2\,p_ip_j)/4 \tag{4.9b}$$

RECOMMENDATIONS ON INTERPRETING THE RESULTS OF DATABASE SEARCHES, ON BINNING, AND ON ESTABLISHING THE UNIQUENESS OF PROFILES

Recommendation 5.1. When the suspect is found by a search of DNA databases, the random-match probability should be multiplied by N, the number of persons in the database.

Recommendation 5.2. If floating bins are used to calculate the random-match probabilities, each bin should coincide with the corresponding match window. If fixed bins are employed, then the fixed bin that has the largest frequency among those overlapped by the match window should be used.

Recommendation 5.3. Research into the identification and validation of more and better marker systems for forensic analysis should continue with a view to making each profile unique.

RECOMMENDATION FOR RESEARCH ON JUROR COMPREHENSION

Recommendation 6.1. Behavioral research should be carried out to identify any conditions that might cause a trier of fact to misinterpret evidence on DNA profiling and to assess how well various ways of presenting expert testimony on DNA can reduce any such misunderstandings.

APPENDIX VII

EXAMPLE DNA CASES

Example DNA profiles from two real forensic cases are presented for use in working through the principles explained in this text. The first case involves a sexual assault where differential extraction is utilized to separate epithelial-rich (female) and sperm-rich (male) components of the case mixture. The second case involves the use of Y chromosome loci to aid in another sexual assault investigation. The electropherograms and case report for both cases were kindly provided by Carll Ladd of the Connecticut State Forensic DNA Laboratory (Meriden, CT). All non-genetic identifiers have been removed to protect the innocent (and the guilty).

CASE (A) – SEXUAL ASSAULT EVIDENCE EXAMINED WITH 13 CODIS AUTOSOMAL STR LOCI

Five evidence samples were submitted as part of this case along with a buccal swab of the victim and a suspect. A description of the evidence and the results of examination are listed below.

Evidence submitted:

#1S1	Cutting from shorts
#4S1	Cutting from sheet
#4S3	Cutting from sheet
#4S4	Cutting from sheet
#6S1	Cutting from underpants
#7	Known swab, 'victim'
#8	Known swab, 'suspect'

Results of examination:

1. DNA was extracted from items #1S1, #4S1, #4S3, #4S4, #6S1, #7, and #8. A differential extraction (see Figure 3.2) was performed on items #1S1, designated #1S1A (epithelial-rich fraction) and #1S1B (sperm-rich fraction); #4S1, designated #4S1A (epithelial-rich fraction) and #4S1B (sperm-rich fraction); #4S3, designated #4S3A (epithelial-rich

fraction) and #4S3B (sperm-rich fraction); #6S1, designated #6S1A (epithelial-rich fraction) and #6S1B (sperm-rich fraction). DNA was purified according to standard Forensic Laboratory protocols (see Chapter 3).

2. The quality and quantity of the DNA obtained from each sample were analyzed by standard Forensic Laboratory protocols (see Chapter 3).

3. Extracted material obtained from the items tested was PCR-amplified (see Chapter 4) by the AmpF*lSTR Profiler Plus and AmpF*lSTR COfiler procedure as described in Forensic Laboratory protocols (see Chapter 5). STR alleles were separated and detected by standard Forensic Laboratory protocols (see Chapters 14 and 15). STR profiles obtained are listed in Table A7.1.

4. Items #1S1, #4S1, #4S3, #4S4, #6S1, #7, and #8 were retained at the laboratory.

5. The results are consistent with the suspect (item #8) being the source of the DNA profiles from items #1S1B (cutting from shorts, sperm-rich fraction), #4S1B (cutting from sheet, sperm-rich fraction), #4S3B (cutting from sheet, sperm-rich fraction), and #6S1B (cutting from underpants, sperm-rich fraction). The results eliminate the victim (item #7) as the source of the DNA profiles from items #1S1B, #4S1B, #4S3B, and #6S1B.

6. The results demonstrate that items #1S1A (cutting from shorts, epithelial-rich fraction) and #4S3A (cutting from sheet, epithelial-rich fraction) are mixtures. The victim (item #7) is included as a contributor to the DNA profiles from items #1S1A and #4S3A. The suspect (item #8) cannot be eliminated as a minor contributor to the DNA profiles from items #1S1A and #4S3A.

7. The results demonstrate that items #4S1A (cutting from sheet, epithelial-rich fraction) and #4S4 (cutting from sheet) are mixtures. The suspect (item #8) is included as a contributor to the DNA profiles from items #4S1A and #4S4. The victim (item #7) cannot be eliminated as a minor contributor to the DNA profiles from items #4S1A and #4S4.

8. The results demonstrate that item #6S1A (cutting from underpants, epithelial-rich fraction) is a mixture. The suspect (item #8) and the victim (item #7) are included as contributors to the DNA profile from item #6S1A.

Please answer the following questions as part of this investigation:

1. Can the suspect be eliminated as a contributor to the evidence samples?

2. Which samples would be most effective to present in court? Should all of them be presented?

3. What is the value of running the female victim's sample in conjunction with the other samples as part of this case?

4. Do all of the female and male fractions from the differential extractions make sense? Is the perpetrator's profile ever seen in the epithelial-rich fraction and if so, what are some possible reasons why? Is the victim's profile ever seen in the sperm-rich fraction and if so, what are some possible reasons why?

Profiler Plus alleles detected

Item #	D3S1358	vWA	FGA	AMEL	D8S1179	D21S11	D18S51	D5S818	D13S317	D7S820
1S1A	16,18	14,16,19	20,21,23, 24	X,Y	12,13,14,15	28,30, 32.2	13,14,18	11	12	9,10,13
1S1B	15,16	14,16	20,21	X,Y	14,15	28,29	14,17	11	12	10,13
4S1A	15,16,18	14,16,19	20,21,23, 24	X,Y	12,13,14,15	28,29,30, 32.2	13,14,17,18	11	12	10,13
4S1B	15,16	14,16	20,21	X,Y	14,15	28,29	14,17	11	12	10,13
4S3A	15,16,18	14,16,19	20,21,23, 24	X,Y	12,13,14,15	28,29,30, 32.2	13,14,17,18	11	12	9,10*
4S3B	15,16	14,16	20,21	X,Y	14,15	28,29	14,17	11	12	10,13
4S4	15,16,18	14,16,19	20,21*	X,Y	12,14,15	28,29*	13,14,17*	11	12	10,13
6S1A	15,16,18	14,16,19	20,21,23, 24	X,Y	12,13,14,15	28,29,30, 32.2	13,14,17,18	11	12	9,10,13
6S1B	15,16	14,16	20,21	X,Y	14,15	28,29	14,17	11	12	INC
7	16,18	19	23,24	X	12,13	30,32.2	13,18	11	12	9,10
8	15,16	14,16	20,21	X,Y	14,15	28,29	14,17	11	12	10,13

COfiler alleles detected

Item #	D3S1358	D16S539	AMEL	TH01	TPOX	CSF1PO	D7S820
1S1A	16,18	11,12	X,Y	9,9.3	8	10,13	9,10
1S1B	15,16	12	X,Y	9.3	8	10,13	10,13
4S1A	15,16,18	11,12	X,Y	9.3	8	10,13	10,13
4S1B	15,16	12	X,Y	9.3	8	10,13	10,13
4S3A	15,16,18	11,12	X,Y	9,9.3	8	10*	9,10*
4S3B	15,16	12	X,Y	9.3	8	10,13	10,13
4S4	15,16,18	11,12	X,Y	9.3	8	10,13	9,10,13
6S1A	15,16,18	11,12	X,Y	9,9.3	8	10,13	9,10,13
6S1B	15,16	12	X,Y	9.3	8	INC	NR
7	16,18	11	X	9,9.3	8	10	9,10
8	15,16	12	X,Y	9.3	8	10,13	10,13

5. If the suspect cannot be eliminated as a contributor to the evidence samples, what is the coincident match probability for the suspect's particular 13-locus STR profile in a U.S. Caucasian, African-American, or Hispanic population? Use Appendix II allele frequencies to determine these profile frequency estimates.

6. What is the likelihood ratio for the suspect having contributed the evidentiary profile versus a randomly selected man from the various U.S. population groups?

7. Can source attribution be declared based on these results following the guidelines used by the FBI Laboratory described in Chapter 21?

8. Would a mixture of the perpetrator and victim have been decipherable without differential extraction? If so, at what loci?

9. Which loci are the most useful in this particular case? Which loci are the least useful?

10. Could there have been more than one perpetrator in this particular case based on the evidence obtained?

11. Considering just sample 1S1A (epithelial-rich fraction) seen in Figure A7.1, which mixture interpretation test would be most useful (see Chapter 22)?

Table A7.1 (above)

*Results with Profiler Plus and COfiler STR kits on Case A evidence items. NR = no result, INC = inconclusive, * = additional minor peak(s) detected < 50 RFU but > 30 RFU.*

Figure A7.1

Profiler Plus green loci for the victim, suspect, and differential extraction female and male fractions from Case A. A mixture with a minor component consistent with the suspect is observed in the epithelial-rich (female) fraction (see third panel). Arrows indicate where stutter filters have removed allele calls.

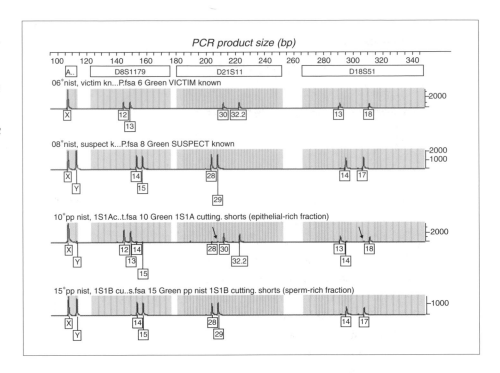

CASE (B) USE OF Y CHROMOSOME STR INFORMATION

Three samples were collected and processed in this case: (1) evidence of the sexual assault in the form of a genital swab from the victim, (2) a buccal swab from the female victim for reference purposes, and (3) a buccal swab from a suspect in the investigation. All three samples were amplified with the Y-PLEX 6 kit (ReliaGene Technologies, Inc.) to detect the Y-STR loci DYS19, DYS389II, DYS390, DYS391, DYS393, and DYS385a/b. Results collected on an ABI 310 Genetic Analyzer are shown in Figure A7.2.

Please answer the following questions as part of this investigation:

1. Examine the Y-STR profiles. Are the results reasonable? Can the suspect be excluded from contributing the crime scene sample?

2. Determine the Y-STR haplotype for the evidence and the suspect and perform a search using these profiles against available online databases to determine the number of times this particular haplotype has been previously observed. See Chapter 9 and Table 9.4 for information on available Y-STR databases.

3. Come up with a match report and statement that might be presented in court regarding this data based upon your searches of available Y-STR databases. Compute the 95% confidence interval with your data.

4. Some peaks are observed in the female victim sample within the DYS390 and DYS391 allele ranges. What are some possible reasons for their occurrence and is the

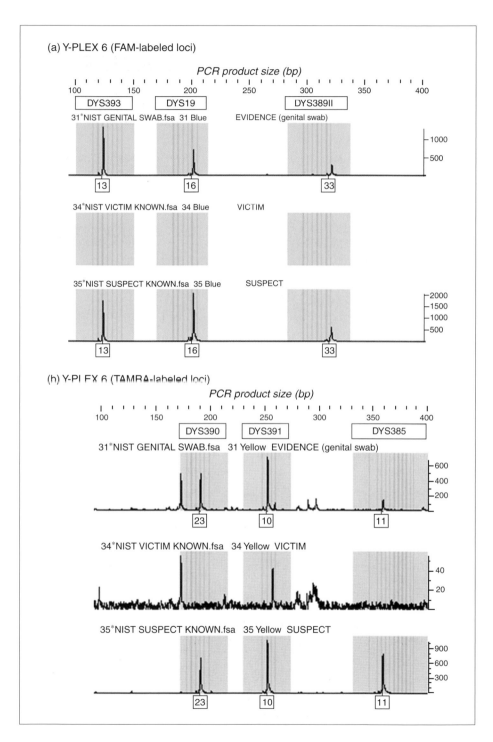

Figure A7.2

(a) Y-PLEX 6 results from FAM-labeled (blue) loci on evidence, victim, and suspect samples in Case B. No PCR products were observed in the female victim sample since it does not contain a Y chromosome.

Figure A7.2

(b) Y-PLEX 6 results from TAMRA-labeled (yellow) loci on evidence, victim, and suspect samples in Case B.

presence of these peaks a concern for this particular case? What kind of issues might the defense raise with the Y-STR data from this case?

5. What are some advantages and disadvantages of Y-STR testing? Suppose the data presented here is the only evidence in this particular case. Would you feel comfortable going to court armed with these Y-STR results? If not, what other kind of information would be helpful in prosecuting the case?

AUTHOR INDEX

SUBJECT INDEX

Figures in Italic and *Tables* in **Bold**
Ap indicates material in Appendices
B indicates material in D.N.A. boxes